WILLIAM F. MAAG LIBRARY
YOUNGSTOWN STATE UNIVERSITY

Biological Flows

World Congress of Biomechanics

Michel Y. Jaffrin Technological University of Compiègne, Compiègne, France
Colin G. Caro Imperial College of Science, Technology, and Medicine, London, England

Biological Flows
Edited by Michel Y. Jaffrin and Colin G. Caro

Biological Flows

Edited by
Michel Y. Jaffrin
Technological University of Compiègne
Compiègne, France

and
Colin G. Caro
Imperial College of Science, Technology, and Medicine
London, England

Plenum Press • New York and London

Library of Congress Cataloging-in-Publication Data

Biological flows / edited by Michel Y. Jaffrin and Colin G. Caro.
      p.   cm.
    "Incorporating selected papers from the Second World Congress of
Biomechanics, held July 10-15, 1994, in Amsterdam, the Netherlands"-
-T.p. verso.
    Includes bibliographical references and index.
    ISBN 0-306-45206-5
    1. Body fluids--Congresses.   2. Fluid mechanics--Congresses.
3. Biomechanics--Congresses.   I. Jaffrin, Michel-Yves. II. Caro,
Colin G. (Colin Gerald) III. World Congress of Biomechanics (2nd :
1994 : Amsterdam, Netherlands)
    [DNLM: 1. Biomechanics--congresses. 2. Body Fluids--physiology-
-congresses.   WE 103 B6138 1995]
QP90.5.B56 1995
612'.014--dc20
DNLM/DLC
for Library of Congress   96-3277
                                                                CIP

Incorporating selected papers from the Second World Congress of Biomechanics,
held July 10 – 15, 1994, in Amsterdam, The Netherlands

ISBN 0-306-45206-5

© 1995 Plenum Press, New York
A Division of Plenum Publishing Corporation
233 Spring Street, New York, N. Y. 10013

10 9 8 7 6 5 4 3 2 1

All rights reserved

No part of this book may be reproduced, stored in a retrieval system, or transmitted in any form or by any means, electronic, mechanical, photocopying, microfilming, recording, or otherwise, without written permission from the Publisher

Printed in the United States of America

PREFACE

Biomechanics has a distinguished history extending at least to the 16th Century. However the later half of this century has seen an explosion of the field with it being viewed as offering exciting challenges for physical scientists and engineers interested in the life sciences, and wonderful opportunities for life scientists eager to collaborate with physical scientists and engineers and to render their scientific work more fundamental. That the field is now well established and expanding is demonstrated by the formation of a World Committee for Biomechanics and the success and large participation in the 1st and 2nd World Congresses of Biomechanics, held respectively in San Diego in 1990 and in Amsterdam in 1994.

With more than 1350 scientific papers delivered at the 2nd World Congress, either within symposia or oral or poster sessions, it would have been out of the question to try to produce comprehensive edited proceedings. Moreover, we are confident that most of the papers have been or will be published in one of the excellent journals covering the field. But we thought that the large amount of effort contributed by the plenary lecturers and the tutorial and keynote speakers of various symposia deserved to be recognised in the form of a specific publication, thus also allowing those unable to attend the presentations to share in the findings. Furthermore, we feel that there is now a need to review aspects of the field.

Being biofluid mechanists ourselves it was natural that we should limit the volume we were editing to biological flows, but we hope that a second volume devoted to biosolid mechanics will follow with other editors.

Very few guidelines were given to the authors. They were asked to write a didactic review of their field incorporating newer findings so that this book could be used for teaching and for reference material. We have been greatly impressed by the quality of the contributions we have received and we are very grateful to the authors for their efforts and especially to Professor Y.C. Fung, Chairman of the World Committee for Biomechanics for his encouragement for this enterprise.

We would also like to thank our secretaries, A.M. Schmitt, M.P. Lenoble, L. Barker and G. Cash for their help.

Of course this book makes no claim to being comprehensive and some important areas have been omitted partly because some invited authors were unable to provide their contributions due to other commitments.

Finally it is hoped that this book will be an incentive for the biomechanics community to attend future World Congresses and that it will help maintain the momentum in the field built up by the first two World Congresses.

Michel JAFFRIN, Compiègne
Colin CARO, London

CONTENTS

1. Biomechanics of Blood Cells: A Historical Perspective 1
   Richard Skalak

2. Mechanics of the Endothelium in Blood Flow 11
   Y. C. Fung and S. Q. Liu

3. New Perspectives in Biological Fluid Dynamics 31
   T. J. Pedley

4. Fluid Mechanics of Arterial Bifurcations 51
   Don P. Giddens, Tongdar D. Tang, and Francis Loth

5. Non-Planar Geometry and Non-Planar-Type Flow at Sites of Arterial Curvature
   and Branching: Implications for Arterial Biology and Disease 69
   C. G. Caro, D. J. Doorly, M. Tarnawski, K. T. Scott, Q. Long, and
   C. L. Dumoulin

6. Computer Simulation of Arterial Blood Flow: Vessel Diseases under the Aspect
   of Local Haemodynamics 83
   K. Perktold and G. Rappitsch

7. Computational Visualization of Blood Flow in the Cardiovascular System 115
   Takami Yamaguchi

8. Biomechanical and Physiological Aspects of Arterial Vasomotion 137
   N. Stergiopulos and J. -J. Meister

9. Architecture and Hemodynamics of Microvascular Networks 159
   T. W. Secomb, A. R. Pries, and P. Gaehtgens

10. Mass Transport through the Walls of Arteries and Veins 177
    M. John. Lever

11. Blood Cross Flow Filtration through Artificial Membranes 199
    M. Y. Jaffrin

12. Cardio-Vascular Interaction Determines Pressure and Flow 227
    N. Westerhof

13. Mechanics of Intramural Blood Vessels of the Beating Heart 255
    F. Kajiya, M. Goto, T. Yada, Y. Ogasawara, and K. Tsujioka

14. Nonlinear Models of Coronary Flow Mechanics 267
    Jos. A. E. Spaan

15. Simulation of Forced Breathing Maneuvers 287
    James J. Shin, David Elad, and Roger D. Kamm

16. Respiratory Mechanics and New Concepts in Mechanical Ventilation 315
    Daniel Isabey, Laurent Brochard, and Alain Harf

17. Biomechanics of Lymph Transport 353
    G. W. Schmid-Schönbein and F. Ikomi

Index 361

BIOMECHANICS OF BLOOD CELLS: A HISTORICAL PERSPECTIVE

Richard Skalak

Department of Bioengineering
University of California, San Diego
La Jolla, California 92093-0412

INTRODUCTION

The notion that all animals and plants are made up of individual living cells and their products is a relatively new concept, dating back to approximately 150 years ago. The application of continuum mechanics to the passive and active behaviors of cells is even much more recent. Detailed analytical and computational solutions describing the deformation of cells and their components has only been carried through extensively in the last two or three decades. In the present paper, the historical development of ideas concerning the cellular nature of all living matter will be outlined first. Secondly, the development of analyses of blood flow will be sketched and thirdly, the application of continuum mechanics to individual blood cells and their component parts will be outlined.

Finally, it should be remarked that the present time is the midst of another revolution in the study of mechanics of cells. This is the introduction of concepts, experimental methods and modeling at the molecular level. Computations of protein folding are well underway. Molecular biology is deriving detailed sequencing of amino acids and base pairs at a rapid rate. In the near future one may expect such information to be utilized in the description and modeling of events in cell mechanics. The modeling of large assemblies of molecules and their incorporation into models of cell structures is not yet extensive, but promises to play a large role in the future.

HISTORY OF CELL THEORY

In the ancient world of Egyptian, Greek and Roman civilizations there was extensive knowledge of human anatomy and there was an active practice of medicine mixed with magic, astrology, and religion (see for example, Singer and Underwood, 1962). But the notion that all living tissues were composed of individual cells was not part of the ancient knowledge or theory. In regard to the physical world, there was a Greek school which hypothesized that all matter is made up of tiny particles, imperceptible to the senses, but of different size, shape and weight. This school of thought contained the concept of atoms of a material. This philosophy held that all physical phenomena and all events ascribed to the soul and supernatural phenomena had, in fact, a physical basis. This philosophy was developed in the writings of Democritis, Epicurus, and Lucretius. Although it encompassed discussion of the nature of inert and living matter with considerable insight, the idea of living cells was not developed.

Lucretius comes close to a cellular concept in the following passage (Lucretius, 1921 translation, p. 139):

> "First, living creatures are sometimes so small
> That even their third part can nowise be seen;
> Judge, then, the size of any inward organ –
> What of their sphered heart, their eyes, their limbs,
> The skeleton? How tiny thus they are!
> And what besides of those first particles
> Whence soul and mind must fashioned be?
> Seest not how nice and how minute?"

The notion that all matter was composed of aggregations of small particles clearly included living matter (Lucretius, 1921 translation, p. 32):

> "Of this homeomeria of things, he thinks
> Bones to be sprung from littlest bones minute,
> And from minute and littlest flesh all flesh,
> And blood created out of drops of blood,

Figure 1. Superficial muscles from the side. From: Andreus Vesalius, De Humani Corpus Fabrica. Basle, 1543.

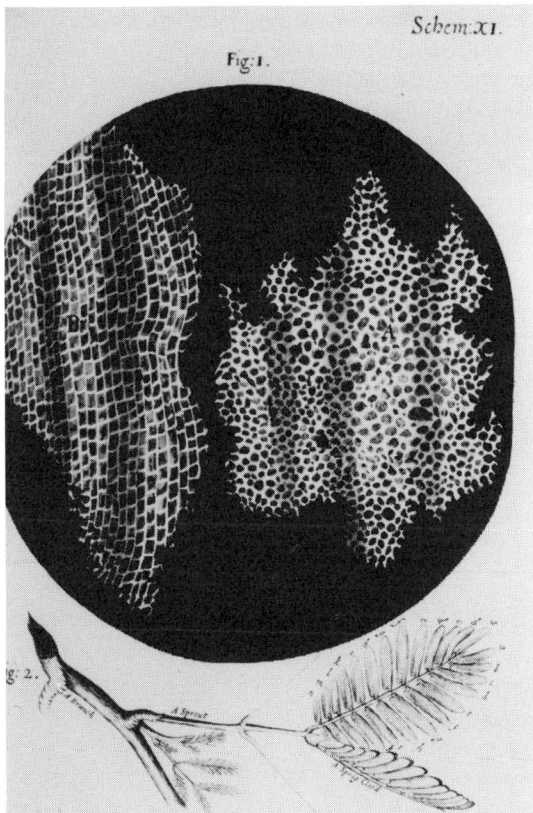

Figure 2. Drawing of a cross-section of cork. From: Thomas Hooke, 1665.

Concerning gold compact of grains of gold,
And earth concerted out of bits of earth."

Aristotle and other Greek philosophers also observed chicken embryo development, but of course, could not discern structures below the resolution of the naked eye.

In the dark ages, there is no progress reported, but with the coming of the Renaissance, the study of anatomy became more detailed, as recorded, for example, in drawings by Leonardo da Vinci (1452-1519) and Andreas Vesalius (1543), Fig. 1. The discovery of the

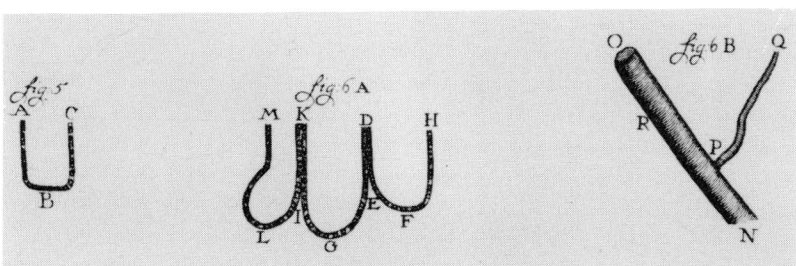

Figure 3. Capillary vessels and blood cells in the tail of an eel. From: Leeuwenhoek, 1688.

cellular structure came shortly after the invention of microscopes. The first illustration of cellular structure and the use of the word cells is usually attributed to Robert Hooke (1635-1703). He observed and sketched the cellular structure of cork (Fig. 2) in his book "Micrographia" in 1665. Individual blood cells, bacteria, muscle fibers and other minute anatomical structures were soon thereafter reported by Leeuwenhoek (1632-1723), Fig. 3. His work marks the beginning of the modern era of direct observation of living cells. Actually his observations of individual red blood cells was preceded by the discovery of frog blood cells by Ian Swammerdam (1637-1680).

In the two centuries following the early microscopist's discovery of cells, the microscopic structures of many different tissues were examined and the relation of cells to tissues was debated and clarified. The internal structures of cells were gradually discovered. In 1833, Robert Brown (of Brownian motion fame) described the cell nucleus in the cells of orchids. Mitochondria of muscle cells were reported by Albert von Kolliker (1817-1905) in 1857. The central role of cells in all living tissues was championed and documented by examination of many types of tissues by Matthias J. Schleiden and Theodor Schwan (1838), Fig. 4. They postulated the basic idea of cell theory that all tissues and their functions derive from cells and their extracellular products. Many investigators contributed to knowledge of cellular structure and function including G.G. Valentin (1810-1883), and J.H. Purkinje (1787-1869) who described cell division before 1838. By the end of the 19th century the cell theory was fully established as the basis of all living animals and plants.

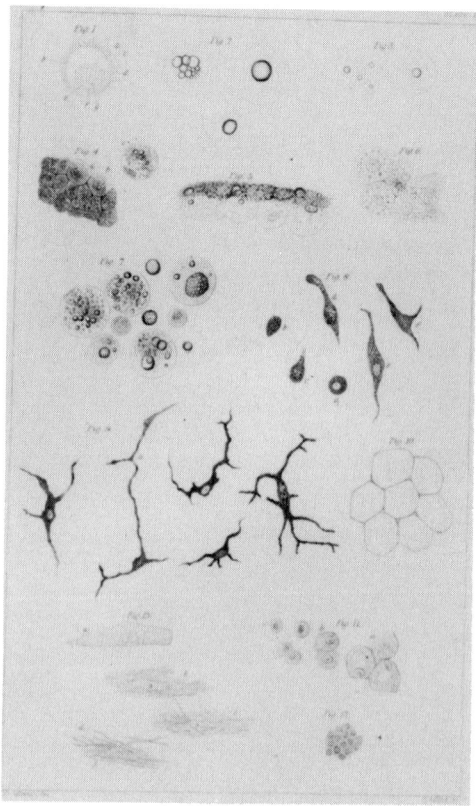

Figure 4. Plate II from: Schwann and Schleiden, 1847.

In the decades following the establishment of cell theory, medical and biological advances were made primarily in physiology, biochemistry, neural biology, respiration, bacteriology, and immunology rather than in the mechanical behavior of cells. During this period Louis Pasteur (1822-1895) introduced his famous vaccines against anthrax (1881) and rabies (1885). Many pathogenic micro-organisms were discovered during the second half of the 19th century and medicine became more scientifically based.

During the same period, the foundations of solid and fluid mechanics were being developed by mathematicians and physical scientists such as Stokes, Navier, Lame, Helmholtz, Lamb, Boussinesq and many others, but largely without pursuing applications in biology and medicine. Mechanical and medical sciences grew with little interchange or collaboration between them. In the first half of the twentieth century many engineering advances involved application of advanced and analytical mechanical analysis. These include the development of large bridges, ships, the automobile and airplanes. The comparable possible uses of analytical continuum mechanics in biology in general and to the biomechanics of cells in particular, received much less attention. In the most recent two or three decades, applications of analytical continuum mechanics to cellular mechanics has begun to be explored more widely and is gradually approaching the sophistication of modern engineering practice. Some examples will be outlined in a subsequent section.

It may be mentioned that the application of mechanics to biological theory had an early and abortive start in the work of Rene Descartes (1596-1650). He developed a completely mechanical theory of physiology which included neural functioning by fluid flow through the nerves. His theory was known as iatrophysics and was later shown to be fallacious in many details and was discarded. A more successful and lasting work was that of Giovanni Alfonso Borelli (1608-1679) who discussed the operation of the musculo-skeletal system in terms of mechanics and levers.

In the 18th and 19th century there were various contributions of mechanics to biology by outstanding scientists such as Leonhard Euler (1707-1783), Thomas Young (1773-1829), Herman von Helmhotz (1821-1894) and Horace Lamb (1849-1934). Their work was on a macroscopic basis and not concerned with individual cells. The paucity of work in biomechanics in this period may be attributed in part to the early failure of Descartes iatrophysics, but more likely is due to two trends that may be associated with this epoch. On one hand, medical and biological innovations were preoccupied with remarkable advances deriving from other fields such as biochemistry, immunology, anesthesia, bacteriology, radiology, and pharmacology which did not require detailed mechanical analysis. Secondly, it was a time when knowledge was expanding so rapidly that few people could be expert in a broad spectrum of scientific and mathematical fields. Medical science and engineering science grew separately. At the present time, a new period of integration is underway.

1. Ovum of a goat, after Krause (Miller's Archiv, 1837, Pl. I, Figure 5).
2. Cells from the yelk-cavity of a nature hen's egg.
3. Cells from the interior of an egg measuring a line and a half in diameter, taken from the ovary of a hen.
4. Portion of the germinal membrane of a mature hen's egg before incubation, viewed from above.
5. Portion of a germinal membrane from a hen's gee after sixteen hours' incubation. It is folded in such a manner that the external surface or serous layer forms the margin.
6. Cells from the serous layer of the same germinal membrane in the neighbourhood of the area pellucida, after separation of the mucous layer.
7. Cells from the mucous layer of the same germinal membrane on the outside of the area pellucida.

8. and 9. Pigment-cells of different kinds and stages of development, from the tail of the tadpole.
10. Cells from the interior of the shaft of a fully developed wing-feather of the raven.
11. Earlier stages of development of the same, from the portion of the shaft of an immature feather which has not as yet become hard.
12. Cell-nuclei, from the same, around which no cells have as yet formed.
13. Flat cells splitting into fibres, from the cortex on the side of the shaft of a raven's feather in progress of formation.

DEVELOPMENT OF BLOOD FLOW THEORY

The history of ideas and discoveries of the circulation of blood has been traced in detail in a book edited by Fishman and Richards (1964). The ancient Egyptians, Greeks, and Romans were familiar with the anatomy of the heart and major blood vessels. They knew that the blood flows in the arteries and veins, but did not realize that blood circulates. In the writings of the famous Greek physician, Galen (131-201) A.D., it was known that the heart valves permit flow in only one direction but it was thought that blood issues and returns to the heart like a tide ebbs and flows rather than circulating.

There are some ancient observations recorded that imply the existence of pulse propagation in the arteries. The earliest known record (3000 B.C.) is in hieroglyphics (see Hamburger, 1939) and describes a trephination. The pulsation of the arteries in the opening of the skull are connected to the beating of the heart. Later, Erasistrathos (280 B.C.) taught that the arterial pulse is originated by the heart and arrives at distal locations a short time later. Capillary blood vessels were not observed as there were no microscopes and their existence connecting arteries and veins was not known.

The existence of capillaries, connected arteries and veins and the idea that blood circulates was postulated before capillaries were actually observed. The discovery of the circulation of blood was made by William Harvey (1628). He estimated the volumetric output of the heart and concluded the blood must circulate from arteries to the veins through "porisities of the flesh." The first observation of such porosities (capillaries) were made in the lung by Malpighi in 1661. Shortly thereafter, Leeuwenhoek reported the observation of blood flow through capillaries in the tail of an eel and remarks that the flow is from arteries to veins. He observed the single file flow of red blood cells in capillaries with clear spaces (plasma) between them and the elongation of the red blood cells as they pass through narrow capillaries.

The first measurements of arterial blood pressures were made by Stephen Hales (1733). He cannulated the carotid artery of a mare tied down to a gate and observed the rise of the blood in the tube. He also estimated the flow and resistance in various parts of the circulation and correctly surmised that the major pressure drop is in the smallest vessels. Hales pointed out that since the length of systole is only one third of the cardiac cycle, about two thirds of the stroke volume must be stored by distension of the arteries. He compared the action of the arteries in maintaining capillary flow during diastole to the action of the air chamber used on fire engines of that time to produce a more even flow of water. This is the origin of the windkessel theory of arterial blood flow.

The equations of motion and continuity which lead to description of wave propagation in arteries were written for inviscid fluid flow by Leonhard Euler in 1775, but were only published posthumously in 1882. Euler was not able to solve these nonlinear equations. Solutions in the context of blood flow were not obtained until 1956 by Lambert, using the method of characteristics which was not known in Euler's time.

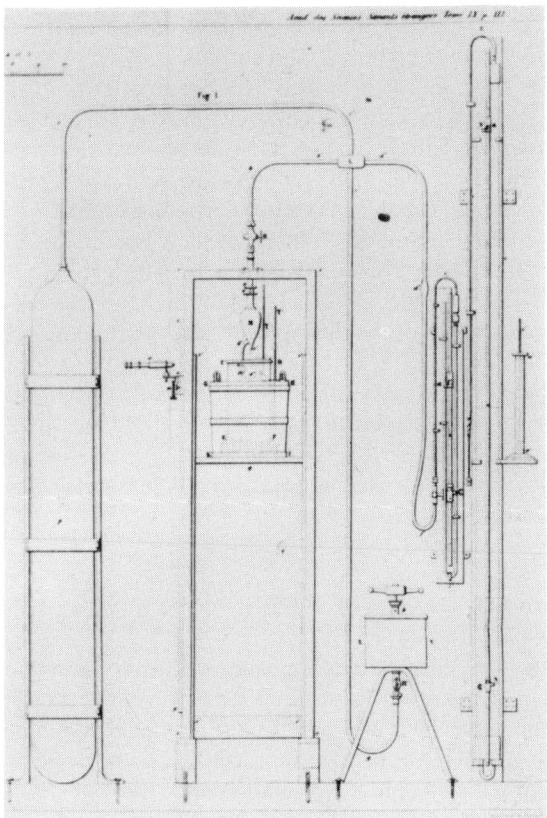

Figure 5. The capillary flow apparatus of Poiseuille. From: Poiseuille, 1846.

Linearized theories of wave propagation were developed extensively during the 19th and 20th centuries. The velocity of wave propagation in the circulation was first derived by Thomas Young (1809) by analogy to the transmission of sound in air. He also estimated the pressure drop throughout the arterial circulation and showed that pressure loss was only a few percent in the large arteries. He also estimated the viscosity of blood was about four times that of water.

Throughout the 19th and 20th centuries, various studies improved the description of pulsatile flow in the arteries and the viscous resistance in small vessels. Both wave propagation and windkessel theories were refined and found useful (see Leake, 1962). The linearized theory of wave propagation in arteries was extensively expounded in the book by McDonald (1974). Nonlinear theory and numerical results are illustrated by Stettler et al., (1987). Extensive review of both linear and nonlinear theories are given in books by Fung (1984) and Pedley (1980).

In regard to the viscous resistance to blood flow, the most famous work is that of J.L.M. Poiseuille (1840). He derived what is now known as Poiseuille's law from his careful experiments in fine glass capillaries, Fig. 5. Poiseuille was a physician and his interest was in the laws of capillary blood flow. His experiments were made with water, alcohol and mercury because blood clotting prevented him from using blood in his apparatus. The analytical derivation of Poiseuille's law was not given until 1858 by Eduard Hagenbach (see Schiller 1933). A history of work related to Poiseuille's law is given by Sutera and Skalak (1993).

Poiseuille's law is accurate for low Reynold flows of Newtonian fluids, but it does not apply to the flow of whole blood accurately due to the particulate nature of blood as a suspension of cells. There are two regions of vessel size in which blood does not behave as a Newtonian fluid. The first range is in vessels between approximately 25 and 300 microns in diameter. It was discovered by Fåhraeus and Lindquist (1931) that the apparent viscosity of blood decreases with tube diameter in vessels below about 300 microns in diameter. This is due to the fact that the blood cells tend to move away from the vessels walls.

The second range is in capillary vessels below about 9 microns in diameter. In this range the blood cells fit more and more tightly into the vessel and the Fåhraeus-Lindquist effect is reversed, i.e., the apparent viscosity increases as the capillary diameter decreases. It is remarkable that since the single file flow was described by Leeuwenhoek in 1668, there was no theoretical consideration of the flow in capillaries containing cells until the suggestions of Prothero and Burton in 1961. A review of capillary flow history is given by Skalak and Chien (1981). The current status of theories of capillary blood flow is summarized by Secomb (1991, 1995). It is of interest to note that in comparing the flows and pressure drops computed for a microvascular network to the experimentally observed values, it appears that capillary resistance in vivo is somewhat larger than predicted from theory and experiments in straight, uniform tubes (Pries et al., 1994).

BIOMECHANICS OF BLOOD CELLS

During the 19th and 20th centuries a great deal of information was developed on the structure and function of many different kinds of cells. However, there was very little consideration given to the mechanical properties of individual cells from the standpoint of theoretical mechanics. In the case of red blood cells, a book by Krogh in 1922, gave a summary of knowledge at that time. Krogh remarks on the great flexibility of red blood cells in passing through filter paper and on their extendibility in narrow capillaries. But no detailed analysis is given. A later book by Ponder (1942) gives a great deal of information on the physiology and pathology of red blood cells, but again without analysis of their biophysical behavior. The first attempt to model the red blood cell as an elastic membrane filled with an incompressible fluid was given by Fung and Tong in 1968. Since then, the properties of the red blood cell have been established with increasing accuracy and detail. The macroscopic behavior of the red blood cell membrane is dominated by the fact that due to its bilayer lipid structure, its surface area is nearly constant. The modulus of elasticity related to isotropic extension of area is five orders of magnitude greater than the shear modulus. The shear modulus is due to the cytoskeleton which is a network of mostly spectrin which is located on the interior surface of the lipid bilayer. The red blood cell membrane also exhibits an appreciable viscosity in shearing motions and it has a small bending elasticity. A summary of these properties may be found in Hochmuth (1987). In treating the red blood cell membrane as a single entity the possibilities of slip between the bilayer leaflets and between the cytoskeletal network and the lipid bilayer are neglected. Recent studies of tethers and micropipette aspiration suggest that such slip does occur (Discher, 1994).

The studies of the properties and flow of red blood cells were historically the first to be undertaken and were relatively straight-forward because of the simple structure of the cell. The next cells to be studied in detail are leukocytes which present a very different and much more complex structure. The leukocyte has much more structure in its interior than a red blood cell in which the interior is fluid. The overall behavior of the leukocyte is viscoelastic and its stiffness is several orders of magnitude greater than that of a red blood cell. The overall viscoelastic behavior of the leukocyte may be modeled as a sphere of a standard solid (Bagge et al., 1971). More detailed models consist of a cortical shell filled with a Maxwell or a

Newtonian fluid. Experiments suggest that the viscoelastic coefficients of the leukocyte cytoplasm are nonlinear both with respect to the extent of deformation and the rate of strain. A recent review of experiments and models of leukocytes is given by Hochmuth (1993).

One of the difficulties of modeling leukocytes and other cells is that the cytoskeleton may be variable in space and time. This is most prominent in the active motion and locomotion of leukocytes. In fact, the continual polymerization and depolymerization of the actin cytoskeleton is essential to cell locomotion. The properties of leukocytes depend on their degree of activation and may account for some of the variation of properties from cell to cell.

The internal events in changes of cytoskeletal structures in the leukocyte and other motile cells have only been established and modeled mathematically in the last few years. A recent summary is given by Stossel (1993). The forward motion of the leading edge of locomoting leukocyte is produced by actin polymerization of the leading edge. The protrusion contains an actin gel which forms a cytoskeleton that is anchored to the substrate over which the cell moves. Toward the end of the cytoskeletal region, the actin gel is depolymerized and contracted. This contraction pulls the main body of the cell forward using the adhered pseudopod at the front end of the cell as an anchor. The different zones of the cell may be clearly visualized in a leukocyte moving in a glass pipette (Usami et al., 1992). Cell locomotion is also important in many situations beside leukocyte migration as in wound repair, embryonic development, and metastasis of cancers. The details of cytoskeletal reorganization, signaling pathways and adhesion mechanisms are currently active areas of research for many different purposes.

CONCLUSION

This brief review is intended to give some historical framework of the development of ideas and mathematical analysis of the mechanics of blood cells and blood flow. The main perspective is that although the cellular nature of blood was first observed over 300 years ago, the detailed analysis of the biomechanics of blood cells has been mostly carried out in the last two or three decades. Current work on blood cells and blood flow is not summarized here since the papers in the remainder of this volume will give a cross-section of the status of such research at the present time.

It remains to remark on future directions of research that are likely to be most promising and profitable from the standpoint of more detailed understanding of cellular events. These directions will most probably involve combinations of biochemistry, molecular biology, and biomechanics. At the present time, molecular structures of proteins and other biological molecules are being developed rapidly and over a wide spectrum of cells and their components (see, for example, Braydon et al., 1995). The studies of protein folding are a leading edge of biological research. Although protein folding itself is mechanically based on the interactions of the component atoms, the studies of cell mechanics from a continuum viewpoint has made little use of the knowledge of detailed molecular structures to date. In the future, the cellular biomechanics based on simulation of molecular substructures may provide a more complete description and understanding of the biomechanics of passive and active behaviors of the many different types of cells.

REFERENCES

Bagge, U., Skalak, R. and Attefors, R., 1977. Granulocyte Rheology. Advances in Microcirculation. 7: 29-48.
Borelli, G.A., 1680. On the Movement of Animals. Translated by P. Maquet, 1989. Springer-Verlag, New York.
Brayden, B.C. and Poljak, R.J., 1995. Structural features of the reactions between antibodies and protein antigens. FASEB J. 9: 9-16.

Discher, D.E., Mohandas, N., Evans, E.A., 1994. Molecular maps of red cell deformation: Hidden elasticity and in situ connectivity. Science 266: 1032-1035.
Euler, L., 1775. Principia pro moto sanquins per arterias determinado. Published post-humously, edited by P.H. Fuss and N. Fuss. Apud Eggers et socios, Petropoli. 2: 814-823, 1862.
Fåhraeus, R. and Lindquist, T., 1931. The viscosity of the blood in narrow capillary tubes. Amer. J. Physiol. 96: 562-568.
Fishman, A.P. and Richards, D.W., 1964. Circulation of Blood: Men and Ideas. Oxford Univ. Press.
Fung, Y.C. and Tong, P., 1968. Theory of sphering of red blood cells. Biophys. J. 8: 175-198.
Fung, Y.C., 1984. Biodynamics: Circulation. Springer-Verlag, New York.
Hales, S., 1733. Statical Essays, Vol. II, Haemastatiks, Reprinted by Hafner Publishing Co., New York, 1964.
Hamberger, W.W., 1939. The earliest known reference to the heart and circulation. The Edwin Smith Surgical Papyrus, circa. 3000 B.C. Amer. Heart J. 17: 259-274.
Harvey, W., 1628. Exercitatio anatomica de moto cordis et sanquinis in animalilus. Translated by C.D. Leake, Thomas Pub., Springfield, Ill., 1958.
Hochmuth, R.M., 1987. Properties of red blood cells. Chap. 12 in: Handbook of Bioengineering, R. Skalak and S. Chien, Eds. McGraw-Hill, New York.
Hochmuth, R.M., 1993. Measuring the mechanical properties of individual human blood cells. J. Biomechanical Engineering. 115: 515-519.
Hooke, T., 1665. Micrographia: Some Physiological Descriptions of Minute Bodies. Martyn and Appleby, London.
Krough, A., 1922. The Anatomy and Physiology of Capillaries. Yale Univ. Press. New Haven, CN.
Lambert, J.W., 1956. Fluid Flow in a Nonrigid Tube. Doctoral Dissertation Series No. 19, 418, University Microfilms. Ann Arbor, Mich.
Leake, C.D., 1962. The historical development of cardiovascular physiology. Handbook of Physiology, Sec. 2. 1: 11-22.
Leeuwenhoek, A.V., 1688. On the circulation of blood. Letter to the Royal Society. Facsimile with introduction by A. Schierbeek. Published by N.B. de Graaf, 1962.
Lucretius, 1921. Translation: Of the Nature of Things. Translated by W.E. Leonard, E.P. Dutton & Co., New York.
Malpighi, M., 1661. De pulmonibus. Observationes Anatomicae. Bologna.
McDonald, D.A., 1974. Blood Flow in Arteries, 2nd Ed. Williams & Wilkins. Baltimore, MD.
Pedley, T.J., 1980. The Fluid Mechanics of the Large Blood Vessels. Cambridge Univ. Press.
Poiseuille, J.L.M., 1846. Recherches experimentales sur le movement des liquides dans les tubes de très-petits diamètres. Mémoires l' Academie Royale des Sciences de l' Institute de France, IX: 433-544.
Ponder, E., 1948. Hemolysis and Related Phenomena. Grune and Stratton, New York.
Pries, A.R., Secomb, T.W., Gessner, T., Sperando, M.B., Gross, J.F. and Gaehtgens, P., 1994. Resistance to blood flow in microvessels in vivo. Circulation Research 75: 904-915.
Prothero, J. and Burton, A.C., 1961. The physics of blood flow in capillaries. Biophys. J. 1: 565-579, 2: 199-212, 2: 213-222.
Schiller, L., 1933. Drei Klassiker der Strömungslehre: Hagen, Poiseuille, Hagenbach. Akad. Verlagsgesellschaft, Leipzig.
Schmid-Schoenbein, G.W., Sung, K.L.P., Tozeren, H., Skalak, R. and Chien, S., 1981. Passive mechanical properties of human leukocytes. Biophys. J. 36: 243-246.
Schwann, T. and Schleiden, M.J., 1847. Microscopical Researches into the Accordance in the Structure and Growth of Animals and Plants. Translated by H. Smith. C. and J. Printers, London.
Secomb, T.W., 1991. Red blood cell mechanics and capillary blood rheology. Cell Biophysics. 18: 231-251.
Secomb, T.W., 1995. Mechanics of blood flow in the microcirculation. To appear in: Biological Fluid Dynamics. C.P. Ellington and T.J. Pedley, Eds., Published by Company of Biologist, Cambridge, UK.
Singer, C. and Underwood, E.A., 1962. A Short History of Medicine. Oxford Univ. Press.
Skalak, R. and Chien, S., 1981. Capillary flow: History, experiments and theory. Biorheology. 18: 307-330.
Stettler, J.C., Niederer, P., Anliker, M., 1987. Nonlinear mathematical models of the arterial system. Chap. 17 in: Handbook of Bioengineering, R. Skalak and S. Chien, Eds. McGraw-Hill, New York.
Stossel, T., 1993. On the crawling of animal cells. Science. 260: 1086-1094.
Sutera, S.P. and Skalak, R., 1993. The history of Poiseuille's Law. Ann. Rev. of Fluid Mech. 25: 1-19.
Usami, S., Wung, S.L., Skierczynski, B.A., Skalak, R. and Chien, S., 1992. Locomotion forces generated by a polymorphonuclear leukocyte. Biophys. J. 63: 1663-1666.
Vesalius, A., 1543. De Humani Corpus Fabrica. Basle, 1543.
Young, T., 1809. On the functions of the heart and arteries. Phil. Trans. Royal Soc. of London. 99: 1-31.

2

MECHANICS OF THE ENDOTHELIUM IN BLOOD FLOW

Y. C. Fung and S. Q. Liu

University of California, San Diego
La Jolla, California 92093-0412

INTRODUCTION

The endothelium is a single layer of confluent endothelial cells covering the entire inner wall of the heart and blood vessels. It is a mono layer (i.e. no cell stays on the back of another) resting on a collagenous basal lamina, with a thickness of from 2 to 5 μm, the higher spots reveling the cell nuclei. The endothelium separates blood from tissue. It's importance to the circulation is somewhat analogous to the skin to a man, the roof to a house, the semipermeable membrane of a reverse osmosis water desalination plant, i.e., it separates; filtrates, and controls the mass transport between the tissue and the blood. It carries enzymes, and manufactures many of them, which control the clotting of blood, dissolution of clots, adhesion of leukocytes or cancer cells, passage of LDL, growth of new blood vessels, as well as inhibition of growth. Molecular biology of the endothelial cells responding to the shear stress of the flowing blood is advancing very rapidly. Paper by P.F. Davies (10, 11, 12), D.F. Dewey, Jr. (13, 55) J.A. Frangos (15), M.A. Gimbrone, Jr. (10, 55, 56). Kuo et. al. (38), Levesque et. al. (40), L.V. McIntire (18, 50), Nerem et. al. (49), Resnick et. al. (56), Sato et. al. (62, 63). Smiesko and Johnson (71), and others have opened up a new vista to biology. It has been found that the locally produced growth factors include the *growth promotors* such as the platelet-derived growth factor (PDGF), (30, 44, 59), and the vascular endothelium-derived growth factor (VEFG), (26, 28, 47, 48, 60), and the *growth inhibitors* such as the endothelium-derived relaxation factor (EDRF) nitric oxide (5, 37, 46). Evidences are mounting that hypertension induces up-regulation of growth promoting substances and their receptors (Sarzani et. al. (59), and down-regulation of growth inhibitory factors EDRF and prostacyclin (Luscher et. al. (43), Panza et. al. (52)). It is established that gene expressions are functions of stress and strain, (Resnick et. al. (56), Shyy et. al. (66), and cell integrity, growth, and differentiation are influenced by stress and strain (Ingber et. al. (31, 32))

With better understanding of the chemical makeup of the endothelial cells, we would like to improve our understanding of their mechanics. We should be able to measure or calculate the stress and strain of every structural molecule inside the cell, because the stress and strain of these molecules influence their mechanical and chemical properties.

In detail, however, the mechanical machinery of cells is still largely unknown. At the present stage of scientific development, it may be helpful to make a theoretical analysis based on a set of alternative hypotheses; i.e., a set of individually plausible, mutually exclusive, but altogether exhaustive kind of hypotheses. The theoretical results will enhance our perspective of nature, and may suggest critical experiments that may finally settle the issues. In the case of the endothelial cells, one set of alternative hypotheses are: either (1), that the cell contents are liquid-like, so that when there is no flow there is no shear stress, or (2), that the cell contents are solid-like. Under hypothesis (1), the shear load of the blood is resisted by the tension in the endothelial cell membrane (together with attached proteins). This case is analyzed in this paper. The alternative hypothesis of a solid interior endothelium is more complex because the mechanical machinery of the cell interior is still controversial, and many plausible structures should be examined. We can only discuss some general features at this time.

Our objective is to derive general results with as few ad hoc assumptions as possible. For example, we examine the endothelium as a continuum, pay attention to the interaction between neighboring cells. A micrographic image of a cross-section of an endothelium may appear either as (A) or as (B) in Fig. 1. In Fig. 1(A), the upper cell membranes of the neighboring cells meet at a junction at an angle. In Fig. 1(B), the neighboring cell membranes meet smoothly at the junctions. They appear to be similar, but we shall show that in case (B), a huge accumulation of tensile stress will occur in the upper cell membranes of the endothelial cells, with disastrous consequences.

In case (A), the stress concentration can be diminished. To achieve complete diminution, so that every cell sustains it's own share of shear load from the flowing blood, without accumulation from its neighbors, an exact difference of the slopes of the upper cell membranes at the cell junctions must be maintained, and this difference is a function of the slope of the side wall relative to the basal lamina. The reasons will be explained in Section 6 below.

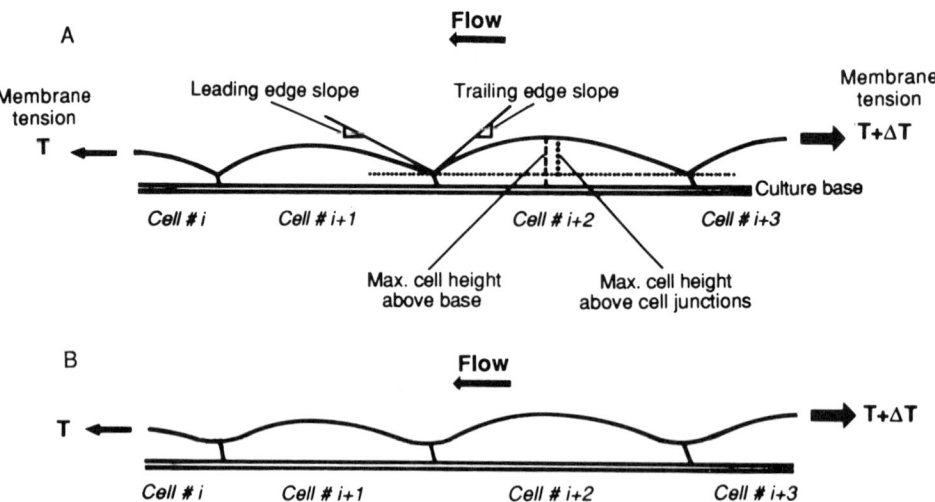

Figure 1. Blood shear induces tension in the upper endothelial cell membranes. There are two possible shapes of the cross-section of an endothelium. (A) The neighboring cell membranes intersect at an angle. (B) The neighbors meet smoothly. The way of tension transmission is very different in these two cases. See text. From (42), by permission.

THE EXTERNAL LOAD

For convenience, we shall call the surface of endothelial cells in an endothelium in contact with blood the **"upper"** cell membrane, the surface in contact with the basal lamina the **"lower"** cell membrane, and the surface in contact with the neighboring endothelial cells the **"side wall."** Figure 2 shows the geometry of the endothelium we shall study.

The upper cell membrane is subjected to the shear stress of the blood at all times in life.
The variations of the endothelial cell morphology, metabolism, and ultra-structure with the shear load have been studied by many authors (7, 23, 41, 51, 57, 58, 67, 68), In 1968, Fry called attention to the existence of a relationship between the shape of the endothelial cells and their nuclei and the shear stress in the blood (19). In 1969, Caro et. al. (4) called attention to a possible connection between arteriosclerosis and the shear stress imparted by the flowing blood. Since then, the subject has been studied vigorously (7, 10 - 13, 49, 62). In human coronary arteries, Giddens et. al. (25) have shown that the axial shear stress varies in the range of 1 to 2 Pa (N/m^2), with a mean value around 1.6 Pa. Kamiya et. al. (34, 35) have observed that the shear stress in flowing blood at the endothelial surface is of the same order of magnitude in all generations of arteries, large and small, including the aorta and capillaries. This, then, is the order of magnitude of the shear load acting on the endothelial cell membrane in contact with the flowing blood.
The exact value will depend on the local condition: entry, exit, branching, flow separation, secondary flow, etc.

The blood pressure acts on the upper cell membrane of the endothelium as an external normal load. The lower cell membrane of the endothelium is adhered to the basal lamina which is collagenous and is a solid. Hence, the boundary conditions of the endothelium on the lower surface are the continuity of displacements. The boundary conditions of each endothelial cell at the side wall are also the continuity of displacements, because it is well known that there are **tight junctions** connecting neighboring cell membranes along the side walls.

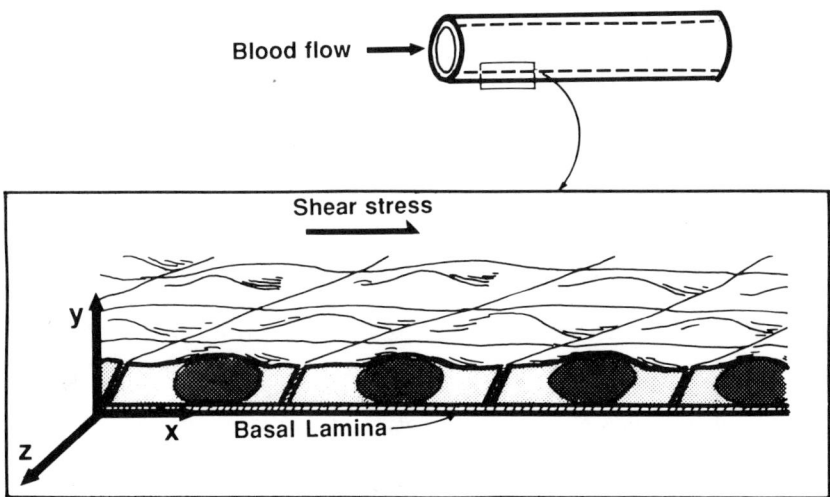

Figure 2. A schematic drawing of an arterial wall showing blood flow in the upper figure, and an enlarged view of the endothelium in the lower figure. The coordinates system shown here is used throughout this paper. From (21), by permission.

The displacements of the basal lamina are determined by the mechanics of the media and adventitia in arteries and veins, (21), and by the contiguous tissue in capillary blood vessels according to the "tunnel in gel" concept of Fung et. al. (17). These tissues are deformed by the blood pressure and shear stress transmitted to them through the endothelium. The stresses in the media and adventitia are much larger than the normal and shear stress in the blood. Chuong and Fung (6), have shown that in a cat thoracic aorta at a blood pressure of 16 kPa (120 mm Hg), a longitudinal stretch ratio of 1.69, and an opening angle of 71° at zero-stress state, the circumferential tensile Cauchy stress is 105 kPa at the inner wall of the media next to endothelium 61 kPa at the outer wall of the adventitia; the longitudinal tensile stress is 95 kPa at the inner wall, 71 kPa at the outer wall; and the radial compressive stress is -16 kPa at the inner wall, and 0 at the outer wall. The maximum shear stress at the inner wall is equal to one-half of the difference between the max principal stress and the min principal stress, (18), i.e., (105 + 16)/2 = 60.5 kPa, acting on a plane which is inclined at 45° to the radial and circumferential axes. This is about 3800 times larger than the shear stress acting on the surface of the endothelium due to blood flow. The maximum shear elsewhere in the media is similarly several thousand times larger than the shear stress of the blood acting on the endothelial cell surface.

The experimental blood pressure and blood shear stress are average values obtained under the assumption that the upper endothelial surface is smooth. Actually, the upper surface of the endothelium is uneven. This waviness is ignored in Poiseuille formula for steady flow, and Womersley formula for pulsatile flow. However, the shear stress acting on the upper cell membrane of the endothelial cells depends on the waviness of the wall. Mathematically, since blood may be considered as an incompressible Newtonian fluid with little error near the vessel wall, the governing equations are the Navier-Stokes equations. For the endothelium problem, the governing equations are similar to those occurring in the theories of peristaltic pumping and atherosclerotic plaques, which have been investigated by Fung and Yih (22), Yin and Fung (74). Lee and Fung (39) used conformal mapping to transform the wavy wall into straight wall. Haldar (27) used power series in the spatial coordinate. Yamaguchi et. al. (73) used Navier-Stokes software. See also Satcher and Dewey (61).

ALTERNATIVE HYPOTHESES ABOUT THE MECHANICAL PROPERTIES OF THE ENDOTHELIAL CELLS[*]

Since little is known about the mechanical properties of the internal parts of the endothelial cell, we are not ready to make a full stress analysis. In this article, we seize on the fact that the vascular endothelial cells in vivo exist in a continuous layer, and ask how do the cells resist the shear force from the blood flow. We ask first what part of the endothelial cell is fluid-like and what part is solid-like. All living cells are viscoelastic. A viscoelastic material is fluid-like if it resists a constant force by flow, as described by a Maxwell model or a generalized Maxwell model; it is solid-like if it resists a constant force by deformation, as described by a Voigt or Kelvin model, or a generalized Kelvin model. In an arterial wall there is no question that collagen and elastin fibers are solids, interstitial fluid is fluid. For the endothelium, we believe that the endothelial cell membrane (with spectrin, actin, tropomodulin, and protein 4.2, 4.1 attached to it) is solid-like because the cells can maintain their shape while they are subjected to a life-long shear force from the flowing blood. It reacts to the shear force with a deformation, not with a flow, hence, is a solid. The contents

---

[*] The following analysis presented in section 3 - 9 is taken from Fung and Liu (21) *J. Biomech. Eng.* (115: 1-12, 1993.)

of the endothelial cell, however, is a composite mixture. Some components, such as the nucleus and actin fibers, may be expected to be solid-like, but the mixture as a whole may be fluid-like. For example, it is known that the content of red blood cells is fluid-like (70), whereas the content of leukocytes obeys the Maxwell model (64). At this time, the rheological property of the content of endothelium is unknown. Hence, we shall make two alternative hypotheses: 1) The content is fluid-like, so that the cell membrane plays a dominant role in maintaining the shape of the endothelium at the steady state. 2) The content is solid-like, so that the cell membrane and the cell content together maintain the endothelium geometry. We investigate the consequences of these alternative hypotheses.

THE HYPOTHESIS OF FLUID-LIKE CELL CONTENT AND *TENSION-FIELD* IN CELL MEMBRANE

Under the hypothesis that the content of the endothelial cells have an overall rheology like a fluid, the shear stress in the cell content vanishes in the steady-state since there is no internal flow. Consequently, the cell membrane is the structure that resists the shear load from the blood flow.

As a structure, the cell membrane (the bilayer together with the attached proteins) is characterized by being very thin. A very thin structure has the special property that it is very easy to buckle, and cannot sustain any significant amount of compressive stress. This low compressive stress offers us a simplification of stress analysis. Explicitly, we assume that:

1. The cell membrane is so thin that it buckles easily and cannot support compression in its plane, and
2. Situation exists in which one of the principal strains in the deformed membrane is positive while the other one is negative or negligible.

A stress analysis based on these assumptions is called a *Tension-Field* theory (9, 54, 72). The word buckling means deformation out of the plane of the membrane, like creating a wrinkle in a piece of paper or cloth by compressing the edges. The critical buckling stress (the minimum uniaxial compressive stress to cause lateral instability and wrinkling of the membrane) limits the compressive force in the orthogonal direction, because beyond the critical condition the membrane can deform further without much additional compressive stress. The value of the critical stress depends on the thickness/length ratio of the membrane and the support condition at the edges. In an analog case of a square plate which is simply supported on all edges of length L, the critical buckling stress is (see Sechler (65), p. 386):

$$\sigma = \frac{\pi^2 E}{3(1 - \upsilon^2)} \left(\frac{t}{L}\right)^2, \tag{1}$$

where t is the thickness of the plate, E is the Young's modulus of elasticity, υ is the Poisson's ratio of the material. The Poisson's ratio of an incompressible material is 0.5. The Young's modulus of an incompressible material is 3 times the shear modulus. The shear modulus of red blood cell membrane is estimated to be 10^3 N/m² according to Skalak et. al. (70). The shear modulus of the endothelial cell membrane is unknown, but may be assumed to be of a similar order of magnitude as that of the red cell. Assuming t = 10 nm, L = 10 μm, E = 3 x 10^3 N/m², υ = 0.5, we obtain the critical buckling stress of 1.3 x 10^{-2} N/m². Comparing this with the estimated tensile stress of 10^3 N/m² acting in the direction of blood flow (see Eq. 8. infra), we see that the buckling stress of the endothelial membrane in the direction perpendicular to the blood flow direction is five orders of magnitude smaller. The exact value

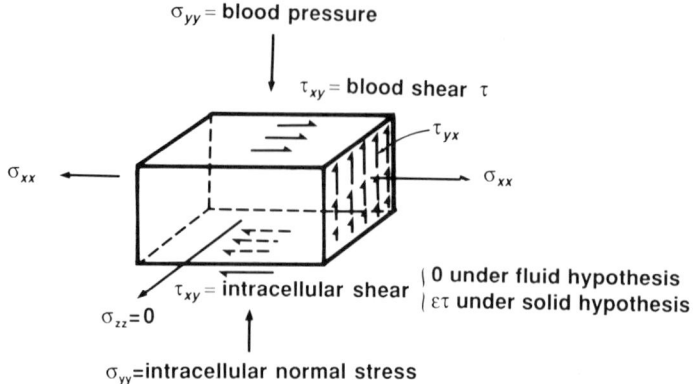

Figure 3. The components of stress acting on a small element of the upper endothelial membrane. The tensile stress σ_{xx} is usually much larger than all the other stresses. σ_{zz} is zero under the tension field assumption. On the top side of the membrane acts the blood pressure and shear. On the underside of the membrane acts the normal and shear stress of the cell content at the interface. Two alternative hypotheses are made with regard to the static stresses in the cell content. See text for details. From (21), by permission.

of the critical buckling stress depends on the shape of the membrane and the edge conditions: widening the width in the direction perpendicular to the load lowers critical stress, narrowing the width increases it, free edge condition lowers the critical stress, clamped edge condition raises it. The postbuckling stress may be somewhat higher than the critical stress, and is a nonlinear function of the amplitude of the wrinkles. But all this will not change the sustainable membrane compressive stress more than one order of magnitude or two. The premise of the *Tension-Field* theory is to say that these small compressive stresses acting in the direction perpendicular to the blood flow can be ignored altogether, and set to zero, in comparison to the tensile stress in the direction blood flow.

The *Tension-Field* theory would fail in any region where both principal stresses or strains are positive. There is no great disaster, however, if *Tension-Field* hypothesis fails, because one loses only a convenient simplification, and can always return to the full continuum mechanics.

TENSILE STRESS IN THE UPPER CELL MEMBRANE UNDER THE FLUID INTERIOR HYPOTHESIS

Fig. 2 shows a flow of blood which causes a shear stress to act on the blood vessel wall. A coordinate system is attached to the vessel wall. Below the blood vessel is shown a schematic drawing of the endothelium, with a longitudinal cross-section in the front. Each upper cell membrane has a system of internal stress and strain as shown in Fig. 3. With reference to a rectangular Cartesian frame of reference (x, y, z) with the x-axis pointing in the direction of the blood flow and y-axis normal to the membrane, the stress tensor has six independent components $\sigma_{xx}, \sigma_{yy}, \tau_{zz}$, and $\tau_{xy} = \tau_{yx}, \tau_{yz} = \tau_{zy} \tau_{zx} = \tau_{xz}$. The magnitude of some components of stress is much larger than that of the others. The normal stress σ_{yy} is equal to the blood pressure on the top side, and to the intracellular static pressure on the under side. The shear stresses τ_{xz}, τ_{yz} are most likely to be very small at steady state if the fluid-mosaic concept of the lipid bilayer promulgated by Singer et. al. (69) is valid. If we accept this concept, then:

Mechanics of the Endothelium in Blood Flow

$$\tau_{zx} = \tau_{yz} = \tau_{xz} = \tau_{zy} = 0. \tag{2}$$

The shear stress τ_{yx} on the upper surface of the upper endothelial cell membrane, which is in contact with the flowing blood, must be equal to the viscous shear stress of the blood, τ. The shear stress τ_{yx} on the bottom of the upper cell membrane depends on the rheological behavior of the cell content. Under the hypothesis that the cell content is fluid-like, then τ_{yx} would be zero if the plasma streaming inside of the cell can be ignored. Plasma streaming is believed to be small. Hence, in the upper cell membrane under the fluid interior hypothesis:

$$\tau_{xy} = \tau_{yx} = \tau \text{ on the surface in contact with blood flow,}$$

$$\tau_{xy} = \tau_{yx} = 0 \text{ on the surface facing cell interior.} \tag{3}$$

Furthermore, according to the *Tension-Field* theory:

$$\sigma_{zz} = 0. \tag{4}$$

Thus, there is left only one normal stress, σ_{xx}, to be considered. The equation of equilibrium of forces in the x-direction acting on an element of the membrane is (see Ref. 18, p. 72):

$$\frac{\partial \sigma_{xx}}{\partial x} + \frac{\partial \tau_{xy}}{\partial y} + \frac{\partial \tau_{xz}}{\partial z} = 0. \tag{5}$$

The third term vanishes according to Eq. 2. Integrating Eq. 5 with respect to y from y = 0 to y = t, the thickness of the membrane, we obtain, by using Eq. 3 and the notation $\overline{\sigma}_{xx}$ for the average value of $\overline{\sigma}_{xx}$ throughout the membrane thickness, the result:

$$t \frac{\partial \overline{\sigma}_{xx}}{\partial x} + \tau = 0, \tag{6}$$

where τ is the shear stress in the blood at the endothelium. Assuming τ to be constant in the length of an endothelial cell and integrating Eq. 6 with respect to x, we obtain:

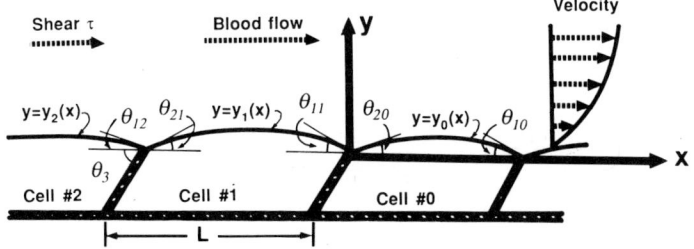

Shape of cells #0, #1, #2. Nomenclature and symbols.

Figure 4. A longitudinal profile of the endothelium showing endothelial cells adhered to a basal lamina. The nomenclature of the cell numbers, dimensions, angles at the junctions, and the equations describing the curve of the upper cell membrane relative to the x-y coordinated (y = y_1 (x) etc.) are given. From (21), by permission.

$$\bar{\sigma}_{xx} = -\frac{\tau x}{t} + (\bar{\sigma}_{xx})_0. \qquad (7)$$

The integration constant $(\bar{\sigma}_{xx})_0$ is independent of x, but can be a function of z. If $(\bar{\sigma}_{xx})_0 = 0$, and $x = -L$ cm, then:

$$\bar{\sigma}_{xx} = \tau L / t. \qquad (8)$$

Thus, we see that if the viscous shear stress of the blood is $\tau = 1$ N/m², the length of the endothelium is 1 cm, the thickness of the endothelial cell membrane is 10 nm, then the tensile stress σ_{xx} is equal to 10^6 N/m2. If the length L is taken to be that of a single cell, of order 10 μm, then $\sigma_{xx} = 10^3$ N/m².

These numerical examples show how significant the length L and the integration constant $(\bar{\sigma}_{xx})_0$ are. A proper treatment of these constants is to consider the transmission of tension between neighboring cells. Note here how large σ_{xx} can be! In Fung and Liu (21), the effect of such a high tension in the upper cell membrane on the shape of the cell nucleus is discussed. Note also, that when the membrane tensile stress σ_{xx} is large, the shear stress in plane sections of the membrane at 45° from the x-axis is also large. Since the maximum shear (called octahedral shear) is equal to $\sigma_{xx}/2$ exactly. Thus, we see that the octahedral shear in the upper cell membrane is many orders of magnitude higher than the blood shear itself.

TRANSMISSION OF THE TENSION IN UPPER ENDOTHELIAL CELL MEMBRANE TO THE BASAL LAMINA THROUGH THE SIDE WALLS

Let us consider an endothelium consisting of a confluent layer of cells whose profiles are schematically drawn in Fig. 4, and consider the role played by the side walls in transmitting some tension in the upper cell membrane to the basal lamina.

With a frame of reference as shown in Fig. 4, assume that the condition is two-dimensional and independent of the coordinate z. The upper membrane of cell No. 1 is described by an equation $y = y_1(x)$, that of cell No. 2 by $y = y_2(x)$, etc. The tensile force

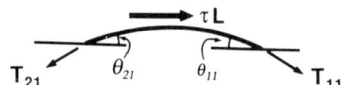

(A) Forces acting on the upper cell membrane of cell #1

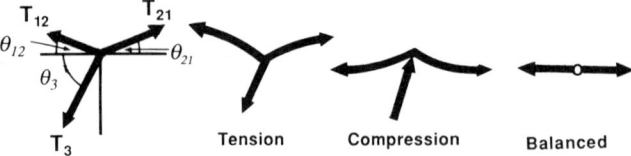

(B) Equilibrium of forces at the junction of cells #1 and #2

Figure 5. (A) The shape of the upper cell membrane of cell No. 1 with the nomenclature of the angles and tensions indicated. (B) The forces acting at the upper junction of cells No. 1 and No. 2. From (21), by permission.

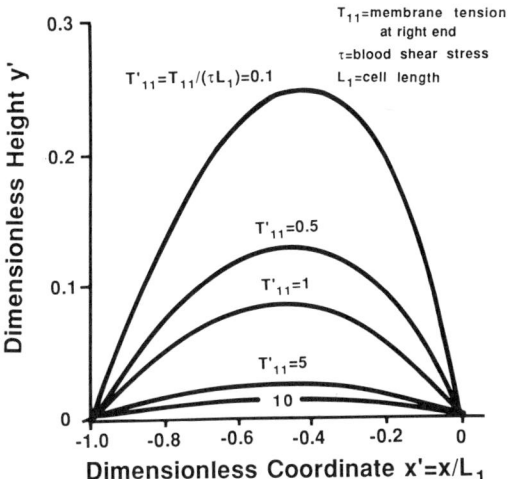

Figure 6. Theoretical shape of the upper cell membrane as a function of the parameters T'_{11} which is equal to the tension per unit width at the right hand end of the cell membrane of cell No. 1 divided by the blood shear force per unit width, τL. The ordinate is dimensionless and is equal to the actual height divided by the product of the length of the cell and the dimensionless pressure parameter p', which is equal to the static pressure difference (also equal to the osmotic pressure difference) divided by the blood shear stress τ. The abscissa is the logitudinal coordinate. From (21), by permission.

per unit length in the upper membrane is equal to the product of the membrane thickness t and the mean stress $\overline{\sigma}_{xx}$, i.e., $t\overline{\sigma}_{xx}$, whose value in cell No. 1 at the junction with cell No. 0 is denoted by T_{11}, that at the junction with cell No. 2 is denoted by T_{21}. The slope of the upper membrane of cell No. 1 at the right end is denoted by $\tan\theta_{11}$, that at the left end is denoted by $\tan\theta_{21}$. The equation of equilibrium of the upper membrane of cell 1 is (see Fig. 5(A)):

$$T_{21}\cos\theta_{21} = T_{11}\cos\theta_{11} + \tau L, \tag{9}$$

where L is the length of the cell No. 1. The balance of forces at the junction of the upper membranes of cell No. 1 and cell No. 2, (Fig. 5(B)) is governed by the equations:

$$T_{21}\cos\theta_{21} - T_{12}\cos\theta_{12} = T_3\sin\theta_3 = T_3\cos\theta_4, \tag{10}$$

$$T_{21}\sin\theta_{21} + T_{12}\sin\theta_{12} = T_3\cos\theta_3 = T_3\sin\theta_4, \tag{11}$$

where the θ's are the angles of inclination indicated in Fig. 5(B), and T_3 is the tension in the side wall. Dividing Eq. 10 by Eq. 11, we obtain:

$$\tan\theta_3 = (T_{21}\cos\theta_{21} - T_{12}\cos\theta_{12})/(T_{21}\sin\theta_{21} + T_{12}\sin\theta_{12}). \tag{12}$$

Squaring both sides of Eqs. 10, 11, adding, and simplifying, we obtain:

$$T_3^2 = T_{12}^2 + T_{21}^2 - 2T_{12}T_{21}\cos(\theta_{12} + \theta_{21}). \tag{13}$$

Since T_{12}, T_{21} are positive, Eq. 11 shows that the membrane tension T_3 in the side wall is positive when θ_{12}, θ_{21} are positive, whereas T_3 becomes negative when θ_{12}, θ_{21} are negative.

According to our *Tension-Field* hypothesis, T_3 cannot be negative (compressive). The smallest value T_3 can have is zero. Setting $T_3 = 0$ in Eqs. 10 and 11, we deduce that:

$$T_{21} \sin(\theta_{21} - \theta_{12}) = 0,$$

$$T_{21} - T_{12} \cos(\theta_{21} + \theta_{12}) = 0. \tag{14}$$

Since T_{12}, T_{21} are positive, the unique solution of Eqs. 12 - 14 is:

$$T_{12} = T_{21}, \quad \theta_{12} = \theta_{21} = \theta_3 = 0, \tag{15}$$

which says that the upper membrane must be flat at the junction and the side wall must be vertical. This is obviously reasonable because if the side wall has no tension, then the two tensile forces in the membranes must pull each other in a single straight line, see the last diagram in Fig. 5(B).

Hence, **the side wall transmits tension in the upper cell membrane to the basement membrane if and only if θ_{12} and θ_{21} are positive, i.e., if and only if the upper cell membrane bulges into the blood stream.** The membrane will bulge outward if there is a static pressure pushing it out. The static pressure in a cell is controlled by the Starling's law for fluid movement across the cell membrane, which states that the rate of outward movement of fluid across the cell membrane, $\dot{m}$, is equal to the product of the coefficient of permeability, k, and the differences of the static pressure p and osmotic pressure π inside the cell (subscript "i") and outside the cell (subscript "o"). Thus, (see Ref. 19, p. 291):

$$\dot{m} = k[p_i - p_o) - (\pi_i - \pi_o)]. \tag{16}$$

At equilibrium, the fluid movement is zero (m vanishes), the static pressure difference balances the osmotic pressure difference:

$$p_i - p_o = \pi_i - \pi_o. \tag{17}$$

The static pressure difference deflects the cell membrane according to Laplace's formula (18, p. 24):

$$t \bar{\sigma}_{xx} \cdot \text{curvature} = p_i - p_o. \tag{18}$$

The curvature of the upper membrane of cell No. 1 shown in Figs. 4 and 5(A), is equal to the negative of the second derivative of the cell membrane surface given by the equation $y = y_1(x)$, if the slope of the surface is sufficiently small. Hence, for small deflection, the differential equation for the cell membrane of the cell No. 1 is, on account of Eqs. 7 and 17:

$$-(T_{11} - \tau x) \frac{d^2 y_1}{dx^2} = p_i - p_o. \tag{19}$$

Introducing the dimensionless variables:

$$y' = y_1/L_1, \quad x' = x/L_1, \quad T' = T_{11}/(L_1 \tau), \quad p' = (p_i - p_o)/\tau, \tag{20}$$

we have

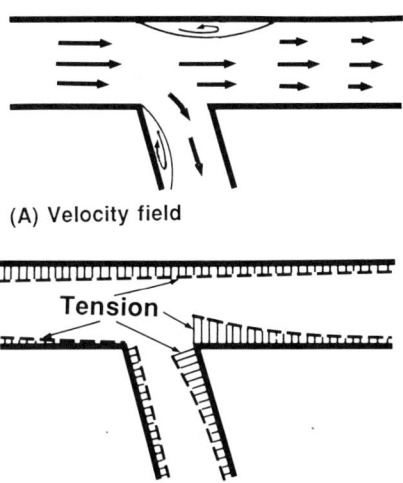

(A) Velocity field

(B) Tension in upper endothelial cell membrane is proportional to the height of the markers from the vessel wall

Figure 7. An illustration of the possible major difference between the distribution of blood shear stress acting on the vessel wall and the distribution of the tensile stress in the cell membrane of the endothelial cells in contact with the blood. The static pressure inside the cell is assumed to be equal to the static pressure of the blood. Under this assumption, the tension in the cell membrane of one cell can be transmitted to the next cell and become accumulated. In the upper figure (A), the velocity profile in a vessel with a branch is shown with two separation regions have secondary flow. The shear stress is proportional to the velocity gradient (not shown). In the lower figure (B), the tensile stress in the upper cell membrane of the endothelial cells is plotted by tick marks perpendicular to the vessel wall. The higher the marker, the larger is the tensile stress. The dotted profile is that of the tensile stress distribution. From (21), by permission.

$$(T' - x') \frac{d^2 y'}{dx'^2} = -p'. \tag{21}$$

The boundary conditions are that the deflection y_1 must vanish at the two ends of the upper membrane (Fig. 4):

$$y' = 0 \text{ when } x' = 0 \text{ and } x' = -1. \tag{22}$$

The solution of Eqs. 21, 22 is:

$$\frac{dy'}{dx'} = -p' \log(T' - x') + c_1$$

$$y' = p'(T' - x')[\log(T' - x') - 1] + c_1 x' + c_2,$$

$$c_2 = -p' T' (\log T' - 1),$$

$$c_1 = -p'(T' + 1)[\log(T' + 1) - 1] + c_2. \tag{23}$$

It is seen that y' is linearly proportional to p', and y'/p' depends only on one variable, T'. Fig. 5 shows the curves of y'/p' vs. x' with T' as a parameter. These show that the angles θ_{11}, θ_{21} are known functions of T_{11}' and p'.

THE CASE OF ZERO STATIC PRESSURE DIFFERENCE, $P_I - P_O = 0$

The case shown in Fig. 1(A) is a special case of zero static pressure difference, $p_i - p_o = 0$. According to Eq. 17, this requires $\pi_i - \pi_o = 0$, so there is no osmotic pressure difference between the content of the endothelial cell and the blood. The case $p_i < p_o$ is unattainable, because it will cause the cell membrane to bulge inward, compressing the side wall (see Fig. 5(B)), causing it to buckle, and returning the cell membrane to a flat configuration, $p_i = p_o$.

When the upper cell membranes of successive endothelial cells are all flat at the cell junctions, then in order to bear the shear stress of the flowing blood, the membrane tension will increase linearly with the distance in a direction opposite to the blood flow. Fig. 7 shows the consequence of this conclusion. The top sketch shows a flow in a blood vessel with a side branch. Two separation zones are shown in which the shear stress is reversed locally. In spite of the change of local shear stress on the vessel wall due to blood flow, in the present case the tensile stress in the upper endothelial cell membrane will grow continuously from one cell to the next as shown in the lower diagram in Fig. 7, which indicates the size of the tensile stress in the upper cell membrane by tick marks perpendicular to the vessel wall. The higher the marks are, the larger is the tensile stress. This sketch is drawn for the case $p_i - p_o = 0$. In this case, the cumulative growth of membrane tension is somewhat mitigated by the reversed flow, but the tension remains high in these regions. On the flow divider, the tensile stress reaches a peak at the apex. This is a mechanism unsuspected before. If the membrane tension is related to the membrane permeability, then this mechanism may be relevant to atherosclerosis (7, 45, 53).

THE HYPOTHESIS OF A SOLID-LIKE CELL CONTENT

Let us now consider the alternative hypothesis that rheologically the content of the endothelial cell is a solid, i.e., it resists a static stress by a deformation from a zero-stress state. Several endothelial cell models have been proposed (33, 51, 63, 67, 68). Use of each model will lead to a specific result. We shall attempt, however, to obtain some results that do not depend on a specific model, but only on the fact that the cell content has a zero-stress state under this hypothesis. The zero-stress state may not coincide with the no-load state; the difference is the residual stress which must be self-equilibrating. Under the shear load of the blood that acts on the upper cell membrane, a system of stress and strain is induced in the cell through the following boundary conditions on the cell content-cell membrane interface, (see Fig. 3):

Normal stress σ_{yy} of cell content = that of cell membrane,

Shear stress τ_{xy} of cell content = that of cell membrane. (24)

Under the solid-interior hypothesis, we shall assume that part of the shear load τ acting on the top side of the upper cell membrane is transmitted through the membrane to its lower side, i.e., at the interface:

$$\tau_{xy} \text{ of cell membrane} = \varepsilon\tau, \quad 0 \leq \varepsilon < 1, \quad (25)$$

where ε is a number between 0 and 1. Then it follows that, at the interface:

$$\tau_{xy} \text{ of cell content} = \varepsilon\tau. \quad (26)$$

The fraction ε is influenced by the condition of integration between the cell membrane and cell content, and by the relative compliance of the cell content and the cell membrane. But, as long as the cell membrane is recognized as an important mechanical part of the cell, ε _ 1, and the role played by the cell membrane can be examined through these boundary conditions.

For the cell membrane, the shear stress is τ on the surface in contact with the blood, and is ετ on the surface in contact with the cell content. The equation of equilibrium of the cell membrane now becomes:

$$\frac{\partial N_x}{\partial x} + (1 - \varepsilon)\tau = 0. \tag{27}$$

where N_x is the membrane tension, $\tau\sigma_{xx}$. Hence, by integration:

$$N_x = -\int_0^x (1 - \varepsilon)\tau\, dx + N_o. \tag{28}$$

If ε is a constant, then:

$$N_x = -(1 - \varepsilon)\tau x + N_o. \tag{29}$$

Hence, with a modification of replacing τ by (1 - ε) τ, all the conclusions reached in the preceding sections remain valid.

A similar argument applies to the side walls of the cells. If we denote the shear stress acting on the side wall due to the solid content by ε't, where ε' is a constant which may differ from ε, then under the solid-content hypothesis the equations of equilibrium of the forces in the cell membranes meeting at a junction of cells remain valid except that T_{12}, T_{21} should be reduced by a factor of (1 - ε) and T_3 should be reduced by a factor of (1 - ε'). The relationships between the T's, p's, o's, and osmotic pressures remain valid with the T's reduced as noted above.

With regard to the *Tension-Field* theory, we note that with a solid interior that supports the cell membrane elasticity, the critical buckling stress of the cell membrane will be higher than that of Eq. 1, and in the postbuckling state the membrane can bear a greater compressive stress. Hence, the *Tension-Field* theory may not apply and we may have to analyze the cell as a shell. However, Kim et. al. (36) has shown that the F-actin stress fibers in endothelial cells are either bundled about the cell periphery, or are aligned with the direction of blood flow. This suggests that the elastic support given to the cell membrane by the stress fibers lies in the flow direction, and not in the direction perpendicular to the flow. Hence, the possible compressive stress in the direction perpendicular to the flow could be very small compared with the tensile stress in the direction of flow, and the simplifying assumption of a *Tension-Field* in the direction of the flow may remain valid.

TURBULENT FLOW

Experiments by Davies et. al. (12) and Helminger et. al. (29) have shown that turbulent flow with a mean correlation length of about 5 times the cell length can cause large increase in cell division and surface cell loss. In a turbulent flow, let the blood pressure acting on the upper endothelial cell membrane be separated into a mean part and a perturbation:

$$p_o = \bar{p}_o + p'_o(x, t), \tag{30}$$

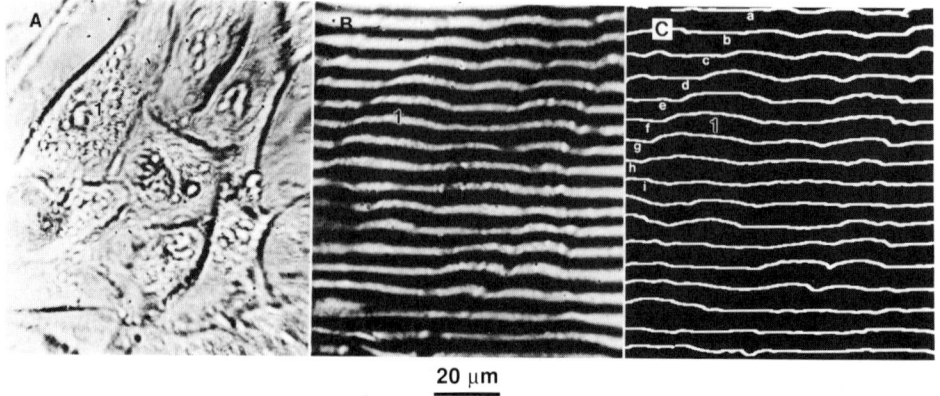

Figure 8. Microscopic images of cultured endothelial cells in a flow chamber at a flow rate that generates a shear stress of about 1 N/m². (A) Ordinary plane image. (B) Images of the same cells as shown in A under an interference microscope. (C) Images of B enhanced by using an Optimas image processing system. Medium osmolarity: 310 mosM. Medium flow direction: right to left. From (42), by permission.

where $\bar{p}_o$ is the mean value of p_o, and $p'_o(x, t)$ is a function of space and time whose average value over a sufficiently long period of time and a sufficiently large area vanishes. Similarly, the surface shear stress from the blood is split into a mean and perturbation about the mean:

$$\tau = \bar{\tau} + \tau'(x, t), \qquad (31)$$

where $\bar{\tau}$ is the mean value of τ and is a constant, whereas τ' is the perturbation about the mean, so that its mean value vanishes. In response to the external load p_o and t, internal stresses are induced in the cell membranes and cell contents. Since there is motion in the cells, there is shear stress in the cell content under both fluid-like and solid-like hypotheses. Hence, we use Eq. 26 to describe the shear stress on the inner wall of the cell membrane. The equation of motion of the upper cell membrane is obtained by adding an inertial force term (mass times acceleration) to the equation of equilibrium, Eq. 27. Under *Tension-Field* theory and using Eq. 31, we have:

$$\frac{\partial N_x}{\partial x} + (1 - \varepsilon)[\bar{\tau} + \tau'(x, t)] = \rho h \frac{\partial^2 u}{\partial t^2}, \qquad (32)$$

where ρ is the mass density of the cell membrane, h is the membrane thickness, u is the displacement of the mass particles of the membrane, and t is time. An estimate of the order of magnitude of the quantities in this equation, with $\bar{\tau}$ about one N/m², ρ about 10^3 kg/m³, h about 10 nm, u about 1 μm, the frequency less than 1000 Hz, we can show that the inertial force term is 5 orders smaller than the shear load $\bar{\tau}$, and can be neglected in Eq. 32. The equation of motion Eq. 32 is reduced to Eq. 27, and the solution is given by Eq 28. Hence, the cell membrane's stress response to the turbulent flow is instantaneous and quasi-static.

Thus, in a turbulent flow, the tensile stress in the upper cell membrane reacts to the pressure $\bar{p} + p'(x, t)$ and shear $\bar{\tau} + t(x, t)$ in the same way as described in Sections 6 to 8.

A major effect of turbulence on the tensile stress in the cell membrane is revealed by this analysis. Suppose that the static pressure difference $p_i - p_o$ fluctuates around a mean value of zero, i.e., $\bar{p}_i - \bar{p}_o = 0$. The instantaneous pressure difference $p_i - p_o$ oscillates between

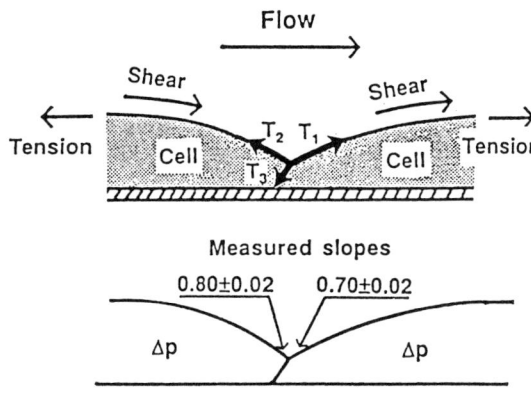

Figure 9. Left: Forces in the endothelial cell membranes. Right: Experimental results of cultured endothelial cells subjected to a flow with a wall shear of about 1 N/m² for 60 min., fluid 310 mosM.

positive and negative values. When $p_i - p_o$ is instantaneously positive, the tension in the upper cell membranes can be transmitted to the basal lamina. When $p_i - p_o$ is instantaneously negative, the transmission in the side walls ceases and the tensile stress from one cell is transmitted to the next and the stress accumulates. The larger the turbulence scale the more severe is the accumulation. This is a kind of off-and-on chain reaction, whose interval, duration, and severity are statistical.

Thus, in a turbulent flow, the tension in the side cell membranes fluctuates, transiently pulling neighboring cells apart if the adhesion of the side wall is not perfect at the junction at all times. The side wall tension also tends to pull the cell away from the basal laminar. In the meantime, the upper cell membrane tension will compress the nuclei. When the cell membrane tension oscillates, the dynamic action will induce an oscillatory motion in the nuclei. These transient events are more severe whenever $p_i - p_o$ can be negative, irrespective of the mean value of $\bar{p}_i - \bar{p}_o$. We suggest that these fluctuations may contribute to the surface cell loss observed by Davies et. al. (12).

A CRITICAL EXPERIMENT[*]

An experimental measurement of the slopes of the upper cell membranes of the endothelium will definitely help. An atomic force microscope may be applied (1, 2), but it is difficult to use it in a flowing fluid. Electron microscopy requires tissue fixation and dehydration, which can induce geometric distortion. Facing these difficulties, we recalled our early work on the determination of the thickness profile of erythrocytes by using a Mach-Zender interference microscope (14, 20), which yielded data with a resolution of 0.02 μm based on physical optics. Hence, we hypothesized that it might yield the endothelium profile.

The experimental details are given in (42). Briefly, a human vascular endothelial cell line was provided by C.-J.S. Edgell (University of North Carolina). The cells were cultured to confluent according to Emeis and Edgell (8). A flow chamber was designed and fabricated (42). The cultured endothelium was subjected to a flow of specific wall shear stress of about 1 N/m² for a specific length of time in a circulating culture medium at 30°C at an osmolarity

---

[*] The material in this section is taken from Liu, Yen and Fung (42), *Proc. Natl. Acad. Sci., USA* 91: 8782-8786, 1994.

of 310 mosM or other specific values. A Leitz interference microscope was used. The principle is described in (3) and (14). A light beam is split into two beams, which separately pass through two identical lens systems in the microscope with a flow chamber containing the cells and culture medium in the path of one beam, and another identical chamber containing the culture medium alone in the path of the other bean. When the two beams are recombined, they interfere and produce a pattern of alternate bright and dark fringes. When the cells are absent, the interference fringes are straight lines. Deviation from a straight line is caused by a light beam passing through cells inducing a phase shift ϕ which is linearly proportional to the product of the thickness of the cell, D, and the difference of the index of refraction of the cell, η_c, and that of the medium, η_m. If λ denotes the wavelength of the light (0.546 mm in our experiment), then the phase shift is given by the formula (3):

$$\phi = (2\pi/\lambda)(\eta_c - \eta_m)D. \tag{33}$$

We used a matching-solution method (3) to measure η_c, and Snell's law to measure η_m. See (3) for details. A typical example is shown in Fig. 8, which refers to a case in a flow with a shear stress of ≈ 1 N/m^2 for 60 min. The flow direction was then reversed, and the same protocol repeated. Further repetitions were made with culture media osmolarity changed by adding NaCL or distilled water.

In Fig. 8, the panel A shows an ordinary image; panel B shows an interference image of the same cells with undisturbed fringes paralleled to the streamlines of flow; panel C shows an image enhanced by using a Optimas software. Further processing and extraction of the data on slope, cell height, the transient effect of osmotic pressure changes, etc., are given in (42).

The measured values of the leading- and trailing-edge slopes were -0.70 ± 0.02 (SE) and 0.80 ± 0.02 (SE), respectively, Fig. 9. The absolute value of the slope at the leading edge of the cells was significantly smaller than that at the trailing edge (P<0.001). The maximal height of the upper cell membrane above the cell junctions was 2.50 ± 0.05 (SE) mm, and the maximum height of the cells above the base was 3.46 ± 0.05 (SE) mm. A reversal of the flow direction induced a reversal of the slopes at these junctions. These results suggest that the shear stress acting on the endothelial cells is responsible for the progressive change in the slopes of the upper cell membrane.

The slope and curvature of the cell membrane indicates that the static pressure inside the endothelial cell is higher than that in the flowing fluid, and that tension exists in the cell membrane. This follows the Laplace formula under the fluid interior hypothesis: tension times curvature equals the static pressure difference across a membrane. Under the solid interior hypothesis, if the term "static pressure" is replaced by the outward normal stress acting on the cell membrane, the Laplace formula remains valid. Thus, we see that the outward bulging implies an excess of the internal hydrostatic pressure over the cytoskeletal stress, and the existence of cell membrane tension.

As analyzed in Ref. (21), the larger the slopes of the cell membranes at the junctions, the smaller may be the transmission of tension from one cell to the next, and less the accumulation of tensile stress due to flow. The smaller the slopes, or the smoother the cell membranes at the junctions, the larger the tension accumulation due to a long line of cells. The slopes we found are relatively small, indicating that the membrane tensile stress accumulation mechanism exists and must be carefully studied in the future.

CONCLUSION

The molecular biology of the endothelial cell has progressed so much that the chemical makeup of the cell is becoming better understood day by day; but the detailed structure of the cell is not yet so clear; whereas the mechanical properties of the structural components in the cell remain practically totally unknown. Hence, we begin the study of cell mechanics by searching for those features that are valid for a wide variety of mechanical properties, i.e., almost independent of these details. We focused on the fact that the endothelium is a continuum of endothelial cells. We found that the accumulation of tensile stress in the upper cell membrane of the endothelium is a major feature. Accumulation in the upstream direction. In arteries, the largest accumulated tensile stress should occur in large arteries and aorta, because they are the most upstream. In veins, the largest accumulated tensile stress should occur in the small veins and venules, because they are most upstream in the venous system.

But cell mechanics needs to go beyond these elementary features. When the structural components of a cell are better known, we must determine their constitutive equations so that a detailed mechanical analysis of the cell can begin. Hence, we are back to a cross road where the determination of the constitutive equations should claim our first priority. For a very complex system, such as the endothelium, it is very unlikely that the constitutive equations can be obtained solely by empirical approach. A practical approach is to propose the form of constitutive equations on theoretical ground, formulate and solve boundary value problems relevant to the experimental determination of constitutive equations on the basis of the proposed ones, perform the relevant experiment, compare the theoretical and experimental results, and learn the necessary improvements. This involves hard work for which high quality is the first requirement. I expect that this is the trend for the next decade.

REFERENCES

1. Barbee, K.A., Davies, P.F., and Lal, R.: Shear stress-induced reorganization of the surface topography of living endothelial cells imaged by atomic force microscopy. *Cir. Res.* 74: 163-171, 1994.
2. Barbee, K.A., Macarak, E.J., Thibault, L.E.: Strain measurements in cultured vascular smooth muscle cells subjected to mechanical deformation. *Ann. Biomed. Eng.* 22: 14-22, 1994.
3. Barer, R. and Joseph, S.: Refractometry of living cells. Part 1, Basic principles. Part 2, The immersion medium. *Q. J. Micros. Sci.* 95: 399-423, 1954 and 96: 1-27, 1955.
4. Caro, C.G., Fitz-Gerald, J.M., and Schroter, R.C.: Atheroma and arterial wall shear- observation, correlation, and proposal of a shear-dependent mass-transfer mechanism for atherogenesis. *Proceed. of Roy. Soc.*, London, [Biol.] 177: 109-159, 1971.
5. Chen R.Y.Z., Chang Ch.H. and P.H. Guth. gastric arteriolar and venular responses to nitrogeneous and nonnitrogenous vasodilating agents in the rats. *Int.J.Microcirc.* 14: 197-203, 1994.
6. Chuong, C.J., and Fung, Y.C.: "On residual stress in arteries," *J. Biomech. Eng.* 108: 189-192, 1986.
7. Curry, F-R.E.: Mechanics and thermodynamics of transcapillary exchange. In Handbook of Physiology, - Cardiovascular System IV, Part I, pp. 309-374, American Physiological Society, Bethesda, MD.
8. Emeis, J.J. and Edgell, C.-J. S.: Fibrinolytic properties of a human endothelial hybrid cell line (Ea. hy 926). *Blood.* 71: 1669-1675, 1988.
9. Danielson, D.A., and Natarajan, S.: Tension field theory and the stress in stretched skin. *J. of Biomech.* 8: 135-142, 1975.
10. Davies, P.F., Dewey, C.F., Bussolari, S.R., Gordon, E.L., and Gimbrone, M.A. Jr.: Influence of hemodynamic forces on vascular endothelial function. In vitro studies of shear stress and pinocytosis in bovine aortic cells. *J. Clin. Invest.* 73: 1121-1129, 1984.
11. Davies P.F., and Tripathi S.C.: Mechanical stress mechanisms and the cell. An Endothelial paradigm. *Circ. Res.* 72: 239-245, 1993.
12. Davies, P.F., Remuzzi, A., Gordon, E.F., Dewey, C.F.,Jr., Gimbrone, M.A., Jr.: Turbulent fluid shear stress induces vascular endothelial cell turnover in vitro. *Proc. Natl. Acad. Sci.* 83: 2114-2117, 1986.
13. Dewey, C.F., Bussolari, S.R., Gimbrone, M.A., and Davies, P.F.: The dynamic response of vascular endothelial cells to fluid shear stress. *J. Biochemical Engineering* 103: 177-185, 1981.
14. Evans, E. and Fung, Y.C.: Improved measurements of the erythrocyte geometry. *Microvasc* 4: 335-347, 1972.

15. Frangos, J.A., Eskin S.G., McIntire L.V., Ives C.L.: Flow effects on prostacyclin production by cultured human endothelial cells. *Science* 227: 1477-1479, 1985.
16. Fry, D.L.: Acute vascular endothelial changes associated with increased blood velocity gradients. *Circ. Res.* 22: 165-197, 1968.
17. Fung, Y.C., Zweifach, B.W. and Intaglietta, M.: Elastic environment of the capillary bed. *Circ. Res.* 19: 441-461, 1966.
18. Fung, Y.C.: *A First Course on Continuum Mechanics*, 3rd ed. Prentice Hall, Englewood Cliff, N.J., 1993.
19. Fung, Y.C.: *Motion, Flow Stress, and Growth.* Springer-Verlag, New York, 1990.
20. Fung, Y.C.: *Biomechanics: Mechanical Properties of Living Tissues.* Springer-Verlag, New York, 1st ed. 1981. 2nd ed. 1993.
21. Fung, Y.C., and Liu, S.Q.: Elementary mechanics of the endothelium of blood vessels. *J. Biomechanical Engineering,* 115: 1-12, 1993.
22. Fung, Y.C. and Yih, C.S.: Peristaltic transport. *J. Appl. Mech.* 35, E: 669-675, 1968.
23. Gau, G.S., Ryder, T.A., and MacKenzie, M.L.: The effect of blood flow on the surface morphology of the human endothelium. *J. of Path.* 131: 55-60, 1980.
24. Geister, A.A.T., M.J. Peach, and G.K. Owen: Angiotensin II induces hypertrophy, not hyperplasia, of cultured rat aortic smooth muscle cells. *Cir. Res.* 62: 747-756, 1988.
25. Giddens, D.P., Zarins, C.K., and Glagov, S.: Response of arteries to near-wall fluid dynamic behavior. *App. Mech. Rev.* 43: S98-S102, 1990.
26. Griffin, S.A., W.C.B. Brown, F., MacPherson, J.C., McGrath, V.G. Wilson, N., Korsgard, M.J., Schelling, H., Fischer, and D. Ganten: Angiotensin II and growth: a link to cardiovascular hypertrophy? *J. Hypertension* 9: 3, 1991.
27. Haldar, K.: Analysis of separation of blood flow in constricted arteries. *Arch Mech, Warszawa,* 43: 103-109,1991.
28. Hamet P., Hadrava V., Kruppa U., and Tremblay J.: Transforming growth factor 1 expression and effect in aortic smooth muscle cells from spontaneously hypertensive rats. *Hypertension* 17: 896-901, 1991.
29. Helmlinger, G., Geiger, R.V., Schreck, S., and Nerem, R.M.: Effects of pulsatile flow on cultured vascular endothelial cell morphology. *J. of Biomech. Eng.* 113: 123-131, 1991.
30. Hsieh H.J., Li N.Q., Frangos J.A.: Shear-induced platelet-derived growth factor gene expression in human endothelial cells is mediated by protein kinase C. *J. Cell Physiol.* 150: 52-558, 1992.
31. Ingber, D.: The riddle of morphogenesis: A question of solution chemistry of molecular cell engineering? *Cell* 75: 1249-1252, 1993.
32. Ingber, D.E. and Folkman,J.: Machanochemical switching between growth and differentiation during fibroblast growth factor-stimulated angiogenesis in vitro: Role of extracellular matrix. *J. Cell Biol.* 109: 317-330, 1989.
33. Johnson, P.C.: *Peripheral Circulation*, John Wiley, New York, 1978.
34. Kamiya, A., Bukhari, R., and Togawa, T.: Adaptive regulation of wall shear stress optimizing vascular tree function. *Bull. of Math. Biol.* 46: 127-137, 1984.
35. Kamiya, A. and Togawa, T.: Adaptive regulation of wall shear stress to flow change in the canine carotid artery. *American J. Physiol.* 239: H14-H21, 1980.
36. Kim, D.W., Langille, B.L., Wong, M.K.K., and Gotlieb, A.L.: Patterns of endothelial microfilament distribution in the rabbit aorta in situ. *Circ. Res.* 64: 21-31, 1989. See also Arteriosclerosis 9: 439-445, 1989.
37. Kishimoto, J., Keverne, E.B., Hardwick, J., and Emson, P.C.: Localization of nitric oxide synthase in the mouse olfactory and vomeronasal system: a histochemical, immunological and in situ hybridization study. *Eur. J. Neurosci.* 5: 1684-1694, 1993.
38. Kuo, L., Davis, M.J., Chilian, W.M.: Endothelium-dependent, flow-induced dilatation of isolated coronary arterioles. *American J. Physiol.* 259: H1063-H1070, 1990.
39. Lee, J.S. and Fung, Y.C.: Flow in locally constricted tubes at low Reynolds number. *J. Appl. Mech.* 37: 9-16, 1970.
40. Levesque, M.J. and Nerem, R.M.: The elingation and orientation of cultured endothelial cells in response to shear stress. *J. Biomech. Eng.* 107: 341-347, 1985.
41. Limas C., Westrum B., Limas C.J.: Comparative effects of hydralazine and captopril on the cardiovascular changes in spontaneously hypertensive rats. *Am. J. Pathol.* 117: 360-371, 1984.
42. Liu, S.Q., Yen, M., and Fung, Y.C.: On measuring the third-dimension of cultured endothelial cells in shear flow. *Proc. Nat. Acad. Sci,* in press.
43. Luscher T.F., Vanhoutte P.M., Raij L.: Antihypertensive treatment normalizes decreased endothelium dependent relaxations in rats with salt-induced hypertension. *Hypertension* 9 (*suppl.* III): III193-III197, 1987.

44. Malek A.M., Gibbons G.H., Dzau V.J., and Izumo S.: Fluid stress differentially modulates expression of genes encoding basic fibroblast growth factor and platelet-derived growth factor B chain in vascular endothelium. *J. Clin. Invest.* 92: 2013-2021, 1993.
45. Markin, V.S., and Martinac, B.: Mechano sensitive ion channels as reporters of bilayer expansion. A theoretical model. *Biophysical J.* 60: 1-8, 1991.
46. Miyahara, K. Kawamoto, K., Yui, Y., Toda, K., Yang, L.X., Hattori, R., Aoyama, T., Yamamoto, Y., Doi, Y. Ogoshi, S., Hashimoto, K., Kawai, C., Sasayama, S., and Shizuta, Y.: Cloning and structural characterizations of the human endothelial nitric-oxide-synthase gene. *Eur. J. Biochem.*, 223: 719-726, 1994.
47. Murphy T.J., Alexander R.W., Griendling K.K., Runge M.S., and Bernstein K.E.: Isolation of a cDNA encoding the vascular type-1 angiotensin II receptor. *Nature* 351: 233-236, 1991
48. Naville, D., Lebrethon, M.C., Hermabon, A.Y., Rouer, E., Benarous, R., Saez, J.M.: Characterization and regulation of the angiotensin II type 1 receptor (binding and mRNA) in human adrenal fasciculata-reticularis cells. *FEBS Lett* 321: 184-188, 1993.
49. Nerem, R.M., and Girard, P.R.: Hemodynamic influence on vascular endothelial biology. Toxic. Path. 18: 572-582, 1990.
50. Nollert, M.U., Diamond, S.L., and McIntire, L.V.: Hydrodynamic shear stress and mass transport modulation of endothelial cell metabolism. *Biotech. and Bioeng.* 38: 588-602, 1991.
51. Palade, G.E., and Bruns, R.R.: Structural modulation of plasmalemmal vesicles. *J. Cell Biol.* 37: 633-649, 1968.
52. Panza J.A., Quyyumi A.A., Brush J.E., Epstein S.E.: Abnormal endothelium-dependent vascular relaxation in patients with essential hypertension. *N. Eng. J. Med.* 323: 22-27, 1990.
53. Pappenheimer, J.R.: Passage of molecules through capillary walls. *Physiol. Rev.* 33: 387-423, 1953.
54. Reissner, E.: Tension field theory. Proceed. of 5th Inter. Cong. of Appl. Mech.: pp. 88-92, 1938
55. Remuzzi, A., Dewey, C.F., Davies, P.F., and Gimbrone, M.A.: Orientation of endothelial cells in shear field in vitro. *Biorheology* 21: 617-630, 1984.
56. Resnick, N., Collins, T., Atkinson, W., Bonthron, D.T., Dewey, D.F., Jr. and Gimbrone, M.A., Jr.: Platelet-derived growth factor B chain promoter contains a cis-acting fluid shear-stress-responsive element. *Proc. Nat. Acad. Sci., USA* 90: 4591-4595, 1993.
57. Repin, V.S., Dolgov, V.V., Zaikina O.E., Novikov I.A., Antonov A.S., Nikolaeva N.A., and Smirnov, V.N.: Heterogeneity of endothelium in human aorta. *Athero.* 50: 35-52, 1984.
58. Rhodin, J.A.G.: Architecture of the vessel wall. In *Handbook of Physiology*, Sec. 2, Vascular Smooth Muscle, ed. by D.F. Bohr, A.P. Samlyo, and H.V. Sparks, Jr. American Physiological Society, Bethesda, MD, Chap. 1: 1-32, 1980.
59. Sarzani R., Arnaldi G., Takasaki I., Brecher P., and Chobanian A.V.: Effect of hypertension and again on platelet-derived growth factor and platelet-derived growth factor receptor expression in rat aorta and heart. *Hypertension* 19 (*suppl* III): III93-III99, 1991.
60. Sasaki, K., Yamano, Y., Bardhan, S., Iwai, N., Murray, J.J., Hasegawa, M., Matsuda, Y., Inagami, T.: Cloning and expression of a complementary DNA encoding a bovine adrenal angiotensin II type 1 receptor. *Nature* 351: 230-233, 1991.
61. Satcher R.L., Jr., and Dewey, C.F.: The distribution of fluid forces on arterial endothelial cells. In "1991 Advances in Bioengineering," American Society of Mechanical Engineers, BED Vol. 20: pp. 595-598, 1991.
62. Sato, M. Levesque, M.J. and Nerem, R.M.: Applications of the micropipet technique to the measurement of the mechanical properties of cultured bovine endothelial cells. *J. Biomech. Eng.* 109: 27-34, 1987.
63. Sato, M., Theret, D.P., Wheeler L.T., Ohsima, N., and Neren, R.M.: Application of the micropipette technique to the measurement of the mechanical properties of cultured porcine endothelial cells. *J. Biomech. Eng.* 109: 27-34, 1987. See also, ibid, 112: 263-268, 1990.
64. Schmid-Schönbein, G.W., Sung, K.L.P., Tözeren, H., Skalak, R., and Chien, S.: Passive mechanical properties of human leukocytes. *Biophys. J.* 36: 243-256, 1981.
65. Sechler, E.E.: *"Elasticity in Engineering,"* John Wiley & Sons, 1945.
66. Shyy Y.J., Hsieh H.J., Usami., and Chien S.: Fluid shear stress induces a biphasic response of human monocyte chemotactic protein 1 gene expression in vasuclar endothelium. *Proc. Nat. Acad. Sci.* USA 91: 4678-4682, 1994.
67. Simionescu, M., Simionescu, N., and Palade, G.E.: Morphometric data on the endothelium of blood capillaries. *J. Cell Biol.* 60: 128-152, 1974.
68. Simionescu, M., Simionescu, N., and Palade, G.E.: Segmental differentiations of cell junctions in the vascular endothelium. The microvasculature. *J. of Cell Biol.* 67: 863-885, 1975. See also, ibid, 68: 705-723, 1976.

69. Singer, S.J., and Nicolson, G.L.: The fluid mosaic model of the structure of cell membranes. *Science* 175: 720-731, 1972.
70. Skalak, R., Tözeren, A., Zarda, R.P., and Chien, S.: Strain energy function of red blood cell membranes. *Biophys. J.* 13: 245-264, 1973.
71. Smiesko, V. and Johnson, P.C.: The arterial lumen is controlled by flow-related shear stress. *News in Physiological Sciences* 8: 34-38, 1993.
72. Wagner, H.: Flat sheet metal girders with a very thin metal web. *Z. Flugtechn. Motor Luft Schiffahrt* 20: 200-314, 1929. Translated into English, NACA TM 604-606.
73. Yamaguchi, T., Hoshiai, K., Okino, H., Sakurai, A., Hanai, S., Masuda, M. and Fujiwara, K.: Presented at 1993 Bioengineering conference, BED Vol 24, ASME, p. 167.
74. Yin, F.C.P., Fung, Y.C.: Peristaltic transport in a circular cylidrical tube. *J. Appl. Mech.* 36: 579-587, 1969.

3

NEW PERSPECTIVES IN BIOLOGICAL FLUID DYNAMICS

T. J. Pedley

Department of Applied Mathematical Studies
University of Leeds
Leeds, LS2 9JT, United Kingdom

ABSTRACT

The subject of biological fluid dynamics is divided into two major parts, internal or physiological fluid dynamics and external fluid dynamics (e.g. swimming and flying) or the interaction of living organisms with their fluid environment. This review will discuss several topics in each category, surveying recent progress and indicating probable growth areas in the near future. The emphasis will be on mechanisms and scientific understanding rather than clinical results. Disproportionately little attention will be paid to cardiovascular fluid dynamics because a disproportionately large number of papers were devoted to it at the World Congress. Otherwise the major topics to be considered are:

(i) Respiratory fluid dynamics - energy loss and pressure drop in airways; forced expiration; gas mixing in airways, especially during high frequency ventilation; surface tension effects in airway closure.

(ii) Peristaltic pumping in the ureter - the conventional concentration on flow within the ureter is now being supplemented by detailed modelling of the contraction of ureteral smooth muscle, against the loads provided by the hydrodynamics, in response to the propagating activation signal.

(iii) Fish swimming - a similar development is taking place in this area: observations of the motion of a fish body can be used not only to compute the time-dependent hydrodynamic forces acting on it, but also to infer the distribution of bending moment along the fish and, with data on the mechanical properties of the tissues, to calculate the forces and rates of contraction of the swimming muscles. The results can be compared with new measurements of muscle properties at different distances along the fish.

(iv) Bioconvection, or spontaneous pattern - formation in dense populations of swimming micro-organisms (certain algae and bacteria in particular). Intriguing experimental observations will be shown and a qualitative explanation given. The need for a stochastic model of random changes in a cell's swimming trajectory will be emphasised.

INTRODUCTION

I was greatly honoured to have been invited to give this plenary lecture. It turned out to be quite appropriate because the week before the World Congress I was chairman of a Symposium with the title of Biological Fluid Dynamics (BFD), held at my University under the aegis of the Society for Experimental Biology (SEB); a book is being published as a result (Ellington & Pedley, 1995). As a result I ought to be able to give you a highly up-to-date and authoritative review of the subject. The problem is selection.

BFD covers both internal, or physiological fluid dynamics (blood flow, airflow in the lungs, excretory flows, mass transfer across and within tissues) and external fluid dynamics, concerned with the interaction of living organisms with their fluid environments (swimming, flying, etc.). Our symposium, and this talk, covers both.

Under physiological fluid dynamics, by far the most research has been directed to the circulation of the blood, with pulmonary fluid dynamics coming a long way second. Some of the main topics of interest are: arterial pulse wave propagation; flow details in complex arterial geometries, because of the link between the distribution of wall shear stress and that of atherosclerosis; venous flow, involving collapse of vessels above the heart; microcirculation, where the blood cannot be treated as a homogeneous fluid. In the lung, there are many interesting fluid dynamic phenomena associated with surface tension effects in the liquid lining layer, as well as gas flow and mixing phenomena.

CARDIOVASCULAR FLUID DYNAMICS

There were enormously many sessions at this Congress on blood flow of one sort or another, and most of the chapters in this book are on that subject, so I am going to discuss it only very briefly. The evidence is by now overwhelming that the distribution of mean shear stress exerted by the flowing blood on an artery wall is causally related to the distribution in that artery of early atherosclerosis. Excellent reviews on the subject are provided by Giddens et al.(1993), Friedman (1993) and Fry (1987). Regions of low mean wall shear stress (WSS), especially if also associated with time-dependent flow separation and hence fluctuations in the magnitude and direction of the WSS, are preferred sites for the intimal thickening which is the precursor of true atheroma (Caro et al. 1971; Zarins et al. 1983; Ku et al. 1985). A clear necessary precondition to understanding the process is a knowledge of the normal distribution of mean and time-dependent WSS in regions which are known to be susceptible to atherogenesis: the insides of bends, the outer walls of bifurcations, etc.

The only point I would like to make here is the following. Unsteady flow at large Reynolds numbers (say 200 - 1000, as in most susceptible arteries) in complex geometries is very complicated. Indeed, the fundamental fluid dynamics is poorly understood. Three-dimensional disturbances, such as curves, branches or indentations, introduce secondary motions and horseshoe vortices, under which the WSS can remain high for a considerable distance downstream of the disturbance (Fukushima & Azuma, 1982). Even in two-dimensional tubes, if non-uniform, unsteadiness in the driving pressure gradient or the wall geometry can lead to regions of flow separation far from the non-uniformity, but associated with high WSS not low as for steady separated flow in two dimensions (Pedley & Stephanoff, 1985; Ralph & Pedley, 1988; Tutty & Pedley, 1993). Moreover, both two- and three-dimensional flows can be non-unique, in that different steady flows can exist in the same tubes with the same driving pressure gradient, depending on the initial conditions (Borgas & Pedley, 1990; Daskopoulos & Lenhoff, 1989). Thus there is great sensitivity to small temporal perturbations as well as small geometric perturbations, making it very difficult to predict flow in one arterial bifurcation from knowledge of another, or even at a particular site in one subject from studies in another subject. Assessing the WSS distribution on the basis of steady and/or two-dimensional intuition will probably be wrong. A full exposition of the above fluid dynamical phenomena is given by Pedley (1995). The message to take away is that accurate prediction of WSS at a given site in a given subject requires totally "customised" representations of the geometry and wave form (not to mention wall elastic properties), which is not currently feasible, computationally, in a time scale short enough to be clinically useful. More insight can probably be gained from studies in idealised geometries which can be reproduced in other laboratories or other computer codes.

RESPIRATORY FLUID DYNAMICS

I choose this as my first major subject because this is where I began, just over 25 years ago. The object then was to learn the nature of air flow in the lung, and see if one could use data on that to infer something of physiological or medical importance.

My colleagues Schroter & Sudlow (1969) had made measurements and visualisations of steady flow in symmetric bifurcations, with dimensionless geometry and dynamic parameters chosen to correspond to middle-sized airways. They observed, and clearly demonstrated, the secondary motions that develop because the airstreams are curved and the effect they have on the axial velocity profiles. Figure 1a reproduces their famous end-on photograph of smoke injected into a daughter tube during steady inspiratory flow, demonstrating the helical streamlines. Figure 1b sketches the effect on the velocity profiles.

When I joined the group we used the measured velocity profiles to compute the rate of viscous energy dissipation in the flow and hence estimate the pressure drop across a bifurcation (Pedley, Schroter & Sudlow, 1970a). In order to extrapolate to the whole lung it was necessary to develop some sort of model to show how the pressure drop would scale with Reynolds number (Re). The data made it apparent that the shear rate or velocity gradient in the flow was greatest in the relatively thin boundary layer on the flow divider - not surprising because the oncoming flow is split there and a new, thin, boundary layer forms

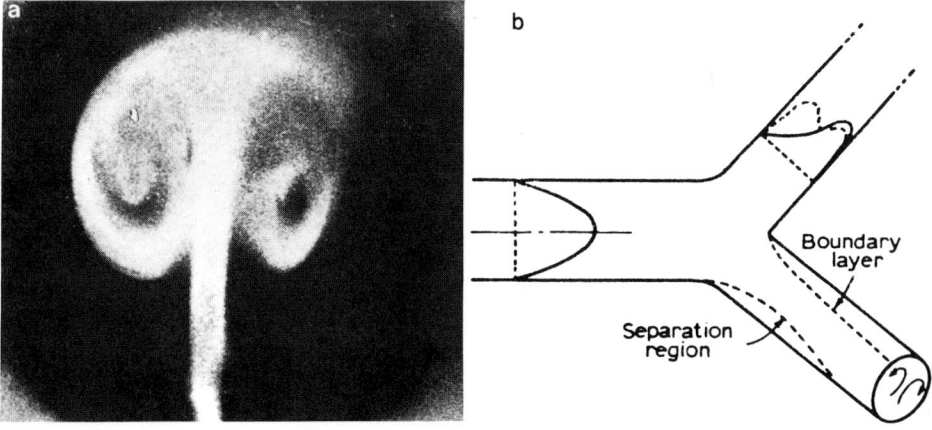

Figure 1. Smoke patterns in the daughter tube of a single bifurcation (end view), showing secondary motions generated during inspiratory flow (from Schroter & Sudlow, 1970). (b) Sketch of flow downstream of a single bifurcation with Poiseuille flow in the parent tube. Direction of secondary motion, new boundary layer and possible separation region shown in lower branch; velocity profiles in plane of junction (solid curve) and in perpendicular plane (broken curve) in upper branch (from Pedley et al. 1971).

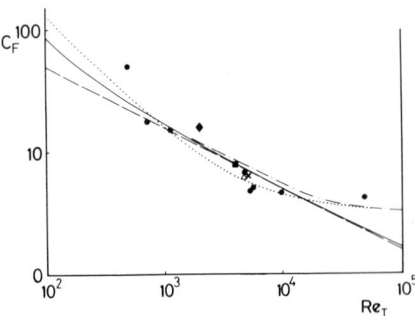

Figure 2. Log-log plot of friction factor (C_F) for inspiratory flow in the lower airways against tracheal Reynolds number (Re_0). *Solid curve*: as predicted from the theory of Pedley et al (1970b); *broken line*: straight line of slope $-\frac{1}{2}$; *dash-dot-curve*: theory modified by Jaffrin & Kesic (1974); *dotted curve*: best fit from a 2-term power series: $C_F = \kappa_1 + \kappa_2/Re_0$. The points are taken from a variety of experimental studies (adapted from Pedley & Kamm, 1991).

as at the entry of a tube, but kept thinner than in a straight tube by the secondary motions caused by curvature. The analogy with entry flow suggested the appropriate scaling, leading to the prediction that the viscous contribution to pressure drop (Δp) would be proportional to $Re^{3/2}$ (or friction factor proportional to $Re^{-1/2}$) as opposed to Re (Re^{-1}) for Poiseuille flow and Re^2 (constant) for turbulent flow (figure 2).

This agreed quite well with the values estimated from Schroter and Sudlow's data, and gave predictions for inspiratory Δp for the whole lung (trachea to alveoli) that agreed quite well with physiological measurement (Pedley, Schroter & Sudlow, 1970b). The model has stood the test of time, undoubtedly helped by the fact that $\Delta p \propto Re^{3/2}$ in fully-developed curved tube flow as well as entry flow. The combination of the two, as in bifurcations, is therefore more or less guaranteed to give the same flow-rate dependence (Pedley & Kamm, 1991).

Those results were for (quasi-) steady inspiratory flow. Rather surprisingly, it has taken over 20 years for something as simple to emerge for steady expiratory flow. In this case the secondary motions and velocity profiles are different and there is not the great simplifying feature of a new, thin boundary layer at each generation. An overall scaling law is thus not theoretically available. However, based on measurements in a model (with rounded corners) Collins et al.(1993) recently concluded that curved tube scaling did work, when measured pressures were fully corrected for the effects of kinetic energy flux. However, studies by Chang's group (Chang, 1989) and by Snyder et al.(1987) have made it clear that

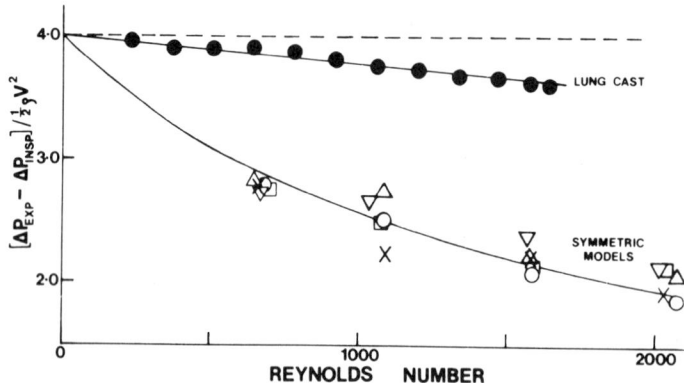

Figure 3. The difference between inspiratory and expiratory viscous pressure drop, corrected for kinetic energy changes, in a cast of a dog lung with smooth walls (closed circles) and in a model made of circular tubes with sharp corners at the joints (other symbols). (From Snyder et al. 1987).

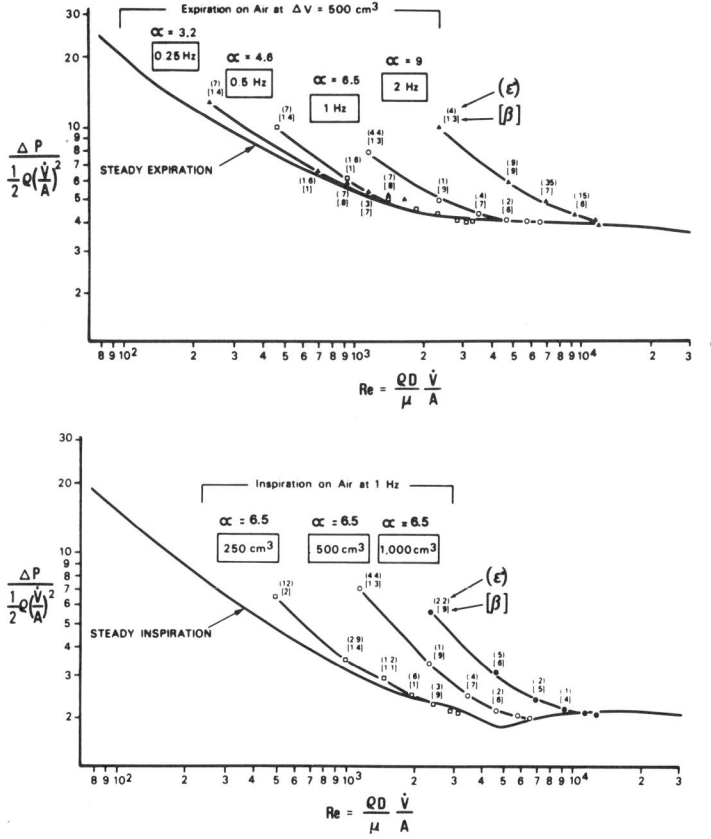

Figure 4. Friction factor vs. tracheal Reynolds number for oscillatory flow in a lung cast: (a) expiratory values at different frequencies; (b) inspiratory values at different tidal volumes. (From Isabey & Chang, 1981).

in sharp-cornered models there is a significant difference between expiration and inspiration, whereas in a human lung cast there is not. These results have yet to be fully explained.

In recent years there has been increased interest in non-quasi-steady oscillatory air flow in the lung, because of the success of High Frequency Oscillation (HFO) in the artificial ventilation of premature babies and animals : that is the airflow at the mouth is oscillated at high frequency (eg 10-15 Hz) and very small amplitude (tidal volumes as low as half the anatomical dead space) (Greenough & Milner, 1987). Here one needs to understand how gas exchange is enhanced as well as determining the pressure distribution in the fragile bronchial trees of such subjects. What progress there has been in the fundamental understanding of gas mixing has been in idealised geometries (e.g. uniform curved tubes), large scale models or adult lungs. Unsteady flow patterns in bifurcations are extremely complex (as mentioned in the context of arteries), especially near the times of flow reversal (Jan, Shapiro & Kamm, 1989). The flow is relatively streamlined at peak inspiration and expiration, so we may hope that the $Re^{-1/2}$ dependence of the friction factor may be valid then, but at other times measurements by Chang et al show a significant frequency-dependence of the friction factor (figure 4; see Isabey & Chang, 1981). No general predictive model is yet available.

The question of how secondary motions and flow oscillations can enhance mixing in airways is usually answered in terms of Taylor dispersion, a process by which shear in a velocity profile can stretch out the interface between fresh gas and old gas longitudinally, inhibited only by transverse molecular diffusion or other mixing processes. Thus the longitudinal mixing, represented by an effective longitudinal diffusivity D_{eff} is reduced as lateral mixing increases. Taylor (1953) and Aris (1956), for example, calculated D_{eff} for dispersion of a solute in steady laminar flow with average velocity $\bar{u}$ in a straight circular tube of radius a, as follows:

$$D_{\text{eff}} = D_{\text{mol}} + \frac{\bar{u}^2 a^2}{48 D_{\text{mol}}}, \tag{1}$$

where D_{mol} is the molecular diffusivity of the solute. The situation is reasonably well understood in steady flow in curved tubes also, but it is so complicated in oscillatory flow that fundamental understanding is still limited. Results for oscillatory flow in a uniform curved tube help to show why.

First consider steady curved tube flow. Here a secondary motion is set up, similar to that depicted in Figure 1, and this tends to distort the axial velocity profile by an amount that depends on the Dean number, κ_D, proportional to the Reynolds number and $(a/R)^{1/2}$, where R is the radius of curvature of the tube centre-line. The secondary motions, of typical velocity V_s, say, might be expected to enhance the lateral mixing and thus reduce D_{eff}, and indeed they do, but not by very much because, although the solute concentration is rapidly smeared out and becomes uniform around a secondary flow streamline, molecular diffusion is still

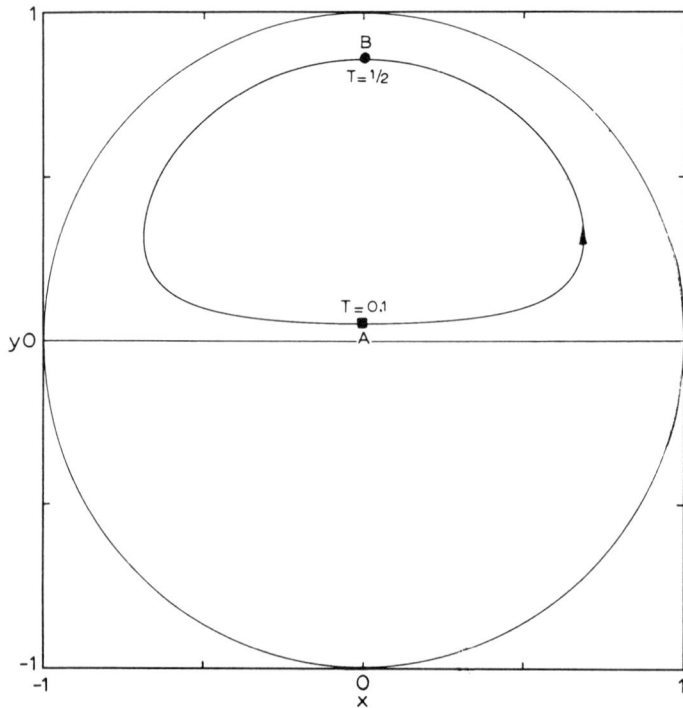

Figure 5. A secondary orbit of period 1 of a fluid element in quasi-steady, oscillatory flow, also of period 1, in a curved tube. If at A (T = 0,1) a particle has a large positive velocity, it will have a small negative one at B (T = ½), leading to resonant dispersion. (From Hydon, 1994b).

required for equilibration between streamlines. Johnson & Kamm (1986) found that, at small values of κ_D, D_{eff} could be reduced only to about 0.2 of the value given in equation (1), in circumstances for which the equilibration *round* streamlines is much faster than that *between* streamlines (i.e. the secondary flow Péclet number, Pe_s, is large, where $Pe_s = V_s a/D_{mol}$.)

In oscillatory flow with zero mean axial velocity, we consider the limit of large Pe_s, or negligibly small molecular diffusivity (more appropriate to liquids than gases). In a straight tube, there would be no longitudinal mixing, because every fluid element would oscillate backwards and forwards with no net displacement. In a curved tube there are more interesting possibilities. Suppose the oscillation frequency to be low, so that the motion is entirely quasi-steady; in particular, the secondary flow streamlines do not change with time although the magnitudes of the velocity components do. Consider fluid elements that lie on that secondary streamline for which the time taken to complete one circuit is equal to the oscillation period. The element which has the greatest positive axial velocity when the flow is maximal in the +x direction (say) will have the lowest negative axial velocity one half-cycle later, when the flow is maximal in the -x direction; one half-cycle later still it will again have greatest positive velocity (Figure 5). Thus this particle will experience a substantial positive displacement each cycle, comparable with its displacement over the same time interval in steady flow. The existence of such resonant conditions mean that the longitudinal dispersion will be significantly greater in a curved tube, over a certain frequency range, than in a straight tube; this remains true when D_{mol} is not zero (Pedley & Kamm, 1988; Eckmann & Grotberg, 1988; Sharp et al. 1991).

At sufficiently high frequencies that the flow is no longer quasi-steady, fluid elements will not in general remain on the same secondary flow streamline and may, even without diffusion, wander across large parts of the tube cross-section. Hydon (1994 a,b) has shown that there are in fact some localised regions ('islands') in the cross-section where elements remain trapped, and can experience resonance such as that described above, but that elements outside the islands experience 'Lagrangian chaos' which is indistinguishable, after a long time, from a diffusive process.

It must be admitted that the influence of such phenomena on gas mixing in the lungs (or solute mixing in blood vessels) is as yet unclear. Important reviews on the topic are those by Grotberg (1994) and Kamm (1995).

All the flows described so far have been discussed on the assumption that the tubes in which the flow takes place are rigid. However, all physiological tubes, such as blood vessels and airways, are in fact elastic. In general, fluid flow in elastic vessels cannot be divorced from the behaviour of the walls. Large changes in cross-sectional area can of course have an important effect on the flow in a tube. Such changes can occur either passively or actively. In the context of the lung, a substantial passive collapse of the large intrathoracic airways occurs during forced expiration, causing flow limitation. Crudely, the reason is that during a strong expiratory effort the pleural pressure (and hence the peribronchial pressure) exceeds atmospheric by a substantial amount, while the pressure in the airways has to fall to the atmospheric level at the mouth, and can become less than the external pressure in larger intrathoracic airways. These may then collapse, especially the largest ones which are not held open by the surrounding parenchyma. The actual mechanism of time-dependent collapse is rather subtle, because fluid-filled elastic tubes can sustain the propagation of waves, but has been successfully modelled by Elad, Kamm & Shin (1994); see Figure 6. Wheezing is also a manifestation of flow-induced oscillation in collapsible airway walls (Gavriely et al. 1989; Grotberg & Gavriely, 1989).

Other physiological examples of collapsible tubes are veins above the heart, arteries under a cuff, the urethra during micturition, etc. There is a small but vigorous community of solid and fluid dynamicists working on such problems. Space does not permit a full discussion here, but reference can be made to the following reviews and recent papers: Shapiro (1977), Kamm (1987),

Kamm & Pedley (1989), Elad & Kamm (1991), Grotberg (1994), Bertram (1995), Heil & Pedley (1995), Luo & Pedley (1995).

Perhaps the newest area of pulmonary fluid dynamics is the very important one of surface tension effects. All airspaces are lined with a thin liquid layer (plus mucus in the larger airways). Various phenomena are governed by it. The one to have been studied most thoroughly is airway closure during expiration and the consequent trapping of gas in the lung. (a) First, a long cylindrical liquid-gas interface is unstable to axisymmetric undulations of wavelength greater than the circumference (a so-called Rayleigh instability), because the surface area (and hence surface energy) is less in the deformed state. If the liquid layer is very thin this might just lead to a corrugated interface, but if there is enough liquid available the disturbance grows until a liquid bridge forms, trapping gas peripherally. Conditions for bridge formation usually do not exist at high lung volumes, because airways are short and the liquid layer thin. However, during expiration both the interface circumference becomes smaller and the liquid layer becomes thicker, so even in healthy subjects gas trapping takes place (Johnson et al, 1991). In certain disease conditions there is more liquid in the lung, so gas can be trapped at higher lung volumes. (b) In other disease states, in which the airways are particularly compliant, an additional, elastic collapse mechanism comes into play. The drop in pressure across the interface means that elastic collapse can also begin (Halpern & Grotberg 1992). (c) Reopening of the airways breaking the liquid bridges - is another aspect of the problem that is currently being studied, both theoretically and experimentally (Perun & Gaver, 1995).

We should note, too, that normal lungs contain a lot of surfactant, so that the surface tension of the liquid layer can be as low as $1/20$ th of that of water. This makes reopening much easier than otherwise, and greatly reduces the pressure required to inflate the lung. Some premature infants are deficient in surfactant, which makes it very difficult for them to breathe at all; the addition of exogenous surfactant is then a very helpful clinical procedure, in addition to the High Frequency Ventilation already discussed. Longitudinal gradients in surfactant concentration lead to gradients in surface tension which can drive a variety of interesting flow phenomena in the liquid layer. For example, drug delivery by aerosols with added surfactant means that peripheral surface tension is greater than where the droplets have impinged, so a fluid flow pulls the drug deeper into the lung. I do not have space to describe any of the interesting research which is proceeding in this area, but refer you to Jensen & Grotberg (1992) Halpern & Grotberg (1993), Otis et al (1993), and once more to Grotberg's recent review (1994).

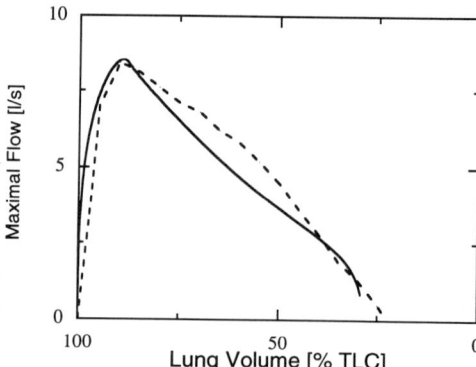

Figure 6. Flow-rate versus expired volume during a forced expiration: broken curve - experimental (from Hyatt & Flath, 1966); solid curve - theoretical model (from Shin, et al. 1995).

THE INTEGRATION OF HYDRODYNAMICS WITH MUSCLE MECHANICS

That is enough on the lung. Every fluid-structure interaction I have described so far involves passive mechanics: elasticity, surface tension, fluid flow. I now want to move on to two totally different areas of BFD which nevertheless have an important feature in common: the wall motions involve active muscular contraction.

Peristaltic Pumping

The first concerns peristaltic pumping of urine in the ureter, whereby waves of contraction are propagated along the smooth muscle in the wall, driving the urine along in boluses from the kidney to the higher-pressure bladder (see Figure 7 and Griffiths, 1989). It is important to understand all aspects of this because there are many disease states in which the peristalsis is impaired or in which fluid from the bladder can make its way back up to the kidneys, possibly carrying harmful bacteria.

The standard fluid dynamical model of peristaltic pumping has considered just the hydrodynamics, given an assumed wall motion (e.g. axisymmetric, sinusoidal waves of small wall-slope on an infinite tube - see Jaffrin & Shapiro, 1971). Output would be flow-rate as a function of the pressure rise, for example. But the input is really the propagating wave of muscle activation, and the wall deformation is part of the output. The muscle contracts in a manner determined by its intrinsic properties (represented by a surface in the space of length,

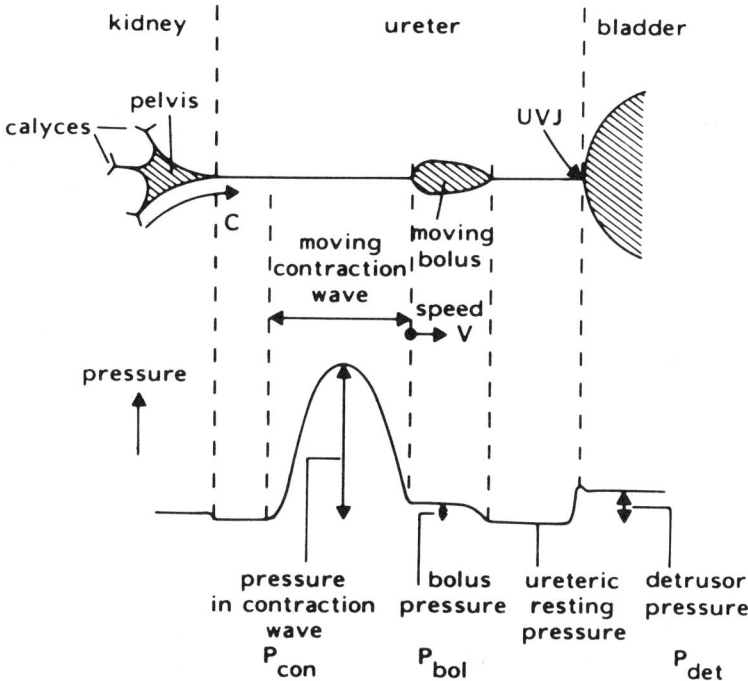

Figure 7. Pressure distribution during transport of an isolated bolus by peristalsis in the ureter (from Griffiths, 1983).

speed of shortening and load against which it is shortening) and the hydrodynamic stresses which the consequent wall motion sets up.

The one person to recognise this in the past, typically, was Y.C. Fung (1971) who published an integrated model of ureteral peristalsis as a chapter in a book, but as far as I know his work has not been followed up. At present there are two people to my knowledge who are seeking to resuscitate the work - R.N. Miftakhov in the United Arab Emirates, whose approach is to try to simulate every feature of the phenomenon, including the electrophysiology and biochemistry of the smooth muscle (see his model of intestinal peristalsis - Miftakhov & Wingate, 1994), and E.O.A. Carew in Leeds, who is trying to develop the simplest possible model. The idea is to postulate an activation wave, put in mathematical descriptions of the smooth muscle contractile properties, and of the visco-elastic properties of the wall tissue, and compute both the hydrodynamics and the corresponding wall motions for comparison with physiological data. An alternative approach is to continue to use the observed wall motions as input, and work back to infer something about the muscle and tissue properties. The problem still is an inadequate knowledge of the contractile properties of the muscle and the mechanical properties of the passive parts of the wall. Further complications lie in the fact that the ureter is of finite length, containing only one or two wavelengths of the contraction wave. This means that the mathematical problem does not reduce to a a steady one in the frame of reference of the propagating wave, but is intrinsically unsteady, as has been recognised in the hydrodynamic context by Li & Brasseur (1992).

Fish Swimming

Peristaltic pumping is one example of the integrated approach to active fluid-structure interactions in biomechanics which I regard as a major new development. Exactly the same integrated approach is being developed by a number of groups in the context of aquatic animal locomotion, in particular fish swimming, and I would like to explain this example in more detail. Again, the traditional approach is to observe how the fish wiggles, perform a hydrodynamic analysis to determine the forces it exerts on the water (via momentum or energy arguments) and compute its swimming velocity, say, and mechanical efficiency. The standard theory is Lighthill's (1970) elongated-slender-body theory, which shows that the mean thrust exerted by an anguilliform or carangiform fish in steady-state swimming at speed U (required to balance the viscous drag) can be computed from a knowledge of conditions at the trailing edge of the tail. With reference to figure 8a, the small-amplitude version of Lighthill's theory uses an energy argument to demonstrate that mean thrust T is given by:

$$T = m(\ell) \left[Uw - \frac{1}{2} w^2 \right]_{x=\ell} \qquad (2)$$

where

$$w = \frac{\partial h}{\partial t} + U \frac{\partial h}{\partial x} \qquad (3)$$

is the lateral velocity of the water next to the fish at location x and time t. Here $h(x,t)$ is the lateral displacement of the fish's central symmetry plane, and $m(x)\Delta x$ is the added mass of the water in a slice of length Δx as it is deflected sideways by the movement of the body with velocity w.

Lighthill's model assumes that the curvature is small and that $w \ll U$, and neglects the effect of the shed vortex wake on the fish body, but it is reasonably accurate and quite

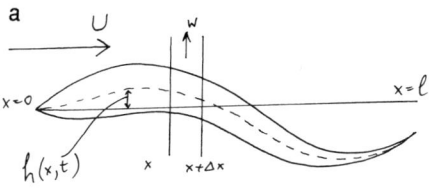

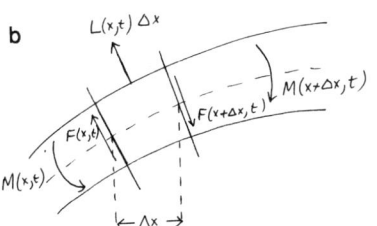

Figure 8. (a) Sketch of a fish swimming steadily at speed U with centre-surface displacement $h(x,t)$; the fluid in contact with the slice between x and $x + \Delta x$ has lateral velocity w. (b) Forces and torques on the slice Δx: bending moment M, beam force F, fluid force L.

versatile - it can, for example, be extended to large amplitude swimming motions and non-steady-state manoeuvres such as starts and turns (Lighthill, 1971; Weihs, 1972, 1973). There is a new, more accurate hydrodynamic theory of fish swimming by Cheng et al.(1991), based on the vortex-ring panel method, which does not assume small curvature, nor neglect the wake, but it is as yet restricted to small amplitude body undulations.

However, what the fish actually does is to send a muscle activation wave down its spinal cord, and the muscles contract according to the hydrodynamic load that they experience. The actual movements of the fish body result from the interaction between the hydrodynamics and the muscle mechanics. An important breakthrough in the analysis of this interaction was achieved by Hess and Videler (1984) by considering the distribution of bending moment down the fish, regarded as an actively bending beam. They considered the force and torque balance on a slice Δx of fish at time t (see Figure 8b) and obtained the following equations relating the displacement $h(x,t)$ to the bending moment $M(x,t)$ via the "beam force" F:

$$m_b(x)\frac{\partial^2 h}{\partial t^2} = -\frac{\partial F}{\partial x} + L \tag{4}$$

$$-\frac{\partial M}{\partial x} - F = 0 \tag{5}$$

where $m_b(x)\Delta x$ is the mass of fish in the slice Δx, and L is the hydrodynamic pressure force, given according to Lighthill's (1970) theory by:

$$-L = \left(\frac{\partial}{\partial t} + U\frac{\partial}{\partial t}\right)(mw). \tag{6}$$

A combination of equations (4) and (5) leads to:

$$\frac{\partial^2 M}{\partial x^2} = -m_h \frac{\partial^2 h}{\partial t^2} - L, \qquad (7)$$

a second order ordinary differential equation for M. This could be used to calculate $h(x,t)$ if we knew the distribution of bending moment. However, the easiest quantity to measure is the displacement of the fish body, $h(x,t)$, using for example the high-speed video recording system of Dr C.S Wardle in Aberdeen (see Wardle & Videler, 1993). The measurements of h can then be used both to compute the hydrodynamic forces, from Equation (6), and to compute the bending moment distribution, by integrating equation (7). The latter would then give information on the contraction of the muscles. [A note of warning should be sounded here, because the "natural" boundary conditions that F and M are zero at both ends of the fish cannot all be satisfied by a solution of Equation (7). A *recoil correction*, consisting of a rigid body translation and rotation (sideslip and yaw) needs to be added to h to ensure that the total torque and lateral component of force on the fish are zero (Lighthill, 1970; Cheng et al. 1994).]

Further work by Cheng, during a six-month fellowship at Leeds, has shown how to subdivide the bending moment M into contributions from the muscle contraction, M_m, and those from the passive (visco-) elastic properties of the tissues. On the basis of simple linear assumptions about the latter, the following equation was deduced:

$$\frac{\partial^2 M_m}{\partial x^2} = \frac{\partial^2}{\partial x^2}\left(\tilde{E}\frac{\partial^2 h}{\partial x^2}\right) + \frac{\partial^2}{\partial x^2}\left(\tilde{\mu}\frac{\partial^3 h}{\partial t \partial x^2}\right)$$

$$+ m_h \frac{\partial^2 h}{\partial t^2} + \left(\frac{\partial}{\partial t} + U\frac{\partial}{\partial x}\right)(mw), \qquad (8)$$

where $\tilde{E}$ is a tissue elasticity parameter and $\tilde{\mu}$ a tissue viscosity parameter. There is not much data yet available on these quantities, but even crude estimates lead to rather good agreement between the time course of M_m deduced from equation (8) and that inferred from experimental observation (Cheng et al. 1995).

The last remark makes it clear why this work is so timely and important. An increasing amount of real data is becoming available, in particular fish species, on (a) the time course of fish muscle activation during steady-state swimming, from EMG measurements (see Wardle & Videler, 1993, 1994) and (b) the contractile properties of fish muscle as a function of oscillation frequency, phase during the cycle (t) and position (x) along the fish (see Altringham et al. 1993). Rapid progress along these lines is being made in two or three laboratories, as could be seen from several presentations at the SEB Symposium in July 1994; see, for example, Bowtell & Williams (1993) and Williams et al (1995), who are developing a fully-integrated description of the swimming lamprey, including electrophysiology and numerical solution of the Navier-Stokes equations; and Daniel (1995), who has made comparably big advances in the study of invertebrates (e.g. jellyfish).

BIOCONVECTION

There is one more story I would like to tell in this brief selection of current problems in BFD. Around half the world's biomass consists of microorganisms living in aquatic environments, many of them forming the oceanic phytoplankton, important for their role both in the food chain and in the global CO_2 balance. Understanding how populations of

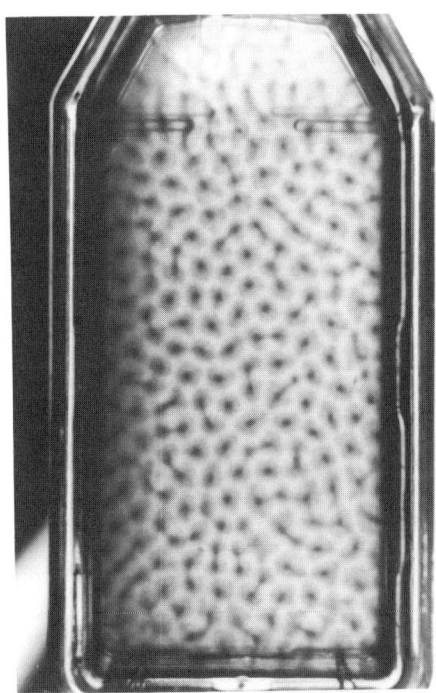

Figure 9. Photograph of a bioconvection pattern in a suspension of swimming algal cells of the species *Chlamydomonas nivalis*. Chamber dimensions 37 × 67 mm; fluid depth 4 mm; mean cell concentration 3 × 10^6 cm^{-3}. (Photograph by courtesy of Prof. J.O. Kessler)

such organisms respond to physical imperatives, or other external influences, is therefore a subject of currently active research. The phenomena I wish to describe have been observed in suspensious of swimming micro-organisms of a variety of types: protozoa, algae and bacteria. What they have in common is that the cells are somewhat denser than the medium in which they swim and, on average, they swim upwards. When a sufficiently concentrated suspension of such upswimmers is placed in a shallow chamber in the laboratory, regular patterns are observed to form spontaneously (Wager, 1911; Platt, 1961; Kessler, 1986; Kessler et al. 1994): see Figure 9 for one example. The patterns arise from a hydrodynamic instability of the initially stationary suspension: the upswimming causes cells to accumulate near the upper surface; the cells being denser than the water, the upper part of the suspension therefore becomes denser than the lower part; when the resulting density gradient exceeds a critical value, the configuration becomes gravitationally unstable and a convective motion sets in, analogous to Rayleigh-Bénard convection in a thermally stratified fluid. The phenomenon has therefore been called *bioconvection* (Platt, 1961). The patterns are particularly easy to observe because they are made visible by variations in concentration of the cells themselves. The pattern changes as the chamber depth is increased.

In collaboration with John Kessler and Nicholas Hill, I have been involved in seeking a quantitative model of the bioconvection process, with a view ultimately to explaining the shapes and dimensions of the observed patterns. Kessler provided the initial impetus by explaining the mechanism by which biflagellate algal cells swim upwards (on average) in the first place: they are bottom-heavy. The cells in question, such as *Chlamydomonas nivalis* or *Dunaliella tertiolecta*, have prolate spheroidal bodies with a pair of swimming flagella at the "front" end of the symmetry axis and with cell contents whose centre of mass is displaced towards the rear (fig. 10). Thus if, in a fluid at rest, a cell's axis is tilted away from the vertical, it will experience a gravitational torque tending to restore it to the vertical, resisted

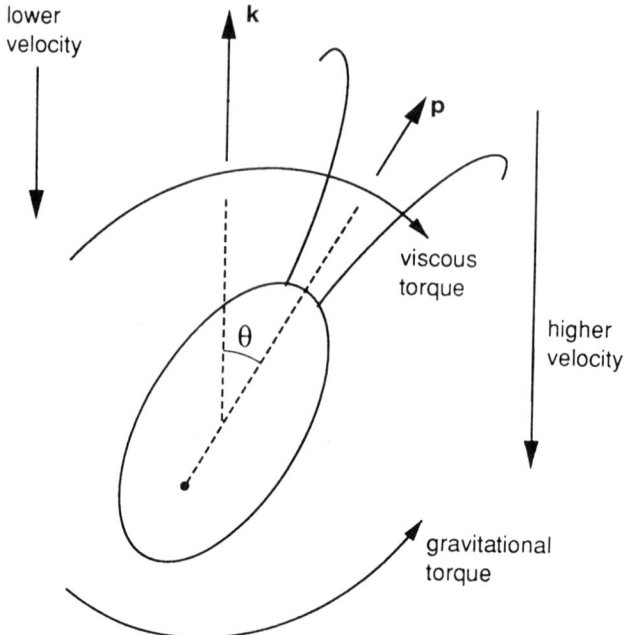

Figure 10. Sketch of a biflagellate algal cell experiencing viscous torque in a vertical shear flow and gravitational torque because of its asymmetric mass distribution, so it swims at an angle to the vertical (cell axis represented by unit vector p, vertical upwards by unit vector k).

by a viscous torque exerted by the fluid. Thus the swimming direction, along the axis, will tend to be vertically upwards too.

When there is an ambient flow, for example a vertical shear flow as depicted in figure 10, the cell will experience a viscous torque, due mainly to the vorticity in the flow, tending to rotate it one way, and a gravitational torque in the opposite direction. The result is that the cell will tend to swim at an angle to the vertical (as long as the vorticity is not too large), in a process known as *gyrotaxis*, and cells will tend to converge on the region of most rapid downflow. Indeed, a proof of the bottom-heaviness hypothesis was Kessler's (1985) observation that the algal cells in question accumulate near the axis in a downwards pipe flow.

Gyrotaxis is also responsible for a new mechanism of instability in a uniform suspension, independent of the development of a density gradient: a blob of fluid containing a greater concentration of cells than its surroundings will fall, thereby setting up a shear flow which, because of gyrotaxis, causes more cells to swim into the blob or its wake, reinforcing

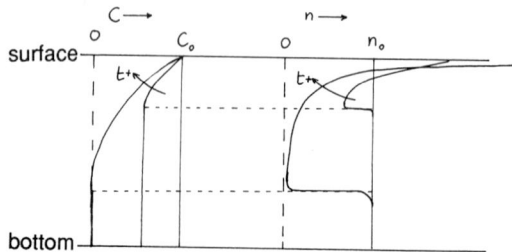

Figure 11. Sketch showing the evolution of cell (n) and oxygen (C) concentrations in a suspension of oxytactic bacterial cells. If the chamber is deep enough the cells at the bottom run out of oxygen, and therefore stop swimming, before they have swum up to the surface.

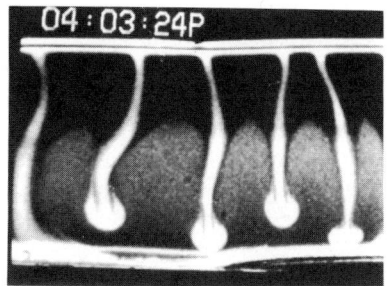

Figure 12. Photograph of a suspension of bacterial cells of the species *Bacillus subtilis* seen from the side in a narrow, vertical chamber. Bioconvective plumes are visible. Chamber depth, 7-8 mm.

the initial fluctuation. The interaction of the two instability mechanisms is probably responsible for the great variety and complexity of observed bioconvection patterns.

The upswimming leading to bioconvection in bacterial suspensions is due not to *gyrotaxis* but to *oxytaxis*, a sort of *chemotaxis* in which cells swim up gradients of oxygen concentration. The intriguing feature in this case is that the bacteria generate their own oxygen gradients by consuming oxygen. Consider a chamber, with its upper surface open to atmosphere, in which the cell (n) and oxygen (C) concentrations are initially uniform at n_0, C_c respectively, C_0 being the oxygen concentration in equilibrium with the atmosphere (Figure 11). The first thing to happen is that the cells consume oxygen, reducing C uniformly except near the free surface where C remains equal to C_0, so a gradient develops. Then the cells in the region of the gradient will swim up towards the surface, leaving a cell-depleted region underneath. This process will continue until one of three conditions obtains: (i) C and n gradients occupy the whole depth of the chamber, and a steady-state is set up; (ii) if the chamber is deep enough the cells at the bottom will run out of oxygen before sensing a significant gradient, so that below the cell-depleted zone will be a cell-rich region of dormant, non-swimming cells and no oxygen; or (iii) the upswimming of cells to the free surface will lead to a sufficiently large density gradient that bioconvection again ensues, as indeed observed (Kessler et al, 1994). In a deep chamber the resulting motion may bring oxygen down to the inactive cells, effectively reventilating them and causing them to start swimming again (fig. 12).

Dark field photography was used, so a lighter shade corresponds to a higher cell concentration. (Photograph by courtesy of Prof. J.O. Kessler). The above has given a concise summary of some of the physical mechanisms responsible for bioconvective pattern formation. I should like now to give a brief flavour of some of the problems that arise in providing continuum models of the phenomena. For a detailed survey of the whole field, see Pedley and Kessler (1992a); for a more discursive treatment, see Pedley & Kessler (1992b); for the most recent theoretical paper, on bacterial suspensions, see Hillesdon et al.(1995).

The basic equations for the fluid motion are the usual continuity and Navier-Stokes equations, with a gravitational body force proportional to the cell concentration, n. The crucial additional equation is the one for n, the cell conservation equation. If the very slow process of gravitational sedimentation is neglected, this takes the following form:

$$\frac{\partial n}{\partial t} = -\nabla \cdot (n\mathbf{u} + n\mathbf{V_c} - \mathbf{D} \cdot \nabla n). \tag{9}$$

That is to say the rate of change of n at a point is equal to the negative of the divergence of the cell flux. There are three terms in the cell flux: the first, $n\mathbf{u}$, corresponds to advection with the fluid velocity $\mathbf{u}$; the second term, $n\mathbf{V_c}$, corresponds to cell transport relative to the fluid by directed cell swimming ($\mathbf{V_c}$ is the average cell swimming velocity in a small fluid element surrounding the point in question); the fluid term, $-\mathbf{D}.\nabla n$, is a cell diffusion term which needs further discussion.

The above description of the mechanisms of gyrotaxis, oxytaxis, etc. was written as if the cell orientation, and hence swimming direction, were deterministically fixed by the torque balance or oxygen gradient. But it is known from direct observation that the micro-organisms in question swim in *random* directions, only *on average* responding to the directing influences of gravity, etc. A single cell changes direction randomly, at random intervals; different cells swim in randomly different directions (what causes the randomness is not currently known: the cells are too big to be significantly affected by ordinary Brownian motion). The diffusive term in equation (9) is a consequence of the randomness, $\mathbf{D}$ being the *cell diffusivity tensor*. Both and $\mathbf{D}$ are consequences of cell swimming. To estimate them we need to consider the probability density function, $f(\mathbf{p})$, for the unit vector $\mathbf{p}$ in which an individual cell swims. If we know $f(\mathbf{p})$ we can calculate $\mathbf{V}_c$ and $\mathbf{D}$ as follows (Pedley & Kessler, 1990):

$$\mathbf{V}_c = \mathbf{V}_s \iint p\, f(\mathbf{p}) d\mathbf{p} = \mathbf{V}_s <\mathbf{p}> \qquad (10)$$

$$\mathbf{D} = \mathbf{V}_s^2\, \tau <(\mathbf{p} - <\mathbf{p}>)(\mathbf{p} - <\mathbf{p}>)), \qquad (11)$$

where V_s is the cell swimming *speed* (independent of direction) and τ is a correlation time. The integral in equation (10), defining the expectation operator $<..>$, is taken over the unit sphere in $\mathbf{p}$-space.

The function $f(\mathbf{p})$ can either be measured (D.P. Häder & N.A. Hill, unpublished) or computed from the solution of an appropriate Fokker-Planck equation. Such a computation has so far been done systematically only for gyrotactic algae (Pedley & Kessler, 1990, 1992a), and then only for flows with very weak vorticity, appropriate to a hydrodynamic instability analysis, or flows in which gravity is swamped by vorticity. Numerical solutions for all vorticity scales are currently being obtained by Dr Hill and his group. A satisfying preliminary result is that, in the absence of bulk flow, the measured and the calculated $f(\mathbf{p})$ agree rather well. The analysis of bacterial oxytaxis is not yet so sophisticated; we have made the rather gross assumptions that $\mathbf{V}_c$ is directly proportional to ∇C (cf Keller & Segel, 1971) and D is isotropic. A further equation is of course required for C. Even with those crude approximations the instability analysis has yielded a number of detailed, qualitative predictions which Prof. Kessler is in the process of testing experimentally.

FLIGHT

Perhaps the main area, within BFD, that I have not touched on at all is that of animal flight. All I can do in the space available is refer you to the pioneering survey of Lighthill (1975, Chapter 8), which inspired much of the work of the last 20 years, and the new demonstration by Ellington (1995) that the principal approximations in the standard models of heavier-than-air aerodynamics - quasi-steady flow over quasi-two-dimensional wings - must be abandoned and replaced with new fluid dynamics which has not yet been developed. That is another major challenge for the near future.

ACKNOWLEDGMENTS

I would like to record my profound thanks first to Sir James Lighthill, FRS, who supervised my initial forays into Biological Fluid Dynamics at Imperial College, London in

1968, and who has taken an active interest in my work ever since, and second to all the colleagues, post-docs and (especially) research students, too numerous to mention individually, who have collaborated with me in exploring the fascinating problems that arise in this splendid subject. Much of the work has been supported by the British Science and Engineering Research Council (now the Engineering and Physical Science Research Council) and, in recent years, by the Wellcome Trust.

REFERENCES

Altringham, J.D., Wardle, C.S. & Smith, C.I., 1993, Myotomal muscle function at different locations in the body of a swimming fish, *J. Exp.Biol.*, **182**: 191-206.

Aris, R., 1956, On the dispersion of a solute in a fluid flowing through a tube, *Proc. R. Soc. Lond., Ser.* A , **235** : 67-77.

Bertram, C.D., 1995, The dynamics of collapsible tubes, in Biological Fluid Dynamics, *S.E.B. Symposium 49*, ed. by C.P. Ellington & T.J. Pedley, London, Company of Biologists.

Borgas, M.S. & Pedley, T.J., 1990, Non-uniqueness and bifurcation in annular and planar channel flows, *J. Fluid Mech.* **214** : 229-250.

Bowtell, G. & Williams, T.L., 1993, Anguilliform body dynamics: a continuum model for the interaction between muscle activation and body curvature, *J. Math. Biol.*, **32** : 83-92.

Caro, C.G., Fitz-Gerald, J.M. & Schroter, R.C., 1971, Atheroma and arterial wall shear: observation, correlation and proposal for a shear dependent mass transfer mechanism for atherogenesis, *Proc. R. Soc. Lond., Ser.* B., **177** : 109-159.

Chang, H.K., 1989, Flow dynamics in the respiratory tract, in *Respiratory Physiology, an Analytical Approach*, ed. by H.K. Chang & M. Paiva, New York, Marcel Dekker, 57-138.

Cheng, J. Blickhan, R., 1994, Bending moment distribution along swimming fish calculated by waving plate theory, *J. Theor. Biol.*, 168: 337-348.

Cheng, J., Pedley, T.J. & Altringham, 1995, A continuous dynamic beam model for swimming fish, in preparation.

Cheng, J., Zhuang, L.-X., & Tong, B.-G., 1991, Analysis of swimming three-dimensional waving plates, *J. Fluid Mech.*, **232** : 341-355.

Collins, J.M., Shapiro, A.H., Kimmel, E. & Kamm, R.D., 1993, The steady expiratory pressure-flow relation in a model pulmonary bifurcation, *J. Biomech. Eng.*, **115** : 299-305.

Daniel, T.L., 1995, Invertebrate swimming: integrating internal and external mechanics, in *Biological Fluid Dynamics, SEB Symposium 49*, ed. by C.P. Ellington & T.J. Pedley, London, Company of Biologists.

Daskopoulos, P. & Lenhoff, A.M., 1989, Flow in curved ducts: bifurcation structure for stationary ducts, *J. Fluid Mech.*, **203** :125-148.

Eckmann, D.M. & Grotberg, J.B., 1988, Oscillatory flow and mass transport in a curved tube, *J. Fluid Mech.*, **188** : 509-527.

Elad, D. & Kamm, R.D., 1991, Modelling a forced expiration, *Comments Theor. Biol.*, **2** : 239-260.

Ellington, C.P., 1995, Unsteady aerodynamics of insect flight, in *Biological Fluid Dynamics, SEB Symposium 49*, ed. by C.P. Ellington & T.J. Pedley, London, Company of Biologists.

Ellington, C.P. & Pedley, T.J., eds, *Biological Fluid Dynamics,* SEB Symposium 49, London, Company of Biologists.

Friedman, M.H., 1993, Atherosclerosis research using vascular flow models : from 2-D branches to compliant replicas, *J. Biomech. Eng.*, **115** : 595-601.

Fry, D.L., 1987, Mass transport, atherogenesis and risk, *Arteriosclerosis*, **7** : 88-100.

Fukushima, T. & Azuma, T., 1982, The horseshoe vortex: a secondary flow generated in arteries with stenosis, bifurcations and branchings, *Biorheology* , **19** 1: 143-154.

Fung, Y.C., 1971, Peristaltic pumping : a bioengineering model, in *Urodynamics : hydrodynamics of the ureter and renal pelvis* , ed. by S. Boyarsky, C.W. Gottschalk, E.A. Tanago & P.D. Zimsking, New York, Academic Press, 177-198.

Gavriely, N., Shee, T.R., Cugell, D.W. & Grotberg, J.B., 1989, Flutter in flow-limited collapsible tubes: a mechanism for generation of wheezes. J. *Appl. Physiol.* **66** : 2251-2261.

Giddens, D.P., Zarins, C.K. & Glagov, S., 1993, The role of fluid mechanics in the localization and detection of atherosclerosis, *J. Biomech. Eng.* **115** : 588-594.

Greenough, A. & Milner, A.D., 1987, High frequency ventilation in the neonatal period, *Eur. J. Pediatr.*, **146** : 446-449.
Griffiths, D.J., 1983, The mechanics of urine transport in the upper urinary tract. II. The discharge of the bolus into the bladder anddynamics at high rates of flow. *Neurourol. Urodynam.* **2** : 167-173.
Griffiths, D.J., 1989, Flow of urine through the ureter: acollapsible, muscular tube undergoing peristalsis. *J. Biomech. Eng.*, **111** : 206-211.
Grotberg, J.B., 1994, Pulmonary flow and transport phenomena, *Annu. Rev. Fluid Mech.* **26** : 529-571.
Grotberg, J.B. & Gavriely, N., 1989, Flutter in collapsible tubes: a theoretical model and wheezes, *J. Appl. Physiol.,* **66** : 2262-2273.
Halpern, D. & Grotberg, J.B., 1992, Fluid-elastic instabilities of liquid-lined flexible tubes, *J. Fluid Mech.*, **244** : 615-632.
Halpern, D. & Grotberg, J.B., 1993, Surfactant effects on fluid-elastic instabilities of liquid-lined flexible tubes: a model of airway closure. *J. Biomech. Eng.*, **115** : 271-277.
Heil, M. & Pedley, T.J., 1995, Post-buckling deformations of a cylindrical shell conveying a viscous flow, in preparation for *J. Fluids Structures.*
Hess, F. & Videler, J.J., 1984, Fast continuous swimming of saithe (Pollachius virens) : a dynamic analysis of bending moments and muscle power, *J. Exp. Biol.*, **109** : 229-251.
Hillesdon, J.A., Pedley, T.J. & Kessler, J.O., 1995, The development of concentration gradients in a suspension of chemotactic bacteria, *Bull. Math. Biol.*, (in press).
Hyatt, R.E. & Flath, R.E., 1966, Relationship of air flow to pressure during maximal respiratory effort in man. *J. Appl. Physiol.,* **21** : 477-482.
Hydon, P.E., 1994a Resonant and chaotic advection in a curved pipe, *Chaos, Solitons & Fractals*, **4** : 941-954.
Hydon, P.E., 1994b, Resonant advection by oscillatory flow in a curved pipe, *Physica D*, **76** : 44-54.
Isabey, D. & Chang, H.K., 1981, Steady and unsteady pressure-flow relationships in central airways, *J. Appl. Physiol.*, **51** : 1338-1348.
Jaffrin, M.-Y. & Kesic, P., 1974, Airway resistance : a fluid mechanical approach, *J. Appl. Physiol.,* **36** : 354-361.
Jaffrin, M.-Y. & Shapiro, A.H., 1971, Peristaltic pumping, *Annu. Rev. Fluid Mech.,* **3** : 13-36.
Jan, D.L., Shapiro, A.H. & Kamm, R.D., 1989, Some features of oscillatory flow in the lung, *J. Appl. Physiol.,* **67** : 147-159.
Jensen, O.E. & Grotberg, J.B., 1992, Insoluble surfactant spreading on a thin viscous film: shock evolution and film rupture, *J. Fluid Mech.*, **240** : 259-288.
Johnson, M. & Kamm, R.D., 1986, Numerical studies of steady flow dispersion at low Dean number in a gently curving tube, *J. Fluid Mech.*, **172** : 329-345.
Johnson, M., Kamm, R.D., Ho, L.W., Shapiro, A.H. & Pedley, T.J., 1991, The nonlinear growth of surface-tension driven instabilities of a thin annular film, *J. Fluid Mech.*, **233** : 141-156.
Kamm, R.D., 1987, Flow through collapsible tubes, in *Handbook of Bioengineering* , ed. by R.Skalak & S. Chien, New York, McGraw Hill, chap. 23.
Kamm, R.D. 1995, Shear-augmented dispersion in the respiratory system, in *Biological Fluid Dynamics, SEB Symposium 49* , ed. by C.P. Ellington & T.J. Pedley, London, Company of Biologists.
Kamm, R.D. & Pedley, T.J., 1989, Flow in collapsible tubes: a brief review, *J. Biomech. Eng.*, **111** : 177-179.
Keller, E.F. & Segel, L., 1971, Model for chemotaxis, *J. Theor. Biol.,* **30** : 225-234.
Kessler, J.O., 1985, Hydrodynamic focussing of motile algal cells, *Nature*, **313** : 218-220.
Kessler, J.O., 1986, The external dynamics of swimming micro-organisms, in *Progress in Phycological Research, Vol. 4.* ed. by F.E. Round, Bristol, Bristol Biopress, 257-307.
Kessler, J.O., Hoelzer, M.A., Pedley, T.J. & Hill, N.A., 1994, Functional patterns of swimming bacteria, in *Mechanics and Physiology of Animal Swimming*, ed. by L. Maddock, Q. Bone & J.M.V. Rayner, Cambridge, Cambridge University Press, 3-12.
Ku, D.N., Giddens, D.P., Zarins, C.K. & Glagov, S., 1985, Pulsatile flow and atherosclerosis in the human carotid bifurcation: positive correlation between plaque localization and low and oscillating shear stress, *Arteriosclerosis,* **5** : 293-302.
Li, M. & Brasseur, J.G., 1993, Nonsteady peristaltic transport in finite-length tubes, *J. Fluid Mech.,* **248** : 129-151.
Lighthill, M.J., 1970, Aquatic animal propulsion of high hydromechanical efficiency, *J. Fluid Mech.,* **44** : 265-301.
Lighthill, M.J., 1971, Large-amplitude elongated-body theory of fish locomotion, *Proc. R. Soc. Lond., Ser. B,* **179** : 125-138.
Lighthill, J., 1975, Mathematical Biofluiddynamics, Philadelphia, S.I.A.M.

Luo, X.-Y. & Pedley, T.J., 1995, A numerical simulation of steady flow in a 2-D collapsible channel, *J. Fluids Structures*, 9: 149-175.

Miftakhov, R. & Wingate, D., 1994, Numerical simulation of the peristaltic reflex of the small bowel, *Biorheology*, **31** : 309-325.

Otis, D.R., Johnson, M., Pedley, T.J. & Kamm, R.D., 1993, The role of pulmonary surfactant in airway closure : a computational study, *J. Appl.. Physiol.*, **75** : 1323-1333.

Pedley, T.J., 1995, High Reynolds numberflow in tubes of complex geometry with application to wall shear stress in arteries, in *Biological Fluid Dynamics, SEB Symposium 49,* ed. by C.P. Ellington & T.J. Pedley, London, Company of Biologists.

Pedley, T.J. & Kamm, R.D., 1988, The effect of secondary motion on axial transport in oscillatory tube flow, *J. Fluid Mech.,* **193** : 347-367.

Pedley, T.J. & Kamm, R.D., 1991, Dynamics of gas flow and pressure-flow relationships, in *The Lung: Scientific Foundations*, ed. by R.G. Crystal & J.B. West, New York, Raven Press, Vol. I, 995-1010.

Pedley, T.J. & Kessler, J.O., 1990, A new continuum model for suspensions of gyrotactic micro-organisms, *J. Fluid Mech.,* **212** : 155.

Pedley, T.J. & Kessler, J.O., 1992a, Hydrodynamic phenomena in suspensions of swimming micro-organisms, *Annu. Rev. Fluid Mech.,* **24** : 313-358.

Pedley, T.J. & Kessler, J.O., 1992b, Bioconvection, *Sci. Progress,* **76** : 105-123.

Pedley, T.J., Schroter, R.C. & Sudlow, M.F., 1970a, Energy losses and pressure drop in models of human airways. *Respir. Physiol.,* **9** : 371-386.

Pedley, T.J., Schroter, R.C. & Sudlow, M.F., 1970b, The prediction of pressure drop and variation of resistance within the human bronchial airways, *Respir. Physiol.,* **9** : 387-405.

Pedley, T.J., Schroter, R.C. & Sudlow, M.F., 1971, Flow and pressure drop in systems of repeatedly branching tubes, *J. Fluid Mech.,* **46** : 365-383.

Pedley, T.J. & Stephanoff, K.D., 1985, Flow along a channel with a time-dependent indentation in one wall: the generation of vorticity waves, *J. Fluid Mech.,* **160** : 337-367.

Perun, M.L. & Gaver, D.P., 1995, An experimental model investigation of the opening of a collapsed untethered pulmonary airway, *J. Biomech. Eng.* 117: 245-253.

Platt, J.R., 1961, 'Bioconvection patterns' in cultures of free-swimming organisms, *Science*, **133** : 1766-1767.

Ralph, M.E. & Pedley, T.J., 1988, Flow I a channel with a moving indentation, *J. Fluid Mech.*, **190**: 87-112.

Shapiro, A.H., 1977, Steady flow in collapsible tubes, *J. Biomech. Eng.,* **99** : 126-147.

Sharp, M.K., Kamm, R.D., Shapiro, A.H., Kimmel, E. & Karniadakis, G.E., 1991, Dispersion in a curved tube during oscillatory flow, *J. Fluid Mech.,* **223** : 537-563.

Shin, J.J., Kamm, R.D. & Elad, D., 1995, Simulation of forced breathing maneuvers, Chapter-15 in this volume.

Snyder, B., Olson, D.E., Hammersley, J.R., Peterson, C.V. & Jaeger, M.J., 1987, Reversible and irreversible components of central-airway flow resistance, *J. Biomech. Eng.,* **109** : 154-159.

Taylor, G.I., 1953, Dispersion of soluble matter in solvent flowing slowly through a tube, *Proc. Ro. Soc. Lond., Ser. A.* , **219** : 186-203.

Tutty, O.R. & Pedley, T.J., 1993, Oscillatory flow in a stepped channel, *J. Fluid Mech.,* **247** : 179-204.

Wager, H., 1911, On the effect of gravity upon the movements and aggregation of *Euglena viridis,* Ehrb., and other micro-organisms. *Phil. Trans. R. Soc. Lond., Ser. B.* , **201** : 333-390.

Wardle, C.S. & Videler, J.J., 1993, The timing of the electromyogram in the lateral myotomes of mackerel and saithe at different swimming speeds, *J. Fish Biol.*, **42** : 347-359.

Wardle, C.S. & Videler, J.J., 1994, The timing of lateral muscle train and EMG activity in different species of steadily swimming fish, in *MMechanics and Physiology of Animal Swimming*, ed. by L. Maddock, Q. Bone & .M.V. Rayner, Cambridge, Cambridge University Press, 111-118.

Weihs, D., 1972, A hydrodynamic analysis of fish turning manoeuvres, *Proc. R. Soc. Lond., Ser. B.,* **182** : 59-72.

Weihs, D., 1973, The mechanism of rapid starting of slender fish, *Biorheology*, **10** : 343-350.

Williams, T.L., Bowtell, G., Carling, J.C., Sigvardt, K.A. & Curtin, N.A., 1995, Interactions between muscle activation, body curvature and the water in the swimming lamprey, in *Biological Fluid Dynamics, SEB Symposium 49,* ed. by C.P. Ellington & T.J. Pedley, London, Company of Biologists.

Zarins, C.K., Giddens, D.P., Bharodvaj, B.K., Sottiurai, V.S., Mabon, R.F. & Glagov, S., 1983, Carotid bifurcation atherosclerosis: quantitative correlation of plaque localization with flow velocity profiles and wall shear stress, *Circ. Res.,* **53 : 502-514.**

4

FLUID MECHANICS OF ARTERIAL BIFURCATIONS

Don P. Giddens, Tongdar D. Tang, and Francis Loth

The Johns Hopkins University
3400 North Charles Street
Baltimore, Maryland 21218

INTRODUCTION

The arterial system carries out its function of distributing blood throughout the body by means of a remarkable network characterized by vessel branching and bifurcation, and fluid dynamic patterns at these sites of flow division can be extremely complex by comparison with flow in unbranching vessel segments. In certain larger and medium-sized arteries, such as the carotid, coronary and aorto-iliac arteries, these sites of branching and bifurcation are associated with the development of atherosclerotic plaques, and hemodynamic factors such as wall shear stress and particle residence time have been implicated as participants in atherogenesis. The complex flow fields which exist in the region of arterial bifurcations are characterized by strong spatial and temporal variations in wall shear and particle trajectory, creating environments considerably different than those found in simple *in vitro* systems, such as Couette or channel flow, that are often employed in the study of fluid dynamic effects on cell function. Understanding the interactions between blood flow and biological behavior of cells in the arterial wall will undoubtedly require a greater knowledge of the response of cells to flow field phenomena that are representative of those occurring in actual arterial bifurcations.

From the theoretical viewpoint of the governing equations of motion, the boundary conditions associated with bifurcating flows are the primary determinants of the flow behavior; hence, the geometry of the bifurcation, the inflow conditions, and the division of flow into the branches all play strong roles in dictating details of the local flow field. This fact can be perplexing to the fluid dynamicist since individual variability is the rule rather than the exception for parameters such as branch angle, ratio of daughter branch to parent vessel area, and flow division. This variability presents a particular challenge when attempting to relate fluid dynamic variables with biological measurements in animal or human vessels.

Considerable attention has been focused in recent years upon the role of hemodynamic wall shear stress, in particular, upon the structure and function of the artery wall, especially as these may relate to localization of atherogenesis or intimal thickening. A number of investigators have demonstrated a correlation between sites of intimal thickening and either low mean wall shear stress or wall shear stress which oscillates in direction during the cardiac cycle. Some of the most extensive correlations have been obtained for the human carotid bifurcation (1,2). This artery, however, is a somewhat unusual bifurcation — the area enlargement in the transition from parent vessel to daughter branches is rather large due to the carotid sinus, the flow division varies considerably during the cardiac cycle, and there is a region of transient flow separation along the outer wall of the sinus which produces a zone of quite low magnitudes of wall shear stress for a considerable portion of the cycle. Thus, in the normal carotid bifurcation the outer wall of the sinus is a region of low mean wall shear, wall shear which oscillates in direction, and long particle residence time; and observations in human vessels demonstrate that atherosclerotic plaques tend to develop in this region much more preferentially than at the flow divider walls. Thus, there is a strong correlation between intimal thickening in the carotid bifurcation and these fluid dynamic factors.

Detailed fluid dynamic studies in representative models of the bifurcation of the left main (LM) coronary artery into the left anterior descending (LAD) and left circumflex (LCX) arteries show several features which are similar to the carotid bifurcation, such as lower wall shear stress along the outer walls than along the flow divider (3). However, the incoming flow waveform and geometry are greatly different than for the carotid bifurcation, and neither flow visualization nor near-wall measurements of velocity profile in these models has demonstrated a flow separation region — yet the left coronary artery is a site of significant atherosclerosis in humans. Thus, although flow separation may be a participant in atherogenesis at some locations, it is clearly not a necessary condition for the initiation of atherosclerosis.

In addition to the natural bifurcations occurring in the arterial system, bifurcations may be created during vascular graft surgical procedures that bypass arterial obstructions. In such cases the specific geometry of the anastomosis of the graft to the host artery is determined by the relative diameters of the graft and artery and by the local anatomy (4). Although to some extent these variables may be controlled by the surgeon, the flow field can vary considerably in the distal anastomosis, depending upon the geometric and flow division characteristics of the individual graft. Intimal hyperplasia is a major cause of graft failure, and fluid dynamic factors associated with graft flow have been implicated as contributing to intimal proliferation (e.g., 5).

In this chapter we discuss our experience with three arterial bifurcations of clinical interest: the bifurcation of the LM coronary artery into the LAD and LCX branches; the bifurcation of the common carotid artery into the internal and external branches; and the distal anastomosis of a vascular graft, which can create a bifurcating flow in the host artery. Each of these bifurcations has its own distinctive flow characteristics, and each may experience significant vascular remodeling which appears to be associated with local flow phenomena and which can lead to pathological consequences. As background, however, we begin with a few brief remarks of a general nature about bifurcating flows.

GENERAL CHARACTERISTICS OF FLOW IN BIFURCATIONS

The fact that flow changes its direction as it moves through the bifurcation means that it acquires a secondary motion in the branches in which fluid moves from the apex toward the outer wall, much as the motion which occurs in flow in a curved tube. Dean (6) was among the first to investigate curved tube flow, and he demonstrated that the nondimensional parameter $K = 2Re^2 a/L$ where Re is the Reynolds number, a is the tube radius and L is the radius of curvature, is a similarity parameter: as K (the Dean number) increases the secondary motion becomes more significant. However, flows in arterial bifurcations, in addition to having secondary flow patterns due to curvature, are also affected by the rapid increase in cross-sectional area which is typically encountered, producing locally adverse pressure gradients that influence near-wall velocity profiles, particularly along the outer walls (i.e., the walls opposite the flow divider).

A number of investigators have studied flows in idealized bifurcations representative of arterial flow (We will not discuss the numerous studies which have addressed branching airway flows in the lungs). For example, Stehbens (7), Ferguson and Roach (8), Zeller *et al.* (9), and Fukushima *et al.* (10) visualized the flow in glass models of arterial bifurcations with constant daughter to parent area ratios and various branching angles. There is a tendency for the secondary, helical patterns to become more pronounced as the bifurcation angle becomes blunter; and flow can indeed separate from the outer wall and reattach downstream, although the separation region is three-dimensional in nature and does not form a closed recirculation region. If the flow rates in the two branches are unequal, separation tends to occur more readily in the branch with the lower flow rate. Furthermore, the likelihood of flow to separate is enhanced as the area ratio (daughter:parent) increases due a the stronger locally adverse pressure gradient as flow enters an area enlargement. Velocity flow fields have been studied in symmetric bifurcations by a number of investigators (e.g., Walburn and Stein (11,12), Kandarpa and Davis (13), Fernandez et al. (14), Patil and Subbaraj (15)), and in-plane velocity profiles are skewed toward the inner or flow divider walls, while in the plane perpendicular to the bifurcation the axial profiles are M-shaped due to helical secondary motion. This results in wall shear stresses being greater at the flow divider than along the outer wall.

To summarize, the flow behavior in arterial bifurcations varies considerably, depending upon the specific geometry, flow division and inflow conditions, but general trends have been identified. Secondary flows exist and increase in strength as the bifurcation angle becomes blunter, and flow separation is enhanced by a larger daughter:parent area ratio and by asymmetric flow division, with the lower flow branch being more prone to separation. Velocity profiles are skewed toward the flow divider, and the wall shear stresses are higher at the flow divider wall than at the outer wall. These general features, while providing intuition to the fluid dynamicist, will vary in their prominence for different arterial bifurcations.

FLOW IN A LEFT CORONARY ARTERY BIFURCATION MODEL

Previous Work

There has been considerable interest in flow in the major coronary arteries largely because of the prevalence of localized atherosclerotic plaques in these vessels and the accompanying clinical significance. Direct studies of flow and velocity profiles in normal human subjects have not been possible due to the lack of adequate noninvasive means of measurement. Nerem *et al.* (16) measured velocity profiles in the LM, LAD and LCX coronary arteries of anesthetized horses with invasive hot film anemometry and showed skewing of flow toward the epicardial surface. Sabbah *et al.* (17) visualized the flow patterns in a model of the left coronary artery of a pig and saw helical flow which was more prominent in the proximal segment of the LCX than in the LAD. Altobelli and Nerem (18) measured velocity profiles in baboon and canine coronary arteries and concluded that flow in the LM had characteristics of a developing flow profile and was skewed toward the epicardial surface. Velocity profiles in the LAD and LCX immediately downstream of the bifurcation were skewed toward the flow divider as well as toward the epicardial surface. None of these investigations demonstrated the presence of flow separation in the LM/LAD/LCX complex, nor did the experimental methods employed allow accurate estimates of wall shear stresses.

Mark et al. (19) used laser doppler velocimetry (LDV) to measure velocity profiles in a model of the human left coronary artery bifurcation in order to determine whether the pulsatile flow was quasi-steady. Under the conditions investigated, a mean Reynolds number of 180 and Womersley parameter of 2.7, unsteady effects were present and an assumption of quasi-steady flow was found to be inadequate. Asakura and Karino (20) employed a particle tracking technique under steady flow conditions and investigated flow patterns and atherosclerotic plaque distribution in human coronary arteries obtained at autopsy. Their experimental methods allowed them to make the human vessels transparent and then to use the actual vessels as laboratory models. These authors concluded that atherosclerotic plaques were located where shear stress was low. However, their particle tracking method did not lend itself to the quantitative study of pulsatile flows.

Geometry and Flow Conditions

Several anatomical studies were found to be useful in developing a representative geometry of the left coronary artery bifurcation. Nerem and Seed (21) measured the dimensions and branching geometry of 19 post-mortem casts of normal human coronary artery trees which included the aortic root. Grottum et al. (22) measured the angles among the branches at the left coronary bifurcation and calculated the radius of curvature of the LAD. MacAlpin et al. (23) measured diameters of the LM, LAD and LCX vessels. We studied 12 post-mortem luminal casts of the left human coronary artery tree prepared at The University of Chicago from the hearts of human subjects who had died from accidental or non-cardiovascular causes. Six of these casts included the aortic root and six did not. Based upon these findings we constructed a model which incorporated the branch angle data as found by Grottum et al. (22) and which modeled the average of all available data for vessel diameter, length and curvature. The dimensions for the model are given in Table 1 and a schematic is presented in Figure 1. In order to provide the appropriate entrance flow conditions, the aortic sinuses were included in the model.

Reliable blood flow data in the normal human coronary arteries are required for a fluid dynamics study, but are difficult to obtain. Although results from various techniques have been published (24-27), data vary according to sex, state of health, and age. For our

Table 1. Dimensions (mm) and angles (degrees) for human left coronary artery bifurcation data and for the upscaled rigid models used in our laboratory study (3).

	In Vivo	Model
<LM,LAD	151	165
<LM, LC	132	117
<LAD,LC	77	78
<LM,AORTA	50	50
Length of LM	11.0	44
Dia. of LM	4.0	16
Dia. of LAD	3.4	13
Dia. of LC	3.0	12
Rad. of curvature of (LAD)	42.8	171
Rad. of curvature of (LC)	39.3	157

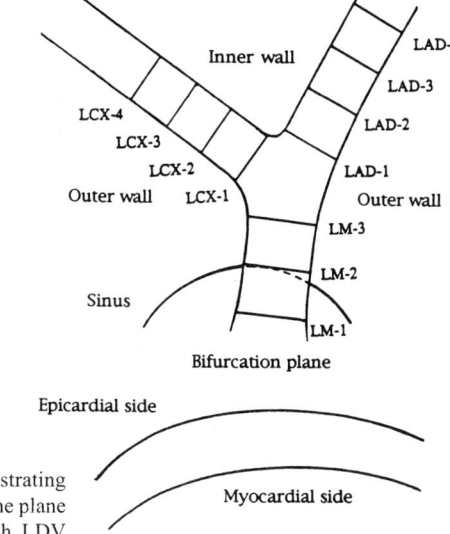

Figure 1. Schematic of the *in vitro* model geometry, illustrating the configuration (a) in the bifurcation plane and (b) in the plane perpendicular to the bifurcation. The stations at which LDV measurements were made are also indicated.

studies we assumed a mean flow rate in the LM of 172 ml/min, which is about 3.5% of the total cardiac output at rest. Flow division data were not found in the literature, so we assumed that the flow rates in the LAD and LCX were such that the mean wall shear stresses in these vessels, as computed by Poiseuille's law, were equal. This assumption was based on the fact that the flow-diameter relationship in normal arteries is such that the mean wall shear stress falls in a fairly narrow range from 10-20 dynes/cm² (28).

The coronary artery flow waveform is considerably different than that in the aorta or peripheral circulation, since blood supply to the heart occurs during diastole. Figure 2 shows the pulsatile waveforms employed in our study, as measured from electromagnetic flowmeters placed in the flow loop. Note that the slightly negative flow often seen at the onset of systole is modeled.

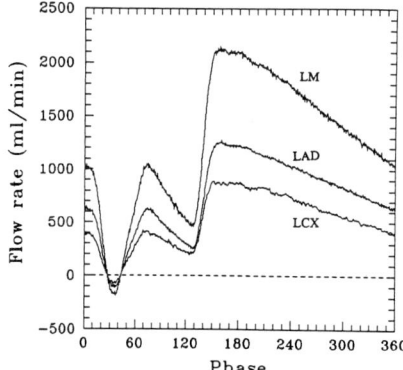

Figure 2. Left coronary artery flow waveforms employed in the experimental study of velocity profiles in the model. Flow rate dimensions are given for the actual laboratory conditions. The corresponding mean flow rate at physiological scales is 172 ml/min.

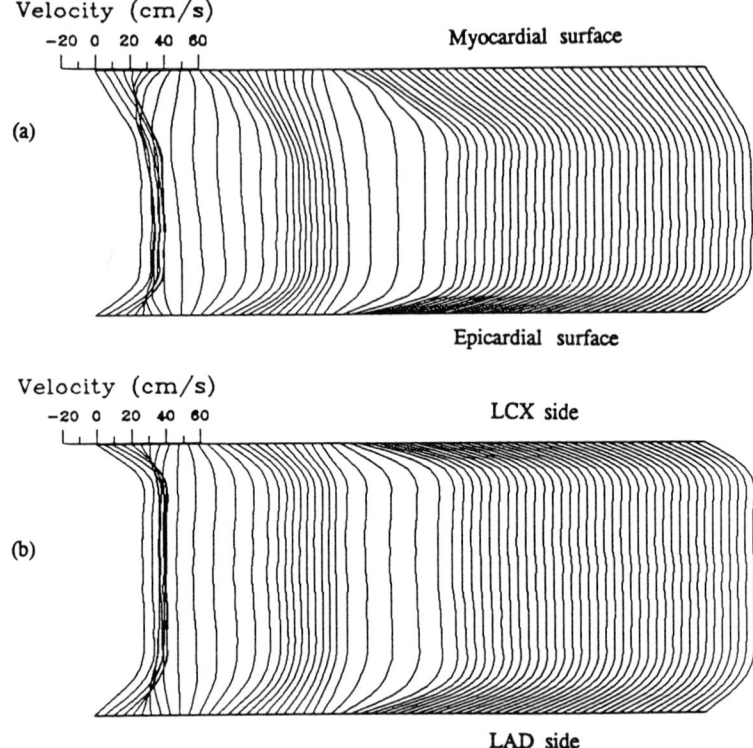

Figure 3. Axial velocity profiles at location LM-1 (a) in the plane perpendicular to the bifurcation and (b) in the bifurcation plane. Profiles are separated by 5.625 degrees in phase during the pulsatile cycle and measured velocities have been scaled to human physiological values in cm/sec.

Velocity Profiles

Velocity profile measurements were performed with a single component LDV system, as described by Tang (3), at the stations shown in Figure 1, both in the plane of the bifurcation and in the plane perpendicular to this. Wall shear stress was estimated from the velocity gradient at the wall and the viscosity coefficient of the working fluid. When scaling to physiological dimensions, a Newtonian blood viscosity coefficient of 4.0 centipoise was assumed.

Figure 3 presents the pulsatile axial velocity profiles at station LM-1: (a) in the plane perpendicular to the bifurcation and (b) in the bifurcation plane. The profiles are separated by 5.625 degrees in phase during the pulsatile cycle, and measured velocities have been scaled to physiological values. The profiles in both planes are blunt in appearance, as is characteristic of an inlet flow, and skewing toward the epicardial surface is readily seen in the plane perpendicular to the bifurcation. Figure 4 presents the axial velocity measurements taken at station LAD-1 using the same format as Figure 3. Profiles are now more developed, but skewing toward the epicardial side is still evident; and the results of the data taken in the bifurcation plane illustrate the higher wall shear stresses which develop along the inner wall, as compared with the outer, and the development of a new shear layer at the flow divider. Axial velocity profiles measured at station LCX-1 are shown in Figure 5. Again, flow is seen to be skewed toward the epicardial surface, and the higher wall shear along the flow divider

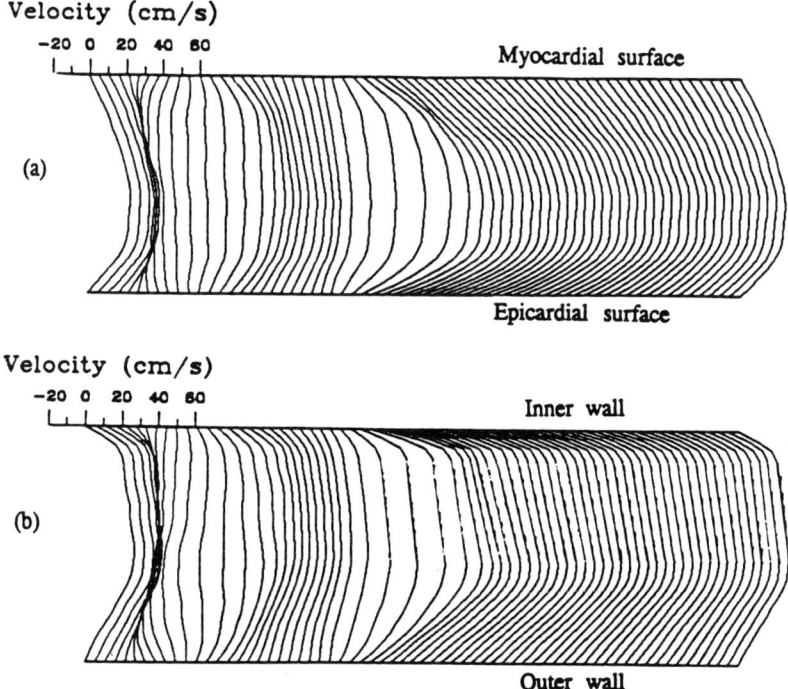

Figure 4. Axial velocity profiles at location LAD-1 (a) in the plane perpendicular to the bifurcation and (b) in the bifurcation plane. Profiles are separated by 5.625 degrees in phase during the pulsatile cycle and measured velocities have been scaled to human physiological values in cm/sec.

side is evident. However, there are strong secondary flows present in the LCX so that the total wall shear stress must be estimated from wall velocity gradients of both the axial and the circumferential velocity components. Tang (3) has reported on a number of axial and circumferential velocity measurements at the various stations identified in Figure 1, and the reader is referred to that reference for additional results on the velocity field.

Wall Shear Stress

The time-varying axial and circumferential wall shear stress components were estimated from the measured near-wall velocity gradients at each of the stations shown in Figure 1. Table 2 presents results for the maximum, minimum and time-averaged mean values of the axial wall shear stress at several selected locations. The wall shear stress tends to be lower on the myocardial as opposed to the epicardial side, and the lowest values occur along the outer wall of the LCX branch, with values along the outer wall of the LAD also being relatively low. We currently do not have comparable data for intimal thickness distribution in the human coronary arteries in order to perform a direct correlation of wall shear stress and intimal thickness. However, the regions of lower wall shear stress fit qualitatively the distribution of atherosclerotic lesions reported by Grottum *et al.* (22).

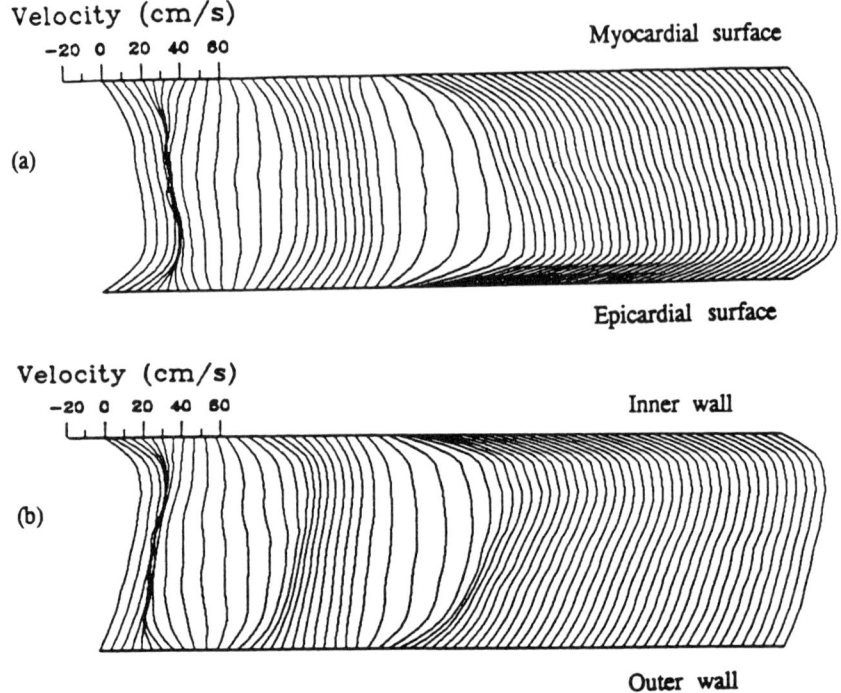

Figure 5. Axial velocity profiles at location LCX-1 (a) in the plane perpendicular to the bifurcation and (b) in the bifurcation plane. Profiles are separated by 5.625 degrees in phase during the pulsatile cycle and measured velocities have been scaled to human physiological values in cm/sec.

General Observations

The velocity and wall shear stress measurements in this model confirmed the trends generally expected from previous studies. The aortic sinuses provided blunt velocity profiles to the LM which, in turn, developed as flow proceeded toward the bifurcation. At each station flow was skewed toward the epicardial side, as expected, due to the curvature of the coronary arteries as they traverse the heart's surface. This resulted in a significantly lower wall shear stress along the myocardial side: the average of the axial component of the mean shear stress along the epicardial surface at all sites where measurements were performed was 19.6 dynes/cm^2 while that along the myocardial side was 11.9 dynes/cm^2. Shear stresses along the outer walls of the bifurcation were lower than along the inner walls, i.e., along the flow divider walls, with the greatest differences being at the proximal stations in the LAD and LCX branches. For example, at the LAD-1 station the mean axial shear stress at the outer wall was 10.2 and at the inner wall was 27.5 dynes/cm^2. For the LCX-1 station, these values were 5.4 and 29.2 dynes/cm^2, respectively.

Although strong secondary flows existed at certain locations, particularly in the proximal segment of the LCX, we were unable to detect any true flow separation, either in the LDV measurements or in the extensive flow visualization studies which accompanied the quantitative measurements. While negative values of wall shear existed, these were associated with the unsteady nature and shape of the flow waveform, and at no time could we identify streamlines or particle pathlines which abruptly diverged from the wall surface.

Table 2. Minimum, maximum and time-averages mean wall shear stress results at selected stations in a model of the human left coronary artery bifurcation. Values were obtained using estimates of the wall velocity gradient and were scaled to physiological values (dynes/cm^2)

Location	Epicardial			Myocardial			Inner Wall			Outer Wall		
	Min.	Max.	Mean	Min.	Max.	Mean	Min.	Max.	Mean	Min.	Max.	Mean
LM-1	-6.5	44	37	-14	28	10	—	—	—	—	—	—
LAD-1	-3.8	35	18	-8.7	22	8.9	-1	54.6	27.5	-9.6	21.4	10.2
LCX-1	-6.9	57	25	-6.7	27	11	-3	54.8	29.2	-11.6	17.8	5.4

The apex of the flow divider was slightly rounded at its tip, a configuration seen consistently in all casts we made of the human coronary bifurcation; and consequently there was a very small stagnation region in the immediate neighborhood that was observed in the flow visualization studies. Otherwise, we saw no zones of flow stagnation or unusually long particle residence time.

It should be noted that the wall shear stress vector changed direction markedly at a number of locations due to the circumferential or secondary flows. Overall, the region where wall shear stress was lowest in magnitude and also changed direction most during the cardiac cycle was along the outer LCX wall in the proximal segment (LCX-1 and LCX-2). If these fluid dynamic variables are causative in atherogenesis, as suggested by correlations in the carotid arteries (1,2), one would expect this site to be the most prone to lesion localization, *at least in those human coronary arteries whose geometry closely approximates that of the* in vitro *model.*

HUMAN CAROTID BIFURCATION

The human carotid bifurcation is unusual among the systemic arteries because of the presence of the carotid sinus in many subjects and because of the low distal resistance of the vascular bed supplied by the internal carotid branch, resulting in a continuously forward flow even during diastole and in a time-varying flow division during the cardiac cycle. The carotid bifurcation is important clinically because of its predilection to develop atherosclerotic plaques which, even though not hemodynamically significant, can lead to transient ischemic attacks and stroke due to plaque ulceration and embolic events.

Model Geometry and Flow Conditions

A model of the carotid bifurcation was developed based on 57 biplanar angiograms of human subjects (29). Flow was studied under steady conditions at mean Reynolds numbers of 400, 800 and 1200 with flow division ratios (internal:external) of 60:40, 70:30 and 80:20 (29,30) and also under pulsatile flow using a physiologic flow waveform with a mean Reynolds number of 300 and peak value of 800 (31). In the pulsatile flow studies an average flow division between the internal and external branches of 70:30 was employed, and the time-varying flow division (as determined in human subjects using noninvasive Doppler ultrasound measurements) was also modelled during the pulsatile cycle. Velocity was measured with a single component LDV system at various sites both in the plane of the bifurcation and perpendicular to the plane. Details of the experimental methods can be found in the references (29-31).

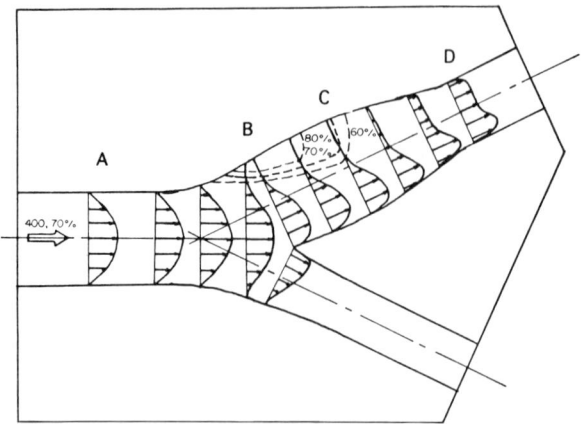

Figure 6. Axial velocity profiles in the bifurcation plane at Re=400 and a flow division ratio of 70:30 into the internal:external carotid branches. The dashed lines in the sinus denote the extent of the region of negative velocity in the profiles for flow divisions of 60:40, 70:30 and 80:20.

Velocity Profiles

Figure 6 presents a schematic of the model geometry, drawn to scale, and of the major flow features observed. Under steady flow a large separation region exists along the outer wall of the carotid sinus, and flow is strongly skewed toward the flow divider. Under pulsatile flow conditions the separation zone is transient, but it does persist over a substantial portion of the cycle because there is continuous forward (and somewhat quasi-steady) flow during diastole as a consequence of the low distal resistance presented to the outflow of the internal carotid branch. However, this zone is not a true recirculation region because of its three-dimensional nature; the separation region contains helical patterns which entrain fluid elements into the region and allow them to exit after a period of time (29,31).

Additionally, Figure 6 contains an overview of the in-plane profiles of the axial velocity component under steady flow at Re = 400 and illustrates the separation region and very strong skewing toward the inner wall (29). The profiles demonstrate that the outer wall experiences quite low shear while the inner wall has a relatively high wall shear environment.

When the model is subjected to pulsatile flow, this picture changes quantitatively but not qualitatively. Figure 7 gives the results of a velocity measurement made near the outer wall of the sinus at a station near the sinus midpoint and compares this to a near-wall velocity measurement made in the internal carotid branch just distal to the tip of the flow divider (See location B in Figure 6). It can be seen that the near-wall velocity along the flow divider is relatively high and always directed distally. In contrast, at the outer wall the behavior is notably different. During systole there is an oscillation in the magnitude and direction of the near-wall velocity, which is reflected in a similar behavior in the wall shear stress. However during the diastolic period, which has approximately constant forward overall flow in the internal carotid, the velocity is nearly zero, indicating flow separation. In fact, it is possible to show that the wall shear stress along the outer wall tends to follow the time derivative of the input flow waveform to a fairly good approximation. The theoretical arguments for this and an experimental demonstration are given by Giddens and Ku (32).

Wall Shear Stress

Table 3 presents pulsatile wall shear stress data as obtained from estimating the wall gradient of the axial and circumferential velocity components at the locations indicated in Figure 6. Minimum, maximum and mean values of the wall shear stress are listed. As expected from the velocity profile behavior, the wall shear stress distribution in the carotid

Fluid Mechanics of Arterial Bifurcations

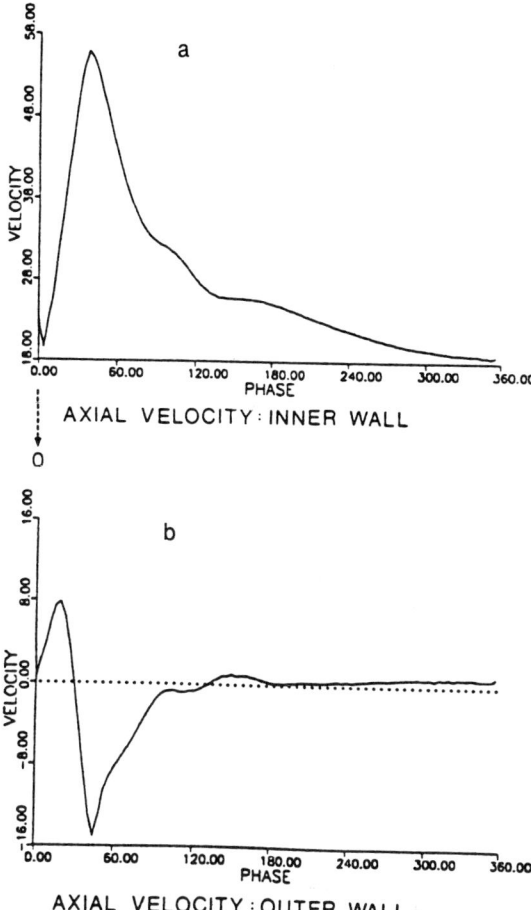

Figure 7. (a) Axial velocity measured near the outer wall of the carotid sinus at section B shown in Figure 6. The low magnitude of the instantaneous velocity can be seen as well as the fact that the velocity during diastole is virtually zero, indicative of a flow separation zone. (b) Axial velocity measured near the flow divider wall (Section B). It should be noted that the origin for the velocity axis is displaced. The relatively high values of velocity during the entire cycle contrast dramatically with the outer wall velocity behavior.

bifurcation model shows the outer wall of the sinus to be a region of low mean shear and of rather small oscillations in magnitude and direction, and the inner wall to be a region of relatively high wall shear which oscillates in magnitude but not in direction. Additionally, the magnitudes of wall shear stress at the side walls in the sinus region are also relatively low, and the shear stress vectors there oscillate in direction during the cycle, indicating that the transient flow separation region is of considerable size. The reader is referred to (29,31) for details on the experimental methods and additional results.

Table 3. Values of minimum, maximum and mean wall shear stress at several locations in the carotid bifurcation model (see Figure 6 for the locations listed).

	Outer Wall			Inner Wall			Side Wall		
Location	Min.	Max.	Mean	Min.	Max.	Mean	Min.	Max.	Mean
A	3	28	7	3	29	7	3	28	7
B	-7	4	-0.5	17	50	26	1	11	3
C	-13	6	-0.7	10	41	17	5	30	8
D	16	49	20	35	109	45	23	70	29

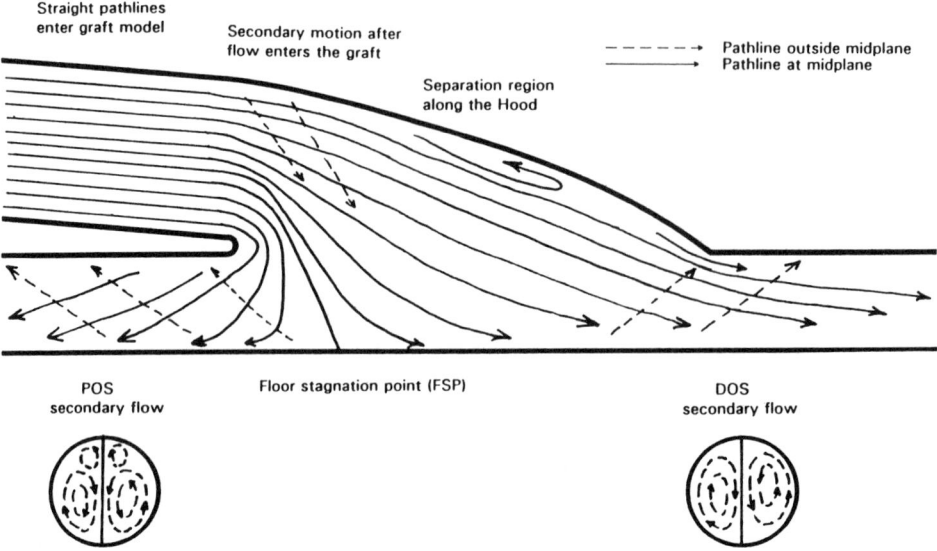

Figure 8. Geometry of a vascular graft model and schematic of general flow patterns which may be observed, depending upon the pulsatile waveform and flow division. POS and DOS refer to the proximal and distal outflow segments, respectively.

General Observations

The flow field in the carotid bifurcation model is rather unlike that in other arterial bifurcations because of the extensive region of transient flow separation. Factors contributing to this are primarily (i) the area enlargement created by the bifurcation itself and by the presence of the sinus and (ii) the common carotid flow waveform which has forward flow at a modest Reynolds number during diastole. These combine to create a substantial period in diastole during which the slower moving fluid which enters the bifurcation encounters an adverse pressure gradient along the outer wall that results in flow separation. This behavior is perhaps unique in the arterial system.

One of the consequences of the separation zone is that a region of long particle residence time is created. This was evident in flow visualization studies which accompanied the LDV measurements and also by the fact that small particles, when introduced into the flow, would remain in the sinus region for several pulsatile cycles before being swept away (31).

Correlations between atherosclerotic intimal thickening in human carotid bifurcations and the measured wall shear stresses showed a strong inverse relationship (1,2), inasmuch as early atherosclerotic plaques were highly preferential in localizing along the outer wall of the proximal internal carotid and this was also a region in which the wall shear was significantly lower, both in the mean and in the maximum values during the cycle, than at other locations. The concept that oscillations in wall shear stress direction might also play a role was discussed in reference (2).

FLOW AND WALL SHEAR IN A VASCULAR GRAFT MODEL

As mentioned previously, under certain situations vascular grafts create a flow bifurcation in the arterial system. For the case of a graft which is implanted to by-pass a

Fluid Mechanics of Arterial Bifurcations

stenotic artery, the distal end of the graft forms an end-to-side anastomosis with the host artery. While the primary goal is to provide blood flow to the circulation distal to the anastomosis, often side branches in the host vessel allow a fraction of the flow to proceed in the proximal direction. A significant clinical problem is that long term patency of vascular grafts may be compromised by neointimal hyperplasia at the distal anastomosis formed with the host artery, particularly for smaller diameter vascular grafts. Thus, knowledge of vascular graft hemodynamics should have value in understanding the mechanisms of graft failure (e.g., 5,33).

Model Geometry and Flow Conditions

The geometry and flow conditions of vascular grafts vary tremendously, so it is difficult to form general conclusions from the study of a single model. The diameters of the graft and artery, the angle of approach, the length of hood, the material of the graft, and the inflow and outflow conditions are each important variables in determining the complex flow field which can occur (4,34,35). We have investigated an *in vitro* model developed from a 5.6 mm (i.d.) PTFE graft implanted in a 3.5 mm canine femoral artery as a model of interest in the study of graft hemodynamics and intimal hyperplasia. Complete details of the development of the model and of the fluid dynamic experiments are given by Loth (34). Both steady flow at representative Reynolds numbers and pulsatile flow using a canine femoral artery waveform were studied with flow visualization and two-component LDV measurements of velocity profiles.

Figure 8 presents a schematic, drawn to scale in its geometry, of the model configuration and of several of the major flow features which can be seen under different conditions. These features change dramatically, depending upon the flow division into the proximal and distal segments and upon whether the flow is steady or pulsatile. Flow in the graft becomes highly three-dimensional as it enters the anastomosis (or bifurcation), and a stagnation point is formed along the floor of the host vessel. For steady flow, this point is stationary, but under pulsatile conditions it moves proximally and distally during the pulsatile cycle. Depending on the flow division, separation may occur along the hood of the graft. This separation zone is enhanced by increasing the flow into the proximal outflow tract; and under pulsatile flow conditions, separation can be either transient or can exist throughout the cycle, depending upon the flow division ratio. Flow accelerates as it enters the distal outflow tract, and

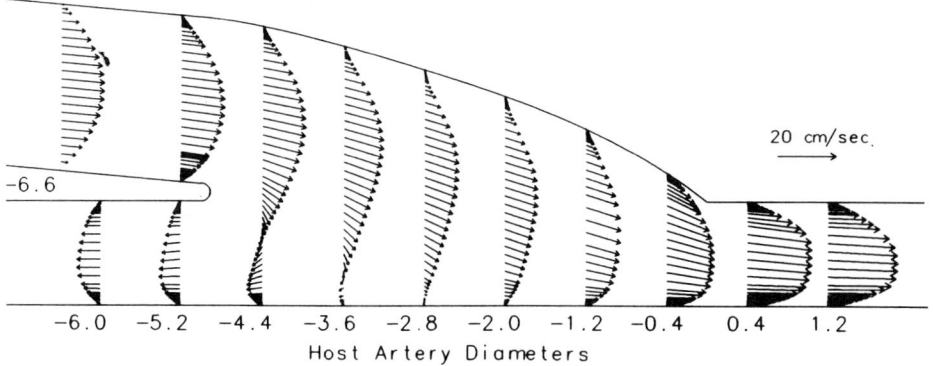

Figure 9. Velocity vectors in the midplane of the model, which is a plane of symmetry, under steady flow conditions. The Reynolds number is 208 and the POS:DOS flow division ratio is 20:80. Velocities have been scaled to canine *in vivo* values (cm/sec).

secondary flow persists for a short distance as flow re-develops. Due to the curvature involved, strong helical flows exist in the anastomosis itself and also as flow enters the proximal outflow segment.

Velocity Profiles

Figure 9 presents the velocity vectors measured in the plane of symmetry of the model under steady flow conditions with a flow division of 20:80 in the proximal to distal outflow tracts and a graft Reynolds number of 208, which corresponds to the mean Reynolds number for a typical canine femoral artery waveform. All values are scaled to *in vivo* conditions. Because of the long entrance length of the graft, flow was fully developed as it approached the anastomosis. The increasing cross-sectional area presented by the anastomosis creates an adverse pressure gradient along the hood, resulting in decreasing values of the wall velocity gradient over much of the hood surface, as can be seen in the figure. Thus, there is a tendency for the hood to experience lower wall shear than other locations, and flow separation can occur along the hood when outflow into the proximal segment is increased. The floor stagnation point can be seen in Figure 9 to be located about half-way along the anastomotic floor, and its position varies strongly with flow division and more gently with Reynolds number. This type of stagnation point flow, where there is direct impingement of flow upon a fairly flat surface, is not present in the normal systemic arterial circulation, although it can occur in the cerebral circulation.

Wall Shear Stress

Figure 10 presents the wall shear stress estimates, obtained from the wall velocity gradient, at locations along the graft hood and host vessel for steady flow at Re=208 and a 20:80 flow division. The shear stress in the proximal segment of the graft (*ca.* 4 dynes/cm^2)

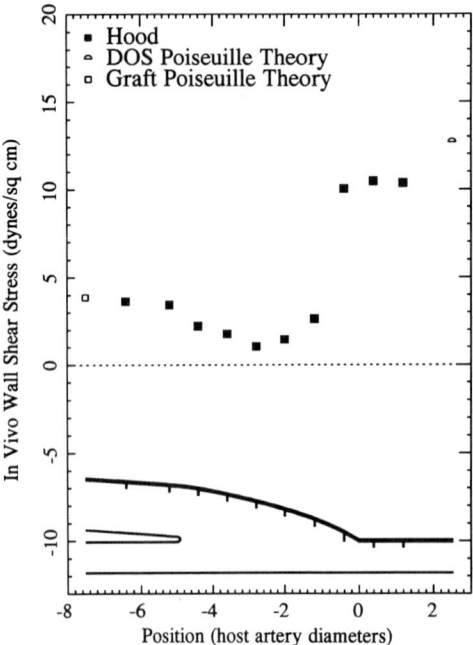

Figure 10. Wall shear stress along the hood of the graft and the host artery at Re=200 and a 20:80 (POS:DOS) flow division ratio. Shear stress values are scaled to canine *in vivo* conditions.

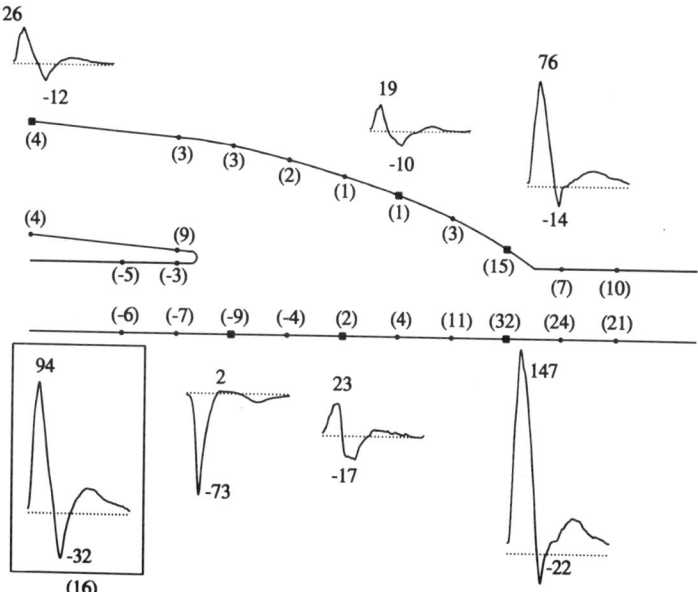

Figure 11. Pulsatile wall shear stress at selected locations in the vascular graft model, scaled to *in vivo* dimensions. The shear stress waveforms shown correspond to locations indicated by (■). Values in parentheses represent time-averaged mean shear stress at the given location. For reference, the Womersly solution for wall shear stress corresponding to the input flow waveform in the normal artery (e.g., without a graft) is shown in the insert at the lower left hand corner of the figure.

is much lower than physiologic (*ca.* 15 dynes/cm^2) because, although the flow rate is normal, the graft diameter is considerably larger than the artery it is supplying. The decrease in wall shear stress along the hood and the rapid rise as flow enters the distal outflow tract with its smaller cross-sectional area can both be readily observed in the data.

Imposing a pulsatile flow changed the velocity field markedly, as expected with a Womersley parameter of 2.7, and of particular interest was the movement of the stagnation point during the cycle. The flow waveform employed had a triphasic nature typical of a femoral pulse. For reference, the Womersley solution for wall shear stress corresponding to the input flow waveform in the normal artery (e.g., without a graft) is given in the insert of Figure 11. Thus, both overall and local flow reversal occurred, and this was reflected in the behavior of the wall shear stress. Figure 11 presents the unsteady shear stress measurements at several selected sites in the model. The overall impression is one of relatively low mean shear and oscillatory shear along much of the hood with more nearly physiological values in the distal outflow segment. Because of the motion of the floor stagnation point, a fairly substantial region of the floor of the host vessel also experiences an oscillating shear stress with a relatively low mean value.

General Observations

The vascular graft model presents an unusual arterial bifurcation whose flow characteristics have elements in common with other bifurcations but which also have notable differences. Because of the geometry of the anastomotic hood, flow enters an enlarging cross-sectional area, and a portion of the fluid makes a sharp turning angle with respect to the incoming stream in order to enter the proximal outflow segment. There are relatively large zones of oscillating and low mean shear along the hood and also along the floor. However, those along the hood are due to the flow reversal in the input waveform and to the tendency for separation to occur on the hood, while the oscillatory shear on the floor is due to movement of a stagnation point during the flow cycle. Hence, although the measurements of wall shear show similar waveform behavior in these two regions, the fluid dynamics mechanisms involved are quite different. We did not observe any indication of turbulence in the measured velocity waveforms under the conditions employed in these experiments. The work of Loth (34) contains additional details of the experimental results.

Measurements of intimal thickening in a series of dogs with similar vascular implants demonstrated consistent intimal growth along the hood and less consistent, though present, intimal thickening along the floor of the host artery (5). Correlation of the fluid dynamics measurements and the morphometric data are in progress.

DISCUSSION

Although there are general fluid dynamics considerations which can be employed to predict, at least qualitatively, the flow behavior in an arterial bifurcation, nonetheless individual bifurcations have features which render them greatly different in their detailed flow fields. In this chapter we have discussed three arterial bifurcations with which we have direct experience, and the reader is also referred to a series of papers by Friedman *et al.* (e.g., 36,37) who have performed extensive studies of coronary arteries and the aortic bifurcation. While there have been a number of correlations between various wall shear measures and intimal thickening, it is safe to state that a universal relationship between specific wall shear behavior and localized intimal thickening has yet to be identified.

The strong correlations between the inverse of low mean wall shear and intimal thickness and between the oscillatory shear index and intimal thickness which were observed in studies of the human carotid bifurcation do not permit a statement as to whether one of these shear measures is more significant than the other, since regions of low mean shear and of oscillating shear coincided and, further, the magnitude of wall shear at the outer wall of the internal carotid was *consistently* low throughout the entire cycle. Additionally, because flows in bifurcations are highly three-dimensional, the presence of secondary flow patterns means that the oscillating wall shear stress vector does not simply oscillate in the proximal and distal orientations (e.g., $+180^0$ to -180^0), but rather can vary over 360^0 in direction (see appendix to reference 2).

We did not observe flow separation in the coronary artery bifurcation, another critically important bifurcation from the standpoint of localization of atherosclerotic plaques. Additionally, Friedman *et al.* (36,37) did not report flow separation in their coronary artery models or their aortic models. Thus, the flow separation found in the carotid artery model — and verified in human subjects using Doppler ultrasound (38) — may be unique in the arterial system. Since bifurcations other than the carotid are prone to atherosclerosis, separation *per se* cannot be the key hemodynamic factor in atherogenesis.

We observed regions of low mean shear and oscillating shear in the vascular graft model — along the hood of the graft and along the floor of the host vessel. However, these regions were created by different fluid dynamics phenomena: the first by reversal of shear due to the shape of the input flow waveform and the tendency for a locally adverse pressure gradient to occur along the hood; the second by the proximal/distal movement of a stagnation point. Our preliminary observations of intimal thickening in a comparable animal model indicate that, while both regions contain localized intimal thickening, it was greater at the hood than along the floor.

The specific geometry and flow division of an arterial bifurcation are crucial to its detailed flow behavior, and care must be taken not to generalize excessively in predicting wall shear at a given site. The velocity field and the wall shear vary considerably over very small spatial dimensions and with time, and particle trajectories are extremely complex. Further studies are needed in which intimal thickness and near-wall flow behavior are examined in animal models, and the effects of mechanical forces on the structure and function of the artery wall must be investigated at cellular and molecular levels. Additionally, the flow field can affect the behavior of blood elements. A platelet or monocyte, for example, can experience a wide range of viscous shear and normal stresses as it travels through a bifurcation; and since interactions of flow with the artery wall will ultimately involve blood-borne cells and molecules as well as forces on endothelial cells, a Lagrangian description of bifurcating flows which includes particle motion will be just as important as the Eulerian description of the velocity field in the attempt to understand vascular biology.

REFERENCE

1. Zarins, C.K., Giddens, D.P., Bharadvaj, B.K., Sottiurai, V.S., Mabon, R.F. and Glagov, S., "Carotid bifurcation atherosclerosis: Quantitative correlation of plaque localization with flow velocity profiles and wall shear stress," *Circulation Research*, Vol. 53, No. 4, 1983, pp. 502-514.
2. Ku, D.N., Giddens, D.P., Zarins, C.K. and Glagov; S., "Pulsatile flow and atherosclerosis in the human carotid bifurcation: Positive correlation between plaque localization and low and oscillating shear stress," *Arteriosclerosis*, Vol. 5, 1985, pp. 293-302.
3. Tang, T. D., "Periodic flow in a bifurcating tube at moderate Reynolds number," Ph.D. Thesis, Georgia Insitute of Technology, 1990.
4. White, S.S., Zarins, C.K., Giddens, D.P., Bassiouny, H., Loth, F., Jones, S.A. and. Glagov, S., "Hemodynamic patterns in two models of end-to-side vascular graft anastomoses: Effect of pulsatility, flow division, Reynolds number and hood length," *Journal of Biomechanical Engineering*, Vol. 115, No. 1, pp. 104-111, 1993.
5. Bassiouny, H.S., White, S., Glagov, S., Choi, E., Giddens, D.P. and Zarins, C.K., "Anastomotic intimal hyperplasia: Mechanical injury or flow induced?" *Journal of Vascular Surgery*, Vol. 15, No. 4, pp. 708-717, 1992.
6. Dean, W.R., "Note on the motion of fluid in a curved pipe," *Journal of Science*, Vol. 20, 1927, pp. 208-223.
7. Stehbens, W.E., "Flow in glass models of arterial bifurcations and Berry aneurysms at low Reynolds numbers," *Quarterly Journal of Experimental Physiology*, Vol. 60, 1975, pp. 181-192.
8. Ferguson, G.G. and Roach, M.R., "Flow conditions at bifurcations as determined in glass models, with reference to the focal distribution of vascular lesions," *Cardiovascular Fluid Dynamics*, Vol. 2, 1972, edited by Bergel, D.H., Academic Press, pp. 141-157.
9. Zeller, H., Talukder, N. and Lorentz, J., "Model studies of pulsatile flow in arterial branches and wave propagation in blood vessels," AGARD, *Fluid Dynamics of Blood Circulation and Respiratory Flow*, 1970.
10. Fukushima, et al., "Characteristics of secondary flow in steady and pulsatile flows through asymmetrical bifurcation," *Biorheology*, Vol. 24, 1987, pp. 3-12.
11. Walburn, F.J. and Stein, P.D., "Velocity profiles in symmetrically branched tubes simulating the aortic bifurcation," *Journal of Biomechanics*, Vol. 14, No. 9, 1983, pp. 601-611.
12. Walburn, F.J. and Stein, P.D., "The shear rate at the wall in a symmetrically branched tube simulating the aortic bifurcation," *Biorheology*, Vol.19, 1982, pp. 307-316.
13. Kandarpa, K. and Davis, N., "Analysis of the fluid dynamic effects on atherogenesis at branching sites," *Journal of Biomechanics*, Vol. 9, 1976, pp. 735-741.
14. Fernandez, R.C., DeWitt, K.J. and Botwin, M.R., "Pulsatile flow through a bifurcation with applications to arterial disease" *Journal of Biomechanics*, Vol. 9, 1976, pp. 575-580.
15. Patil M.K. and Subbaraj, K., "Finite element analysis of two dimensional steady flow in model arterial bifurcations," *Journal of Biomechanics*, Vol. 21, 1988, pp. 219-233.
16. Nerem, et al. "Hot film coronary velocity measurements in horses," *Cardiovascular Research*, Vol. 10, 1976, pp. 301-313.

17. Sabbah, N.N., Walburn, F.J. and Stein, P.D. "Patterns of flow in the left coronary artery," Journal of Biomechanical Engineering, Vol. 106, 1984, pp. 272-279.
18. Altobelli, S.A. and Nerem, R.M., "An experimental study of coronary artery fluid mechanics," *Journal of Biomechanical Engineering*, Vol. 107, 1985, pp. 16-23.
19. Mark, F.F. et al. "Nonquasi-steady character of pulsatile flow in human coronary arteries," *Journal of Biomechanical Engineering*, Vol. 107, 1985, pp. 24-28.
20. Asakura, T. and Karino, T. "Flow patterns and spatial distribution of atherosclerotic lesions in human coronary arteries," *Circulation Research*, Vol. 66, 1990, pp. 1045-1066.
21. Nerem, R.M. and Seed. W.A. "Coronary artery geometry and its fluid mechanical implications," *Fluid Dynamics as a Localizing Factor for Atherosclerosis*, 1983, edited by Schettler, G., Springer-Verlag Press.
22. Grottum, P., Svindland, A. and Walloe, L., "Localization of atherosclerotic lesions in the bifurcation of the main left coronary artery," *Atherosclerosis*, Vol. 47, 1983, pp. 55-62.
23. MacAlpin, R.N., Abbasi, A.S., Frollman, J.H., and Eber, L., "Human coronary artery size during life," *Radiology*, Vol. 108, 1973, pp. 567-576.
24. Knoebel, S.B., McHenry, P.L., Stein, L. and Sonel, A., "Myocaridal blood flow in man as measured by a coincidence counting system and a single bolus of RbCl," *Circulation*, Vol. 36, 1967, pp. 187-196.
25. Pitt, A. Friesinger, G.C. and Ross, R.S. "Measurement of blood in the right and left coronary artery beds in humans and dogs using the Xenon technique," *Cardiovascular Research*, Vol. 3, 1969, pp. 100-106.
26. Ross, R.S., Ueda, K., Lichtlen, P.R., and Rees, J.R., "Measurement of myocardial blood flow in animals and man by selective injection of radioactive inert gas into the coronary arteries," *Circulation Research*, Vol. 15, 1964, pp. 28-41.
27. Cohen, A., et al., "The quantitative determination of coronary flow with a positron emitter (Rubidium-84)," *Circulation*, Vol. 32, 1965, pp. 636-649.
28. Giddens, D.P., Zarins, C.K. and Glagov ,S., "Response of arteries to near-wall fluid dynamic behavior," Applied Mechanics Reviews, Vol. 43, No. 5, 1990, pp. S98-S102.
29. Bharadvaj, B.K., Mabon, R.F. and Giddens, D.P., "Steady flow in a model of the human carotid bifurcation: Part I - flow visualization," *Journal of Biomechanics*, Vol. 15, No. 5, 1982, pp. 349-362.
30. Bharadvaj, B.K., Mabon, R.F. and Giddens, D.P., "Steady flow in a model of the human carotid bifurcation: Part II - laser doppler anemometer measurements," *Journal of Biomechanics*, Vol. 15, No. 5, 1982, 363-378.
31. Ku. D.N., and Giddens, D.P., "Pulsatile flow in a model carotid bifurcation," Arteriosclerosis, Vol. 3, 1983, pp. 31-39.
32. Ku D.N. and Giddens, D.P., "Pulsatile flow visualization in a carotid bifurcation model," 34th Annual Conference on Engineering in Medicine and Biology, Houston, TX, September 1981.
33. LoGerfo, F.W., Quist, W.C., Nowak, M.D., Crawshaw, H.M. and Haudenschild, D.D., "Downstream anastomotic hyperplasia," *Annals of Surgery*, 197, No. 4, April 1983, p. 479.
34. Loth, F., "Velocity and wall shear measurements inside a vascular graft model under steady and pulsatile flow conditions," Ph.D. Thesis, Georgia Institute of Technology, 1993.
35. Figueras C., Jones, S.A., Giddens, D.P., Zarins, C., Bassiouny, H.S. and Glagov, S., "Relationships between flow patterns and geometry in end-to-side anastomotic grafts," BED-Vol. 20, *Advances in Bioengineering*, ASME, , 1991, pp. 225-257.
36. Friedman, M.H., Bargeron, C.B., Deters, O.J., Hutchins, G.M. and Mark, F.F., "Correlation between wall shear and intimal thickness at a coronary artery branch," *Atherosclerosis*, Vol. 68, 1987, pp. 27-33.
37. Friedman, M.H., Bargeron, C.B., Hutchins, G.M., Mark, F.F. and Deters, O.J., "Hemodynamics measurements in human arterial casts and their correlation with histology and luminal area," *Journal of Biomechanical Engineering*, Vol. 102, 1980, pp. 247-251.
38. Ku, D.N., Giddens, D.P., Phillips, D.J., and Strandness, D.E., Jr., "Hemodynamics of the normal human carotid bifurcation: in vitro and in vivo studies," *Ultrasound in Medicine and Biology*, Vol. 11, No. 1, 1985, pp. 13-26.

5

NON-PLANAR GEOMETRY AND NON-PLANAR-TYPE FLOW AT SITES OF ARTERIAL CURVATURE AND BRANCHING: IMPLICATIONS FOR ARTERIAL BIOLOGY AND DISEASE

C. G. Caro,[1] D. J Doorly,[1] M. Tarnawski,[1] K. T. Scott,[1] Q. Long,[1] and C. L. Dumoulin[2]

[1] Centre for Biological and Medical Systems
Imperial College
London SW7 2BX, United Kingdom
[2] GE Corporate Research and Development Center
Schenectady, New York

INTRODUCTION

There has been interest in arterial mechanics over centuries, but this has mainly been confined to the blood pressure, the volume flow rate of the blood and pulse wave propagation (see chapter by Skalak in this volume). Rindfleisch, in 1872, proposed that sites in arteries which experience 'the full stress and impact of the blood' are prone to atheroma. However, it is only within the past 20-30 years that interest has developed in the details of the arterial velocity field.

This interest has come about in part from the finding that the velocity field prominently influences the biology of vessels, including their metabolism, structure and dimensions (Kamiya and Togawa, 1980; Frangos et al, 1985; Yoshida et al. 1988; Henderson, 1991; Davies & Barbee, 1994). In addition, it has come about from recognition that the velocity field can influence the development of arterial disease.

There is now substantial support for the view that atherosclerosis develops preferentially in regions in arteries where the wall shear stress is low on average and undergoes large variations of direction during the cardiac cycle (Caro et al. 1971; Kjaernes et al. 1981; Friedman et al. 1981; Zarins et al. 1983; Schettler et al. 1983; Yoshida et al. 1988; Liepsch, 1994). Moreover, there is increasing support for the view that the velocity field can influence the development of disease at sites of vascular surgery. Several workers report that intimal hyperplasia occurs preferentially at arterial bypass grafts where the wall shear stress is low (Dobrin et al. 1989; Ojha, 1994; White et al. 1993; Jones et al. 1994; Ethier et al. 1994;

Perktold et al. 1994). The disease also occurs at arteriovenous fistulae created for example for chronic haemodialysis, but there has been less study of the relationship of its distribution to the velocity field (Pflugbeil et al, 1994; Fillinger et al. 1990).

Since arterial geometry strongly influences the local blood flow pattern, it is essential to understand both the geometry and the flow. Arteries evidently branch and are often curved. It is also found there is variation of arterial geometry between individuals (Friedman et al. 1983) and that the geometry is affected by skeletal and other motion, such as that of the heart (Caro, 1992; Pao et al. 1992).

The loci generated by tracing the cross-sectional centre of an artery and any branches in the flow direction may be used to establish the geometry of the curvature and branching. It is planar if the space curves corresponding to the loci lie in a plane; otherwise it is non-planar. At any point on a space curve, the local tangent vector and a vector along the radius of curvature, define a plane (called the osculating plane in differential geometry). Torsion measures the rate at which the orientation of this plane varies along the curve. Thus, both curvature and torsion are required to describe a non-planar curve, such as a helix, whereas only curvature is required for a planar curve. Pao et al. (1992) have used both parameters to quantify the time-dependent geometry of the canine coronary arteries. Branching complicates the geometric description, because it allows for discontinuities in the parameters where curves join. For example, at a branch or bifurcation, the curves corresponding to the joining arterial segments may each be planar, but with respect to differently oriented planes. Here the value of torsion is zero apart from at the join point, where the non-planarity arises from the discontinuity in orientation of the planes.

Most studies of the flow in arterial bends and branches are based on models and assume planar geometry (Schettler et al.1983; Yoshida et al. 1988; Liepsch, 1990; Mosora et al. 1990; Liepsch, 1994). In studies which consider non-planar geometry attention is mainly focused on specific locations: aortic arch (Caro, 1971; Farthing, 1979; Paulsen, 1983; Frazin et al. 1990; Yearwood and Chandran, 1984; Kilner et al. 1993; Hoydu et al. 1994), branching of superficial coronary arteries (Batten and Nerem, 1982; Altobelli and Nerem, 1985; Sabbah et al. 1984), bifurcation of aorta (Moore et al. 1994), branching of femoral artery (Back et al. 1985), distal femoral artery (Scholten and Wensing, 1995), carotid syphon (Perktold, 1988). These studies reveal (or model) local flow patterns different from those expected with planar geometry (Berger et al. 1983; Liepsch, 1990; Mosora et al. 1990; Lou and Yang, 1992).

Arterial geometry is recognised to be complicated (Karino et al. 1994; Sabbah et al. 1984) but it does not appear to have been proposed previously that non-planarity is commonly required to describe the curvature and branching of the arteries and the associated flow. Our interest in this subject arose from a crude model study in which we introduced a bend upstream of and non-planar to an existing bend. We observed swirling-type flow and improved clearance of material from the system and, possessing a longstanding interest in arterial mechanics, we were led to ask about the occurrence of non-planar curvature and branching in the arterial system. Our interest has subsequently extended to the parameters which describe the geometry, and the details and biological significance of non-planar-type flow.

In the work which we report, we have examined the geometry of the human aorta and the rabbit aorta using casts. In addition, we have used magnetic resonance (MRI) techniques to measure the geometry and flow pattern in arteries in a small group of human subjects, and in simple models. The MRI studies were carried out with a 1.5 Tesla scanner (GE Medical Systems, Milwaukee, WI); imaging sequence parameters are given. The human subjects were healthy and in the age range 20-32 yr, and the studies were undertaken with Ethical Committee approval.

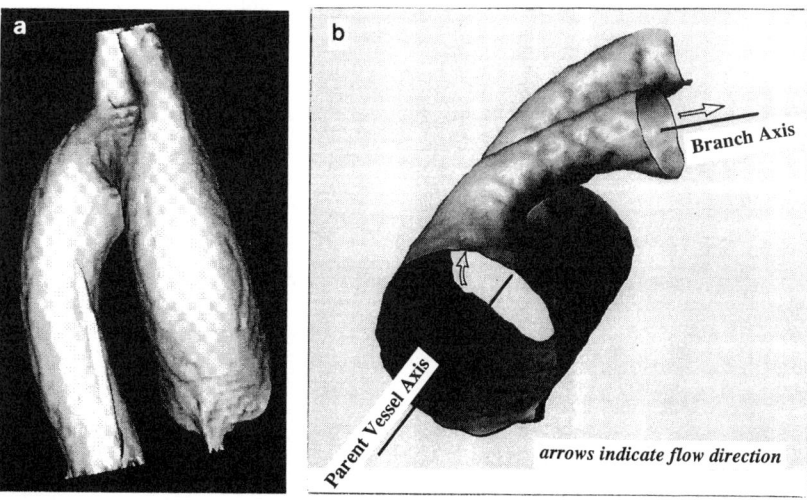

Figure 1. Reconstructed views of cast of human aorta. Fig 1a is a projected view of the aortic arch; the arch has helical-type curvature and there is curvature of the origins of the branches. (There is a streaking artefact over the descending aorta, caused by the reconstruction technique). Fig 1b shows the origins of the branches of the abdominal aorta; the branching geometry can be expected to force the flow to follow a path with helical curvature, indicated by the arrows.

MEASUREMENTS OF NON-PLANAR ARTERIAL GEOMETRY AND FLOW

Studies in Casts

The cast of the human aorta and the rabbit aorta were prepared according to the method of Tompsett (1967). The human aortic cast was made of 'Woods metal' alloy and silicone rubber moulds were made of the arch and abdominal segment. The moulds were filled with dilute copper sulphate solution and images were acquired using a 3-D gradient echo sequence (TR 33 ms, TE 9 ms, flip angle 10°) to obtain contiguous 1 mm slices through the mould with an in-plane resolution of 0.6 mm. The slices were processed using the AVS package (Advanced Visual Systems Inc) to produce 3-D data sets which could be displayed as rotated 2-D projections to demonstrate the non-planar nature of the aortic arch and abdominal aorta.

The curvature of the aortic arch was approximately helical. The slices through the aortic arch and abdominal aorta showed curvature of the origins of the major branches (Figs 1a,b). The aortic bifurcation in the human cast, unlike in vivo (Moore et al. 1994; Caro et al. 1994) was essentially planar. The rabbit cast showed approximately helical curvature of the aortic arch. In addition, there was curvature of the orgins of major branches of the aortic arch and abdominal aorta, and non-planarity of the trifurcation.

In Vivo Studies

Aortic Bifurcation MR angiographic images were obtained at the aortic bifurcation in 6 subjects, using a thick-slice 2-D phase contrast gradient echo sequence (TR 33 ms, TE 9 ms, flip angle 20°) gated to the subject's ECG in order to reduce pulsatile blood flow artefacts. The images are 2-D projections through a 28 cm thick section of tissue in coronal

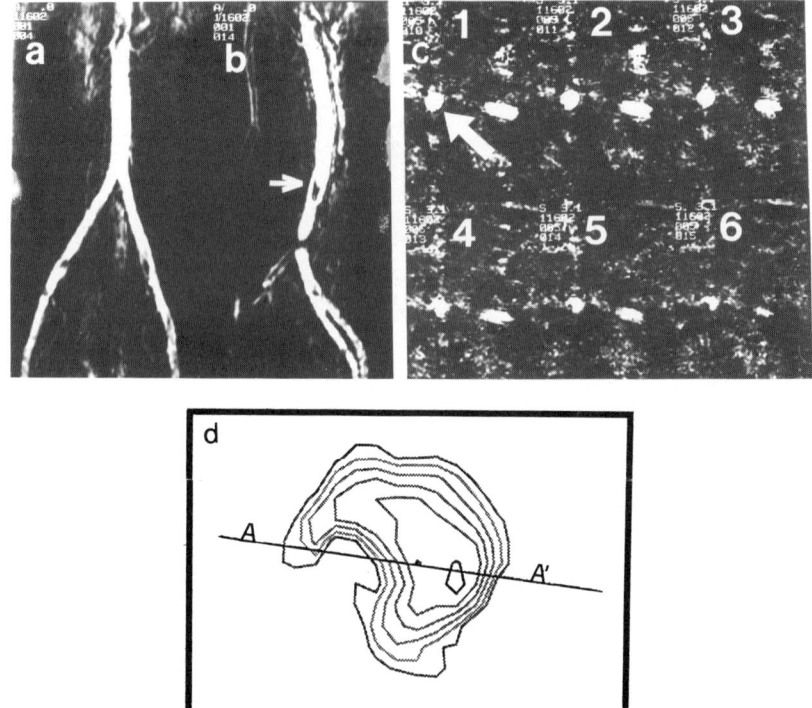

Figure 2. Shown are (a) coronal and (b) sagittal MRI views of the aortic bifurcation of a healthy human subject (arrow indicates level of bifurcation). In (c) and (d) are seen the distribution of axial velocity in the right common iliac artery of the same subject (arrowed). The six phase images (1-6) in (c) were acquired at approximately 30 ms intervals starting 250 ms after the ECG R-wave. Shown in (d) is the axial velocity contour plot from phase image 4. The velocity distribution is rotated out of the 'plane' of bifurcation (line A-A') and is asymmetric.

and sagittal planes, having an in-plane resolution of 1.25 mm. All images were encoded for a maximum velocity of 100 cm/s.

Velocity images were obtained at the aortic bifurcation in the 6 subjects who were studied angiographically, using a 2-D cine phase contrast sequence (TR 30 ms, TE 12.6 ms, flip angle 30°, maximum velocity 80 cm/s). The 1 cm excitation slab was located 2 cm downstream of the aortic bifurcation and rotated about two independent axes so as to be normal to the axis of a common iliac artery, rather than to projections of the axis.

The angiographic studies revealed the aortic bifurcation to be non-planar in all the subjects (Figs 2a,b), the non-planarity arising from curvature (convexity anterior) of both the abdominal aorta and the bifurcation. Although the curvature was mild (aortic radius/radius of curvature typically 1/20) flow effects were found. There was variation of the axial velocity distribution during the cardiac cycle (Fig 2c) with the development of pronounced secondary motion. The secondary motion causes a crescentic distribution of axial velocity but, unlike in a planar bifurcation, the crescents were asymmetric and rotated out of the

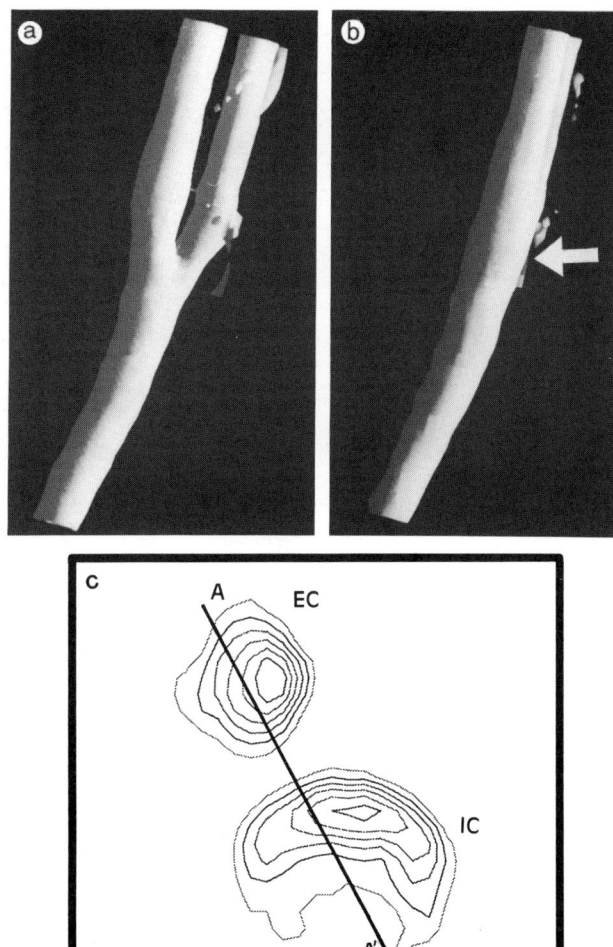

Figure 3. Reconstructed views showing (a) coronal and (b) sagittal MRI projections of the bifurcation of the left common carotid artery of a healthy human subject. Arrow denotes location of bifurcation. The sagittal projection shows curvature in the 'plane' of bifurcation, which renders the bifurcation non-planar. (Artefacts are seen, caused by the reconstruction technique). Shown in (c) are axial velocity contour plots for internal (IC) and external EC) carotid arteries, acquired 1 cm downstream of the bifurcation. The axial velocity distribution is seen to be rotated out of the 'plane' of bifurcation (A-A') and to be asymmetric.

'plane' of bifurcation (Fig 2d). The angle of rotation (measured looking upstream) was about 70° anti-clockwise for the right common iliac artery and 75° clockwise for the left.

Carotid Artery Bifurcation MR angiograms of the carotid bifurcation were obtained in 4 subjects, using a 2-D time-of-flight gradient echo sequence (TR 47 ms, TE 8.7 ms, flip angle 45°) with venous saturation to acquire 64 contiguous 1.5 mm slices with an in-plane resolution of 0.8 mm. The images were processed using AVS (as described above) to obtain both coronal and sagittal views of the bifurcation. Axial velocity images were obtained, using the 2-D cine phase contrast sequence (TR 33 ms, TE 7.5 ms, flip angle 45°), 1 cm downstream of the carotid bifurcation in 2 of the 4 subjects studied angiographically. The measurements were encoded for a maximum velocity of 60 cm/s with a resolution of 0.5 mm.

The carotid bifurcation appeared non-planar in all the subjects. Both the left and right common carotid arteries ran axially, ie in the long axis of the body, as well laterally towards the carotid bifurcation. In contrast, the daughter vessels, the internal and external carotid arteries, had essentially axial courses at their origins (Fig 3a,b). Fig 3c shows velocities measured at the left common carotid bifurcation in one subject. The axial velocity distribution in the carotid sinus was asymmetric and rotated out of the 'plane' of bifurcation during

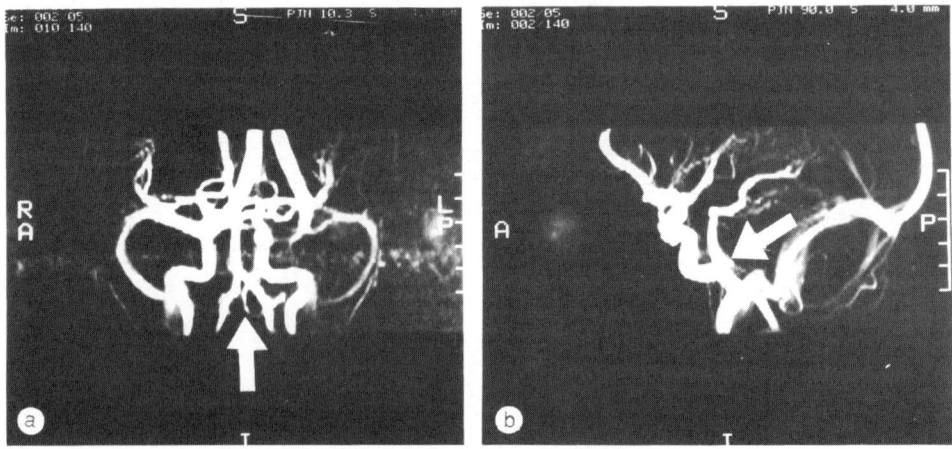

Figure 4. MR angiograms of the intra-cranial vessels of a healthy human subject. The coronal view (a) shows the coalescence of the vertebral arteries to form the basilar artery. The sagittal view (b) shows curvature in the 'plane' of coalescence, which renders the coalescence non-planar. Arrow denotes location of coalescence.

part of the cardiac cycle. The angle of rotation (measured looking upstream) was about 45° clockwise. As noted, both features are consistent with non-planarity. Others have observed curvature of the terminal common carotid artery and helical-type flow in carotid bifurcation models (Masawa et al, 1994).

Vertebral Artery Coalescence. Coronal and sagittal views of the basilar and vertebral arteries were obtained in 3 subjects using a 3-D phase contrast gradient echo sequence TR 24 ms, TE 8.7 ms, flip angle 20°, maximum velocity 50 cm/s to acquire 60 contiguous 1.5 mm slices with an in-plane resolution of 0.9 mm. Reconstruction of these views is carried out automatically on the GE Signa MR system. In all the subjects, the coalescence of the vertebral arteries to form the basilar artery was non-planar (Fig 4a,b).

Model Flow Studies

Non-Planar Tube Phantom We constructed a phantom from 15 mm internal diameter glass tubing, which comprised two successive planar bends (75° arc, radii/radii of curvature 1/10) arranged approximately orthogonally.

Tube cross-section at the two bends was slightly elliptical (major/minor axis approximately 1.2, minor axis in plane of curvature) (Fig 5a). The flow (of dilute copper sulphate) was steady and laminar (Reynolds number 850) and was fully developed at the entrance to the first bend. The phantom was imaged in a standard quadrature head coil. Two flexible tubes of different diameter were positioned 90° apart along the length of the phantom. They were filled with dilute copper sulphate and served as indicators of its orientation.

Thick slice localising 2-D phase contrast angiograms were obtained (TR 33 ms, TE 9 ms, flip angle 20°, 256x128, FOV 32 cm, Venc 15 cm/s, GRASS, scan time 1.5 min). In addition, thin slice oblique 2-D phase contrast angiograms were obtained to localise a plane perpendicular to the tube axis (TR 33 ms, TE 10 ms, flip angle 30°, GRASS, slice thickness 20 mm, 256x128, NEX 20, scan time 1.5 min). Velocity measurements were made with a 3-D phase contrast sequence, using a fast gradient echo sequence with two spatial dimensions and a third dimension converted to a velocity dimension by the inclusion of a bipolar

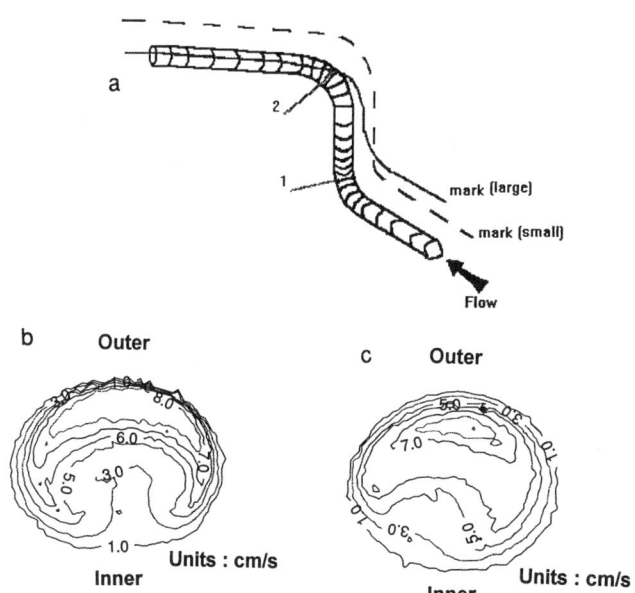

Figure 5. (a) shows the non-planar curved phantom, the location of the marker tubes along it and the positions of the two stations used for MR velocity measurements. (b) shows the axial velocity distribution at station 1 (planar bend). The velocity distribution is symmetrical about the plane of curvature. (c) shows the axial velocity distribution at station 2 (second bend). The axial velocity distribution is rotated out of the plane of curvature and is asymmetric.

gradient. Each image showed the same spatial position, but a specific velocity range corresponding to image velocity contours in the axial (z) direction. (TR 44.1 ms, TE 31.6 ms, flip angle 30°, FOV 10 cm, 256x128, 0,4 mm resolution, slice thickness 3 mm, velocity sensitivity 1 cm/s, 32 images 1-16 representing -16 cm/s to -1 cm/s, 17-32 representing +1 to +16 cm/s, NEX 4, scan time 12 min).

At the first bend (Fig 5b) the distribution of axial velocity was skewed symmetrically about the plane of curvature, with a substantially steeper near-wall velocity gradient at the outer than inner wall of curvature. At the second bend (Fig 5c) the distribution of axial velocity was asymmetric and rotated out of the plane of curvature. There were roughly 'diametrically' opposed regions of high and low near-wall velocity gradient, but the gradient was substantially more uniform 'circumferentially' than at the first bend.

Non-Planar Bifurcation Phantom. For a planar bifurcation, the velocity distribution in the daughter tubes would be symmetrical about this plane, with crescent-shaped contours of higher axial velocity pointing towards each other, ie on the outside of the bend seen by the flow in each tube as it flows from parent to daughter. The measurements show however that the non-planarity results in skewing of the velocity distribution, with the higher velocities rotated away from this plane. The velocities near the inside of the bend, and the near wall velocity gradient are higher than in the planar case, so the shear stress is higher in this region.

The phantom comprised a symmetrically bifurcating glass tube, with a 60° degree included angle between the daughter vessels (Fig 6a), representing a simplification of the geometry of the aorta and the common iliac arteries in the region of the aortic bifurcation. The internal diameter of the parent vessel was 13.5 mm and that of both the daughters was

9 mm. The parent and daughter vessels were curved about a common axis, with radius of parent vessel/radius of curvature approximately 1/20. This degree of curvature was commonly observed in the in-vivo studies described above. Non-planarity thus arises with respect to the flow in the daughter tubes as the combination of curvature in one plane and bifurcation (implying curvature) in another plane. For this symmetrical case, at a given point

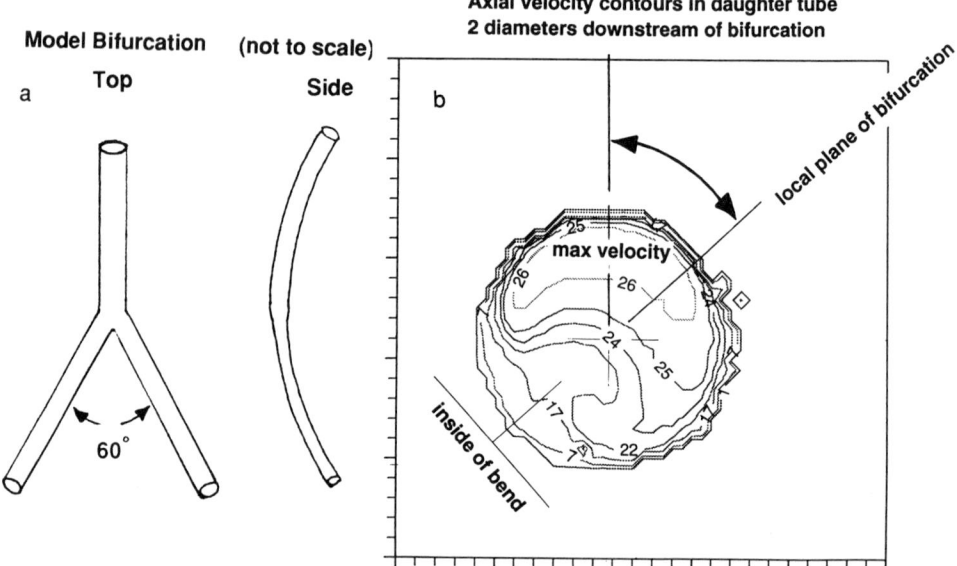

Figure 6. (a) Sketch (not precisely to scale) showing curved symmetrical model bifurcation, with 60° degree included angle between the daughter vessels. The curvature of the bifurcation is similar to that of the abdominal aortic bifurcation measured in-vivo in several subjects. (b) Contours of axial velocity in daughter tube at 2 (daughter tube) diameters downstream of the bifurcation. The plane which is tangent to the centrelines of both the daughter vessels at this point is the local plane of the bifurcation (it passes through the line joining the centres of the daughter vessels at right angles to the paper).

downstream of the bifurcation the local plane of the bifurcation can be defined as that which is locally tangential to the centre lines of the daughter vessels.

Measurements of steady flow (at Re 1300) in the daughter tubes, with a fully developed velocity profile at the entrance to the curved section upstream, revealed the characteristic crescentic velocity distribution typical of a bifurcation, with crescentic contours of high streamwise velocity displaced towards the outside of the bend (Fig. 6b). The crescents were however rotated out of the 'plane' of bifurcation, as found in vivo, and the velocity profile showed the same skewing of the low and high shear regions seen at the second bend in the non-planar tube phantom study. This steady flow model does not, of course, include all the conditions which pertain in vivo, but the similarity of the results to those found in vivo suggests that non-planarity is the cause of the asymmetrical velocity distribution observed in vivo.

Indicator Dispersion Studies

As a further means of investigating the effects of non-planarity on flow, we have studied the dispersion of indicator in steady flow in curved-tube arterial models (Johnson et al, 1994). The models were constructed from 8.0 mm internal diameter polyvinylchloride tubing. They consisted of a 50 cm straight section leading to a 50 cm test section, which took different forms in different experiments: I - straight for 37.5 cm before 90° arc planar bend (radius/radius of curvature 1/20); II - straight for 25 cm before planar S-bend (bends as in I); III - as in II but bends in orthogonal planes. The flow was laminar (Reynolds number 900) and fully developed at the entrance to the model. 0.5 ml of indicator (0.4 g/l methylene blue) was injected in about 0.5 sec at the upstream end of the test section, via a 0.1 cm outside diameter catheter with side holes and a sealed end. The 'foot-to-foot' transit time of the indicator was measured with photodiodes at either end of the test section.

The mean values (SD) (n=10) for I, II and III were respectively: 2.07 s (0.04), 2.78 s (0.12), 3.03 s (0.04). Lengthening of the transit time implies reduction of the range of velocity within which the indicator travels, as the result of mixing. Models I and III are respectively analogous to planar and non-planar branching. The results are consistent with non-planarity having increased mixing and the circumferential uniformity of wall shear over that present with planar geometry.

DISCUSSION

We consider firstly the methods. Although the casts were prepared in situ and at physiological transmural pressure, there could have been distortion post-mortem. Such distortion could explain why the bifurcation of the human aorta was essentially planar in the cast, whereas it was non-planar in vivo. It should also be noted that the in vivo studies involved only a small number of subjects.

In the MR measurement techniques, there is averaging within voxels which will give rise to problems of spatial resolution. However, the resolution was adequate to determine the overall geometry of the vessels and hence to assess non-planarity. The use of 1 cm slices limits the accuracy of measurement of velocity, because of intra-voxel averaging of velocities. Nevertheless, the images showed the rotation and asymmetry of the velocity distribution expected with flow in non-planar geometries (Batten and Nerem, 1982; Kao, 1987). Therefore, intra-voxel velocity dephasing would not seem to have limited the velocity measurements significantly. The velocity contours shown in the figures are unscaled, but could be scaled given additional information or the use of additional imaging sequences (Dumoulin et al. 1993). The 2-D cine phase contrast sequence used acquires data continuously during the cardiac cycle and the data are retrospectively reconstructed to create the images. Some time averaging will therefore occur. However, previous experience with the technique combined with experience gained with the use of more accurate (but lengthier) imaging sequences in phantoms, suggests that this would not have seriously limited the measurements.

The present studies confirm findings by others that the curvature of the aortic arch and, usually, the bifurcation of the aorta are non-planar. In addition, they show that the bifurcation of the carotid arteries and the coalescence of the vertebral arteries to form the basilar artery are non-planar. Moreover, they suggest that the origins of major branches from the aortic arch and the abdominal aorta are non-planar; the flow at these sites has not yet been measured. It has been recognised previously that the branching of superficial coronary arteries and the femoral artery is non-planar. Therefore, the curvature and branching of larger arteries appears commonly to be non-planar. Furthermore, the non-planarity appears to be approximately helical.

The studies also show that the flow pattern just downstream of the aortic and carotid bifurcations differs from that associated with planar bends and branches (Berger et al. 1983; Lou & Yang, 1992). Instead, it resembles the flow pattern seen at the branching of the coronary arteries (Batten and Nerem, 1982; Moore et al. 1994) and in the non-planar phantoms. It resembles too the pattern computed for the carotid syphon (Perktold, 1988). It would appear therefore that the non-planar branching and curvature of arteries is associated with a distinctive flow pattern.

Flow in helical pipes is widely used in engineering applications, but has been little studied theoretically (Manlapaz & Churchill, 1980; Wang, 1981; Germano, 1982; Kao, 1987). The secondary motion in fully developed flow in a helical pipe, like that in a planar bend, consists of a pair of vortices. However, it can be much distorted, with one vortex being squeezed into a narrow region (Kao, 1987). Such asymmetry of the axial distribution of velocity was observed in this work downstream of both the aortic and carotid bifurcations and in the non-planar phantoms.

The flow was mainly measured within and just downstream of non-planar regions, but the need to measure the flow upstream, in the transition region and further downstream, is recognised. Relevant studies by others include those of Yearwood and Chandran (1984), Frazin et al.(1990), Kilner et al.(1993), Hoydu et al.(1994), Moore et al.(1994), Sabbah et al.(1984), Stonebridge and Brophy (1991), Masawa et al.(1994) and Scholten and Wensing, (1995). In several of these studies, swirling or helical-type flow is reported downstream of a non-planar region. However, it is necessary to distinguish between the symmetric counter-rotational vortex pattern seen at planar bends and branches, sometimes called Dean motion (Dean, 1927; 1928), and the asymmetric distribution of velocity observed with non-planar geometry. It can be expected that there will be increased mixing and a circumferentially more uniform wall shear with non-planar than planar geometry. Our indicator dispersion studies, reported above, gave results consistent with these expectations (Johnson et al. 1994). Furthermore, in the tube phantom study, the near-wall velocity gradient was circumferentially more uniform with non-planar than planar geometry.

The apparent commonness of non-planar curvature and branching and of associated non-planar-type flow in the arterial system, raises a number of questions. In respect of vessel biology, it directs attention to the mechanisms which underlie non-planarity. It is likely that these involve anatomical factors, but also possible that they involve fluid mechanical factors, including wall shear. A potentially related question concerns the biological significance of non-planar-type flow. Adaptive regulation of wall shear stress in arteries to change of flow (Kamiya & Togawa, 1980) is consistent with the design of vessels being optimised, within limits, for the role of transporting blood over long periods of time. However, the change in that study was effected by change of vessel diameter, and we are unaware of any comparable study involving non-planarity. It should be added that the distribution of non-planarity in the arterial tree, and elsewhere in the vascular system, is unknown.

In the light of current understanding, it can be anticipated that the flow patterns observed in vivo will also influence the development and distribution of arterial disease (Schettler et al. 1983; Yoshida et al. 1988; Liepsch, 1990; 1994). Fox et al.(1982), Masawa et al. (1994) and Scholten and Wensing (1995) have reported a helical distribution of atherosclerotic lesions downstream of non-planar regions, consistent with a helical distribution of low wall shear. The observed arterial flow patterns may also have implications for vascular surgery; graft-related intimal hyperplasia appears to affect preferentially regions which experience low wall shear (Dobrin et al. 1989; Ojha, 1994; White et al. 1993; Jones et al. 1994; Ethier et al. 1994; Perktold et al. 1994).

Several fluid mechanical questions arise from the study, which merit investigation. Whilst the work has examined the non-planar geometry of arteries at a number of locations and has shown apparently strong sensitivity of flow to non-planarity, as yet the conditions upstream and downstream, at higher Reynolds numbers, and with unsteady flow have not been examined. The flow in arteries varies considerably within the physiological range. Therefore, if the design of arteries is optimised, within limits, for the role of transporting blood over long periods of time, it is of interest to examine the sensitivity to geometry over a range of flow conditions, to determine how well the design works. Lastly, it is permissible to speculate whether an understanding of the influence of the three dimensional geometry of arteries on the flow could have relevance beyond vascular function, for example in the design of general piping systems.

ACKNOWLEDGMENTS

L. Q. is a Royal Society Royal Fellow, People's Republic of China. We acknowledge help from Drs. J. Mestel and P. Cashman and Mr. N. Watkins, and collaborations with Mr K. Robinson, and Prof. B. Hillen, Drs. J. Ravensbergen and J. Krijger and Mr. P. Wensing, Utrecht University, The Netherlands. We thank St. Mary's NHS Trust for access to the MR scanner and BUPA Medical Foundation, BT Charity, 3M and Huntleigh Nesbit Evans Healthcare for support.

REFERENCES

Altobelli, S.A., and Nerem, R.M., 1985, An experimental study of coronary artery fluid mechanics, *Biomech. Eng.* 107:16-23.
Back, M.R., Cho, Y.I., and Back, L.H., 1985, Fluid dynamic study in a femoral artery branch casting of man with upstream main lumen curvature for steady flow, *J. Biomech. Eng.* 107:240-248.
Batten, J.R., and Nerem, R.M., 1982, Model study of flow in curved and planar arterial bifurcations, *Card. Res.* 16:178-186.
Berger, S.A., Talbot, L., and Yao, L.S., 1983, Flow in curved pipes, *Ann. Rev. Fluid Mech.* 15:461-512.
Caro, C.G., Fitz-Gerald, J.M., and Schroter, R.C., 1971, Atheroma and arterial wall shear: observation, correlation and proposal of a shear dependent mass transfer mechanism for atherogenesis, *Proc. Roy. Soc. B* 177:l09-159.
Caro, C.G., Dumoulin, C.L., and Graham J.M.R., 1992, Secondary flow in the human common carotid artery imaged by MR angiography, *J. Biomech. Eng.* 114:147-149.
Caro, C.G., Doorly, D.J., Tarnawski, M., Scott, K.T., Johnson, M.J., and Dumoulin, C.L., 1994, Non-planar curvature and branching of arteries, *J. Physiol.* 475.P:60P.
Davies, P.F., and Barbee, K.A., 1994, Endothelial cell surface imaging: insights into hemodynamic force transduction, *NIPS* 9:153-155.
Dean, W.R., 1927, Note on the motion of fluid in a curved pipe, *Phil. Mag.* 4:208-223.
Dean, W.R., 1928, The stream-line motion of fluid in a curved pipe, *Phil. Mag.* 5:673-695.
Dobrin, P.B., Litooy, F.N.,and Endean, E.D., 1989, Mechanical factors predisposing to intimal hyperplasia and medial thickening in autogenous vein grafts, *Surgery* 105:393-400.
Doorly, D.J., Tarnawski, M., Caro, C.G., Scott, K.T., and Cybulski, G., 1994, Unsteady arterial flow, distensibility and velocity distribution measurement techniques, *Clin. Haemorheol.* 14:426.

Dumoulin, C.L., Doorly, D.J. and Caro, C. G., 1993, Quantitative measurement of velocity at multiple positions using comb excitation and Fourier velocity encoding, *Magn. Reson. Med.* 29:44-52.

Ethier, C.R., Zhang, X.D., Karpik, S.R., Johnston, K.W., and Steinman, D.A., 1994, Flow patterns in hooded vs. non-hooded anastomoses. *2nd World Congress of Biomechanics*, Amsterdam, II:259b.

Farthing, S., and Peronneau, P., 1979, Flow in the thoracic aorta, *Card. Res.* 13:607-620.

Fillinger, M.F., Reinitz, E.R., Schwartz, R. A., Resetarits, D.E., Paskanik, A.M., Bruch, D., and Bredenberg, C.E., 1990, Graft geometry and venous intimal-medial hyperplasia in arteriovenous loop grafts. *J. Vasc. Surg.* 11:556-566.

Fox, B., James, K., Morgan, B., and Seed, W.A., 1982, Distribution of fatty and fibrous plaques in young human coronary arteries, *Atherosclerosis* 41:337-347.

Frangos, J.A., Eskin, S.G., McIntyre, L.V., and Ives, C.L., 1985, Flow effects on prostacyclin production by cultured endothelial cells, *Science* 227:1477-1479.

Frazin, L.J., Lanza, G., Vonesh, M., Khasho, F., Spitzzeri, C., McGee, S., Mehlman, F.D., Chandran, K.B., Talano, J., and McPherson, D., 1990, Functional chiral asymmetry in descending thoracic aorta, *Circulation* 82:1985-1994.

Friedman, M.H., Hutchins, G.M., Bargeron, C.B., Deters, O.J., and Mark, F.F., 1981, Correlation between intimal thickness and fluid shear in human arteries, *Atherosclerosis* 39:425-436.

Friedman, M.H., Deters, O.J., Mark, F.F., Bargeron, C.B., and Hutchins, G.M., 1983, Arterial geometry affects hemodynamics: a potential risk factor for atherosclerosis, *Atherosclerosis* 46:225-231.

Germano, M., 1982, On the effect of torsion on a helical pipe flow, *J. Fluid Mech.* 125:1-8.

Henderson, A.H., 1991, Endothelium in control, *Brit. Heart J.* 65:116-125.

Hoydu, A.K., Bergey, P. D., and Haselgrove, J.C., 1994, A MRI bolus tagging method for observing helical flow in the descending aorta, *Magn. Reson. Med.* 32:794-800.

Hutchins, G.M., 1983, Arterial geometry affects hemodynamics: a potential risk factor for atherosclerosis, *Atherosclerosis* 46:225-231.

Johnson M.J., Caro, C.G., Scott, K.T., Doorly, D.J., Tarnawski, M., and Dumoulin, C.L., 1994, Indicator dispersion in planar and non-planar artery models, *J. Physiol.* 475.P:11P.

Jones, S.A., Giddens, D.P., Loth, F., Kajiya, F., Morita, I., Hiramatsu, O., Ogasawara, Y., Tsujioka, K., and Zarins, C.K., 1994, In-vivo measurements of blood flow velocity profiles in canine ilio-femoral anastomotic bypass grafts. *2nd World Congress of Biomechanics*, Amsterdam, II:167b.

Kamiya, A., and Togawa, T., 1980, Adaptive regulation of wall shear stress to flow change in the canine carotid artery, *Am. J. Physiol.* 239:H14-H21.

Kao, H.C., 1987, Torsion effect on fully developed flow in a helical pipe, *J. Fluid Mech.* 184:335-356.

Karino, T., Takeuchi, S., Kobayashi, N., and Abe, H., 1994, Vascular geometry, flow patterns and preferred sites for aneurysm formation in human intracranial arteries, *Second World Congress of Biomechanics, Amsterdam* II:259a.

Kjaernes, M., Svindland, A., Walloe, L., and Wille, S., 1981, Localisation of early atherosclerotic lesions in an arterial bifurcation in humans, *Acta Pathologica et Microbiologica Scandinavica A* 89:35-40.

Kilner, P.J., Yang, G. Z., Mohiaddin, R.H., Firmin, D.N., and Longmore, D.B., 1993, Helical and retrograde secondary flow patterns in the aortic arch studied by three-directional magnetic resonance velocity mapping, *Circulation* 88:2235-2247.

Liepsch, D., (ed)., 1990, *Biofluid Mechanics: Blood flow in large vessels. Proc. 2nd Int. Symp.* Berlin: Springer-Verlag.

Liepsch, D., (ed)., 1994, *Biofluid Mechanics: Proc. 3rd Int. Symp.*, Dusseldorf: VDI-Verlag.

Lou, Z., and Yang, W-J., 1992, Biofluid dynamics at arterial bifurcations, *Crit. Rev. Biomed. Eng.* 19:455-493.

Manlapaz, R.I., and Churchill, S.W., 1980, Fully developed laminar flow in a helically coiled tube of finite pitch, *Chem. Eng. Commun.* 7:57-78.

Masawa, N., Glagov, S., and Zarins, C.K., 1994, Quantitative morphologic study of intimal thickening at the human carotid bifurcation: I. Axial and circumferential distribution of maximum intimal thickening in asymptomatic uncomplicated plaques, *Atherosclerosis* 107:137-146.

Moore, J.E., Maier, S.E., Ku, D.N., and Boesiger, P., 1994, Haemodynamics in the abdominal aorta: a comparison of in vitro and in vivo measurements, *J. Appl. Physiol.* 76:1520-1527.

Mosora, F., Caro, C.G., Krause, E., Schmid-Schonbein, H., Baquey, C., and Pelissier, R., (eds)., 1990, *Biomechanical Transport Processes*, New York: Plenum Press.

Ojha, M., 1994, Wall shear stress temporal gradient and anastomotic intimal hyperplasia, *Circ. Res.* 74:1227-1231.

Pao, Y.C., Lu, J.T., and Ritman, E.L., 1992, Bending and twisting of an in vivo coronary artery at a bifurcation, *J. Biomech.* 25:287-295.

Paulsen, P.K., and Hasenkam, J.M., 1983, Three-dimensional visualization of velocity profiles in the ascending aorta in dogs, measured with a hot-film anemometer, *J. Biomech.* 16:201-210.

Perktold, K., Florian, H., Hilbert, D., and Peter, R., 1988, Wall shear stress distribution in the human carotid siphon during pulsatile flow, *J. Biomech.* 21:663-671.

Pflugbeil, G., Troster, J., Liepsch, D., and Maurer, P.C., 1994, Special fluid dynamic profiles in different models of arterio-venous fistulas: consequences for patients on dialysis. *Proc. 3rd Int. Symp.* Dusseldorf: VDI-Verlag, 11-14.

Rindfleisch, E., 1872, *A Manual of Pathological Histology, Vol 1*, London, New Sydenham Society.

Sabbah, H.N., Walburn, F.J., and Stein, P.D., 1984, Patterns of flow in the left coronary artery, *J. Biomech. Eng.* 106:272-279.

Schettler, G., Nerem, R.M., Schmid-Schonbein, H., Morl, H., and Diehm, C., (eds), 1983, *Fluid Dynamics as a Localising Factor for Atherosclerosis,* Berlin: Springer-Verlag.

Scholten, F. G., and Wensing, P. J. W., 1995, Atherogenesis in the distal part of the femoral artery: a functional anatomical study of local factors, *PhD Thesis*, University of Utrecht.

Stonebridge, P.A., and Brophy, C.M., 1991, Spiral laminar flow in arteries? *Lancet* 338:1360-1361.

Tompsett, D.H., 1967, Casts of the vessels of the head and vertebral column, *Br. J. Surg.* 54:719-723.

Wang, C.Y., 1981, On the low Reynolds-number flow in a helical pipe, *J. Fluid Mech.* 108:185-194.

White, S.S., Zarins, C.K., Giddens, D.P., Bassiouny, H., Loth, F., Jones, S.A., and Glagov, S., 1993, Hemodynamic patterns in two models of end-to-side vascular graft anastomoses. *J. Biomech. Eng.* 115:104-111.

Yearwood, T.L., and Chandran, K.B., 1984, Physiological pulsatile flow experiments in a model of the human aortic arch, *J. Biomech.* 15:683-704.

Yoshida, Y., Yamaguchi, T., Caro, C.G., Glagov, S., and Nerem, R.M., (eds), 1988, *Role of Blood Flow in Atherogenesis,* Tokyo: Springer-Verlag.

Zarins, C.K., Giddens, D.P., Bhjaradvaj, B.K., Sotiurai, V.S., Mabon, R.F., and Glagov, S., 1983, Carotid bifurcation atherosclerosis: quantitative correlation of plaque localisation with flow velocity profiles and wall shear stress, *Circulation Research*, 53:502-514.

6

COMPUTER SIMULATION OF ARTERIAL BLOOD FLOW

Vessel Diseases Under the Aspect of Local Haemodynamics

K. Perktold and G. Rappitsch

Institute of Mathematics
Technical University Graz
A-8010 Graz, Austria

INTRODUCTION

Since the late sixties it has been increasingly accepted that haemodynamic factors are of importance in the initiation and development of atherosclerotic lesions, and the role of blood flow dynamics as a localizing factor in the genesis of atherosclerosis has provided considerable impetus for the investigation of arterial flow phenomena during the last two decades. Clinical observations have proven that atherosclerosis, a disease of large and medium size arteries, has a pattern which is of a local nature whereby regions of branchings and of sharp curvatures are sites of enhanced predilection. From the fluid dynamical point of view these are zones where the flow is highly disturbed. Because of the influence of the geometry on the detailed characteristics of the pattern of blood flow in arteries, the focal nature of atherosclerosis and its relationship to vessel geometry provides indirect evidence for a role of blood flow in the development of the disease. The vascular geometry may be a risk factor (Friedman et al., 1983, Nerem, 1984), one which could be genetically passed on from one generation to the next, and thus, could be a part of the family history of a human being.

In order to understand the role of arterial fluid dynamics in the genesis of atherosclerosis, it is necessary to know the relationship of the pattern of blood flow to the disease pattern. This requires an understanding of the detailed characteristics of flow in arteries and its relationship to vessel geometry. Many important basic questions on global and local blood flow have been already treated in the monograph "Blood Flow in Arteries" by McDonald (1974) (first published 1960). Further books containing pioneer work in the area of arterial flow have been published by Patel and Vaishnav (1980), Pedley (1980) and Fung (1981). In Pedley's book general mathematical analysis of arterial blood flow is extensively treated. Experimental methods and measurements in cardiovascular blood flow with emphasis on clinical applications are published by Sugawara et al. (1989).

Today extensive documentation on experimental and numerical flow studies in arteries is available. Frequently the flow in curved tubes simulating blood flow in curved arteries and the flow in arterial bifurcation models have been carried out. A review of experimental and numerical studies of arterial bifurcation flow up to 1991 has been published by Lou and Yang (1992). As a result over the last 30 years much has been learned about arterial fluid dynamics and about the relationship between the genesis of the disease with the locally irregular flow field. Accordingly to recent findings, flow separation and flow recirculation, low and oscillating wall shear stress and enhanced particle residence time are decisive factors. On the basis of Caro's low shear theory (Caro et al., 1971) it is widely accepted that an important effect depending on the flow is the locally disturbed mass transfer in separation and stagnation areas. Zarins et al. (1983), Ku et al. (1985), Glagov et al. (1988) and Friedman (1989) concluded that intimal thickening and atherosclerosis develop largely in regions of low wall shear stresses.

However, in order to investigate the complex mechanisms broad knowledge of detailed arterial flow phenomena is necessary. Modern techniques as ultrasound Doppler and laser Doppler anemometry are available. An alternative approach to detailed arterial flow characteristics employs computer simulation in vessel models. The computer methods are very useful in supporting experimental methods and often enable the determination of flow variables which are difficult to obtain in experiments, e.g. wall shear stress is of considerable physiological interest, however, it is difficult to measure this quantity. Recently the high level of computer technology and the development of efficient numerical methods have permitted the consideration of relevant physiological three-dimensional model geometries and the calculation under physiological pulsatile flow conditions including improved rheological assumptions and distensible vessel walls. Simplified two-dimensional geometric idealizations yield basic information, but an appropriate analysis requires the consideration of three-dimensional geometries, because two-dimensional models are unable to show important effects such as secondary motion.

The mathematical description of local arterial flow uses the time-dependent, three-dimensional, incompressible Navier-Stokes equations for Newtonian and non-Newtonian inelastic fluids where the complex blood rheological behaviour is approximated with a shear thinning model on the basis of experimental viscosity data in oscillatory blood flow. The flow calculations are performed with our recently developed pressure correction method applying finite elements. The basis of the numerical procedure is the application of an orthogonal decomposition theorem for a vector field on the Navier-Stokes equations. In the algorithm an equation system for an auxiliary vector field, the pressure equation and an equation for the divergence free velocity are solved. In the case of compliant models the transient wall calculations employing an incrementally linearly elastic constitutive relation and a geometric non linear thin shell model are performed using the finite element program package ABAQUS. In a coupled approach the wall equations and the flow equations subject to the moving boundaries are solved. The approach uses an iteration procedure at each time level in the time-dependent calculations.

The numerical studies are concerned with axial and secondary velocity, flow detachment and reattachment and the corresponding separation zones, the wall shear stress distribution and with the particle residence time. These effects will be illustrated using specific problems. First the velocity patterns in a multiply curved non-planar tube model simulating the carotid siphon are analysed. The study concentrates on the patterns of the secondary motion. The next example concerns the flow in saccular aneurysms, where the investigation focuses on the flow activity in the aneurysms. Two further demonstrations refer to the flow in the carotid artery bifurcation, where the influence of different geometries and the influence of vessel compliance on flow characteristics and wall shear stress have been analysed. One investigation concerns the flow phenomena in a bypass configuration. The final example illustrates pulsatile convective mass transfer in a curved artery model.

MATHEMATICAL FLOW MODEL

In the artery segments of interest here convective local separated viscous flow with recirculation regions can be expected. Thus, in detailed flow analysis rigorous simplifications cannot be applied, and one must therefore look to the most complete mathematical model in fluid dynamics. The governing equations describing the laminar flow of a viscous fluid are the equations of momentum conservation and of mass conservation. Specific conditions in arterial blood flow are irregular geometry, pulsatile nature of the flow, non-Newtonian fluid behaviour and compliant vessels.

The basic kinematic equations express the incompressible mass and momentum balance. In stress-divergence form in absence of body forces the governing equations are given by:

$$\nabla \cdot \mathbf{u} = 0 \tag{1}$$

$$\rho\left(\frac{\partial \mathbf{u}}{\partial t} + (\mathbf{u} \cdot \nabla)\mathbf{u}\right) = \nabla \cdot \mathbf{S} - \nabla p \quad \text{in} \quad G \subset R^n, \; n = 2 \text{ or } 3, \tag{2}$$

where $\mathbf{u}$ is the n-component flow velocity vector, p is the pressure, ρ is the constant fluid density; $\mathbf{S}$ is the extra stress tensor, G denotes a bounded regular domain of the n-dimensional space.

The constitutive equations specify the extra stresses expressing the rheological properties of the fluid. Assuming a Newtonian fluid the extra stresses are linear functions of the rate of deformation tensor $\mathbf{D}$:

$$\mathbf{S} = 2\mu \mathbf{D} \tag{3}$$

$$\mathbf{D} = \frac{1}{2}\left(\mathbf{L} + \mathbf{L}^T\right), \quad \mathbf{L} = (\nabla \mathbf{u})^T \tag{4}$$

μ denotes the constant dynamic viscosity of the fluid, $\mathbf{L}$ is velocity gradient tensor. In fully developed (Poiseuille-) flow the occurring extra stress is defined by , where is called the shear rate and is the velocity gradient in the direction perpendicular to the flow axis. Basic flow types are discussed by Mc. Donald (1974).

For a non-Newtonian inelastic fluid the extra stresses are defined in terms of the rate of deformation tensor and of the apparent viscosity μ:

$$\mathbf{S} = 2\mu \, (D_{II}; p_1,\ldots,p_m) \, \mathbf{D} \tag{5}$$

The extra stress components are non-linear functions of an appropriate tensor quantity, the second invariant of the rate of deformation tensor and of specific parameters. Essentially the apparant viscosity is shear rate dependent. Considering fully developed (parabolic) flow the tensor quantity is related to the scalar shear rate .

In the mathematical treatment of Newtonian and non-Newtonian inelastic fluids it is important that the extra stresses can be inserted into the momentum equation:

$$\rho\left(\frac{\partial u_i}{\partial t} + u_j \frac{\partial u_i}{\partial x_j}\right) - \frac{\partial}{\partial x_j}\left(\mu\left(\frac{\partial u_i}{\partial x_j} + \frac{\partial u_j}{\partial x_i}\right)\right) + \frac{\partial p}{\partial x_i} = 0, \qquad i,j = 1,2,3. \tag{6}$$

Considering the incompressible mass conservation the classical Navier-Stokes equations for Newtonian fluids in the variables velocity and pressure (using the standard Laplacian symbol) may be expressed as:

$$\rho\left(\frac{\partial u_i}{\partial t} + u_j \frac{\partial u_i}{\partial x_j}\right) - \mu \Delta u_i + \frac{\partial p}{\partial x_i} = 0$$

$$\frac{\partial u_j}{\partial x_j} = 0, \qquad i,j = 1,2,3. \tag{7}$$

In the case of variable viscosity the generalized Navier-Stokes equations can be derived:

$$\rho\left(\frac{\partial u_i}{\partial t} + u_j \frac{\partial u_i}{\partial x_j}\right) - \mu \Delta u_i - \frac{\partial \mu}{\partial x_j}\left(\frac{\partial u_i}{\partial x_j} + \frac{\partial u_j}{\partial x_i}\right) + \frac{\partial p}{\partial x_i} = 0$$

$$\frac{\partial u_j}{\partial x_j} = 0, \qquad i,j = 1,2,3. \tag{8}$$

In Eqns.6-8 Einstein's summation convention is applied. When a subscript is repeated in a term summation is indicated. These forms of the Navier-Stokes equations are referred to as the primitive equations since they are written in terms of the primitive variables velocity components and pressure. Alternative forms use the streamfunction and the vorticity transport as dependent variables. The stable calculation of the velocities and especially of the pressure from these variables is not simple.

The equations describe the flow in a model with rigid walls. For compliant models with moving boundaries a modification is required. In the Arbitrary Lagrangian-Eulerian method (ALE) for fluid-structure interaction by Hughes et al. (1981) a correction of the convective velocity is applied. The ALE description of the momentum equations as the governing equations for the fluid motion is:

$$\rho\left(\frac{\partial u_i}{\partial t} + (u_j - \hat{u}_j)\frac{\partial u_i}{\partial x_j}\right) - \frac{\partial}{\partial x_j}\left(\mu\left(\frac{\partial u_i}{\partial x_j} + \frac{\partial u_j}{\partial x_i}\right)\right) + \frac{\partial p}{\partial x_i} = 0, \qquad i,j = 1,2,3. \tag{9}$$

In the concept of the ALE method is material velocity (flow velocity) and describes the velocity of the points of the flow domain.

The Navier-Stokes equations may be conventially normalized or non-dimensionalized using a characteristic (reference) length, a characteristic (reference) velocity and the kinematic viscosity. Using the normalizations:

$$x_i^* = x_i / L_0, \quad u_i^* = u_i / U_0, \quad p^* = p / \rho U_0^2, \quad t^* = t U_0 / L_0 \qquad (10)$$

Eqn. 7 becomes

$$\frac{\partial u_i^*}{\partial t^*} + u_j^* \frac{\partial u_i^*}{\partial x_j^*} - \frac{1}{Re} \Delta u_i^* + \frac{\partial p^*}{\partial x_i^*} = 0$$

$$\frac{\partial u_j^*}{\partial x_j^*} = 0, \qquad i,j = 1,2,3, \qquad (11)$$

where all quantities are non-dimensionalized, and $Re = L_0 U_0 / \nu$ is the Reynolds number. In the case of oscillatory or pulsatile flow in addition a dimensionless frequency parameter, the Stokes number, is introduced $\eta = \sqrt{4\pi L_0^2 f / \nu}$, where f is the frequency.

Because the viscous stress field is often important in arterial flow investigations the components of the symmetric stress tensor for the incompressible three-dimensional (rectangular Eulerian co-ordinate system (x,y,z)) and for axisymmetric (r,z,θ) flows studied are demonstrated explicitely. The stresses in terms of constant or variable viscosity and of velocity gradients are:

$$\tau_{xx} = 2\mu \frac{\partial u}{\partial x}, \quad \tau_{xy} = \mu\left(\frac{\partial u}{\partial y} + \frac{\partial v}{\partial x}\right), \quad \tau_{xz} = \mu\left(\frac{\partial u}{\partial z} + \frac{\partial w}{\partial x}\right)$$

$$\tau_{yy} = 2\mu \frac{\partial v}{\partial y}, \quad \tau_{yz} = \mu\left(\frac{\partial v}{\partial z} + \frac{\partial w}{\partial y}\right), \quad \tau_{zz} = 2\mu \frac{\partial w}{\partial z}; \qquad (12)$$

$$\tau_{rr} = 2\mu \frac{\partial u}{\partial r}, \quad \tau_{rz} = \mu\left(\frac{\partial v}{\partial r} + \frac{\partial u}{\partial z}\right), \quad \tau_{zz} = 2\mu \frac{\partial v}{\partial z}, \quad \tau_{\theta\theta} = 2\mu \frac{u}{r}. \qquad (13)$$

The transport of macromolecules and dissolved gases in the vessel lumen and to the vessel wall is a problem of high interest. The corresponding convective diffusion process can be mathematically described with the transport equation:

$$\frac{\partial c}{\partial t} + u_j \frac{\partial c}{\partial x_j} = D \frac{\partial}{\partial x_j}\left(\frac{\partial c}{\partial x_j}\right), \quad j = 1,2,3, \qquad (14)$$

where c is solute concentration (or tension of a dissolved gas) and D is the diffusion coefficient of the solute. The characterization of the transport process uses the Peclet number defined as $Pe = U_0 L_0 / D$. For high Peclet numbers in convection dominated physiological diffusion processes ($Pe \approx 10^5$ for oxygen transport, $Pe \approx 10^7$ for albumin transport) stabilization of the numerical algorithm is necessary (Rappitsch and Perktold, 1995).

BOUNDARY CONDITIONS

The Navier-Stokes equations can be solved for the velocity and the pressure given appropriate boundary and initial conditions. For the flow problems of interest here, the boundary of the flow domain partitions into two sections with different types of boundary conditions. At the one boundary section (inflow boundary and vessel wall) the time-dependent flow velocity is prescribed:

$$u_i(t) = g_i(t), \quad i=1,2,3, \quad t>0. \tag{15}$$

At the inflow boundary mostly physiologically correct inflow velocity data are not available, and often fully developed velocity profiles (Womersley solutions) corresponding to the prescribed pulse wave form are assumed. These profiles are calculated as long straight tube profiles. In relatively straight artery segments (e.g. common carotid artery) fully developed velocity profiles are acceptable.

At the remaining boundary section, outflow boundary or "artificial" outflow boundary located downstream of the flow domain of interest, the condition describing surface traction forces often can be assumed. The condition is modelled mathematically as:

$$\left(-p\delta_{ij} + \mu(\frac{\partial u_i}{\partial x_j} + \frac{\partial u_j}{\partial x_i})\right) n_j = h_i(t), \quad i,j=1,2,3, \quad t>0, \tag{16}$$

where $n_j, j = 1,2,3$, are the components of the outward pointing normal unit vector at the outflow boundary. In distensible wall calculations the pressure at the outlet boundary is prescribed as normal force $\sigma_n = -p(t)$,, the tangential force $\sigma_t = 0$.

Under the assumption of rigid vessel walls traction free outflow is assumed. Generally the specification of outflow boundary conditions is difficult for finite flow domains. Stress conditions are useful at artificial boundaries which are located downstream the domain of interest. In order to make use of idealized conditions, the inflow and outflow cross-sections are chosen far enough upstream and downstream of the region of interest, so that the flow is uniform at these sections.

Appropriate initial conditions require the specification of the flow velocities at the initial time $t=t_0$, where the condition:

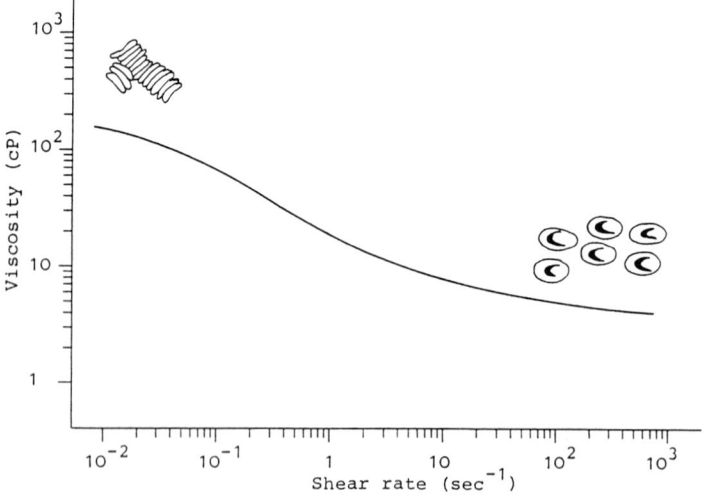

Figure 1. Apparent viscosity as function of the shear rate.

$$\frac{\partial u^0_j}{\partial x_j} = 0, j = 1,2,3, \text{ is satisfied.}$$

BLOOD RHEOLOGY

Blood is a colloidal fluid where particles are suspended in the plasma. The major determinants of blood viscosity are plasma viscosity and red blood cell behaviour (concentration, deformability and aggregation). Blood has complex non-Newtonian behaviour (Thurston, 1979).

In large arteries where relatively high shear rates occur blood behaves nearly as a Newtonian fluid, and often numerical studies in large arteries apply Newtonian simplification with a constant representative viscosity. An improvement can be made including the shear rate dependent viscosity. The shear thinning effect of blood can be mathematically modelled using an appropriate shear thinning model. In Fig. 1 the apparent viscosity of blood against the shear rate is illustrated.

The aggregation of red blood cells causes an elevation of blood viscosity at low shear rates. Disaggregation of the reversible formations yields a viscosity decrease at high shear rates.

In our investigation the shear thinning behaviour according to the modified Cross model is applied:

$$\mu(\dot\gamma) = \mu_\infty + (\mu_0 - \mu_\infty) \frac{1}{(1+(\lambda\dot\gamma)^b)^a}. \tag{17}$$

For the application of a shear thinning model in the description of the complex rheological behaviour in pulsatile blood flow appropriate viscosity data are necessary. In this study experimental data measured in oscillatory blood flow by Liepsch et al. (1991) are used for the numerical fitting of the five occurring parameters in Eqn. 17.

The comparison of Newtonian and of non-Newtonian results shows a minor quantitative influence of the shear thinning behaviour in large arteries where the shear rates are relatively high (Perktold et al., 1991b). A further improvement of the rheological assumptions requires the consideration of viscoelastic properties, although an essential influence in local haemodynamics cannot be expected. In the mathematical modelling of blood viscoelasticity the Oldroyd-B model as constitutive relation is suggested in literature (Perktold et al., 1994c).

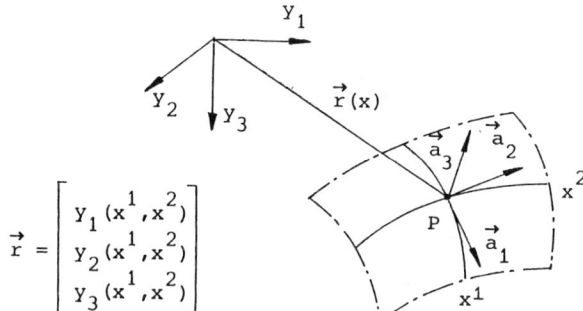

Figure 2. Shell middle surface, covariant tangential base vectors $\vec{a_1}$, $\vec{a_2}$ to the curvilinear co-ordinate lines defining a tangential plane.

$$\vec{r} = \begin{bmatrix} y_1(x^1, x^2) \\ y_2(x^1, x^2) \\ y_3(x^1, x^2) \end{bmatrix}$$

MATHEMATICAL WALL MODEL

Basic concepts of stress analysis are explained using a thin wall tube. In the demonstration the intravascular pressure and longitudinal tethering force are the primary sources of stress in the blood vessel wall (Patel and Vaishnav, 1980). In some simplified cases vascular segments may be treated as homogeneous circular cylindrical tubes of uniform thickness.

Using certain approximations the estimation of stresses in blood vessels may be calculated from simple equilibrium considerations. The stresses occurring in a blood vessel segment under action of an intravascular pressure p and a longitudinal force F are the circumferential, longitudinal and radial stresses.

From a diametral longitudinal plane bisecting of the tube the circumferential stresses can be found as:

$$2 S_\Theta L h = 2 p R_i L \quad \rightarrow \quad S_\Theta = p R_i / h \tag{18}$$

where is the inner tube radius and h is the wall thickness.

The longitudinal stresses are determined considering a vessel cut by a transverse plane perpendicular to the vessel axis. The equlibrium yields:

$$2 \pi R h S_z = F + p \pi R_i^2 \quad \rightarrow \quad S_z = p(R/2h - 1/2) + F/2\pi R h \tag{19}$$

Assuming $F=0$ Eqns.18-19 express the well known fact that the magnitude of the circumferential stresses is twice the longitudinal stress.

The circumferential and the longitudinal stresses are the dominant forces in this simple tube mechanics under internal pressure and tethering force. The third force component, the radial stress, is of minor importance. The corresponding expression cannot be derived easily (Patel and Vaishnav, 1980).

Eqns.18-19 are very basic formulas in vascular mechanics, and limitations should be carefully kept. In the case of irregular geometries as bifurcations the simplifications cannot be applied, and the description of the vessel wall mechanics requires improved models. An essential improvement is to model the artery wall as thin shell structure.

Shell Mechanics

A shell is a three-dimensional continuous medium for which one dimension, the thickness, is small with respect to the two others. The basic idea of shell theories is an appropriate transformation and the integration over the thickness and to get a two-dimensional model, formulated on the middle surface of the shell representing an approximation of the three-dimensional model. (Pioneers of shell theory are: Kirchhoff, Koiter). The shell mathematics uses tensor notations, and the mathematical formalism is not simple (Koiter and Simmonds, 1973).

Knowing the physical characteristics of the material, the initial configuration C, the distribution of the applied forces and the boundary conditions, the shell problem is to determine the displacement of the points of C. From the displacements the strains and the stresses at any point of the new configuration C* can be calculated. The equations of equilibrium represent the total potential energy, this is the loading energy and the strain energy. The analysis uses geometric non-linear shell theory and incrementally linearly elastic wall behaviour. The viscous flow problem and the shell problem generally cannot be solved exactly and the application of numerical methods is required.

NUMERICAL SOLUTION

The most important numerical methods for the approximate solution of partial differential equations problems are the finite difference methods and the finite element methods. One of the significant advantages of finite element methods is the high flexibility, and thus, it is one of the most efficient methods of dealing with irregular shaped boundaries. Finite element methods yield accurate results for many practically important problems, and extensive results have been published. A bibliography of finite element applications in biomechanics is given by Mackerle (1992). The numerical treatment involves deriving the desired form of the equations, choosing a suitable mesh system, developing the finite element form of the equations and solution procedure and determining the stability of the resultant algebraic scheme.

Solution procedures for solving the Navier-Stokes equations are discussed by many authors (Temam, 1979, Girault and Raviart, 1980). The development of accurate, stable and efficient algorithms is still an important mathematical reseach topic. If the Navier-Stokes equations are solved in primitive variable form, the determination of the pressure is part of the solution procedure and is available when the solution is achieved. Investigations show that the accuracy of the primitive variable solutions is very sensitive to the convergence tolerance used for the pressure solver.

Flow Calculations

The numerical algorithm for the Navier-Stokes equation is based on our newly developed Galerkin finite element method, where operator splitting is applied. The splitting employs a mathematical theorem which enables the decomposition of a $(L_2(G))^n$, $n = 2$ or 3, vector field into orthogonal subspaces H and $H^\perp$ (Girault and Raviart, 1980):

$$(L_2(G))^n = H \oplus H^\perp, \quad for \quad n = 2 \ or \ 3. \tag{20}$$

The spaces H and $H^\perp$ for $n=3$ are characterized as:

$$H = \{curl\, \vec{\varphi} \,|\, \vec{\varphi} \in (H^1(G))^3 \wedge \gamma_v curl\, \vec{\varphi} = 0\}$$
$$H^\perp = \{\nabla q \,|\, q \in H^1(G)\} \tag{21}$$

$H^1(G)$ is a special Hilbert space, γ_v denotes the normal trace operator at the boundary ∂G of the domain. The decomposition defines an orthogonal projection and can be applied to the variational formulation of the incompressible Navier-Stokes equations. The application results in a fractional step method, whereby the flow velocity and the uncoupled pressure are corrected (Perktold and Rappitsch, 1994).

The method, of which the decomposition of a sufficiently regular vector field in a divergence-free and a rotation-free field is the fundamental, has been developed on the basis of Chorin's method where finite differences are used (Chorin, 1968). Chorin's method has been developed further and modified by several authors (Donea et al., 1981, Gresho et al., 1984a, 1984b). The advantage of the method is an uncoupling of the occurring variables resulting in an effective calculation algorithm for the very time-consuming three-dimensional flow problem.

The decomposed variational equation system is approximated on a chosen finite element subspace. The elements applied here are isoparametric bricks with trilinear interpolation for the velocity components and constant pressure. The approximation functions in an element (e) then take the form:

$$u_i^e(x,y,z;t) = \sum_{s=1}^{8} N_s(x,y,z) u_{i,s}^e(t), \quad i=1,2,3, \quad p^e(t) = M p_c^e(t) \tag{22}$$

where N_s, $s = 1,...,8$, and M are the element interpolation functions for the velocity components and for the pressure ($M=1$); u_{is}^e, $s = 1,...,8$, and p_c^e are the unknown velocity node values and the unknown pressure element centre value.

Using implicit Euler backward differences for the time derivatives and finite element discretization of the rigid flow domain and of the non-Newtonian inelastic flow equations at a discrete time level leads to the following iterative calculation algorithm (the symbol "→" indicates the vector of node values):

$\Delta t = t^{n+1} - t^n$ is time step, $n = 0,1,...,N-1$ is time level, $m = 0,1,...,M-1$ is iteration level. $\vec{x}_w^n$ is wall position, $\vec{u}^n$, p^n, μ^n are velocity including the wall velocity $\vec{u}_w^n$, pressure and viscosity at the preceding time level $t^n = \Delta t n$.

$$\vec{x}_w^{n+1,0} = \vec{x}_w^{n,M}(=\vec{x}_w^n), \quad \vec{u}_w^{n+1,0} = \vec{u}_w^{n,M}(=\vec{u}_w^n), \quad \vec{u}^{n+1,0} = \vec{u}^{n,M}(=\vec{u}^n), \quad p^{n+1,0} = p^{n,M}(=p^n),$$

$$\mu^{n+1,0} = \mu^{n,M}(=\mu^n).$$

1. Calculate the auxiliary vector field $\vec{u}^{n+1/2,m+1}$ from the linearized system:

$$\left[C + \Delta t K1(\vec{u}^{n+1,m}) + \Delta t K2(\mu^{n+1,m}) \right] \vec{u}^{n+1/2,m+1} = C\vec{u}^{n,M} + \Delta t \hat{Q} p^{n+1,m} \tag{23}$$

$$\vec{u}^{n+1/2,m} = \vec{u}^{n+1} \quad at \quad \partial G1$$

2. Calculate the pressure correction $q^{n+1,m+1}$:

$$\overline{Q}^T C_d^{-1} \overline{Q} q^{n+1,m+1} = -\frac{1}{\Delta t} Q^T \vec{u}^{n+1/2,m+1} \tag{24}$$

3. Calculate the divergence-free velocity field $\vec{u}^{n+1,m+1}$:

$$\vec{u}^{n+1,m+1} = \vec{u}^{n+1/2,m+1} + \Delta t \, C_d^{-1} \overline{Q} q^{n+1,m+1} \tag{25}$$

4. Calculate the updated pressure $p^{n+1,m+1}$:

$$p^{n+1,m+1} = p^{n+1,m} + q^{n+1,m+1} \tag{26}$$

5. Calculate the non-Newtonian viscosity $\mu^{n+1,m+1}$:

$$\mu^{n+1,m+1} = \mu_1 + (\mu_0 - \mu_1) \frac{1}{\left[1 + (2\lambda\sqrt{D_{II}^{n+1,m+1}})^b\right]^a} \tag{27}$$

C and C_d are consistent and concentrated (lumped) mass matrix, $K1$ is linearized convection matrix, $K2$ is linearized diffusion matrix, Q and Q^T are gradient and divergence matrix. $\overline{Q}$ and $\overline{Q}^T$ are modified with respect to the velocity boundary conditions. The matrix $\hat{Q} = C C_d^{-1} Q$ reflects the application of the concentrated mass matrix in the pressure equation (Eqn.24). The pressure matrix $\overline{Q}^T C^{-1} \overline{Q}$ is symmetric and positive definite. No explicit boundary conditions are necessary for the pressure equation.

The validation of the scheme has been performed in different Newtonian calculations (Hilbert, 1987; Perktold, 1987; Perktold et al., 1991c; Perktold et al., 1991d) and non-Newtonian calculations (Perktold et al., 1991b; Perktold and Resch, 1991). Recently the method

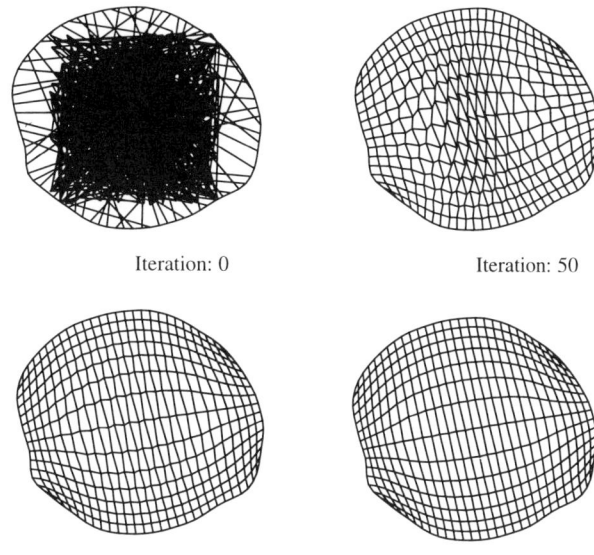

Figure 3. Iterative development of a two-dimensional mesh starting with a random node distribution.

has been modified with respect to the consideration of a streamline upwind technique published by Brooks and Hughes (1982). This enables the stable calculation of higher Reynolds number flow. The resulting equation systems are solved applying our recently developed modified bi-conjugate gradient algorithm for non-symmetric matrices. Essential features are the application of an optimum pre-conditioning technique (incomplete LU-factorization) and the development of a compact storage of the sparse finite element matrices (Hofer and Perktold, 1995).

Shell Calculation

The strain energy in the variational formulation is approximated by a middle surface integral. The integrand consists of a quadratic expression in the components of the strain

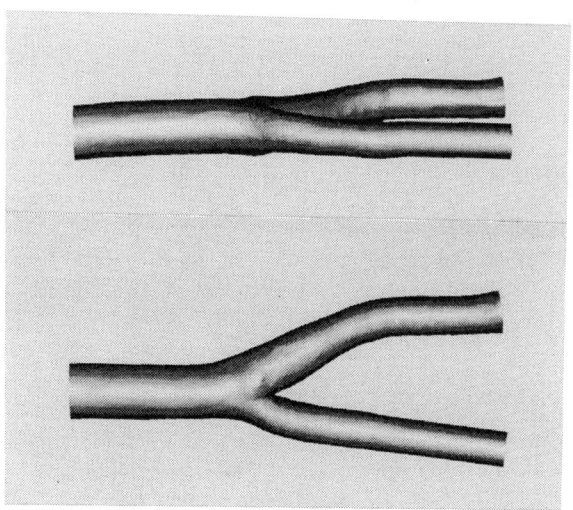

Figure 4. Anatomically realistic computer model of the carotid artery bifurcation. (From Perktold et al., 1994d).

tensor and in the tensor of curvature changes. The former expresssion defines the extensional energy per unit area, the latter represents the bending energy per unit area. In the numerical solution the principle of minimum potential energy is applied with respect to the chosen finite element subspace. A review of linear and non-linear finite element shell models has been given by several authors (Ernst, 1981; Gallagher, 1973).

To compute approximative solutions for the wall deformation and the intramural stresses the finite element program package ABAQUS, Version 5.3 (Hibbit et al., 1993) has been used. The shell load is the total inner wall pressure (Navier-Stokes pressure and the outflow pressure condition) and the wall inertial force. Viscous forces are not considered. The selected element type is S4RF, a four node doubly curved, shear flexible element with reduced integration and hourglass control using five degrees of freedom: three displacement components and two in-surface rotation components. This element is especially suited for large displacements.

Expressing the fluid-structure interaction the wall motion and the fluid motion are iteratively coupled at a time step

1. Calculate the wall position $\vec{x}_w^{n+1,m=1}$ and the wall velocity $\vec{u}_w^{n+1,m+1}$ from the total wall pressure $p_w^{n+1,m}$ and the vessel wall mass forces applying the finite element package ABAQUS.
2. Update the vessel geometry and the finite element mesh and interpolate the velocities and the pressure at the updated nodes. Update the convective velocity $\vec{u}^{n+1,m+1} - \hat{u}^{n+1,m+1}$ taking into account the mesh velocity $\hat{u}^{n+1,m+1}$ which is determined by an appropiate distance function.
3. Update the occurring matrices with respect to the current data.

The transient shell calculations yield the deformation vector in each surface node and the principle stresses. Using the deformation vector the flow domain is updated in each step. Calculation of the wall velocity in each surface node yields the velocity boundary conditions in the next Navier-Stokes step. From the surface node movement the position and the velocity of the inner nodes can be calculated.

Finite Element Mesh Generation

An important step in the application of finite element methods is the generation of the computational mesh. Essential for the successful application of the finite element method is the efficient definition of a finite element subdivision of the geometric domain with appropriate computational regularity. Commercial software packages are available and can be used efficiently for many problems. In the finite element modelling of complex three-dimensional domains often it is necessary to develop special purpose grid generators. In general to generate the numeric data necessary to define a model surface experimental surface data are necessary. The basis in our recently developed finite element grid generator is the optimisation of local grid properties, as smoothness and orthogonality of the mesh (Kumar and Kumar, 1988). The local grid properties are expressed in a corresponding function which controls the properties. The grid function at a grid point P is defined as:

$$F = a_s f_s + a_0 f_0 + a_b f_b \qquad (28)$$

where f_s is the potential energy of elastic springs connecting the central point P and the neighbouring grid points of a macro. Each inner point P is arranged as the common corner point in a macro consisting of eight neighbouring isoparametric bricks. f_0 describes the orthogonality of specified vectors in the macro. f_b controls the orthogonality of the grid near

Computer Simulation of Arterial Blood Flow

the surface, a_s, a_o, a_b are suitable weights. The minimisation of F at a Point P with respect to the co-ordinates (x,y,z) of P yields a set of equations:

$$\frac{\partial F}{\partial x} = 0, \quad \frac{\partial F}{\partial y} = 0, \quad \frac{\partial F}{\partial z} = 0 \qquad (29)$$

The generation method then consists in choosing the initial grid and updating it by minimizing the specified weighted local grid properties iteratively. Selecting the parameters requires some experience. Fig.3 demonstrates the iterative development for a two-dimensional grid starting with a random node distribution.

The generator is especially suited in modelling of anatomically realistic arterial bifurcations. The strong geometry dependence of the flow field and of related quantities is the motivation for the development of anatomically realistic computational models. Fig. 4

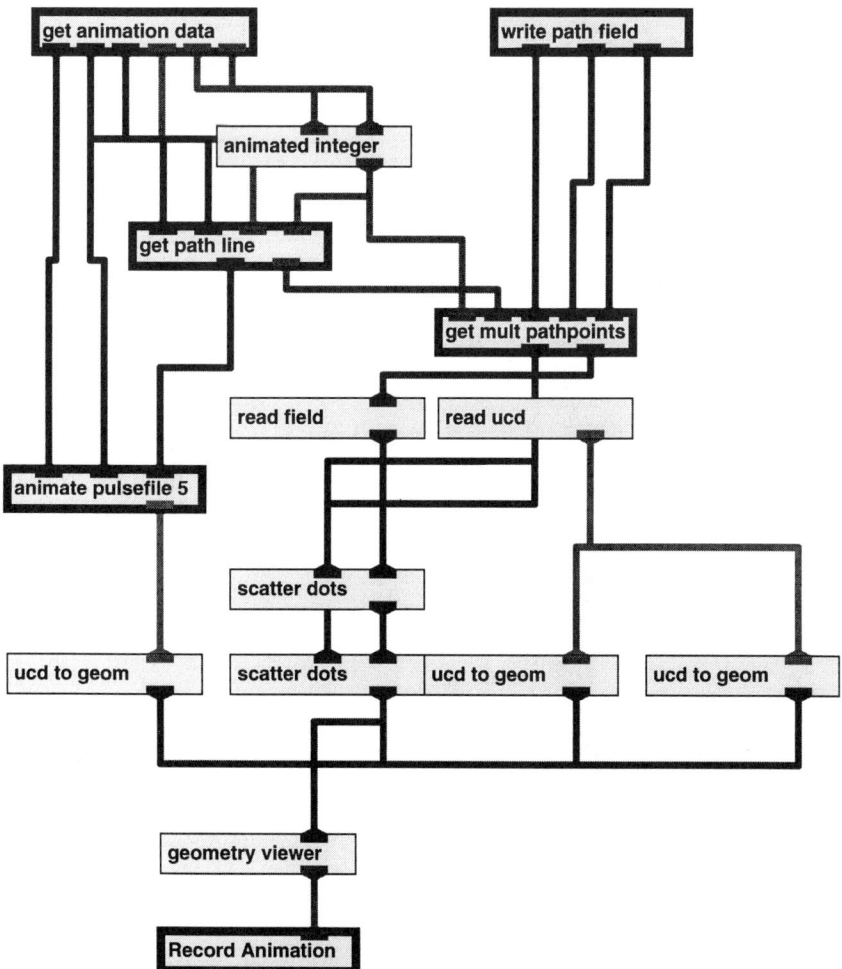

Figure 5. AVS-network of moving particle animation.

shows the generated computational model surface of a carotid artery bifurcation based on experimental data.

DISCUSSION OF THE MODEL ASSUMPTIONS

Wall distensibility causes wall motion and wave phenomena; the wave is reflected at discontinuities. In our model local reflection is not included in the model. The most essential discontinuities in the arterial system are the vessel bifurcations. According to Pedley (1980), Reuderink (1991) local reflection at bifurcations is small. In the range of essential frequencies (frequencies in the Fourier analysis of the pulse wave) having essential amplitudes high transition and small reflection occur. Reflection at the periphery is important and expressed in the pressure pulse waveform and in the flow rate pulse waveform as well as in the flow division ratio.

In the developed model incrementally linearly elastic behaviour of the vessel wall has been assumed. The influence of this rigorous simplification on local flow and stress pattern has been proved appying a pressure loaded cylinder having different mechanical properties: linearly elastic, hyperelastic and viscoelastic behaviour. The hyperelastic calculation uses a strain energy potential based on uniaxial stress-strain data by Patel and Vaishnav (1980). The viscoelastic constitutive equation is defined as:

$$\tau = \int_0^t G_R(t-s)\dot{\gamma}(s)ds = G_0\gamma - \int_0^t \dot{G}_R(s)\gamma(t-s)ds. \tag{30}$$

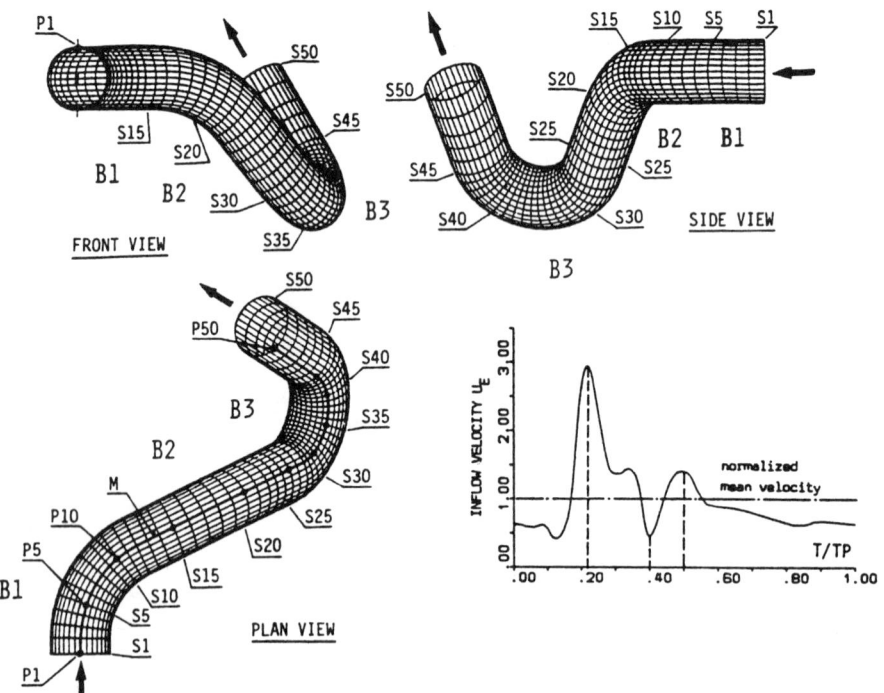

Figure 6. Front view, plan view and side view of a carotid siphon model.

The parameters of the Prony series expansion of the relaxation modulus:

$$G_R(t) = G_\infty + \sum_{i=1}^{N} G_i e^{-t/\tau_i} \qquad (31)$$

are fitted using experimental relaxation test data by Fung (1981).

The mechanical data for large arteries have been taken from references (Reneman et al., 1985). Comparison of the radial displacement demonstrates minor influence of hyperelasticity and viscoelasticity on local flow characteristics. The modelling of the vessel wall properties is essential in global investigations as pulse wave propagations (Pedley, 1980, Horsten et al., 1989, Stergiopulos, 1994). In local calculations the simplification is acceptable.

ADVANCED POST-PROCESSING TECHNIQUES

The paths of fluid particles and the particle residence times are demonstrated by means of numerical flow visualization. In a first post-processing step the paths of specified particles are calculated. In the simple model each particle follows a certain trajectory. Hence, for each point $\mathbf{x}$ of the flow domain there is a path $\mathbf{s}(t)$ representing the trajectory of $\mathbf{x}$. For different trajectories $\mathbf{s}(\mathbf{x},t)$ denotes the path by $\mathbf{x}$, with the initial condition $\mathbf{s}(\mathbf{x},t_0) = \mathbf{x}$. Let $\mathbf{u}$ denote the time-dependent fluid velocity, then the path of a fluid particle is described by the following initial value problem:

$$\frac{d}{dt}\mathbf{s}(\mathbf{x},t) = \mathbf{u}(\mathbf{s}(\mathbf{x},t),t),$$

$$\mathbf{s}(\mathbf{x},t_0) = \mathbf{s}_0. \qquad (32)$$

The system of first order ordinary differential equations is solved using a predictor-corrector procedure. The particle path calculations are carried out for specified particles which enter the flow field and start moving at the specified time. The observation of the particles covers the time up to the exit of the computational flow domain.

The calculated particle paths are used for the production of an animation sequence applying the visualization package AVS. The time-dependent visualization requires the development of own program modules. The network of the moving particle animation consisting of AVS-modules and own modules (bold marked fields) is shown in Fig.5.

The single frames of the animation sequence are recorded on a laser-disk from which the video film is produced. Further animations concern special representations of the computational geometry, the dynamics of the axial flow velocity profiles, the transient development of reversed flow and the vessel wall deformation. The animation technique allows an impressive demonstration of three-dimensional dynamic processes.

The wall shear stress as a quantity of primary physiological interest is derived from the flow velocity. The shear stresses in the flow field are described with the extra stress tensor (Eqns. 3-4). For incompressible fluids and no-slip condition at the wall the shear stress is:

$$\tau_w = \mu \left.\frac{\partial \mathbf{u}_t}{\partial \mathbf{n}}\right|_{wall} \qquad (33)$$

where denotes the tangential velocity, **n** is the normal unit vector at the vessel wall. The time-dependent wall shear stress is calculated from the finite element velocity field near the wall.

NUMERICAL RESULTS AND DISCUSSION

The computer simulation of arterial blood flow has been carried out in models of different arterial segments. This chapter outlines recent results of local arterial flow studies.

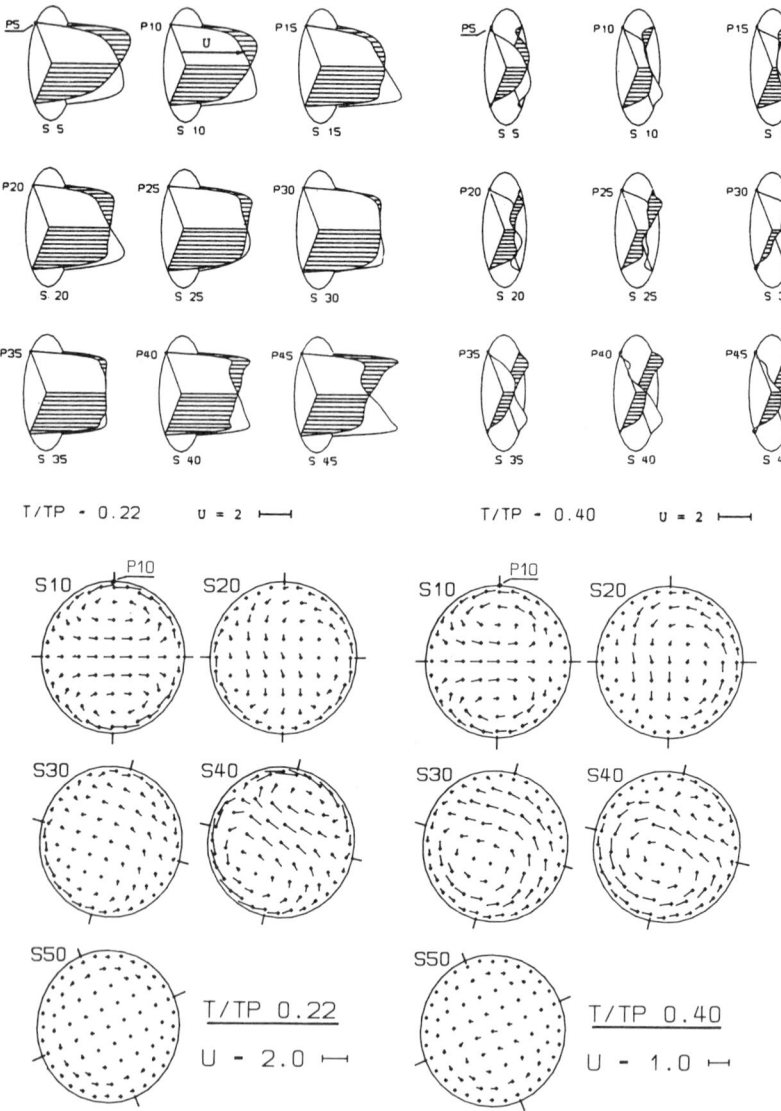

Figure 7. Axial velocity profiles and secondary velocity vector field at different flow cross sections (as indicated in Fig.6) for the pulse phase angles of flow maximum and flow minimum.

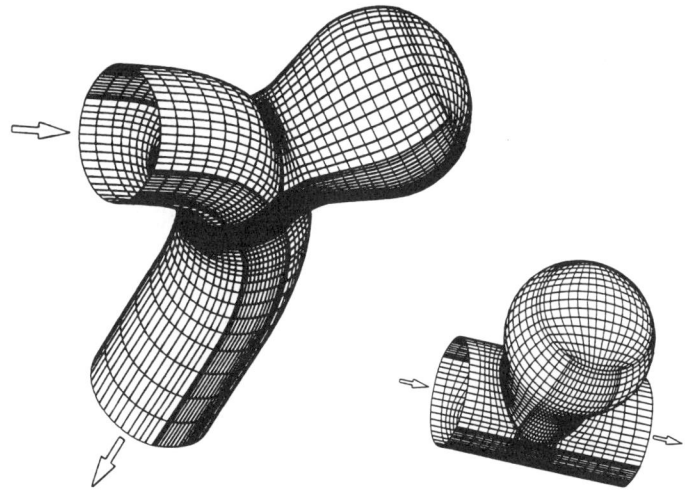

Figure 8. Three-dimensional representation of a curved vessel segment with a lateral aneurysm (first model) and a straight vessel segment with a lateral aneurysm (second model).

Analysis of Pulsatile Blood Flow in a Carotid Siphon Model

The carotid siphon is a multiply curved segment of the human internal carotid artery located at the base of the skull. The segment consists of several non planar bends which immediately follow each other; here three of these bends are considered. The carotid siphon is a typical curvature site in the human circulatory system where the formation of early lesions occurs.

In Fig.6 a front view, plan view and a side view of the model segment are shown. The specified cross sections are denoted by S1,...,S5,....,S50. The marked point P1 at the boundary of S1 and the generating line M, as indicated, gives the position and the orientation

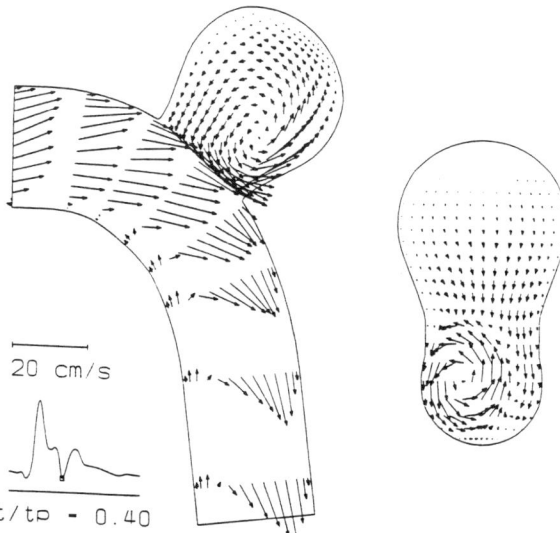

Figure 9. Velocity vector field in the curvature plane and secondary velocity vector field at minimum flow rate in the first model.

of the flow cross-sections for presentation of results. The Newtonian calculation is performed for a Reynolds number of $Re=200$ and a Stokes number of $\eta = 12.2$ The investigation concentrates on the effect of multiple non-planar bending. The numerical results are illustrated for the axial velocities and for the secondary velocity.

For fluid flow through a curved passage a transverse (secondary) velocity component occurs in the plane perpendicular to the axis of the tube. As a result of the no-slip condition, the velocity of the fluid particles in the core of the tube is higher than those of particles near the wall, the centrifugal force affecting the particles moving along the bend is therefore larger for the core particles. Fluid in the tube centre is swept to the outside of the bend and fluid near the wall moves towards the inside. Flow phenomena in curved tubes have been analysed in detail by Pedley (1980) and Berger et al. (1983).

The axial flow field and the secondary flow field are displayed at different cross sections in Fig.7. At S10, which is located in the zone of the downstream end of the first bend B1, the secondary motion is well developed and symmetrical, the flow changes its direction as the second bend B2 is approached. The cross-section S20 is situated at the downstream end of B2, the cross sections S30 to S50 belong to the bend B3. In the third

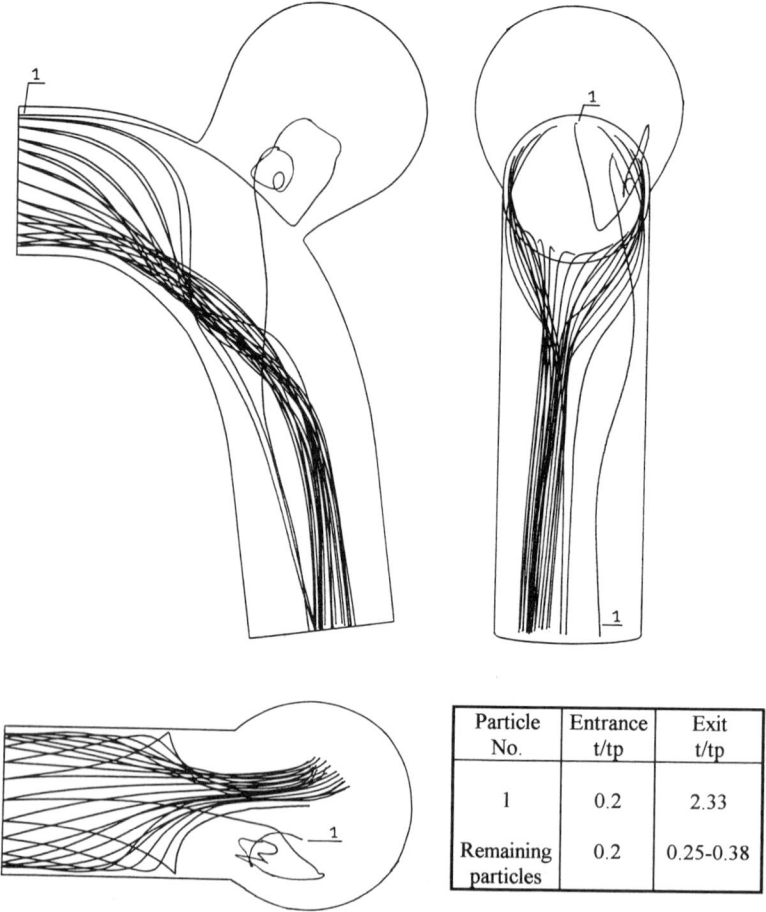

Particle No.	Entrance t/tp	Exit t/tp
1	0.2	2.33
Remaining particles	0.2	0.25-0.38

Figure 10. Particle paths of single fluid particles in the first model. The particles enter the flow field near the wall.

Computer Simulation of Arterial Blood Flow

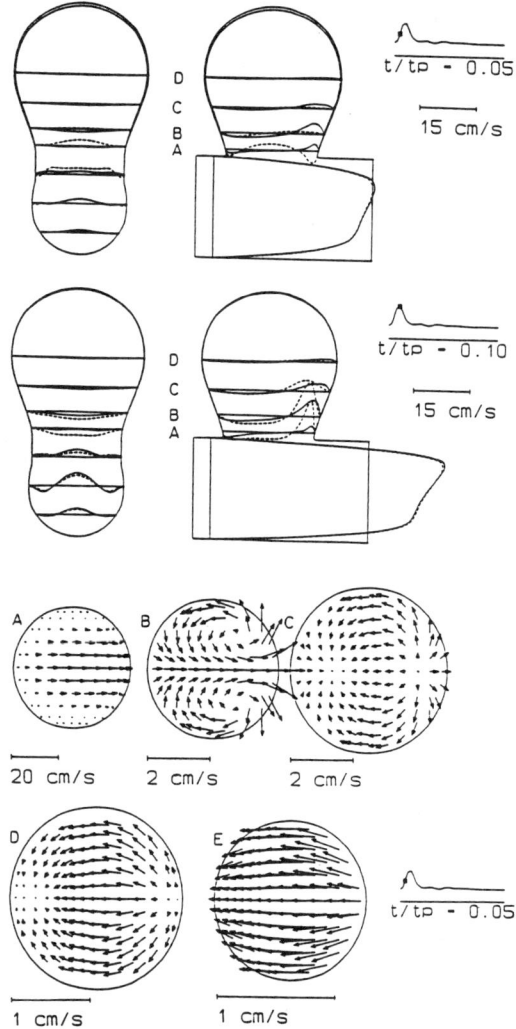

Figure 11. Velocity profiles at accelerated flow and peak flow rate (elastic wall: solid line, rigid wall: dashed line) and secondary velocity vector field in the elastic wall model at accelerated flow.

bend strong secondary circulation arises. Axial velocity profiles are shown across perpendicular vessel diameters. At the inflow cross-section S1 time dependent flow is assumed to be fully developed. Secondary motion influences the distribution of axial velocity. Results on haemodynamic characteristics in the carotid siphon model have been published by Perktold et al. (1987), Perktold et al. (1988).

Pulsatile Flow Characteristics in Saccular Aneurysms

The investigation contributes to the analysis of flow activity in lateral saccular aneurysms under different conditions. Information on flow activity of aneurysms is of special interest in the context with the breaking off the parts of a thrombus and the washing out of these parts from the aneurysm into the main flow stream. Thus, the danger of embolism arises in the small downstream vessels. In the first model the parent vessel is curved and the aneurysm assuming rigid walls is placed at the outer side of the curvature. In the second model the parent vessel is straight; the aneurysm wall is assumed to be rigid and distensible. The calculations are carried out under non-Newtonian inelastic flow conditions. According to the study by Löw et al. (1993) the mean reference Reynolds number $Re = 200$ and the reference Stokes number $\eta = 9.85$.

The representation of the results concentrates on the flow velocity vector field and on particle paths in the first model and on secondary motion in the second model. The axial velocity vector field in Fig.9 shows recirculating flow in the aneurysm. The corresponding secondary velocity vector field shows asymmetric recirculation and relatively low velocity in the aneurysm directed toward the orifice.

The low flow activity within the aneurysm is also demonstrated by the paths of single particles entering the flow field at the same time equally spaced near the wall (Fig.10). Only one particle enters the aneurysm and remains there for more than two periods.

Velocity patterns in the distensible model are shown in Fig. 11. The comparison with rigid wall results illustrates significant differences during the systolic phase.

The investigation shows a complex velocity field in the aneurysms whereby the flow activity is relatively small in all cases under consideration. The flow velocity in the

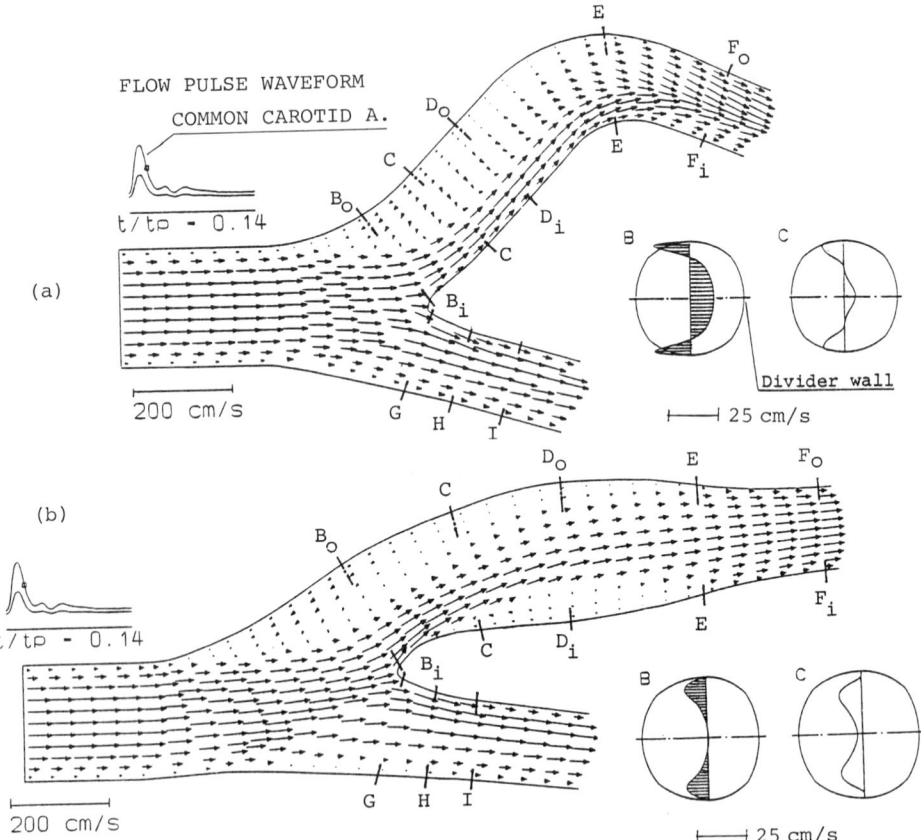

Figure 12. Flow velocity vector field in the branching plane and secondary velocity profiles during systolic deceleration; wide angle carotid model (a) and acute angle model (b).

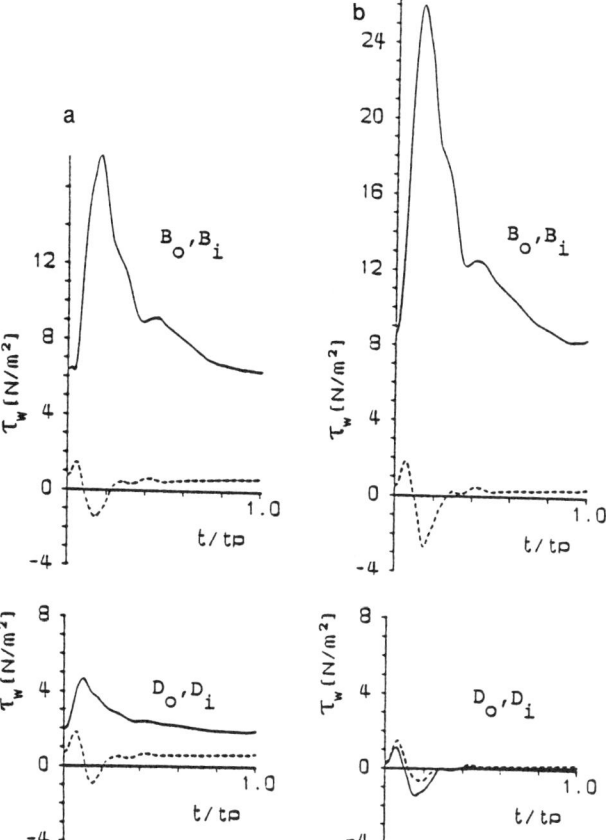

Figure 13. Wall shear stress during the pulse cycle at the outer and the inner internal carotid wall at proximal sinus location B_o, B_i and at sinus mid location D_o, D_i (locations according to Fig.12), wide angle model (a), acute angle model (b), solid line: inner wall, broken line: outer wall.

aneurysms is an order of magnitude smaller than in the parent vessel. The numerical flow visualization illustrates this behaviour also. The consequence is high particle residence time in the aneurysm. This fact favours the thrombus formation (Steiger, 1990). In the distensible wall model globally increased but still low flow activity can be observed.

Flow Phenomena in Carotid Artery Bifurcation Models

An arterial segment in which lesions preferentially occur is the human carotid artery bifurcation with the carotid sinus. A considerable amount of experimental and numerical research work has been performed to study the flow dynamics in this segment (compare references in Perktold et al. 1994a). Here results concerning the flow field and the wall shear stress distribution are shown and discussed in two models which differ in the bifurcation angle and in the shape of the bifurcation region and the sinus. The study originally published by Perktold et al. (1994a) concentrates on flow phenomena in the carotid sinus. The bifurcation models considered here are essentially based on findings by Desyo (1990) on carotid bifurcation height and positive correlation with the bifurcation angle size and typical locations of deposits and stenosis. The two models under consideration are a wide angle carotid with a relatively narrow sinus width and an acute angle carotid with an enlarged sinus diameter. Simplified rigid vessel wall and non-Newtonian inelastic blood behaviour are assumed. The calculations have been carried out under the physiological flow pulse waveform and flow division ratio internal to external carotid described by Ku et al. (1985). The

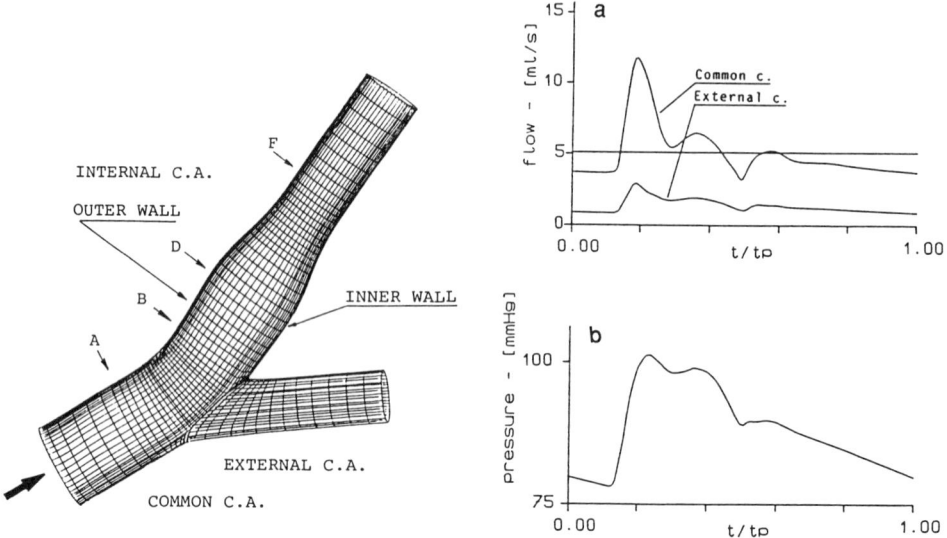

Figure 14. Three dimensional representation of the carotid artery bifurcation (model by Ku et al., 1985); (a) flow pulse waveform in the common carotid artery and the external carotid artery, (b) pressure pulse waveform at the internal carotid outlet (data by Osenberg (1991)).

mean flow rate in the common carotid is 5.1 *ml/s*. The mean (diameter defined) reference Reynolds number $Re = 300$, the pulse frequency is 80 strokes per minute. Fully developed velocity profiles corresponding to the common carotid velocity pulse waveform are prescribed at the inflow boundary as time-dependent boundary conditions.

Fig.12 shows the flow velocity vector field in the branching (symmetry) plane and the secondary flow patterns during decelerated flow. Significantly different flow separation and recirculation in the sinus is demonstrated. Separated flow occurs at the outer (non-divider) sinus wall in both models, inner (divider) sinus wall separation can be seen only in the acute angle model with the wide sinus. The secondary motion in bifurcations results from the curvature and the branching effect. The comparison of the secondary velocity profiles in the wide angle and in the acute angle model demonstrates different patterns. In the wide angle model at proximal sinus location (B) in the flow cross-section centre strong secondary

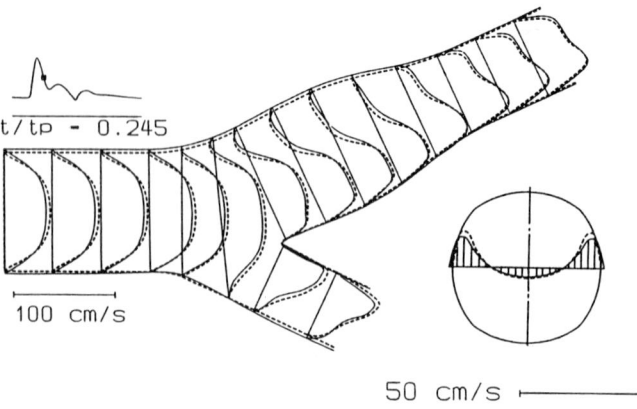

Figure 15. Axial flow velocity profiles at the symmetry plane during systolic deceleration phase and secondary velocity profile at sinus maximum diameter location (solid line: distensible model, dashed line: rigid model.)

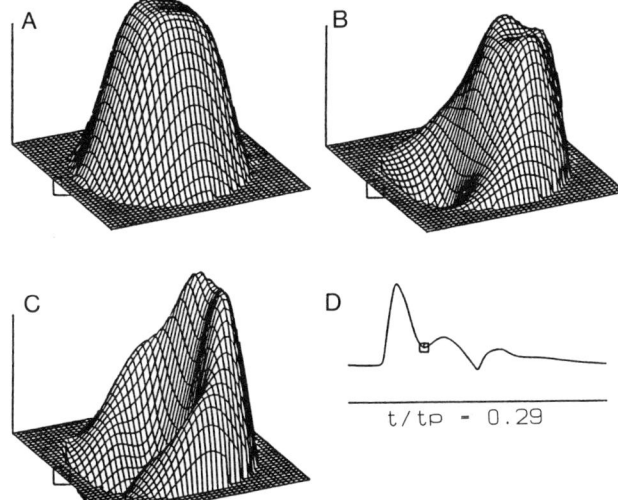

Figure 16. Three-dimensional representation of axial flow velocity at specified cross sections for the pulse phase angle of systolic flow minimum. Three-dimensional representation of axial flow velocity at specified cross sections for the pulse phase angle of systolic flow minimum.

flow is directed towards the divider; in the acute angle model only low secondary velocity which is directed away from the divider can be observed. This differences result from the different curvature and from the different position of the apex related to the common carotid axis.

Wall shear stress at the outer and the inner internal wall during the cardiac cycle at proximal sinus location B_0, B_i and at sinus maximum diameter location D_0, D_i is demonstrated in Fig.13. It can be seen that the wall shear stress maximum is significantly higher in the acute angle model near the flow divider (location). At the outer sinus wall location B_0 and D_0 the shear stress is relatively low. Negative shear stress occurring at the outer wall of the wide angle model and at the outer and inner wall of the acute angle model indicates temporal flow separation.

The study showed that the fluid dynamic variables are significantly affected by the geometry of the bifurcation model. An aspect of clinical importance is described by Desyo (1990), who reports that in bifurcations with wide angles, atherosclerotic plaques are typically located in the region of the beginning of the internal artery and in the sinus, whereas in lower level bifurcations with acute angles typical atherosclerotic processes can be detected

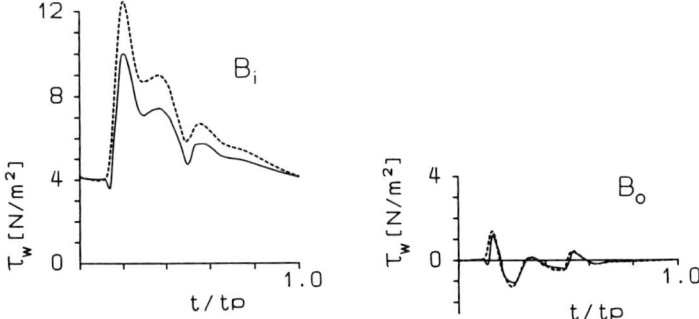

Figure 17. Wall shear stress during the pulse cycle at the proximal sinus location; inner wall B_i, outer wall B_o; solid line: distensible model, dashed line: rigid model.

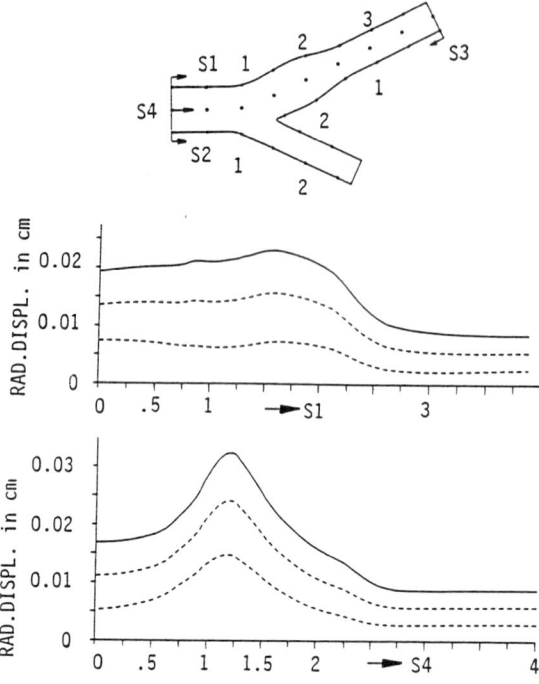

Figure 18. Normal displacement in *cm* along the generating lines S1 (common-internal (outer) wall), and S4 (common-internal side wall) related to maximum pressure load p_{max} (solid line), $1/3 p_{max}$ and $2/3 p_{max}$ (dashed lines).

at the terminal end of the common carotid artery. These different locations of plaques reflect the different positions of the zones of flow separations as shown here. Further results concerning the geometry-dependency of the flow field and the wall shear stress distribution in the carotid artery bifurcation are demonstrated in Perktold and Resch (1990), Rindt et al. (1990) and Perktold et al. (1991a). Additional numerical studies confirming that local flow structure is mainly determined by the local geometry have been performed by Yamaguchi et al. (1990), He (1993).

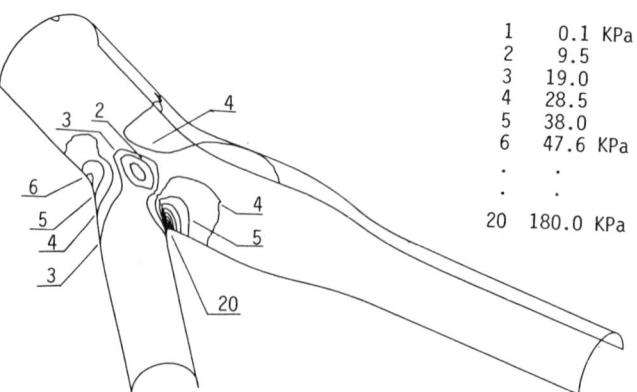

1	0.1 KPa
2	9.5
3	19.0
4	28.5
5	38.0
6	47.6 KPa
.	.
.	.
20	180.0 KPa

Figure 19. Maximum principal stress contours at the inner surface at the pulse phase angle of maximum pressure load p_{max}.

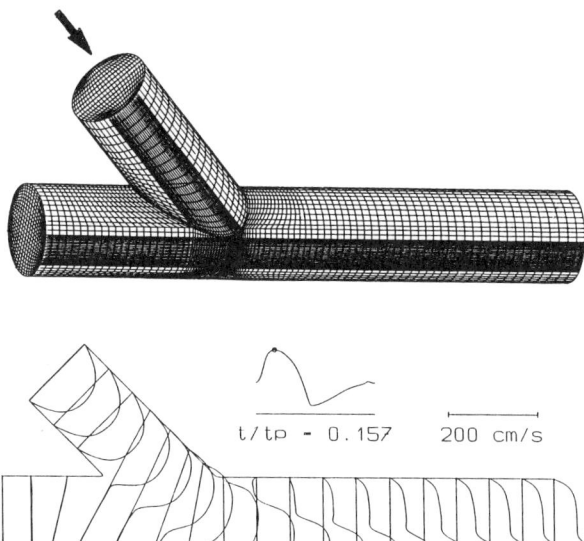

Figure 20. Bypass anstomosis model. Axial velocity profiles in the branching plane at peak flow rate.

Vessel Mechanics and Blood Flow Dynamics in a Compliant Carotid Artery Bifurcation Model

This demonstration primarily deals with the interaction of pulsatile flow and vessel wall distensibility in a compliant carotid bifurcation model. The effect of wall distensibility on flow separation and recirculation and on wall shear stress distribution has been analysed. The distensible results have been compared to results in a corresponding rigid wall model. Further results concern the intramural stress distribution at maximum pressure load. The used flow pulse wave form and the internal carotid outlet pressure pulse wave form are shown in Fig.14. The incremental elasticity parameter describing the wall material loaded with the incremental pressure from the diastolic pressure level to the peak systolic pressure has been determined from experimental displacement data and corresponding pressure data by Reneman et al. (1985) and is $E = 3.61*10^6 \, dynes/cm^2$ Results as outlined here have been published by Perktold and Rappitsch (1993), Perktold and Rappitsch (1995).

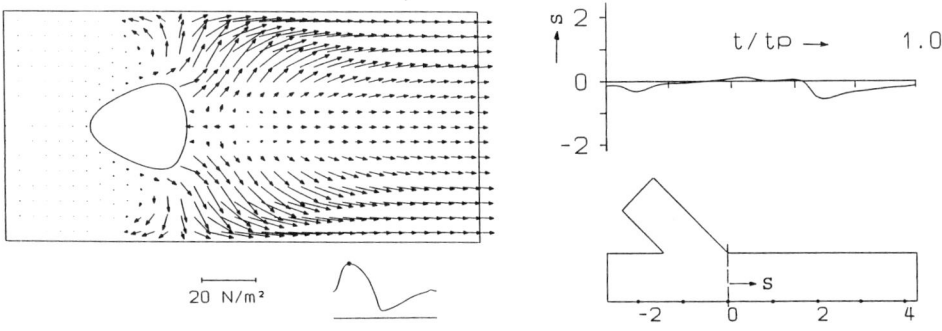

Figure 21. Wall shear stress vector field at the artery wall (plane representation of the wall) at peak systole and locations of the flow stagnation point on the artery floor during the pulse cycle.

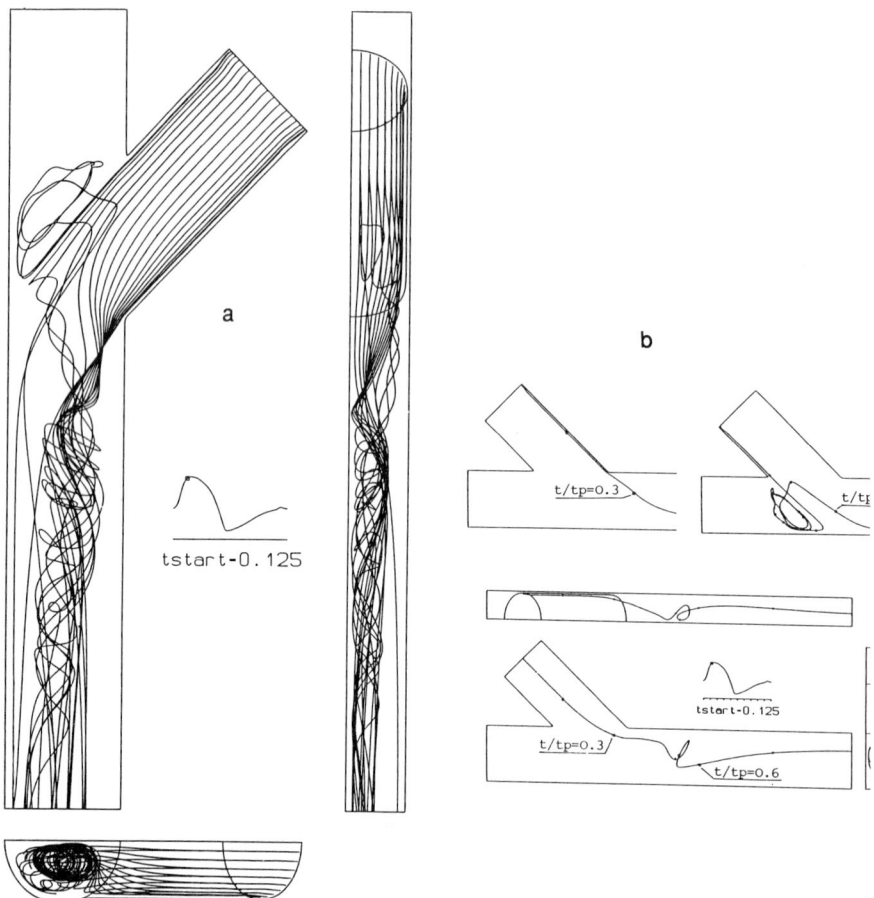

Figure 22. Fluid particle path representation and particle residence time; (a) Propagation of a particle front, (b) Paths of single particles and corresponding residence time.

Axial velocity profiles in the branching plane (symmetry plane) of the model and the secondary velocity profile at sinus mid location during systolic deceleration phase are shown in Fig.15. The comparison with rigid wall results (dashed line) indicates an increase of the internal axial flow velocity in the distensible model during systolic deceleration phase where the vessel lumen contracts. The distension of the vessel wall leads to less flow separation and flow recirculation and to a decrease in magnitude of the reversed flow velocity in the region of the outer carotid sinus wall.

The wall shear stress distribution during the pulse cycle at different levels at the outer internal wall B_0, and at the inner wall (internal divider wall) B_i is shown in Fig.17. Negative shear stresses indicate reversed flow. In the compliant model (solid line) the wall shear stress is lower than in the rigid model (dashed line). The relative decrease $|\tau_{w,\,dist} - \tau_{w,\,rig}|/|\tau_{w,\,dist}|$ is up to 25 % for high values.

The vessel wall mechanics is analysed by means of the normal displacements and the principal stresses. Fig.18 shows the normal vessel displacements along generating lines related to maximum pressure load pmax, $1/3 p_{max}$ and $2/3 p_{max}$. Fig.19 is a contour plot of the maximum principal stresse at the inner surface of the wall at maximum pressure load. The

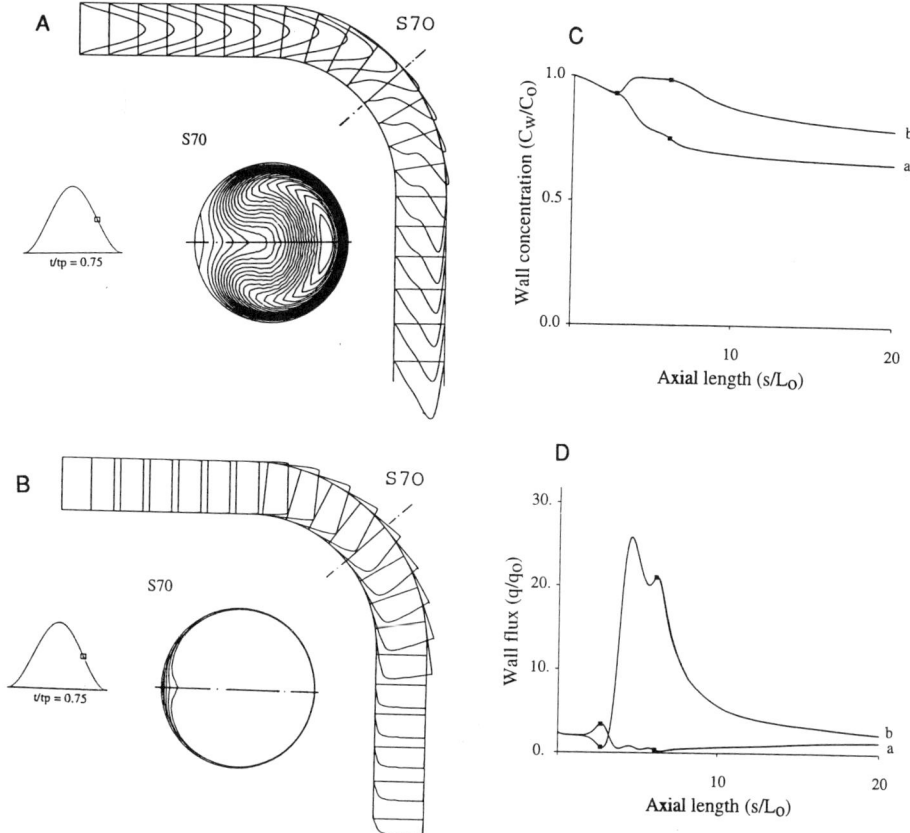

Figure 23. (A) Normalized axial velocity profiles and axial velocity contour lines during decelerated flow. (B) Normalized solute concentration profiles at the curvature plane and concentration contour lines. (C) Normalized solute concentration at the inner (a) and at the outer wall (b) of the bend. (D) Normalized wall flux through the inner wall (a) and through the outer wall (b).

stress concentration factor defined as the ratio of the local stress to the uniform stress in the common carotid is 6.3 in the apex region. In addition to the locally high stresses, steep stress gradients can be observed.

The study shows that the distensibility of the vessel wall in the physiological range affects the flow field quantitatively. This has also been confirmed in the studies by Anayiotos (1990), Reuderink (1991). Primarily the wall shear stress magnitude is reduced in the distensible model. Whether this influence is important in atherogenesis is not clear. Further studies are necessary to obtain a sufficient understanding of the influence of the wall distensibility in atherogenesis.

The comparison of non-Newtonian and Newtonian results for the axial velocity profiles and the secondary motion in the rigid carotid artery bifurcation model shows only minor differences. For non-Newtonian flow the maximum velocity and the magnitude of reversed flow decrease slightly. The wall shear stress results generally illustrate an increase in the non-Newtonian case (about 10-12%), only at levels B and D at the inner wall the behaviour is contrary. In general, wall shear stress is increased in the non-Newtonian case as has been analysed by Perktold et al. (1991b).

Fluid Dynamic Characteristics in Vascular Graft Anastomoses.

Local haemodynamics in anastomoses is of essential importance for success in bypass surgery. According to clinical observations (Trubel et al., 1994) post operative intimal hyperplasia and thrombus formations occur in regions of anastomoses where abnormal flow patterns can be observed such as whirls with increased particle residence time and low pulsatile wall shear stress. Therefore the surgical vessel reconstruction should be made considering optimum reduction of these effects. The optimum flow in arterial anastomoses lessens the risk of post operative thrombosis. In our recent study a systematic numerical analysis of the haemodynamics in distal anastomoses has been carried out (Perktold et al., 1993, Perktold et al., 1994b). Fig.20 shows the extremely skewed axial velocity profiles and flow separation in the artery near the toe and on the bottom.

The shear stress vector field at the artery wall is displayed in Fig.21. In the plane wall representation the artery is cut open along the floor. The plot ilustrates the variation of the shear stress vector direction. At positions where the vectors have an upstream directed component reversed flow occurs. The locations of the weak migrating stagnation point on the artery floor are also shown in Fig.21. The distance of the stagnation point from the reference plane through the downstream intersection of the tubes perpendicular to the artery axis is drawn over the normalized time of the pulse period.

Fig.22 demonstrates the propagation of fluid particles. The particles under consideration enter the flow field near the graft wall. It can be seen that the motion of many particles within the artery takes place under strong influence of the secondary motion. The representation demonstrates the extremely disturbed flow field and very different particle residence times. The flow of blood particles in different arterial segments and its relation to thrombogenesis has been experimentally studied by Goldsmith (1972), Motomiya and Karino (1984). The vessel wall mechanics in a bypass anastomosis model has been analysed by Rappitsch et al. (1993).

Arterial Mass Transfer in Curved Tubes

The study of mass transfer in an arterial vessel and through the vessel wall is important in the understanding of atherogenetic processes (Friedman and Ehrlich, 1975, Back et al., 1977, Caro et al., 1987, Patel and Vaishnav, 1980). Here the oxygen transport in a curved tube artery model is analysed where shear dependent permeability of the vascular wall is taken into account. According to Perktold et al. (1995) the applied parameters are: mean Reynolds number $Re=300$, Womersley number $\alpha = 4.1$, Peclet number $Pe=381.000$, Dean number $\kappa = 106$.

The investigation shows the spatial and temporal variation of the concentration of dissolved oxygen in the convective flow field and demonstrates a strong effect of the wall shear stress distribution on the resulting wall flux. The main characteristic is the substantial wall flux reduction through the inner wall of the curvature. Further studies concerning the convective diffusion of macromolecules (albumin) will be carried out in near future.

CONCLUSION

During the last decades numerical methods were applied to study blood flow phenomena with increased detail. The numerical simulations allow a fine resolution of the flow field, both in space and time, and thus, a very detailed description of the flow and of related quantities can be provided. The changing of governing parameters in the numerical simulation can be done easily, and thus, parameter studies can be carried out economically.

To simulate accurately physiological conditions which are difficult to consider in the experimental techniques and to isolate different factors in the investigation numerical methods are advantegous.

In general, finite element modelling can be employed under two different points of view. In one approach commercial program packages can be applied whereas in the other case own developments are required. The two finite element applications have specific advantages and drawbacks. In general the application of commercial program packages is efficient and allows many researchers the treatment of complex numerical problems. The present high level of software-engineering guarantees broad possibilities in the analysis. Packages for the solution of various types of problems are available.

In commercial software packages standard pre- and post-processing is often included (e.g. the generation of simple meshes, the graphic representation of default output variables). Problems may arise from the fact that the user has insufficient information about the applied algorithms and therefore, the interpretation and the criticism of results is often difficult. In parameter controlled programs the quantitative influence of chosen parameter values on the results is often not known to the user. A typical example is the upwind parameter. Modifications of the code are practically impossible. The pre- and post-processing routines available are often not sufficient, and therefore own extensions are necessary, but further post-processing beyond the routines offered is difficult and often only restricted possible.

Own developments are time-consuming and require a great amount of knowledge in numerical mathematics and computer science. The major advantages are the high flexibility in the sense that special modifications allow the consideration of various problem specific facts. The knowledge of the numerical algorithms and of the structure of the code allows the implementation of special purpose extensions for the description of additional effects.

In our biological flow studies essentially we apply our own extensive mathematical and program developments. Our recently developed procedures comprise a finite element mesh generator for complex three-dimensional domains, a Newtonian and non-Newtonian inelastic Navier-Stokes equation solver including a conjugate gradient solver for sparse non-symmetric matrices, a Postscript post-processing graphic package and animation and visualization tools. The occurring mechanical shell problem, as an aspect in our analysis, is solved using the commercial program package ABAQUS. The interaction of both program units is controlled through a UNIX shell-script for program scheduling, data conversion and input-output handling.

The application of the developed software on biological flow has been demonstrated for some specific problems of large artery blood flow. Dynamical phenomena in viscous flow which are important in the genesis of atherosclerosis are investigated under different conditions including non-Newtonian effects and vessel compliance. The results form a basis for an extended interpretation of arterial flow phenomena and for further studies including convective mass transport.

ACKNOWLEDGEMENT

Tis study is supported by the Austrian Science Foundation, Project-No. P9071-TEC, P10494-TEC, Vienna, Austria.

REFERENCES

Anayiotos, A., 1990, Fluid Dynamics at a Compliant Bifurcation Model. PhD-Thesis, Georgia Inst. of Technology.

Back, L.H., Radbill, J.R. and Crawford, D.W., 1977, Analysis of oxygen transport from pulsatile, viscous blood flow to diseased coronary arteries of man. J. Biomechanics, **10**, 763-774.
Berger, S.A., Talbot, L. and Yao, L.S., 1983, Flow in curved pipes. Annual Reviews in Fluid Mechanics, **15**, 461-512.
Brooks, A.N. Hughes, T.J.R., 1982, Streamline upwind/Petrov-Galerkin formulations for convection dominated flows with particular emphasis on the incompressible Navier-Stokes equations. Comp. Meth. in Appl. Mech. and Eng., **32**, 199-259.
Caro, C.G., Fitz-Gerald, J.M. and Schroter, R.C., 1971, Atheroma and arterial wall shear: Observation, correlation and proposal of a shear dependent mass transfer mechanism of atherogenesis. Proc. Roy. Soc. Lond., **177**, 109-159.
Caro, C.G. and Parker, K.H. 1987, The effect of hemodynamic factors on the arterial wall. In Atherosclerosis-Biology and Clinical Science (Edited by Olsson, A.G.), pp. 183-195. Churchill Livingstone, Edinburgh.
Chorin, A.J., 1968, Numerical solution of the Navier Stokes equations. Math. Comput., **22**, 745-762.
Desyo, D., 1990, Radiogrammetric analysis of carotid bifurcation: hemodynamic-atherogenetic repercussions on surgical practice, In: Biofluid mechanics (Edited by D.Liepsch), 45-56, Springer Verlag.
Donea, J., Giuliani, S., Laval, H. and Quartapelle, L., 1981, Solution of the unsteady Navier-Stokes equations by a finite element projections method. In: Computational techniques in transient and turbulent flow (Edited by Taylor, C. and Morgan, K.), Pineridge Press, 97-132, Swansea.
Ernst, L.J., 1981, A geometrically nonlinear finite element shell theory. Thesis, Laboratory of Engineering Mechanics, Delft University of Technology, The Netherlands.
Friedman, M.H. and Ehrlich, L.W., 1975, Effect of spatial variations in shear on diffusion at the wall of an arterial branch. Circulation Research, **37**, 446-454.
Friedman, M.H., Deters, O.J., Mark, F.F., Bargeron, C.B. and Hutchins, G.M., 1983, Arterial geometry affects hemodynamics-a potential risk factor for atherosclerosis. Atherosclerosis, **46**, 225-231.
Friedman, M.H., 1989, A biological plausible model of thickening of arterial intima under shear. Arteriosclerosis, **9**, 511-522.
Fung, Y.C., 1981, Biomechanics - Mechanical Properties of Living Tissues. Springer Verlag, New York Heidelberg Berlin.
Gallagher, R. H., 1973, Finite element analysis of geometrically nonlinear problems. In Theory and Practice in finite element structural analysis (Edited by Y. Yamada and R.H. Gallagher), Tokyo Univ. Press.
Girault, H.G. and Raviart, P.A., 1980, Finite element methods for Navier-Stokes equations. Springer Verlag, Berlin.
Glagov, S., Zarins, C., Giddens, D.P. and Ku, D.N., 1988, Hemodynamics and atherosclerosis. Arch. Pathol. Lab. Med., **112**, 1018-1031.
Goldsmith, H.L., 1972, The flow of model particles and blood cells and its relation to thrombogenesis. Prog. Hemost. Thromb., **1**, 97-112.
Gresho, P.M., Chan, S.T., Lee, R.L. and Upson, C.D. , 1984a, A modified finite element method for solving the time-dependent incompressible Navier-Stokes equations. Part 1: theory. Int. J. Num. Meth. Fluids, **4**, 557-598.
Gresho, P.M., Chan, S.T., Lee, R.L. and Upson, C.D. , 1984b, A modified finite element method for solving the time-dependent incompressible Navier-Stokes equations. Part 2: applications. Int. J. Num. Meth. Fluids, **4**, 619-640.
He, X., 1993, Numerical simulations of blood flow in human coronary arteries. Ph.D.Thesis, Georgia Institute of Technology, Atlanta, Ga., USA.
Hibbit, Karlsson & Sorensen Inc., 1993, ABAQUS (Version 5.3). 1080 Main Street, Pawtucket, R. I. 02860.
Hilbert, D. ,1987, Ein Finite Elemente-Aufspaltungsverfahren zur numerischen Lösung der Navier-Stokes Gleichungen und seine Anwendung auf die Strömung in Rohren mit elastischen Wänden. Diss., TU-Graz.
Hofer, M. and Perktold, K., 1995, Vorkonditionierter konjugierter Gradienten Algorithmus für große schlecht konditionierte unsymmetrische Gleichungssysteme. Supplement Volume ZAMM **75** SII, S641-S642.
Horsten, J.B.A.M., Steenhoven, A.A. van and Dongen, M.E.H. van, 1989, Linear propagation of pulsatile waves in visco-elastic tubes. J. Biomechanics, **22**, 477-484.
Hughes, T.J.R., Liu, W.K. and Zimmermann, T.K., 1981, Lagrangian-Eulerian finite element formulation in incompressible viscous flows, Comp. Methods Appl. Mech. Engrg., **29**, 329-349.
Motomiya, M. and Karino, T., 1984, Flow patterns in the human carotid artery bifurcation. Stroke, **15**, 50-56.
Koiter, W.T. and Simmonds, J. G., 1973, Foundations of shell theory. In Proc. 13th Int. Congress Theor. Appl. Mech., pp. 150-176, Springer 1973.

Ku, D.N., Giddens, D.P., Zarins, C.K. and Glagov, S., 1985, Pulsatile flow and atherosclerosis in the human carotid bifurcation. Arteriosclerosis, **5**, 293-302.
Kumar, A. and Kumar, N.S., 1988, A new approach to grid generation based on local optimisation, In: Numerical Grid Generation in Computational Fluid Mechanics (Edited by S. Sengupta, J. Häuser, P.R. Eiseman, J.F. Thompson), Pineridge Press, Swansea, U.K., 177-184.
Liepsch, D., Thurston, G. and Lee, M., 1991, Studies of fluids simulating blood-like rheological properties and applications in models of arterial branches. Biorheology, **28**, 39-52.
Lou, Z. and Yang, W.J., 1992, Biofluid dynamics at arterial bifurcations. Critical Reviews in Biomed. Eng., **19**, 455-493.
Löw, M., Perktold, K. and Raunig, R., Hemodynamics in rigid and distensible saccular aneurysms: A numerical study of pulsatile flow characterisitcs.
Mackerle, J., 1992, Finite and boundary element methods in biomechanics: a bibliography (1976-1991). Engineering Computations, **9**, 403-436.
McDonald, D.A., 1974, Blood Flow in Arteries, Arnolds, London.
Nerem, R.M., Cornhill J.F., 1980, The role of fluid mechanics in atherogenesis. ASME J.- Biomech. Eng., **102**, 181-189.
Nerem, R.M., 1984, Atherogenesis: hemodynamics, vascular geometry, and the endothelium, Biorheology, **21**, 565-569.
Osenberg, H.P., 1991, Simulation des arteriellen Bluflusses - Ein allgemeines Modell mit Anwendung auf das menschliche Hirngefäßsystem. Diss., ETH-Zürich 9342, IBT Zürich.
Patel, D.J. and Vaishnav, R.N., 1980, Basic hemodynamics and its role in disease processes. University Park Press, Baltimore.
Pedley, T.J., 1980, The Fluid Mechanics of Large Blood Vessels. Cambridge University Press, U.K.
Perktold, K., Florian, H. and Hilbert, D., 1987, Analysis of pulsatile blood flow: a carotid siphon model. J. Biomcd.Eng., **9**,46-53.
Perktold, K., 1987, On numerical simulation of three-dimensional physiological flow problems. Ber. Math. Stat. Sektion, Forschungsges. Joanneum Graz, Nr. 280/5.
Perktold, K., Florian, H., Hilbert, D. Peter, R.O., 1988, Wall shear stress distribution in the human carotid siphon model during pulsatile flow. J. Biomechanics, **21**, 663-671.
Perktold, K. and Resch, M., 1990, Numerical flow studies in human carotid artery bifurcations: basic discussion of the geometric factor in atherogenesis. J. Biomed. Eng., **12**, 111-123.
Perktold, K., Peter, R.O., Resch, M. and Langs, G., 1991a, Pulsatile non-Newtonian flow characteristics in three-dimensional carotid bifurcation models: numerical study of flow phenomena under different bifurcation angle. J. Biomed. Eng., **13**, 507-515.
Perktold, K., Resch, M. and Florian, H., 1991b, Pulsatile non-Newtonian flow characteristics in a three-dimensional human carotid artery bifurcation model. Transactions of the ASME-J. Biomech. Eng., **113**, 464 464- 475.
Perktold, K., Nerem, R.M. and Peter, R.O., 1991c, A numerical calculation of flow in a curved tube model of the left main coronary artery. J. Biomechanics, **24**, 175-189.
Perktold, K., Resch, M. and Peter, R.O., 1991d, Three-dimensional numerical analysis of pulsatile flow and wall shear stress in the carotid artery bifurcation. J. Biomechanics, **24**, 409-420.
Perktold, K. and Resch, M., 1991, Numerische Lösung der Navier-Stokes Gleichungen für Verallgemeinerte Newtonsche Fluide. Z. angew. Math. Mech., **71**, 5, T416-T419.
Perktold, K., Tatzl, H. and Schima, H., 1993, Computer simulation of hemodynamic effects in distal vascular graft anastomoses. ASME-BED-Vol. 26, (Edited by Tarbell, J. M.), 91-94, ASME, New York.
Perktold, K. and Rappitsch, G., 1993, Numerical analysis of arterial wall mechanics and local blood flow phenomena. In: Advances in Bioengineering, ASME-BED-Vol. 26, (Edited by Tarbell, J. M.), 127-131, ASME, New York.
Perktold, K. and Rappitsch, G., 1994, Mathematical modeling of local arterial flow and vessel mechanics. In: Computational methods for fluid-structure interaction (Edited by Crolet, J. and Ohayon, R.), 230-245. Pitman Research Notes in Mathematics Series 306, Longman Scientific &Technical, J. Wiley and Sons, New York.
Perktold, K., Thurner, E. and Kenner T., 1994a, Flow and stress characteristics in compliant carotid artery bifurcation models. Medical & Biological Engineering & Computing, **32**, 19-26.
Perktold, K., Tatzl, H. and Rappitsch, G., 1994b, Flow dynamic effects of the anastomotic angle: a numerical study of pulsatile flow in vascular graft anastomoses models. Technology and Health Care, **1**, 197-207.
Perktold, K., Leuprecht, A. and Peter, R.O., 1994c, Problems in applications of viscoelastic models to the numerical analysis of arterial blood flow. In: Biofluid Mechanics (Edited by Liepsch, D.), VDI-Verlag, Reihe 17: Biotechnik, Nr. 107, 471-476.

Perktold, K., Rappitsch, G. and Liepsch, D., 1994d, Flow and wall shear stress in the human carotid artery bifurcation: computer simulation under anatomically realistic conditions, In: Advances in Bioengineering 1994 (Edited by Askew, M.J.), ASME-BED-Vol. 28, 437-438, ASME, New York.

Perktold, K. and Rappitsch, G., 1995, Computer simulation of local blood flow and vessel mechanics in a compliant carotid artery bifurcation model. J. Biomechanics (in press).

Perktold, K., Rappitsch, G. and Pernkopf, E., 1995, Arterial mass transport in curved tubes: a computational study of convective diffusion processes including shear dependent wall permeability. In: Proc. ASME - Summer Bioengineering Conference 1995, Beaver Creek.

Rappitsch, G., Perktold, K. and Guggenberger, W., 1993, Numerical analysis of intramural stresses and blood flow in arterial bifurcation models. In: Computational Biomedicine (Edited by K.D. Held, C.A. Brebbia, R.D. Ciskowski and H. Power), 149-156. Computational Mechanics Publications, Southampton Boston.

Rappitsch, G. and Perktold, K. 1995, Computer simulation of convective diffusion processes in large arteries. J. Biomechanics (accepted).

Reneman, R.S., van Merode, T., Hick, P. and Hoeks, A.P.G., 1985, Flow velocity patterns in and distensibility of the carotid artery bulb in subjects of various ages. Circulation **71**, 500-509.

Reuderink, P., 1991, Analysis of the flow in a 3D distensible model of the carotid artery bifurcation, Thesis, Eindhoven Institute of Technology, The Netherlands.

Rindt, C.C.M., Van Steenhoven, A.A., Janssen, J.D., Reneman, R.S. and Segal, A., 1990, A numerical and experimental analysis of the flow field in a three-dimensional model of the human carotid bifurcation. J. Biomechanics, **23**, 461-473.

Steiger, H.-J., 1990, Pathophysiology of Development and Rupture of Cerebral Aneurysms. Acta Neurochirurgica, Supplementum 48, Springer-Verlag, Wien, New York.

Stergiopulos, N., Pythoud, F. and Meister, J.-J., 1994, Analysis of the Forward and Backward Running Waves in Arteries: The Role of Nonlinear Wall Elasticity. In: Biofluid Mechanics (Edited by Liepsch, D.), VDI-Verlag, Reihe 17: Biotechnik, Nr. 107, 727-732.

Sugawara, M., Kajiya, F., Kitabatake, A. and Matsuo, H. (Editors), 1989, Blood Flow in the Heart and Large Vessels. Springer-Verlag, Tokyo, Berlin, New York.

Temam, R. (1979), Navier-Stokes Equations. North Holland, Amsterdam.

Thurston, G.B., 1979, Rheological parameters for the viscosity, viscoelasticity and thixotropy fo blood, Biorheology, **16**, 149-162.

Trubel, W., Moritz, A., Schima, H., Raderer, F., Scherer, R., Ullrich, R., Losert, U. and Polterauer, P., 1994, Compliance and formation of distal anstomotic intimal hyperplasia in dacron mesh tube constricted veins used as arterial bypass grafts. ASAIO Journal 1994, M273-M278.

Yamaguchi, T., Nakano, A., Hanai, S., 1990, Three dimensional shear stress distribution around small atherosclerotic plaques with steady and unsteady flow, In: Biomechanical Transport Processes (Edited by Mosora, F. et al.). Plenum Press, New York, 173-182.

Zarins, C.K., Giddens, D.P., Bhardvaj, B.K., Sottiurai, V.S., Mabon, R.F. and Glagov, S., 1983, Carotid bifurcation atherosclerosis - quantitative correlation of plaque localization with flow velocity profiles and wall shear stress. Circulation Research, **53**, 502-514.

7

COMPUTATIONAL VISUALIZATION OF BLOOD FLOW IN THE CARDIOVASCULAR SYSTEM

Takami Yamaguchi

Tokai University
Department of Bio-Medical Engineering
School of High-Technology for Human Welfare
317 Nishino, Numazu, Shizuoka 410-03, Japan

INTRODUCTION

Fluid flow, such as blood flow and air flow, is a sine qua non mechanical phenomenon for maintaining the life of animals particularly vertebrates. Among them, blood flow is of vital concern not only from the view point of normal physiological conditions but also with respect to various disorders. It is, however, noteworthy that we can not clearly separate the physiological role and the pathological behavior of blood flow because the pathological process begins under normal physiological conditions. In other words, pathological phenomena should be regarded as being seamlessly continuous with the physiological state [1]. This is particularly true for some vascular diseases which start and develop under a strong influence of blood flow [2]. Atherosclerosis is representative among these diseases and is very important because its development finally results in the diminution and cessation of blood flow to crucial organs, particularly to the brain and the heart [3]. In westernized or industrial societies, death directly or indirectly caused by atherosclerosis usually occupies the top of the mortality statistics.

Some general discussions on the studies of blood flow placing stress on the computational fluid dynamics (CFD) method and the visualization based on the computation are given in the present article. We would like to describe the state of the art and the near future technology of computational visualization which is becoming available in the field of biological flow studies in these years. In the course of discussion, we have to depict some basic problems in the studies of blood flow, common to both experimental and computational ones, with respect to the physiological condition and pathological consequences. Computational bio-fluid mechanics directed to these fundamental problems undoubtedly represents one of the most promising directions in the field of biological flow studies, just as in various fields of engineering where fluid dynamics plays the leading role.

Biological Flow, Edited by Michel Y. Jaffrin and Colin Caro
Plenum Press, New York, 1995

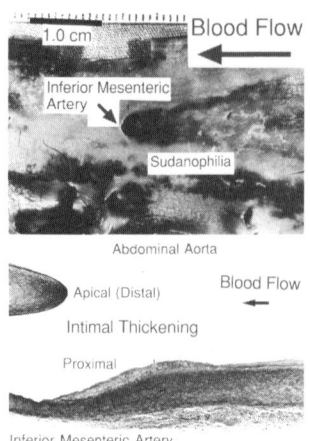

Figure 1. Spatial distribution of the early atherosclerotic lesion by macroscopic (top) and by microscopic (bottom) views. Macroscopically, there is a specific (fan-shaped) pattern of sudanophilic region (stained by Sudan IV dye, which shows strong affinity for lipids) on the upstream wall of the abdominal aorta of 20-year-old subject (who died of an accident.) From the longitudinal section of this region (bottom), we can see localized intimal thickening both on the proximal and on the apical (or distal) wall of branching. They show remarkably different morphological structure. The proximal location where a loose, cell rich intimal thickening grows is where the atherosclerotic plaque later develops. It is easily understood from these two figures that the 2D section is merely showing a limited part of the phenomenon, and 3D investigation is mandatory.

Pathophysiological Significance of Blood Flow

Vascular diseases such as ischemic heart disease and cerebral strokes are known to be strongly influenced by blood flow in terms of their onset and advancement [1-3]. Most of these cardio- and cerebro-vascular diseases develop as sequelae of atherosclerosis, which has been postulated to start at vessel locations where the wall shear stress is expected to be low [4]. Though the so-called low wall shear stress regions of the artery are not yet exactly located in the three dimensional (3D) flow field, no one doubts that there are certainly specific locations prone to atherosclerosis [5].

There are seemingly two major reasons why the wall shear stress, among several mechanical phenomena, interests many investigators. Firstly, atherosclerosis is predominantly a disease of the intima of the arteries from the pathological view point. Mechanical forces and effects other than the wall shear stress tend to affect all the layers of the arterial wall. Secondly, the wall shear stress, when it is viewed as a transport phenomenon of momentum, is similar to other transport phenomena of mass, heat, etc. The transport phenomena are thought to be very important because atherosclerosis has been traditionally looked upon as the disease of accumulation of excess lipids in the arterial wall.

It is well known that the wall shear stress is determined by the flow just adjacent to the wall. Consequently, its distribution shows markedly different patterns in the complex configuration of arteries. Arteries, in fact, have very complex structures, including branching, curvature, constriction, dilatation, etc. Blood flow in the artery itself is, moreover, unsteady with a complex time course. Such a complex nature contributes to the whole flow field in a completely non-linear way. This is because the physics of the fluid flow, as is described by its governing equations, the Navier-Stokes equation, is highly non-linear. The whole flow field consequently may result in a surprisingly deviated one from what is expected by linear superposition.

Computational Fluid Dynamics

As we previously reported [6-23], CFD is a powerful tool for investigating the complex nature of blood flow. Although there are many difficult problems in the CFD study of blood flow, for example, its unsteadiness, the non Newtonian viscosity of blood, etc, it is gradually becoming a matter of computing power to overcome them. We are nowadays able to carry out computations assuming the complex nature of the blood flow using an extremely complex model based on a real geometry of the blood vessels.

Flow related phenomena, such as heat and mass transfer, and any derived parameters of mechanics, such as the wall shear stress exerted by blood flow, are easily computed in CFD. This is particularly important in studies of the role of blood flow in atherogenesis because a combination of spatial as well as temporal variations of these parameters seem to play an influential role in the process. These four dimensional (4D) (that is, three spatial dimensions plus time) phenomena can be fully analyzed only by the computational method at the present stage of investigation.

Importance of Geometry

Though the potential of CFD is very high, its actual applications are still limited mostly to arterial and cardiac flow models based on rather simple idealistic geometries. As is frequently pointed out, the most significant factor which affects the flow structure [24] and hence the diseases [25-26] is the 3D configuration of the flow field. Therefore the reality of the computational model is crucial for correlating fluid mechanics to physiological and pathophysiological phenomena such as atherogenesis [11,14]. Fortunately, similarities between the real flow field and a model can be pursued in a flexible way in the fluid mechanical studies, particularly in CFD. We can introduce any arbitrary scaling systems as far as they are consistent, so that we can analyze models of microscopic scale to very large scale by the same method.

Recent advances in computer technology have enabled us to achieve more realistic modeling. We found that vascular casts are useful for the study of complex vascular structures [14, 27]. Using these casts as a starting point for the realistic modeling, an accurate comparison can be made of the CFD results with those obtained by blood flow measurement. Three-dimensionally reconstructed images obtained by non-invasive methods such as CT, MRI, and ultrasound methods can also be applied to computational mechanics studies.

Significance of the Visualization

As is discussed above, we need to understand the effect of the 3D structure of the vasculature to study blood flow. This is essential not just to know the fluid motion, but eventually for the understanding of the influence of the flow on the wall. It is particularly important for the study of blood flow with respect to the vascular disorders. Computational fluid dynamics incorporating computer graphics technology is presently the only available method to allow such an analysis. The 4D nature of the real flow can be understood by the animation technique in which consecutive changes in the 3D distribution of computed values are displayed using sophisticated rendering techniques with color graphics. Not only velocity components and pressure, but also a variety of derived parameters, for example, the wall shear stress and its spatial derivatives can be displayed from a variety of viewpoints on the computer screen as well as on video equipment.

COMPUTATIONAL METHODS IN PHYSIOLOGICAL FLOW STUDIES

General Consideration

Computational analysis of the flow can be conceptually divided into several steps or stages as shown in Fig. 2. Before conducting the computation, we need to model the real flow mathematically. The first step of the mathematical modeling is an approximation. The full Navier-Stokes equation is also an approximation, in a sense, because it is based upon the continuum hypothesis, neglecting molecular behavior of the fluid. The word approximation here is, however, not used in this sense. Further reductions in the full Navier-Stokes equation, such as reducing it to the Euler equation dropping viscosity, two-dimensionalization, etc., is meant in the present context. We also need flow modeling. Turbulence modeling, for instance, using k-ε two equations, is also applied if necessary.

Based upon these approximations and modeling, we formulate partial differential equations. Since the formulated partial differential equations are for continuous variables, we need to discretize the equations for numerical simulation on a computer. In discretization, we utilize various methods called numerical schemes. By these processes, we finally obtain algebraic equations which can be applied to the numerical models of the flow field. A computer program called "pre-processor" is usually used to define computational models and to set necessary initial and boundary conditions.

A computer program that performs the main body of CFD computation is usually called a "solver", and it produces numerical results. The result itself is just a list of numbers which represents computed fluid mechanical parameters such as velocity, pressure, and some other variables. A "post-processor" is a program to convert this computed result into physically or physiologically meaningful data and to visualize them.

From the bio-fluid dynamics point of view, we are interested in the physiological and pathophysiological interpretations of the computed result. We therefore use and trust commercial packages of programs which are based on necessary theoretical formulations and can concentrate the pre- and post-processing of the computation. This is nowadays an ordinary approach to CFD studies in most of the engineering fields too. Needless to say, we have to know what we are doing, in other words, the governing physical process and equations describing the phenomena when we conduct CFD analysis.

Governing Equations

Basic Considerations: As is frequently mentioned, there are several basic problems in applying the Navier-Stokes equations to blood flow. Firstly, it is assumed that the fluid is Newtonian, i.e., the viscosity does not vary with the flow condition. This is not strictly true for blood, since it is known to show shear dependent viscosity particularly when the shear rate is very low [29]. However, the viscosity becomes relatively constant if the shear rate exceeds a certain limiting value. Fortunately, the shear rate estimated in larger arteries usually

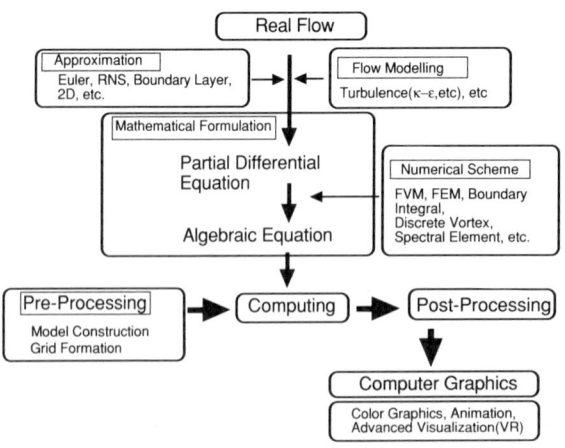

Figure 2. Scheme of the steps of the CFD studies. Translated and modified from H. Daiguji, in Numerical Fluid Dynamics (Suuchi Ryutai Rikigaku) (in Japanese) Chapter 1, Eds M.Yasuhara and H. Daiguji, University of Tokyo Press, Tokyo, 1992, p.4. [28]

exceeds this threshold. Therefore, we can safely neglect the non-Newtonian viscosity when we consider blood flow in large arteries[30]. Next, a frequently asked question relates to the deformation, particularly that due to the non-linear stress-strain relationship of the vessel wall. The arterial and venous wall deforms according to the transmural pressure as well as the wall shear stress. The pressure primarily changes by the pulsatile nature of blood flow and secondarily by the fluctuations of the disturbed velocity field, such as vortices and turbulence. They are unfortunately very difficult to incorporate into currently available computational methods to their full extent. Intensive studies on this subject have been performed [31,32].

Equations of Motion and Continuity. As is discussed above, our usual analysis can be based upon the Navier-Stokes equations of motion, and the continuity equation for the incompressible Newtonian fluid as the following [33];

$$\rho \frac{D\mathbf{v}}{Dt} = -\nabla p + \mu \nabla^2 \mathbf{v} \qquad (1)$$

$$(\nabla \cdot \mathbf{v}) = 0 \qquad (2)$$

Where ρ is the density, $\mathbf{v}$ is the velocity, $D\mathbf{v}/Dt$ is the substantial velocity time derivative with 3D partial derivative components of both time and distance in 3D, p is the pressure and μ is the viscosity.

Equations of Transport of Scaler Variables: In the computation, we can integrate the transport equations for heat and mass. Since the mass transport in the liquid is particularly convection dependent, the distribution of the solute masses can be computed based on the velocity field obtained by the solution of the equations of motion. In the heat transfer, natural convection may be important even in a liquid such as blood, and should be modeled in that case. Nevertheless, the temperature distribution is small in the living system and the heat production within the fluid can be neglected so that we do not usually have to take the interference of heat on the velocity field into account.

Turbulence. It is well known that turbulence, if it occurs under the same global flow conditions, induces significantly larger wall and intra-fluid shear stresses. There is a strong possibility that turbulence in the living system affects the various pathophysiological phenomena, such as atherosclerosis. Turbulence has been shown to exist in the larger arteries in the cardiovascular system [34-38]. This is particularly true when the flow rate is increased, for example, in an exercise condition. Turbulence modeling has been successfully applied to the computation of biological fluid flow and its effect on the wall. One of these approaches is the so-called k-ε approach, in which two averaged properties of turbulent flow are introduced into the equations of motion and are traced in the course of computations [39]. It is noteworthy that the introduction of such parameters is in a sense arbitrary or *a priori* in terms of the recognition of transition to turbulence. Therefore the transition is not predicted by this method, but the average behavior of turbulence can be computed.

Numerical Method

Pros and Cons of Three Different Methods: In general, currently available techniques in CFD analysis fall into three or four classes [40-41]. They are the finite difference or the finite volume type method, the finite element approach, the boundary element method, and so forth. The finite difference or the finite volume method was historically first developed and has been available for many years. It is established that these methods give stable results with relatively less computation time than is required by other methods. However, the model construction is less flexible than the finite element method later developed. Although the finite element method allows greater flexibility to build the computational domain, much more time and memory space is necessary for the computation. The boundary element method is mainly used for a class of flow problems (potential flow) which is not always relevant to blood flow. We have been using a finite volume method to compute the fluid flow alone. As pointed out above, relatively short computational time is spent to obtain stable results in the finite volume methods, and this point is important because we are using workstation class computers for their greater flexibility.

It is recognized that the moving wall type problem can not be properly dealt with by the finite volume type approach and may be completely impossible. In most efforts so far reported to solve the coupled equations for the wall motion and the fluid motion, a finite element type method is utilized[31]. This is partly because the solid mechanics problems have been mostly solved by finite element methods.

Boundary Conditions: Since CFD analysis falls into a category of boundary value problems, we need three to four sets of boundary conditions. They are, the inlet condition, the wall condition, the outlet condition, and conditions to define symmetry of the model. The fluid and solid interfaces such as the one between the blood and the arterial wall is usually assumed as a non-slip rigid boundary. In some special cases, the wall may be dealt with as a permeable wall, which allows a certain flow across it. In an analysis of the coupled motion of the fluid and the wall, the wall moves according to the normal and shear stresses exerted by the fluid. Even in this case, a no-slip condition on the wall is usually assumed.

If the models are like a pipe, the inlet flow boundary is set as if there is an infinitely long inlet part before the model starts by assuming some theoretical fluid velocity distribution at the inlet cross section. However, in unsteady flow cases, there are some problems to be considered. When the basic frequency is very low and the flow can be regarded as virtually steady, the flow velocity profile may be parabolic. This is the case when the Womersley's alpha parameter is very small, and is sometimes irrelevant in the larger arteries and airways. In this case, we do not know precisely the inlet boundary condition so that a flat profile becomes the only choice. Another possibility is to compute appropriate flows to define the inlet boundary condition. This approach will be useful particularly when we try to build a general model of the circulatory system. We will discuss more this relatively new idea, naming it a multi-level CFD model of the flow tracts in the living system.

Different consideration is necessary for the outlet condition, especially when the unsteady flow is calculated. In the unidirectional steady flow case, the outlet condition is rather simple. What we usually do is just assume zero pressure and zero gradients for all the dependent variables, such as velocity components and pressure. Nevertheless, in the unsteady case, the pressure gradient can be reversed anywhere in the computing domain including the outlet cross section. In such a case, we need to know what is happening outside the computing model.

Importance of the Pre-Post Processing [42]

Pre-Processor: Since we always try to tackle with flow tracts with complex configurations the pre-processor is very important. It is frequently estimated that the pre-processing, that is the model building, requires more than 90% of the effort of computational analysis. An expertise is still necessary to build a complex model, particularly when it is 3D, even by using the most advanced CAD based pre-processor system. There are two types of difficulties in pre-processing. The first one is to define the complex surface shape of the model. This task has been very much facilitated by introducing 3D CAD software. The second difficulty is to define the computational mesh inside the flow tract. The computational mesh should be carefully located by paying attention to the predicted flow distribution [43,44].

Post-Processor: There are several issues that should be discussed under this heading. First, it is important to convert and process the numerical values, since we can not necessarily be content with the usual output of the computational fluid mechanical simulation. Secondly, the graphics representation is mandatory to understand the flow itself and its effect on the vascular wall. Another issue is related to the data compatibility between the CFD system and other systems to post-process it, which we do not discuss here.

Three dimensional velocity and pressure values are usually obtained from the computation. The velocity itself is undoubtedly important to understand the flow field, and is directly calculated in the computational study. The virtue of computational methods is that the obtained data can be easily re-processed in many ways. We always make color graphic mapping of the distribution of the wall shear stress and its derived parameters, such as a plain average in time, a RMS (root mean square) value, etc. This is usually difficult to carry out by using the experimental data due to limitations of measurements. Temporal as well as spatial correlation between these derived parameters and other mechanical or physiological parameters can also be processed using the computed results. By doing this, we can search the physiologically significant characteristics with respect to the normal and abnormal state of the living system.

Visualization Technique

Computer Graphics: In usual scientific publications, color printing is prohibitively costly and most graphics are represented in black and white drawings or half-tone photos. However, in advanced computer graphics, color representations are easier than the graphics of the black and white line drawings especially in 3D graphics. This is particularly true if we need to build specialized graphics system to visualize the CFD results.

As has been repeatedly stressed in the foregoing, we are interested in true 3D distribution of a certain class of physical properties, such as the wall shear stress in the complex 3D configuration. In order to restitute all information, we need to visualize the geometry of the computational domain and the computed results in the same frame at the same time. Most modern computer graphics tools, such as GL, PHIGS, etc. are based on color graphics particularly on the polygonal drawing with its lighting and shading. This means that the surface of a 3D object is easily rendered with various special effects.

The Virtual Reality Technology: The CRT display used for computer graphics is flat and can only show pseudo 3D effects by using various techniques for representing the perspective. It is, therefore, difficult, if not impossible, to comprehend the true 3D spatial location of physical objects by a static graphical drawing. To overcome this, the virtual reality technology is one of most promising but not widely used modalities. Here, we use the phrase "virtual reality" to mean a technology to give an interactively controllable or modifiable graphically represented imaginary environment which gives a user a sense of immersion as if he/she was in an imaginary world.

In the fields of radiology and surgery, as a means to visualize medical image data, virtual reality is now widely used. However, these applications are not coupled with computational mechanics. So far it is used to visualize the image data obtained by Xray CT, MRI, or ultrasound equipment. Very limited attempt has been made to visualize the compu-

Figure 3. A part of 3D representation of the flow field computed in a model abdominal aneurysm. Velocity vectors are rendered as 3D arrows and the wall shear stress is mapped in a color representation of the wall surface. When this is displayed in a stereoscopic display, the angle of view, the depth of the viewpoint, etc, can be modified by the movement of the eye gaze (through the detected head position).

tational results in the bio-fluid mechanics field. The complex nature of the 3D flow structure in the normal and diseased artery is a most suitable problem to which to apply the virtual reality technology. It is especially so because the human capability for the 3D perception.

Another possibility related to the use of VR technology in CFD is the real time "steering" of the computation. This means that the computational process is controlled during its course by an operator watching the results through graphical representation. This technique is particularly promising in bio-fluid mechanics because the complex conditions governing the fluid dynamics of blood flow could only be fully studied in a hybrid, i.e. human and computing, interacting approach such as this.

To actualize virtual reality, there are three important component technologies, including a stereoscopic display, an interactive computer graphics software, and a high-speed graphics computer. Combining these resources, we can produce an imaginary or "virtual' world in which the computational results are displayed and even "touched" by an observer. An example is shown in Fig. 3. In this application, a part of the flow velocity field in a model abdominal aneurysm [8,19] is rendered in 3D color graphics with parallax and we can see the 3D structure of the velocity vectors stereoscopically.

The technology is thought to be very useful to grasp the complex 3D structures of the flow field and the fluid-solid interactions in the living system.

Animation Technique: The animation technique is widely used to present physical phenomena in four dimensions, i.e., spatial 3D plus time 1D. This is very important in understanding unsteady flow such as that in the cardiovascular system and the respiratory system. In animated figures, it is easy to visualize 3D constructions because they provide a sense of depth due to the differential movement of an object in the view volume. Animation is therefore a crucial technique nowadays to facilitate the understanding of computational results not only in unsteady flow cases but also in steady flow cases with complex geometry.

Ideally, the computed physical values and derived parameters should be displayed as they are in the real space and the real time. By this means, we would view the system from arbitrarily chosen angles with different rendering techniques at will. Unfortunately, this is not possible at the moment for two reasons. Firstly, the amount of data becomes enormous if we try to store it for direct display, easily exceeding several gigabytes, and causes shortage of the disk space and takes a very long access time. Secondly, the rendering speed of graphic workstations is still slow even if improved very much recently. Therefore, we need to record the rendered images on a recording device in a frame-by-frame mode. Although it is not

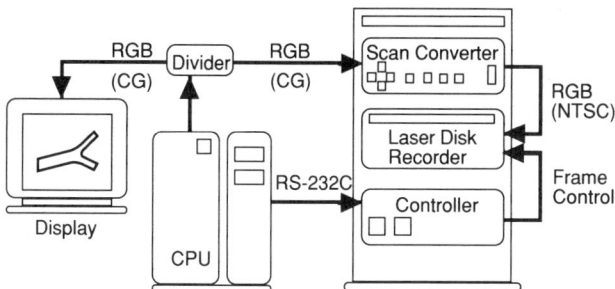

Figure 4. An animation recording system used in our laboratory. A rendered computer graphics frame is displayed in the computer display and the high frequency video signal is also transferred to a scan converter. The scan converter transforms the frame definition from CG standard to usual broadcasting standard (NTSC, for example). Converted RGB signal is fed to a write-once laser disk recorder which is controlled by a controller connected to the computer via a serial communication channel. Thus, an animation creating program renders a frame, sends a signal to the recorder and a frame of the animation is recorded on a disk within a second.

possible to alter the conditions of the rendering, such as view angles, thresholds of physical parameters, etc, through this technique, it is still a very powerful means to understand the nature of any kind of flow fields computed by CFD method.

To make animation video from a computer graphics output, there are two major methods. The conventional method we have been using is to convert the graphics output video (RGB) signal into any broadcast format (NTSC, PAL, etc.) and record it on a frame-to-frame video (tape or disk) recording equipment. Usually, the graphics output of computers has more scan lines than that of the usual television video signal, and hence the video signal of the computer graphics display contains much higher frequency. Therefore we need to convert the scan rate by special equipment called a scan converter. After this conversion, the video signal is recorded in three separate channels as red, green and blue (RGB) or in a composite channel according to various broadcasting systems. When recording a frame, we can use a video-tape recorder which can record a frame, or a disk type recorder. Though most home-use video recorders can not be used to make rigorous frame-to-frame recordings, the merit of using the video recorder is its low cost. However the video-recorder is sometimes painfully slow. To record a video frame (i.e. for 1/30 second), the tape must be rolled back every time (pre-rolling), for about 5 seconds, and the total time is nearly 1 minute for just taking a frame. At this speed, it will take 25 hours to take an animation of a length of 1 minute. When it was just to make a video for a demonstration purpose, one day for a minute did not matter. However, we need a handier method to utilize animation as a tool for analyzing the computed results in four dimensions. In this context, a laser disk type recorder is mandatory to cut the recording time. Only one second is necessary to record one video frame if a laser video disk recorder is used and it can be controlled digitally in a integrated program for the visualization(Fig. 4). By using this method, we can compare the minute spatial distribution alterations due to the combination of the geometrical configuration and the time course of the blood flow.

Another modern method is to utilize digital video recording technology which is available even on personal computers. The rendered image is stored in a digital video format on computer disks and can be displayed at any time or even transferred in a compressed format (eg. MPEG) on a communication channel. This is undoubtedly the future trend with the usage of computer networks, such as the internet. In this technology, no expensive

Figure 5. Canine left ventricle cast made by methacrylate resin in maximally dilatated state. Grids marked on the surface of the cast were used to locate the pin sensor of the 3D coordinate measuring machine employed to measure the spatial shape of the cast.

specialized hardware such as a scan converter is necessary and various information compression methodologies under vigorous development will facilitate this trend very much.

SOME RESULTS OF COMPUTATIONAL VISUALIZATION OF BLOOD FLOW

In the following section, some examples of our computational studies are shown, namely, the intraventricular flow fields of the heart, the 3D wall shear stress distribution in arterial bifurcations, and the flow adjacent to cultured endothelial cells. They are shown here with the intention of presenting both ends of a wide spectrum of computational studies.

Intraventricular Flow Field

Realistic Modeling of the Left Ventricle: First we present results from our series of studies dealing with the intraventricular flow fields in the heart. This study is characterized in two ways. First, this is based on an attempt to incorporate the uncoupled wall motion into the CFD. Secondly, the model used in this study is a realistic one based on a measurement of a cast of the real heart.

In order to compute the intra-cardiac flow field we made some casts in the left ventricular chamber of the heart, which is shown in Fig. 5. This cast was made by injecting a methacrylate resin into the left ventricle of a dog. To reconstruct the true 3D configuration, it is necessary to obtain the 3D coordinate values. A coordinate measuring machine was used to measure and to digitize the surface of the cast. The measured coordinate values were digitally sent to a workstation where they were recorded in a data file.

Figure 6 shows the canine left ventricle representation constructed by the measured surface shape of the cast. The coordinate values of the vascular structure geometric patterns were inserted into the SCRYU 3D flow simulation package used to compute complex flow patterns [11].

Computational Modeling of Ventricular Wall Motion: The computation code used is the SCRYU flow simulation package by Software Cradle Ltd. (Osaka, Japan). This system performs a finite volume integration of the 3D Navier-Stokes equations based on the

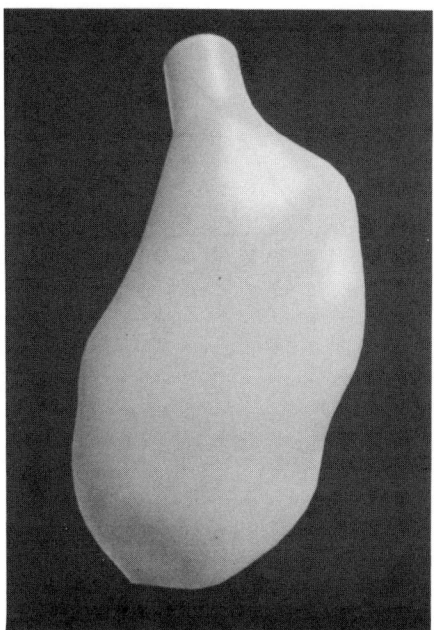

Figure 6. The smoothed volume representation of the inner cavity of the canine left ventricle in the surface rendering mode. This is a part of an animation video, in which the normal and the infarcted wall motion was simulated by using specially developed computational methods.

SIMPLE algorithm with a body-fitted coordinate (BFC) system [43,44]. Velocities and pressure values were computed. The computational grids were made using our own user-developed system and the preprocessor system of the SCRYU package, which is able to build BFC grids.

To deal with the intraventricular flow as that induced by the voluntary wall motion (i.e. contraction of the heart muscle), we solved the fluid motion and the wall motion in an uncoupled manner assuming the following.

First the time course of the left ventricular wall change was assumed to follow a function of time taken from established cardiac volume measurements without any reactions from the intraventricular fluid. Another important assumption is that the movement of the ventricular wall can be regarded as equivalent to the flow across the ventricular wall within the short duration of time in which the wall does not move significantly. Then the wall movement can be substituted by the flow across the wall [20].

We assumed that the time scale of the movement of the contracting ventricular wall is much slower than the time scale of the temporal resolution of the computation. In other words, the left ventricular wall moves much slower than the fluid particle nearby moves, and could be assumed quasi-steady for the time scale of the velocity development used. By assuming this, we can introduce two different time scales, namely the "flow time scale" and the "cardiac wall time scale". What we assumed is that the "cardiac wall time scale" is much larger than the "flow time scale". If we assume this, we can compute the evolution of the velocity field using a stationary computational grid in each "cardiac wall time step". Usually we used the "flow time scale" of an order of magnitude of 10^{-4} seconds and the "cardiac wall scale" of 10^{-2} seconds. We tested a couple of combinations of these two time scales and obtained basically the same results, which partly justifies the assumption.

Results of Systolic and Diastolic Flow Computations: Using the developed model, we simulated various phases of flow in the left ventricle of the heart, including the normal heart [16], the heart with abnormal wall motion due to myocardial infarction, intracardiac

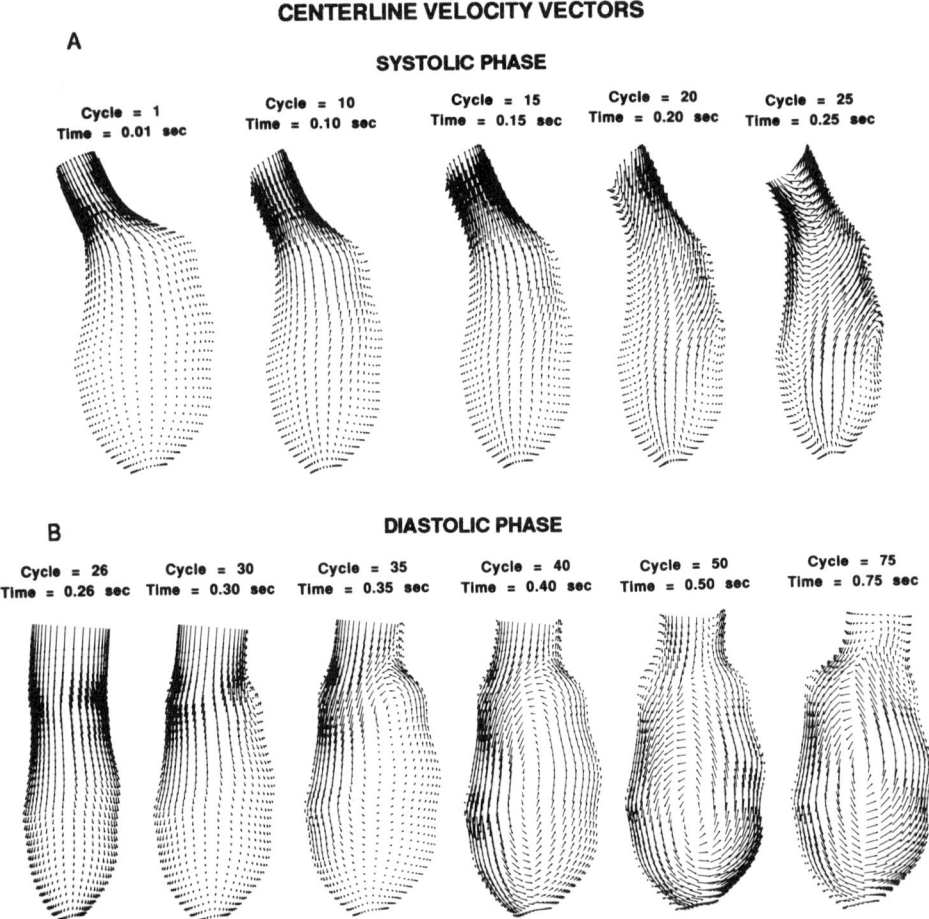

Figure 7. Intraventricular longitudinal velocity distribution pattern of the normal heart with end systolic ejection fraction 60%. Panel A : systolic flow. Panel B: diastolic flow. Reprinted from Biorheology 32, T.W. Taylor and T.Yamaguchi, "Flow patterns in 3D left ventricular systolic and diastolic flows determined from computational fluid dynamics", Copyright(1995) with kind permission from Elsevier Science Ltd, The Boulevard, Langford Lane, Kidlington, OX5 1GB, UK.

systolic blood flow under the condition of aortic stenosis [15], and the interactions of the diastolic filling and the systolic ejection flow under normal physiological condition [23] and in heart failure. Here we show an example of the intraventricular flow pattern in the normal heart at systole (Figure 7 A) and in diastole (Figure 7 B). Cross-sectional velocity vectors were also computed at systole (Figure 8 A) and at diastole (Figure 8 B).

Significance of the Computations of Intraventricular Flow: By introducing different time scales for the wall motion and the fluid motion induced by the wall motion, the computation of intraventricular flow fields was shown to be effectively conducted. This was possible because the muscle of the left ventricular wall exerts a much larger power than that needed to give the momentum to the blood before the real ejection. In this context, the interaction of the blood and the ventricular wall can be regarded as unilateral and there is a very small reaction from the fluid to the muscle This assumption enabled us to simulate the

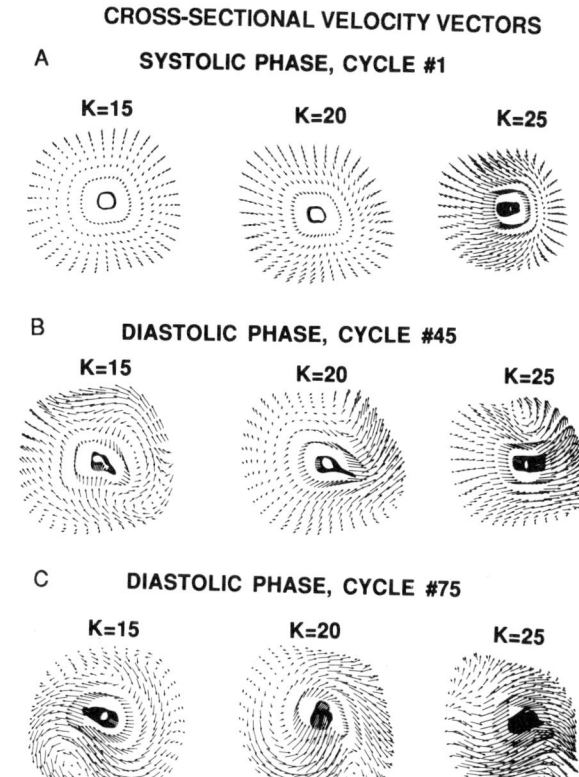

Figure 8. Intraventricular cross sectional velocity distribution pattern of the normal heart with end systolic ejection fraction 60%. Panel A : systolic flow. Panel B,C: diastolic flow. Reprinted from Biorheology 32, T.W. Taylor and T.Yamaguchi, "Flow patterns in 3D left ventricular systolic and diastolic flows determined from computational fluid dynamics", Copyright(1995) with kind permission from Elsevier Science Ltd, The Boulevard, Langford Lane, Kidlington, OX5 1GB, UK.

ventricular wall motion by replacing it by the flow across the wall. This method can be widely applied to different situations in the living system.

It is also noted that the construction of a realistic model whose configuration is based on real vascular structures was established in the studies shown here. This is a necessary technique for correlating fluid mechanics to the atherogenic and other vascular disease processes. This technique coupled with our computational method makes it possible to compare simulation results with measurements by various methods such as ultrasound imaging and MRI.

The Wall Shear Stress Distribution in the Bifurcation of Arteries

Purpose of the Study: Since atherosclerosis is known to occur at specific locations especially in its early stage, various possible phenomena that can cause such non-uniform spatial distribution have been intensively investigated. The biochemical properties of the blood are essentially uniform within a particular segment of the artery, for example in the bifurcation, so that they were omitted from serious considerations. Physical properties, such as pressure and wall shear stress can produce very fine spatial differences within a localized zone of artery. They have, therefore, been suspected to be the cause of the spatial distribution of the lesions, and hence to be underlying mechanisms of the diseases per se.

Most of previous studies on the flow and its mechanical influence on the arterial wall have been conducted using conventional 2D flow analysis. This is partly because the 3D configuration of the bifurcation is very complicated to model, and the true 3D unsteady computation needs enormous computation time. Experiments have been extensively carried

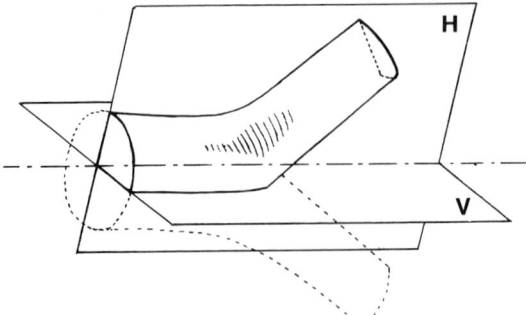

Figure 9. Schematic drawing of the parametrically defined bifurcation model of arteries. The bifurcation is divided into 4 parts by the horizontal (denoted H) plane of symmetry and the vertical (denoted V) plane of symmetry. By this method, a boundary fitted coordinate model is naturally defined.

out regarding this type of flow field [45-47]. Unfortunately, however, interpretations of the results were sometimes affected by traditional understandings of the 2D flow field. This is why we conducted extensive 3D studies of the bifurcational flow fields.

Model Design: We constructed a parametrically defined smooth-walled 3D bifurcation model using a blockwise design. As shown in Fig. 9, the 3D bifurcation was divided into 4 parts, each of which represents a quarter of the mother pipe continuously connected to a half of the daughter pipe. By such division, we can reduce 3D complex geometry of the bifurcation into rather a simple continuous body. Ratio of the diameters, curvature of the connection wall, and the density of the computational mesh inside the model can be parametrically defined. We can think of a few levels in the symmetrical structure of this type of model. If the branch angle is symmetrical and the diameter ratio between either daughter pipe and the mother pipe is symmetrical (i.e. the same), only a quarter of the model is enough to compute the flow inside the bifurcation. On the other hand, half of the bifurcation must be connected (i.e. two quarters should be combined), if a kind of asymmetry, such as different branch angles between two daughter vessels is introduced. Further asymmetry can be introduced to model the real arterial bifurcation.

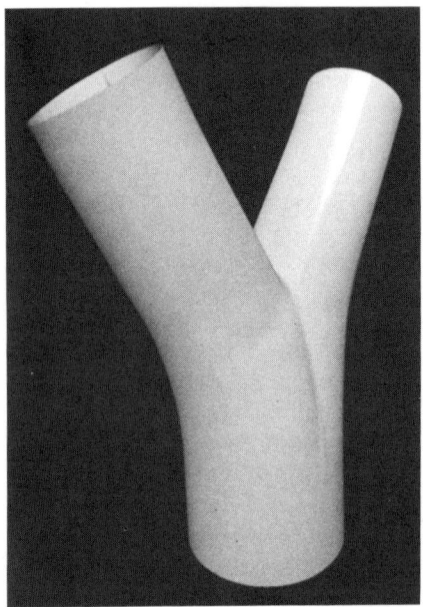

Figure 10. Surface rendering of a bifurcation model. Only the outer surface is rendered for a full bifurcation, that is, composed of 4 quarter models connected together.

In Fig. 10, a symmetrical model is shown by the surface rendering computer graphics. The transitional walls between the daughter and the mother, and between the daughters are smooth and no abrupt irregularities are found. In our models, the bifurcational fork wall is parametrically defined in comparison to the previously reported models [48].

Computational Results: By extensive computational studies covering different configurations, Reynolds numbers, and unsteadiness parameters, we found some very interesting phenomena. The lateral wall of the bifurcation, the observed atherosclerosis prone region, is not the position where the wall shear stress is low under any definition, such as instantaneous, average, or RMS. The definitions are as follows; the instantaneous wall shear stress is the one computed at every time step of the CFD computation at each wall element. The average is a time average over a cardiac cycle calculated at each wall element. The RMS value means the statistical standard deviation corresponding to the average above defined over a cardiac cycle. The lowest shear stress was observed 90 degree off the lateral wall, that is to say, the hip of the bifurcation region as shown in Fig. 11. This clearly shows the necessity of taking into account true 3D nature of the flow field to investigate the effects of the flow on the vessel wall. The same offset of the so-called low wall shear stress region was also observed by other investigators[49]. The lateral wall of the bifurcation is where the spatial as well as the temporal variance of the wall shear stress is strong and these fluctuations of the wall shear stress may be a determining factor of the early or precursory change of atherosclerosis [50,51].

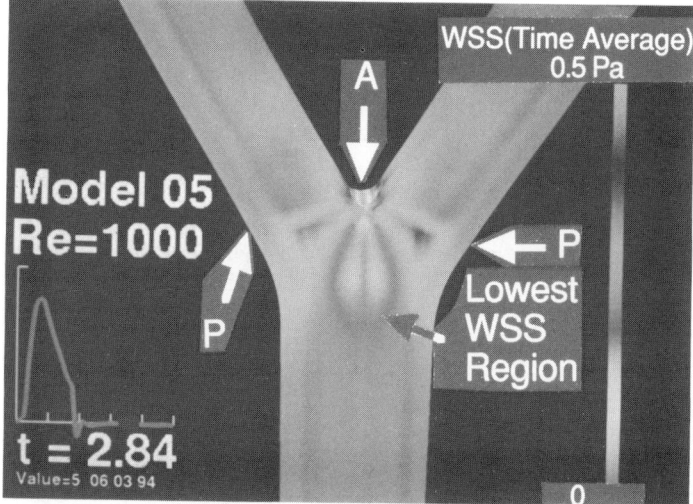

Figure 11. Spatial distribution of the time average of the wall shear stress (WSS) computed for a symmetric bifurcation. The average WSS value for a cardiac cycle is shown. Reynolds number calculated from the peak velocity at inlet cross section and the diameter of the mother pipe was 1,000. Velocity pattern used is also shown in the left corner. A: Apex of the bifurcation (WSS is always very low). P: Proximal regions which are prone to atherosclerosis. The lowest WSS region is noted by an arrow which resides 90 degree off the lateral wall. The root mean square (RMS) value of WSS was also the lowest in the same region as the average. (The original is a color graphics drawing).

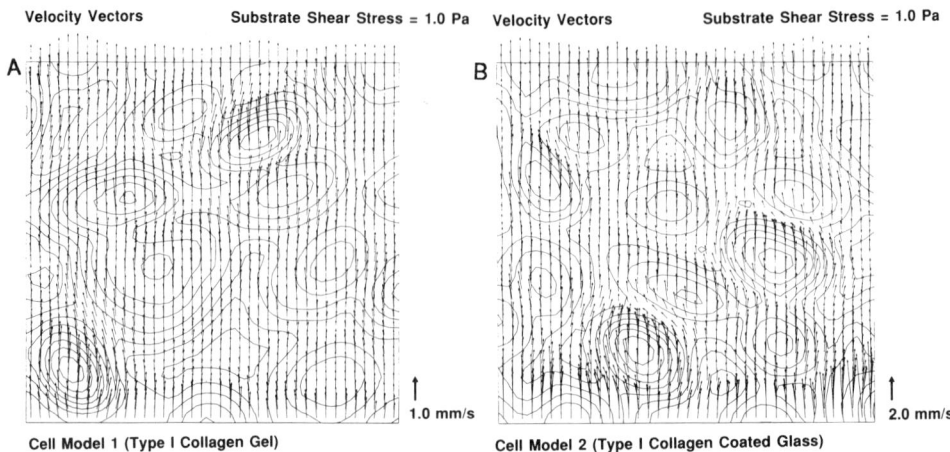

Figure 12. Velocity vectors calculated for confluent cell models. Panels A and B show vector distributions calculated for model 1 (monolayer formed on a collagen gel substrate) and model 2 (monolayer formed on a collagen coated glass substrate).

The Cellular Surface Flow and the Wall Shear Stress Distribution

Purpose of the Study: Vascular endothelial cells in situ have long been known to respond to blood flow and the active role of blood flow in the regulation and remodeling of blood vessels is currently of great interest [52]. To overcome difficulties in doing experiments using blood vessels, investigators have designed in vitro devices for studying the function and morphology of endothelial cells exposed to fluid flow.

When living cultured endothelial cells are examined by a light microscope, their 3D appearance can be easily appreciated. For example, when cells are cultured on a rigid substrate such as glass or plastic, the center part of a cell is significantly elevated from the substrate due to the presence of the nucleus, namely, the nuclear bulge. We found that the size of the nuclear bulge depended upon the nature of the substrate surface since cells cultured on a collagen gel had smaller nuclear bulges than those cultured on a solid surface. The diameter of the cells ranged from 50 to 70μm. Thus, the aspect ratio (height/diameter) of these endothelial cells was of the order of 10^{-1}. This aspect ratio should not be overlooked from the fluid mechanical viewpoint, and we decided to investigate the effects of cell shape on the flow pattern over the cell surface and on the consequent wall shear stress exerted by flow on the cell surface by CFD method [7,13,22].

Modeling of Confluent Endothelial Cells: Two types of confluent cell models were considered: one with cells cultured on type I collagen gel substrate, and the other with those cultured on type I collagen coated glass substrate. Cells on the collagen gel (model 1) express certain features similar to those of the endothelium *in vivo*..

Optically sliced serial images with a thickness of 0.4 micrometers for the model 1 and 0.9 micrometers for the model 2 were taken under no flow condition. The edge of each slice was traced manually to produce a contour drawing, and the contour data were fed into the modeling process developed for the purpose.

Results of Computation and the Wall Shear Stress Distribution: Velocity vectors calculated for confluent cell models are shown in Fig. 12. Panels A and B show vector distributions calculated for model 1 (monolayer formed on a collagen gel substrate) and model 2 (a monolayer on the glass surface coated with collagen), respectively. We computed the flow fields under some flow conditions. The main flow that followed the cell surface went around the nuclear bulge. Consequently, fluid which passed over the peak of the nuclear bulge came from higher layers of the flow with higher velocity.

The wall shear stress distribution is shown in Fig. 13. Panel A is for the model 1 and Panel B is for the model 2. At the top of the nuclear bulge, shear stress values were two to three times larger than those found on the low areas of the cell surface.

Cell heights were significantly different between the cells cultured on the collagen gel (model 1) and those cultured on the glass surface (model 2). However, the magnitude of shear stress did not differ much between these models. It should be realized that the height of the nuclear bulge was not the only difference between the two models. For example, cell density, configuration and alignment were clearly different, although we did not quantitatively evaluate them in the present study. We have conducted similar computations using more idealistic configurations and reported some preliminary results [53].

CONCLUDING REMARKS

General Structure of Near Future of Bio-Fluid Dynamics

One of our goals of computational studies of physiological flow is to form a system to reconstruct the living system in the computational space and test physiological phenomena in full detail without difficulties which we usually encounter in direct experimental studies. To bring about this task we need to establish a comprehensive model of the physiological flow field. Let us take the blood flow as an example. The circulatory system consists of numerous inter-connected structures. From a mechanical view point, it would be appropriate to start from the heart, because it is the only energy source in the whole circulation. As is well known, the blood expelled from the heart continuously flows from tracts of larger scales to those of smaller scale, and eventually to the capillary network. The length scales of different parts of the circulation is shown in Table 1 in terms of the order of magnitude. The order of the significant length scale varies over a range of 10^{-1} - 10^{-8}. This range is prohibitively wide for the currently available computational methods, and will continue to

Table 1. Representative scales and required resolutions estimated along the streamwise direction in the cardiovascular system

Streamwise Scales (m)

	Heart	Aorta	Artery	Arteriole	Capillary EC Surface
Length Scale	10^{-1}	10^{-2}	10^{-3}	$10^{-4} \sim 10^{-5}$	10^{-6}
Required Resolution Overall	10^{-2}	10^{-3}	10^{-4}	$10^{-5} \sim 10^{-6}$	10^{-7}
Near Wall	10^{-3}	10^{-4}	10^{-5}	$10^{-6} \sim 10^{-7}$	10^{-8}
Streamwise Running (xD)		10^{0}		10^{1}	10^{2}

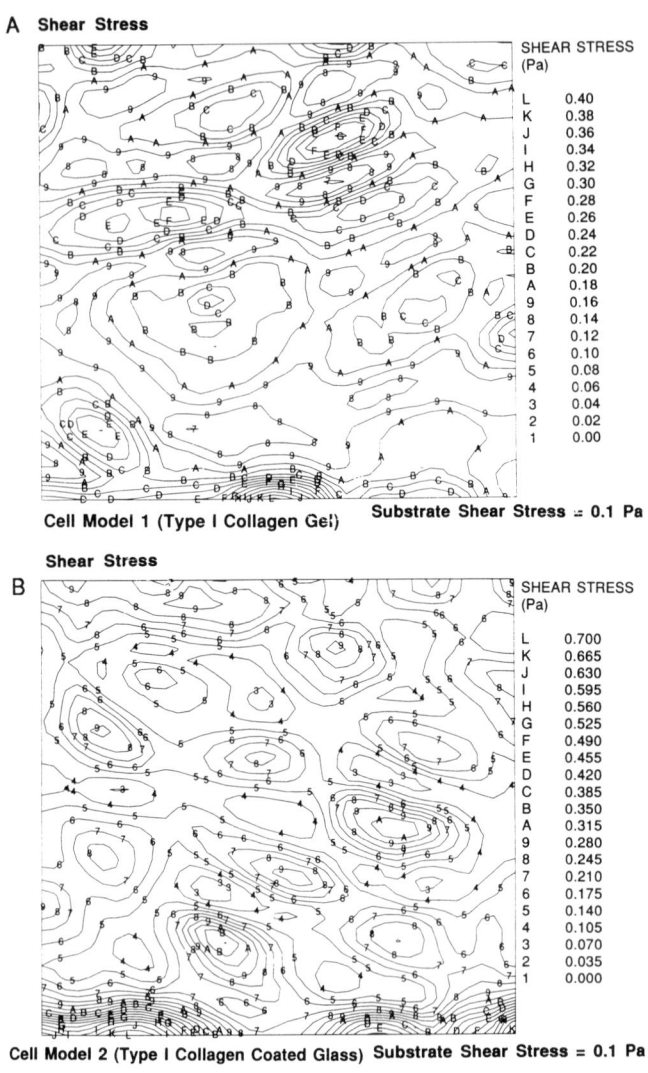

Figure 13. The wall shear stress distribution is shown. Panel A is for the model 1 and Panel B is for the model 2. Wall shear stress distribution patterns again could not be distinguished between the low and the high flow conditions except for the values, so that the results obtained under the low flow condition are shown.

be so in the foreseeable future. We therefore need to divide the system, model its parts separately, perform computation for the divided model, and integrate the obtained results to understand the whole mechanical behavior of the system.

If we divide and model the circulatory system into different parts, we can expect that the model of each part will have appropriate resolution requirements as also shown in Table 1. For example, the aorta has an overall length scale of 10^{-2} m, so that required overall resolution in terms of length would be one order less, that is 10^{-3} m, and 10^{-4} m for near wall phenomena. This range of resolution corresponds to the maximum resolution currently available in the CFD studies. It is, of course, not only the length scale that determines the requirement of resolution. The representative velocity should be taken into account. Other-

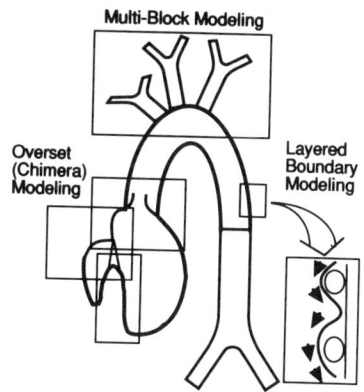

Figure 14. A scheme of multi-block layered boundary CFD modeling. The whole cardiovascular system is divided into many blocks along the flow direction. In the figure, the heart, the aorta, and some bifurcation structures are separately modeled and are integrated. In the heart, some important substructures such as the papillary muscle, valves are modeled in finer resolution and they are overset in the whole heart model (Chimera modeling). Finally, the near wall phenomena on the wall of arteries due to different level of unevenness or irregularities are modeled by a layered approach, for example by considering endothelial cell configuration.

wise, the near wall phenomena may not be fully resolved by the distribution of the computational grid according to the consideration of length scales. For covering this difficulty, it is a usual maneuver to introduce some parametric modeling of the characteristics of the near wall behavior of the fluid. This is typically performed in turbulence modeling by assuming the so-called log law in which the near wall distribution of the flow is modeled by a logarithm.

We can manage these difficulties by introducing two different types of multiplicity of the CFD modeling, namely in the streamwise levels and in the expansion of the boundary. The former has already been explained partly. The whole system is divided into several segments along the flow direction, so that the segments provide the inflow and the outflow condition to each other. The latter is to expand some part of the flow field to give finer resolution so that the contribution of the smaller structures to the global fluid dynamics is fully estimated. As shown in Fig. 14, there are two slightly different methods. Firstly, relatively large substructures can be modeled by the overset or chimera grid techniques. In this technique, a detailed computational model is constructed for a substructure and this is set in the global model exchanging the values of the flow field parameters with each other. Secondly, a model of very near wall phenomena can be constructed to give a lumped parametric contribution to the global flow from a finer structure of the wall. We call it a layered boundary approach.

Conclusion

In this article, general discussions were given on the relevance of the CFD studies of blood flow for correlating its physiological state and the pathological consequences. We emphasized that the extremely complex nature of blood flow including its susceptibility to the geometrical configuration of blood vessels particularly under unsteady conditions should be carefully dealt with. In this sense, the CFD method is the most promising method to reveal the nature of the blood flow and its effects on various physiological and pathological phenomena in the living system.

In the CFD study, the modeling is undoubtedly the most important yet the most time-consuming part. From a series of our attempts to model and to obtain the flow field in the cardiovascular system, three examples were shown in the present article. The construction of a realistic model whose configuration is based on macro- and microscale vascular structures was shown to be necessary for correlating fluid mechanics to the physiological an disease processes. This was thought to allow a finer detail of flow understanding than is presently available.

Another aspect stressed in this article was the visualization of the CFD results. We need to understand the effect of three dimensional structure of the vasculature. This is not only to know the fluid motion in the blood flow, but also for the understanding of the influence of the flow on the wall, particularly for the study of the blood flow with respect to vascular disorders. The four dimensions of the real flow can be understood by the animation technique in which the consecutive changes in the three dimensional distribution of computed values are displayed using 3D rendering with color graphics.

In understanding those graphically presented computational results, we can naturally rely upon the human ability of pattern recognition. The only characteristic in which we are undoubtedly superior to computing machines is, and will forever be, the intuitive ability for pattern recognition. In order to fully utilize this ability, we need to visualize the computational results from different viewpoints, various thresholds of rendering of computed values, and with different temporal scales. By these rendering techniques, we can depict the subtle changes of spatial parameters and can discuss the underlying physics and physiology. This will be accomplished by a combination of novel hardware including the virtual reality technology and the sophisticated 3D rendering software.

It should be noted that the available computer power is still unsatisfactory to examine very complicated flow fields in the cardiovascular system. However, the rapid advancement of computer technology has made it possible to compute fairly complicated flow fields even by a workstation. Therefore the use of a realistic model in conjunction with sophisticated visualization of the flow field should become a mandatory part of an investigation relating fluid dynamic factors to normal and pathological behavior of vascular tissues and cells.

ACKNOWLEDGMENTS

A part of the studies represented in this article was supported by the Research Grant for Cardiovascular Diseases (3A-3) from the Ministry of Health and Welfare of Japan, the Grant-in-Aid for Scientific Research on Priority Areas, "Biomechanics of Structure and Function of Living Cells, Tissues, and Organs" (#04237101), Grant #04454537, and Grant #07680954 from the Ministry of Education, Science and Culture of Japan.

REFERENCES

1. Yoshida, Y., Wang, S., Yamane, T., Okano, M., Oyama, T., Mitsumata, M., Suda, K., Yamaguchi, T., and Ooneda, G., 1990, Structural differences of arterial walls which are either vulnerable or resistant to atherosclerosis, Acta Medica et Biologica 38:1-19.
2. Caro, C.G., Pedley, T.J., Schroter, R.C., and Seed ,W.A., 1978, The mechanics of the circulation, Oxford University Press, Oxford, 341-346.
3. Woolf, N., 1982, Pathology of Atherosclerosis, Butterworth Scientific, London, 187-216.
4. Caro, C.G., Fitz-Gerald, J.M., and Schroter, R.C., 1971, Atheroma and arterial wall shear observation, correlation and proposal of a shear dependent mass transfer mechanism for atherogenesis, Proc. Roy. Soc. London (Biol) 177:109-159.
5. Yamaguchi, T., Hanai, S., Oyama, T., Mitsumata, M., and Yoshida, Y., 1986, Effect of blood flow on the localization of fibrocellular intimal thickening and atherosclerosis at the young human abdominal aorta-inferior mesenteric artery branching (In Japanese), Recent Advances in Cardiovascular Disease 7:97-108.
6. Yamaguchi, T., Nakano, A., and Hanai, S., 1990, Three dimensional shear stress distribution around small atherosclerotic plaques with steady and unsteady flow, In: Mosora, F., Caro, C.G., Krause, E., Schmid-Schönbein, H., Baquey, C., and Pelissier, R., (Eds.) Biomechanical Transport Process, Plenum Press, New York,173-182.

7. Sakurai, A., Nakano, A., Yamaguchi, T., Masuda, M., and Fujiwara, K., 1991, A computational fluid mechanical study of flow over cultured endothelial cells, Advances in Bioengineering, BED-20:299-302.
8. Taylor, T.W., and Yamaguchi, T., 1992, Three-dimensional simulation of blood flow in an abdominal aortic aneurysm using steady and unsteady computational methods, Advances in Bioengineering, BED-22:229-232.
9. Yamaguchi, T., and Taylor, T.W., 1992, A parametrically defined computational fluid mechanical model for the study of the flow in arterial bifurcations, Advances in Bioengineering, BED-22:237-240.
10. Yamaguchi, T., and Taylor, T.W., 1992, Computational fluid mechanical study of the coronary spasm, Advances in Bioengineering, BED-22:333-340.
11. Taylor, T.W., and Yamaguchi, T., 1992, Three-dimensional graphics and computational model construction of vascular chambers using physiological cast measurements, Advances in Bioengineering, BED-22:469-472.
12. Taylor, T.W., Okino, H., and Yamaguchi, T., 1993, Three dimensional analysis of left ventricular ejection using computational fluid dynamics, Bioengineering Conference BED-24:136-139.
13. Yamaguchi, T., Hoshiai, K., Okino, H., Sakurai, A., Hanai, S., Masuda, M., and Fujiwara, K., 1993, Shear stress distribution over confluently cultured endothelial cells studied by computational fluid mechanics, Bioengineering Conference BED-24:167-170.
14. Sakurai, A., Yamaguchi, T., Okino, H., Hanai, S., and Masuda, M., 1993, A method for formulating realistic mathematical models based on arterial casts for the computational fluid mechanical studies on arterial flow and atherosclerosis, Journal de Physique III 3:1551-1556.
15. Taylor, T.W., Okino, H., and Yamaguchi, T., 1993, The effects of supravalvular aortic stenosis on realistic three-dimensional left ventricular blood ejection, Biorheology 30:429-434.
16. Taylor, T.W., Okino, H., and Yamaguchi, T., 1993, Realistic three-dimensional left ventricular ejection determined from computational fluid dynamics, Advances in Bioengineering BED-26:119-122.
17. Yamaguchi, T., 1993, A computational fluid mechanical study of blood flow in a variety of asymmetric arterial bifurcations, Frontiers of Medical and Biological Engineering 5:135-141.
18. Yamaguchi, T., and Taylor, T.W., 1993, Computational fluid mechanical study of the blood flow with moving walls in the cardiovascular system, Theoretical and Applied Mechanics, 42:331-338.
19. Taylor, T.W., and Yamaguchi, T., 1994, Three-dimensional simulation of blood flow in an abdominal aortic aneurysm - steady and unsteady flow cases, Journal of Biomechanical Engineering 116:89-97.
20. Taylor, T.W., Okino, H., and Yamaguchi, T., 1994, Three-dimensional analysis of left ventricular ejection using computational fluid mechanics, Journal of Biomechanical Engineering 116:127-130.
21. Yamaguchi, T., and Taylor, T.W., 1994, Some moving boundary problems in computational bio-fluid mechanics. In: Crolet, J.M., and Ohayon, R., (eds) Computational methods for fluid-structure interaction, Longman Scientific & Technical, Harlow, U.K. 306:198-213.
22. Yamaguchi, T., 1994, Maximum wall shear at the nuclear bulge of endothelial cells and their alignment along the blood flow - a computational fluid mechanical study -, Advances in Bioengineering, BED-28:347-348.
23. Taylor, T.W., and Yamaguchi, T., 1995, Flow patterns in three-dimensional left ventricular systolic and diastolic flows determined from computational fluid dynamics, Biorheology 32:107-117.
24. Friedman, M.H., Deters, O.J., Mark, F.F., Bargeron, C.B., and Hutchins, G.M., 1983, Arterial geometry affects hemodynamics. A potential risk factor for atherosclerosis, Atherosclerosis 46:225.
25. Friedman, M.H., Brinkman, A.M., Qin, J.J., and Seed, W.A., 1993, Relation between coronary artery geometry and the distribution of early sudanophilic lesions, Atherosclerosis 98:193-199.
26. Masawa, N., Glagov, S., and Zarins, C.K., 1994, Quantitative morphologic study of intimal thickening at the human carotid bifurcation: I Axial and circumferential distribution of maximum intimal thickening in asymptomatic, uncomplicated plaques, Atherosclerosis 107:137-146.
27. Vesely, I., Eickmeier, B., and Campbell, G., 1991, Automated 3-D reconstruction of vascular structures from high definition casts, IEEE Trans. Biomed. Eng. 38:1123-1129.
28. Daiguji, H., 1992, Numerical Fluid Dynamics (Suuchi Ryutai Rikigaku) (in Japanese) Chapter 1, (Eds.) Yasuhara, M., and Daiguji, H., University of Tokyo Press, Tokyo, 1992, 4.
29. Fung, Y.C., 1993, Biomechanics: mechanical properties of living tissues, 2nd Ed., Springer-Verlag, New York, 66-72.
30. Perktold, K., Resch, M., and Florian, H., 1991, Pulsatile non-Newtonian flow characteristics in a three-dimensional human carotid bifurcation model, J. Biomechanical Engineering 113:464-475.
31. Perktold, K., Thurner, E., and Kenner, T., 1994, Flow and stress characteristics in rigid walled and compliant carotid artery bifurcation models, Med. Biol. Eng. Comput. 32:19-26.
32. Reuderink, P., 1991, Analysis of the flow in a 3D distensible model of the carotid artery bifurcation. Thesis, Univ. Eindhoven.

33. Bird, R.B., Stewart, W.E., and Lightfoot, E.N., 1960, Transport Phenomena, John Wiley & Sons, New York, 80-81.
34. Yamaguchi, T., Kikkawa, S., Yoshikawa, T., Tanishita, K., and Sugawara, M., 1983, Measurement of turbulence intensity in the center of the canine ascending aorta with a hot-film anemometer, J. Biomechanical Engineering, 105:177-187.
35. Yamaguchi, T., and Parker, K.H., 1983, Spatial characteristics of turbulence in the aorta, Annals of New York Academy of Sciences, 404:370-373.
36. Yamaguchi, T., Kikkawa, S., and Parker, K.H., 1984, Application of Taylor's hypothesis to an unsteady convective field for the spectral analysis of turbulence in the aorta, J.Biomechanics 17:889-895.
37. Yamaguchi, T., Kikkawa, S., Tanishita, K., and Sugawara, M., 1988, Spectrum analysis of turbulence in the canine ascending aorta measured with a hot-film anemometer, J.Biomechanics 21:489-495.
38. Hanai, S., Yamaguchi, T., and Kikkawa, S, 1991, Turbulence in the canine ascending aorta and the blood pressure, Biorheology 28:107-116.
39. Bradshaw, P., Cebeci, T., and Whitelaw, J.H., 1981, Engineering Calculation Methods for Turbulent Flow, Academic Press, London, 37-57.
40. Peyret, R., Taylor, T.D., 1983, Computational Methods for Fluid Flow, Springer-Verlag, New York, 18-140.
41. Fletcher, C.A.J., 1988, Computational Techniques for Fluid Dynamics, Vol I, Springer-Verlag, Berlin, 98-162.
42. Nakahashi, K., and Fujii, K., 1995, Grid Generation and Computer Graphics (Computational Fluid Dynamics Series 6; (Ed.) Murakami S.) (In Japanese), University of Tokyo Press, Tokyo, 1-134.
43. Thompson, J.F., Warsi, Z.U.A., and Mastin, C.W., 1982, Boundary-fitted coordinate systems for numerical solutions of partial differential equations—A review, J. Comp. Physics 47:1-108.
44. Thompson, J.F., Warsi, Z.U.A., and Mastin, C.W., 1985, Numerical Grid Generation Foundations and Applications, Elsevier Science Publishing, New York.
45. Ku, D.N., and Giddens, D.P., 1987, Laser Doppler anemometer measurements of pulsatile flow in a model carotid bifurcation, J. Biomechanics 20:407-421.
46. Fukushima, T., Homma, T., Azuma, T., and Harakawa, K., 1987, Characteristics of secondary flow in steady and pulsatile flows through a symmetrical bifurcation, Biorheology 24:3-12.
47. Fukushima, T., Homma, T., Harakawa, K., Sakata, N., and Azuma, T., 1988, Vortex generation in pulsatile flow through arterial bifurcation models including the human carotid artery, J. Biomechanical Engineering 110:166-171.
48. Thiriet, M., Pares, C., Saltel, E., and Hecht, F., 1992, Numerical simulation of steady flow in a model of the aortic bifurcation, J. Biomech. Engng. 114:40-49.
49. Perktold, K., Rappitsch, G., and Liepsch, D., 1994, Flow and wall shear stress in the human carotid artery bifurcation: computer simulation under anatomically realistic conditions, Advances in Bioengineering BED-28:437-438.
50. Zarins, C.K., Giddens, D.P., Bharadvaj, B.K., Sottiturai, V.S., Mabon, R.F., and Glagov, S., 1983, Carotid bifurcation atherosclerosis: quantitative correlation of plaque localization with flow velocity profiles and wall shear stress, Circ. Res. 53:502-514.
51. Ku, D.N., Giddens, D.P., Zarins, C.K., and Glagov, S., 1985, Pulsatile flow and atherosclerosis in the human carotid bifurcation: positive correlation between plaque location and low and oscillating shear stress, Arteriosclerosis 5:293-302.
52. Davies, P.F., and Tripathi, S.C., 1993, Mechanical Stress Mechanisms and the cell : an endothelial paradigm, Circ. Res. 72:239-245.
53. Yamaguchi, T., 1994, Deformation and alignment of arterial endothelial cells along blood flow (a computational fluid mechanical study), (In Japanese) Trans. Japan Soc. Mech. Engineers B 60:3665-3671.

8

BIOMECHANICAL AND PHYSIOLOGICAL ASPECTS OF ARTERIAL VASOMOTION

N. Stergiopulos and J. -J. Meister

Biomedical Engineering Laboratory
Swiss Federal Institute of Technology
Lausanne, Switzerland

INTRODUCTION

Muscular arteries possess the ability to alter their geometry and apparent mechanical properties by changing the degree of muscular tone. Arterial muscular tone alterations may result from external stimuli (i.e., neural or humoral factors, changes in pressure or flow, etc.) though spontaneous oscillations in muscular tone, in the absence of variations in external stimuli, are also observed. These are usually manifested by low-frequency oscillations in the arterial diameter which, by analogy to small resistance arteries and arterioles, can be termed arterial vasomotion.

An example of vasomotion recorded noninvasively in the human radial artery is shown in Figure 1. For the radial artery the amplitude of the diameter oscillations is typically in the order of 50 to 100 microns, which is several times higher than the diameter excursion due to pulse pressure (typically 20 microns). Such low-frequency components are not observed in mean pressure or the RR intervals. Therefore, the large scale diameter oscillations were thought to be due to spontaneous variations in the smooth muscle tone.

Spontaneous variations in the muscular tone in small, resistance type arteries and arterioles have been known to exist for many years (Johansson and Bohr, 1966). Because dynamic alterations in the intraluminal diameter of small vessels is one of the primary mechanisms regulating blood flow, vasomotion in small arteries and arterioles has been the subject of considerable investigation (Funk and Intaglietta, 1983; Intaglietta, 1991; Oude Vrielink et al., 1989 ; Oude Vrielink et al., 1990).

In contrast to vasomotion in the microvasculature, arterial vasomotion is a phenomenon not widely known or acknowledged. Muscular arteries are typically considered as vessels capable of changing their caliber, but, under basal conditions, they were commonly assumed to remain rather inert. It is characteristic that in his classic paper "Description of the Myogenic Hypothesis", Folkow (1964) wrote "...(vasomotion) is normally present in the precapillary resistance and sphincter sections...By contrast, the smooth muscles in larger vessels...seem to exhibit little or no intrinsic activity...".

Biological Flow, Edited by Michel Y. Jaffrin and Colin Caro
Plenum Press, New York, 1995

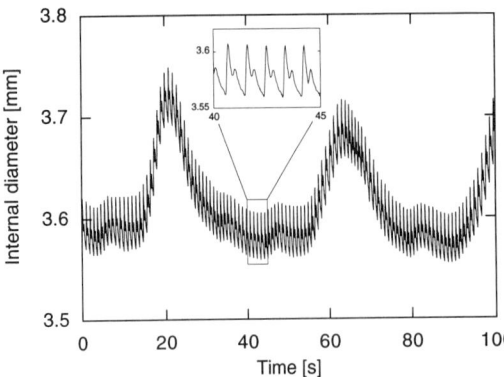

Figure 1. Noninvasive in vivo recordings of the radial artery diameter on a healthy subject under resting conditions. There is clear evidence of slow (~0.023 Hz), high-amplitude vasomotion upon which diameter variations due to pulsatile pressure (70 bpm) are superimposed.

Perhaps the main reasons why arterial vasomotion is a poorly recognized phenomenon is that it is difficult to observe noninvasively and no evident physiological significance to this phenomenon can be ascribed. In small arteries and arterioles vasomotion has a direct impact on perfusion (Secomb et al., 1989), influences filtration (Intaglietta, 1981) and can be thought of as a mechanism for keeping resistance vessels in a continuous state of "alert". Large muscular arteries however, with a diameter larger than 300 microns, have negligible effects on resistance and the control of peripheral perfusion. Thus, at a first glance and even when as pronounced as the example shown in Figure 1, arterial vasomotion may appear to lack physiological significance. Results from studies on arterial vasomotion, however, have shown that vasomotion has a profound effect on the apparent mechanical properties of the arteries. Furthermore, there is evidence that arterial vasomotion is influenced by the local mechanical environment (circumferential and intimal shear stresses) and it has been suggested that vasomotion patterns may relate to the physiological or pathological state of the artery itself.

In this chapter we will discuss arterial vasomotion and its relation to the properties and physiology of muscular arteries. A short introduction to the physiology and biomechanics of muscular arteries will be given first. The core of the discussion will contain a) recent experimental findings and b) theoretical models of arterial vasomotion. Special attention will be given to the role of mechanical forces and biomechanical parameters on vasomotion. The largest part of the material presented is based on original contributions from our laboratory and from collaborators.

BASIC PHYSIOLOGICAL CONSIDERATIONS

In this section we will review briefly the basic physiological and biomechanical aspects of muscular arteries in relation to vasomotion.

Arterial and Smooth Muscle Mechanics

The mechanical properties of arterial smooth muscle are, for most purposes, adequately described in terms of a Hill's "Maxwell" type model (Figure 2a). This simple phenomenological model allows us to relate primary and easily measured mechanical parameters, such as stress (force), strain (length), displacement (diameter) and velocity of displacement, to its three global components: the parallel elastic component (PEC), the series elastic component (SEC), and the contractile component (CC). Sometimes, a fourth compo-

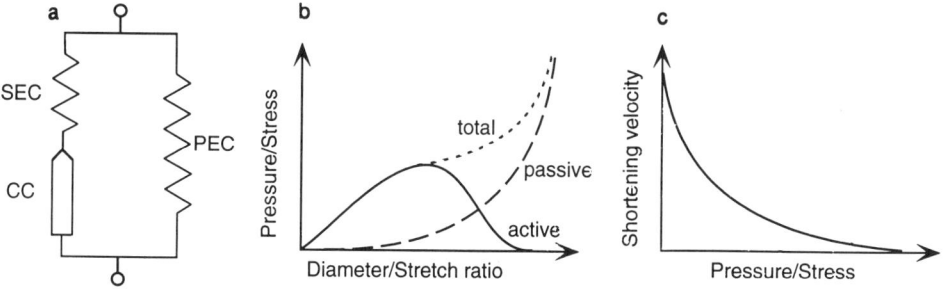

Figure 2. (a) Schematic of a Maxwell model, with the parallel elastic component (PEC) arranged in parallel to a series combination of the contractile component (CC) and the series elastic component (SEC). (b) Qualitative drawing of active (solid line), passive (dashed line) and total (dotted line) stress-stretch ratio curves. (c) Qualitative drawing of muscle length (vessel radius) shortening velocity versus stress (pressure) curve.

nent, called the parallel viscous component (PVC), is added in parallel to PEC to account for internal viscous dissipation effects (Gonzalez-Fernandez and Ermentrout, 1994). Although the Maxwell model was originally proposed to describe the properties of striated or smooth muscle, such types of model are often extended to describe the behavior of the whole vessel (Mulvany and Aalkaer, 1990).

The parallel elastic component describes the response of the muscle/vessel in the absence of muscular tone. In vascular tissues the relationship between stress and stretch (or, equivalently, pressure and diameter) is nonlinear, with the material becoming increasingly stiffer at higher strains. This passive response of an unstimulated muscle/muscular artery is represented by the dashed line in Figure 2b.

The series elastic element couples the contractile mechanism with the intracellular and extracellular structures. Its existence is justified by the additional elasticity observed when the contractile mechanism is engaged. The effect of SEC is best seen at lower strains, below the point of engagement of the PEC (Murphy, 1980). The SEC is characterized by exponential stress-strain curves which appear to be very elastic in vascular tissues. This is quite important because it suggests that under isometric contractions the CC will experience significant shortening (Murphy, 1980).

When stimulated, the contractile component develops force and /or shortens. The static properties of the CC are usually described by the active stress-stretch ratio curves developed under maximum stimulation and under isometric conditions. As shown schematically in Figure 2b, active stress (solid line) is calculated by subtracting passive stress from total stress (dotted line). Active stress shows a clear biphasic dependence on length (stretch ratio), reaching a maximum at a certain optimal length. The bell-shape curve is asymmetric, with a steeper fall after the maximum point. This particular shape is thought to be a consequence of the sliding filament theory (Huxley, 1980; Murphy, 1980). The dynamic properties of the CC are characterized by velocity of shortening-stress (pressure) curves. These appear to be approximately hyperbolic, as shown in Figure 2c.

Smooth Muscle Contraction Mechanisms

The contractile activity of the smooth muscle depends on the level of intracellular free calcium concentration, $[Ca^{2+}]i$. Calcium elicits the phosphorylation of myosin light chains leading to formation of crossbridges between actin and myosin with potential stress development or shortening of muscle length. At equilibrium, the dependence of the degree of phosphorylation on calcium appears to be a sigmoidal one (Rembold and Murphy, 1990).

For a given smooth muscle length, the developed stress depends on the actual number of cross-bridges formed.

The mechanisms for $[Ca^{2+}]_i$ regulation are: 1) transmembranal flux through voltage-dependent channels, receptor operated channels, stretch-activated ion channels, and leakage, 2) release or uptake from intracellular or membrane-bound stores, and 3) extrusion of calcium by the ATP-driven calcium pumps and the Na/Ca-exchange mechanism (Mulvany, 1983).

Origin of Vasomotion

Spontaneous diameter oscillations in blood vessels have been known to exist for more than a century (Jones, 1852). However, the exact mechanisms that cause and control vasomotion are still unknown. In 1964, Folkow hypothesized that rhythmic contractions originate at specialized groups of smooth muscle cells (pacemaker cells) which have a surprisingly low and unstable membrane potential and which possess the ability gradually to build up the potential and subsequently depolarize completely spreading the excitation and contraction to neighboring cells. The frequency of the pacemaker activity is determined by the rate of membrane depolarization which, in turn, is positively correlated with the degree of distention. Thus, there is a close link between vasomotion and the myogenic response, the latter referring to an intrinsic property of arteries which causes them to contract when pressure is raised and to relax when pressure is lowered (Bayliss, 1902; Folkow, 1964). As a consequence, at higher pressures (higher hoop stress) the frequency of spontaneous contractions increases.

The existence of vascular pacemakers was confirmed by Hermsmeyer (1973). Co-lantuoni et al. (1985) suggested that vasomotion originates in pacemaker cells located in the vicinity of bifurcations and spreads in the upstream and downstream direction. Thus vasomotion in a given vessel arises from the combined effect of vasomotion waves originating at upstream and downstream bifurcations. The frequency of oscillations increases towards more peripheral vessels.

A popular theory for the origin of vasomotion is based on the dynamics of transmembranal Ca^{2+} and K^+ ion flux, and specifically on the interaction between voltage-dependent Ca^{2+} channels and voltage-calcium-dependent K^+ channels (Gonzalez-Fernandez and Ermentrout, 1994; Mulvany, 1983). Calcium influx induces depolarization of the cell membrane and elicits the opening of the K^+ channels. Potassium outflux tends to repolarize the membrane. Any inherent delays between the Ca^{2+} and K^+ flux control mechanisms could lead to oscillations. This theory is supported by observations showing that oscillations are abolished by calcium and potassium blockers and enhanced by calcium agonists (Demey et al., 1988).

EXPERIMENTAL STUDIES ON ARTERIAL VASOMOTION

In Vivo Studies of Vasomotion in Human Arm Arteries

Noninvasive measurements of the arterial diameter and wall thickness were performed in the arteries of the human arm by means of a recently developed, high-precision ultrasonic echotracking device (NIUS 2, Omega Electronics, Bienne, Switzerland). Arm arteries, and especially the radial and ulnar arteries, are ideal for this study because they are muscular arteries and sufficiently superficial to facilitate ultrasonic measurements. Vasomotion was observed also in the brachial and digital arteries (Porret et al., 1995). Pressure was measured simultaneously with a photoplethysmographic device (Finapres, Ohmeda, Denver,

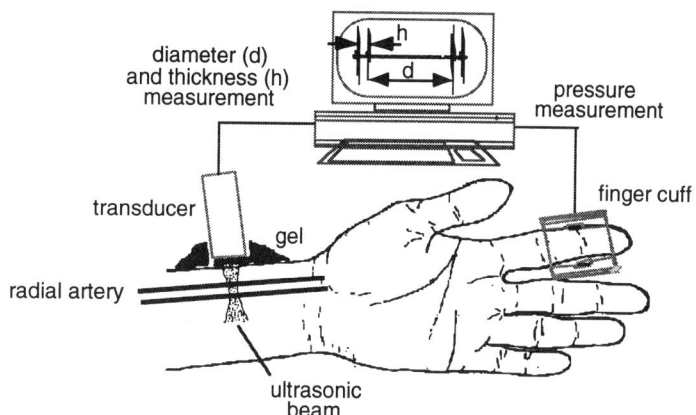

Figure 3. Schematic of the experimental setup used for noninvasive, simultaneous in vivo measurements of arterial diameter, wall thickness, pressure and flow in the human arm arteries. Diameter probe is placed approximately 5 cm proximal to the wrist. Drawing adopted from Tardy (1992).

CO) in the nearby digital artery while flow was measured by means of a continuous Doppler system (Doptek, Montpelier, France). A drawing of the experimental setup is shown in Figure 3. For details about the experimental setup and the devices used see the article by Tardy et al. (1991). With such an experiment it is possible to obtain a long-term record of the diameter as well as the pressure and wall hoop stress. Thus, one may study not only vasomotion, but also compute, on a cycle-to-cycle basis, the influence of vasomotion on the apparent mechanical properties of the artery.

Influence of Vasomotion on the Apparent Mechanical Properties: The simultaneous measurement of diameter, thickness and pressure permits the determination of mechanical properties of the arterial wall, such as the distensibility and incremental modulus of elasticity at any instant and at every pressure level (Tardy et al., 1991). Figure 4 shows the influence of vasomotion on the distensibility-pressure curves of a human radial artery measured in vivo. Distensibility, D, is defined as

$$D = \frac{1}{A}\frac{\partial A}{\partial p} \tag{1}$$

where A is the cross-sectional area of the artery and p is the transmural pressure. Distensibility-pressure curves 1, 2 and 3 in Figure 4b were obtained using the area-pressure curves of corresponding heart cycles 1, 2 and 3, shown in Figure 4a. Heart cycle (1), in the trough of the vasomotion wave, yields an area-pressure curve representative of an artery at a relatively constricted state, which results in a less distensible artery (curve 1 in Figure 4b). Likewise, at the peak of the vasomotion wave (point 3 in Figure 4a), the artery is relatively relaxed and this results in a more distensible artery, as curve 3 in Figure 4b shows. High-pass filtering can eliminate vasomotion but does not change the effect on the apparent mechanical properties, because these depend on the instantaneous slopes of the pressure and diameter signal which, in general, are several orders of magnitude higher than the slow-varying vasomotion signal. The influence of spontaneous changes in muscular tone on the distensibility can be significant. For the example of Figure 4 and at 90 mm Hg the variation in distensibility within the vasomotion wave (trough to peak) was in the order of 70%.

The significance of the results presented in Figure 4 is clear: the apparent mechanical properties of muscular arteries, even under resting conditions, can be quite strongly influenced by spontaneous changes in muscular tone. Thus, an important, and often overlooked, limitation to the assessment of the mechanical properties of muscular arteries in vivo, arises

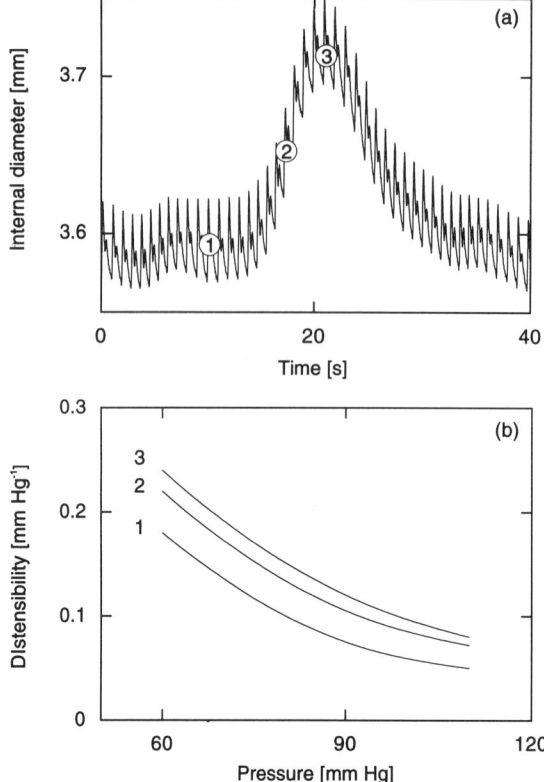

Figure 4. Effect of vasomotion on the distensibility-pressure curves of a human radial artery measured in vivo. Curves 1, 2 and 3 shown in (b) are calculated using diameter and pressure signals of corresponding cardiac cycles shown in (a).

from the lack of reference point with regard to the contractile state of the arterial smooth muscle.

Propagation and Cross-Correlation of Vasomotion in the Human Arm: As mentioned earlier, several studies have advocated the propagative nature of vasomotion in small arteries and arterioles (Colantuoni et al., 1985). Smooth muscle cell contractions are thought to originate in pacemakers cells and subsequently spread from cell to cell causing a slowly propagating vasomotion wave.

To examine whether conduit arterial vasomotion is also able to propagate simultaneous in vivo measurements of the diameter at two locations of the radial artery were obtained (Porret et al., 1995). The experimental setup is similar to the one depicted in Figure 2, except that a second diameter probe was placed in the same radial artery, 5 cm proximal to the control probe. The simultaneous diameter recordings would be used to a) assess the correlation between the vasomotion patterns in the two sites and b) examine whether there exists a consistent phase shift between the two signals.

A typical set of simultaneous mean diameter (averaged over the heart cycle) recordings at the two radial sites is given in Figure 5a. A remarkable correlation between the two diameter signals is observed. At both sites, mean diameter oscillated with relatively constant fundamental frequency of 0.016 Hz. The correlation coefficient, obtained simply by linear regression between the two diameter signals was, for this particular case, 0.92. In general, for all subjects studied (n = 5), the correlation coefficient was larger than 0.90. Therefore, vasomotion in two closely spaced points within the same artery, was always strongly correlated.

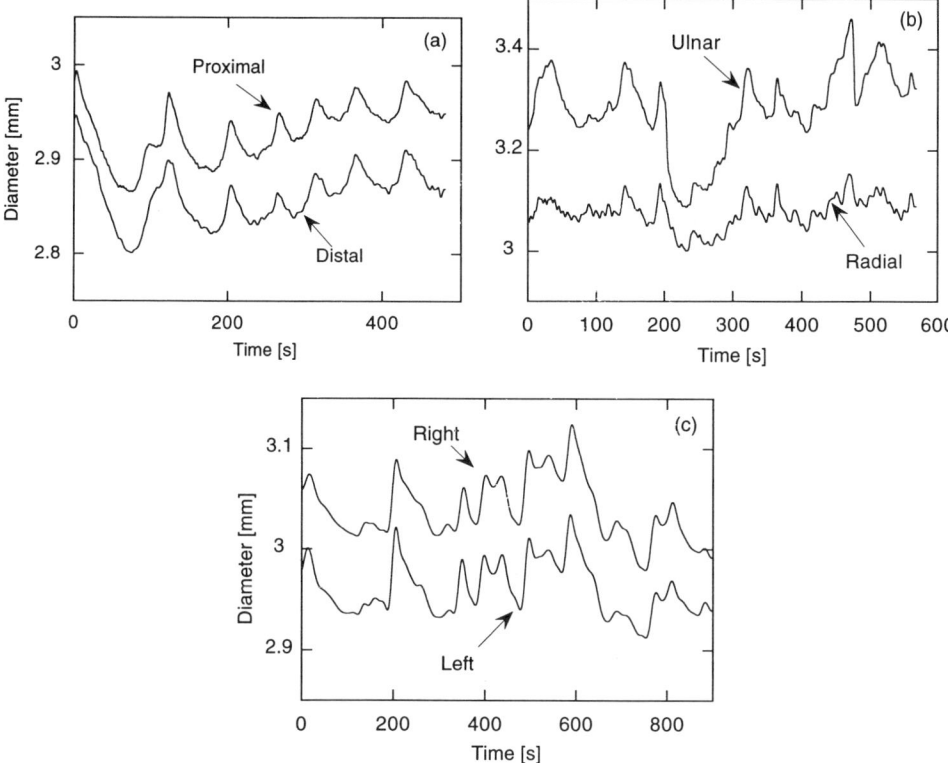

Figure 5. Simultaneous diameter recordings in two sites of the human arm arteries. a) ipsilateral measurements on the radial artery, b) ipsilateral measurements on the radial and ulnar artery, c) contralateral measurements on the left and right radial artery.

To find whether there was a consistent time lag between the two mean diameter signals, a cross-correlation analysis was performed. In all cases (n = 5) the analysis yielded a time delay less than 2 seconds. The delay was not consistent in magnitude and direction and could change sign even for the same subject. If vasomotion were propagative in nature and similar to the one found in arterioles, one would expect consistent patterns in the phase shift between the two mean diameter signals, which was not the case here. Furthermore, there were cases where the signals, although strongly correlated, exhibited basic frequencies which were not the same and progressively diverged with time (Porret et al., 1995).

Porret et al. examined also the correlation between mean diameter signals registered simultaneously in a) the radial and ulnar artery of the same arm (ipsilateral measurements), and b) the left and right radial artery radial (contralateral measurements). Both radial and ulnar artery measurements were done at approximately 5 cm from the wrist. A typical set of simultaneous mean diameter tracings in the radial and ulnar artery is shown in Figure 5b. Diameter fluctuations are more irregular than those presented in Figures 1, 4a and 5a, but irregular vasomotion was as common as the regular one. For the particular case shown in Figure 5b, the amplitude of irregular diameter fluctuations in the ulnar artery was twice as large as the one in the radial artery. However, there was a remarkable synchronicity and similarity in the shape of vasomotion in the two sites. In all cases studied (n = 5), ulnar and radial arteries exhibited similar vasomotion patterns.

A high degree of correlation was observed in vasomotion measured simultaneously in the right and left radial arteries. A characteristic example is shown in Figure 5c, where, again, despite the fairly irregular type of vasomotion, there is a striking similarity and synchronicity in the vasomotion patterns. All fine details in the mean diameter variation are present in both radial arteries at approximately the same time and with roughly the same amplitude. The correlation coefficient was consistently higher than 0.9 for all cases (n = 7) studied.

Considering the high degree of correlation and synchronicity in the vasomotion in the radial and ulnar arteries of both arms, a theory of propagation of the vasomotion waves, similar to the one for arterioles, seems rather unlikely. Similarly, theories suggesting that vasomotion results from local instabilities in the dynamics of smooth muscle cells seem to contradict the data, although one could argue that chaotic vasomotion eventually becomes synchronous because even weakly coupled nonlinear oscillators may exhibit entrainment (Griffith, T. M., personal communication, 1994). The observations, and especially the similarity between left and right radial artery vasomotion, support the notion of a global regulating mechanism. However, Hayoz et al. (1993) found no correlation between mean radial diameter and the R-R signal in six healthy subjects, and thus concluded that the

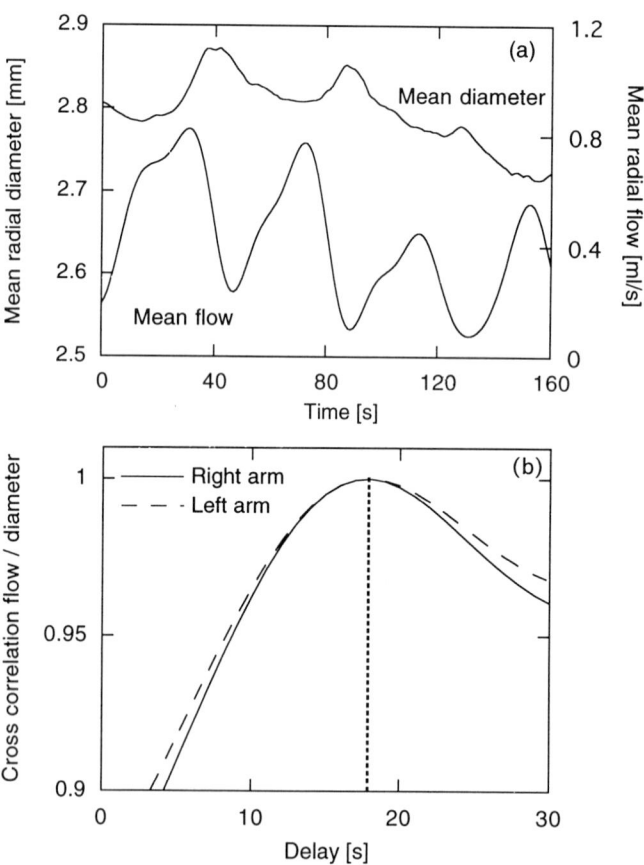

Figure 6. (a) Mean diameter and mean flow measured simultaneously in the left radial artery. (b) Plot of the cross-correlation function between mean diameter and mean flow for the left and radial arteries.

spontaneous contractile activity of the smooth muscle does not result from phasic neural or humoral triggers.

Vasomotion and Flow: In the search for a global regulatory factor for arterial vasomotion, Hayoz et al. (1995) measured simultaneously radial artery diameter and flow. In many cases, but not all, there was a strong similarity in the mean flow and mean diameter signals. The two signals were not synchronous, rather, mean flow led mean diameter by 15-20 seconds. An example of mean radial flow and mean radial diameter tracings is shown in Figure 6a. Mean flow exhibits low-frequency oscillations with approximately the same frequency as mean diameter and with significant amplitude, which in the example of Figure 6a are in the order of 100% within each vasomotion cycle. Since mean pressure remained relatively constant, the observed large scale flow variations were due to analogous variations in peripheral resistance.

The time delay between mean flow and mean diameter was determined by the maximum of the cross-correlation function (Figure 6b). For the example of Figure 6 the maximum was at a positive 18 seconds delay, which means that, in the average, mean flow precedes diameter by 18 seconds. Hayoz et al. (1995) hypothesized that mean diameter variations are merely a dynamic response of the vessel following alterations in the level of mean flow (or shear stress). Thus, what is observed as spontaneous arterial vasomotion is, perhaps, an artery "forced" to oscillate following the variations of intimal wall shear stress. EDRF release, now known as NO synthase, is thought to be the primary mechanism for the signal transduction (14). In vivo experiments in animals have shown that arteries respond to changes in flow, usually by vasodilatation when flow is increased and by vasoconstriction when flow is decreased (29). The characteristic time delay between a step change in flow and the diameter response is in the order of 10 seconds, which compares favorably with the time delay of Figure 6.

In Vitro Studies: Effect of Pressure on Frequency and Amplitude of Vasomotion

Vasomotion can be studied on in vitro preparations of small arteries and arterioles, under controlled perfusion conditions (Achakri et al., 1995; Busse et al., 1982; Osol and Halpern, 1988). The advantage of in vitro studies is the facility in measuring certain quantities (diameter, thickness, mechanical properties, etc.) as well as the independent control of important parameters such as pressure, flow, chemical environment, etc. A major disadvantage is that, for most cases, vasomotion is induced by the addition of vasoactive agents, such as norepinephrine or other calcium agonists, in the superfusion or perfusion solution, and thus the correspondence to in vivo vasomotion may be questionable.

In this section we will discuss the effect of pressure on the frequency and amplitude of vasomotion. The discussion will be based on the findings of an in vitro study by Achakri et al. (1995), although similar experiments have been carried out in the past by other investigators before (Busse et al., 1982; Oude Vrielink et al., 1989; Oude Vrielink et al., 1990). The perfusion tests were run on a Living System Inc. dual perfusion system (LSI, Vermont, USA). For details of the experimental setup and procedures, see the original article by Achakri et al. (1995).

The tests were performed on segments of rat femoral (mean diameter = 770 mm) and on rat mesenteric arteries (mean diameter = 390 mm) perfused with Tyrode's solution at a constant rate of 100 ml/min. Vasomotion was induced and maintained by constrictor agents (NE 10^{-6} M for mesenteric and NE 10^{-6} M + Bay K8644 10^{-7} M for femoral arteries). Figure

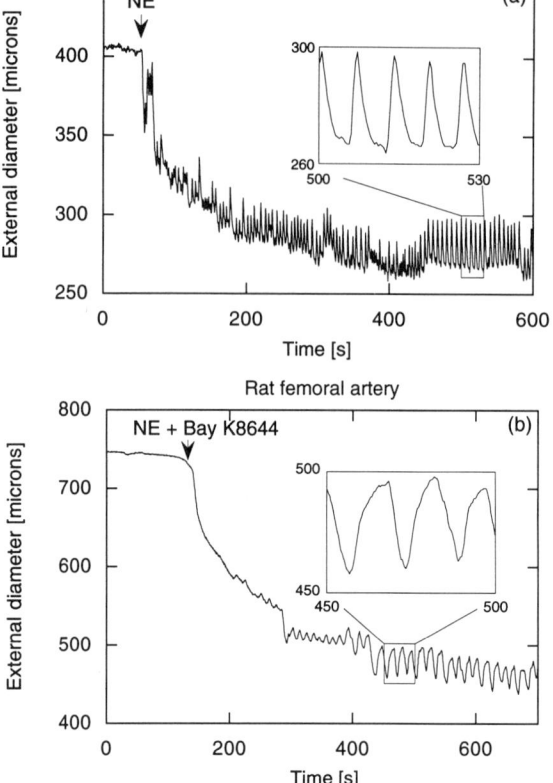

Figure 7. Initial contractile response and development of vasomotion as induced by NE in the mesenteric artery (a) and by NE + Bay K8644 in the femoral artery (b). Note the difference in vasomotion patterns shown in the magnified inserts: the mesenteric artery shows fast relaxations followed by slower contractions, whereas the femoral artery exhibits fast contractions followed by slower relaxations.

7 shows the contractile response and subsequent development of vasomotion induced in the two types of arteries.

To study the effect of pressure on frequency and amplitude of vasomotion, intraluminal pressure was increased or decreased between 0 and 120 mm Hg in a stepwise manner by means of adjustable-height distal fluid columns. A typical example of an experiment in a mesenteric artery showing the evolution of vasomotion following step changes in pressure (increasing sequence) is shown in Figure 8a. Analogous measurements for the femoral, but with a decreasing sequence of pressure, are shown in Figure 8b. The results were reproducible when the sequence (increase or decrease) of pressure changes was reversed.

The dependence of vasomotion frequency and amplitude on pressure is shown in Figures 9a and 9b. Vasomotion frequency shows a monotonic increase with pressure, with a tendency to level off at higher pressures. The linear regression coefficient was $r = 0.83$ and $r = 0.72$ for the rat mesenteric and rat femoral arteries, respectively. These results agree with previous findings (Busse et al., 1982; Harder, 1984; Osol and Halpern, 1988), all of which show a positive correlation between vasomotion frequency and pressure. Although the functional form of the frequency-pressure curves seems to be the same for the two types of arteries, the absolute value of frequency, for a given pressure, appears to be inversely correlated to the arterial diameter. Similar findings were reported by Funk and Intaglietta (1983) for the small arteries and arterioles of the hamster skin fold.

The amplitude of vasomotion exhibits a biphasic (skewed bell-shape) dependence on pressure (Figure 9b), which is quite similar for both arterial groups when vasomotion

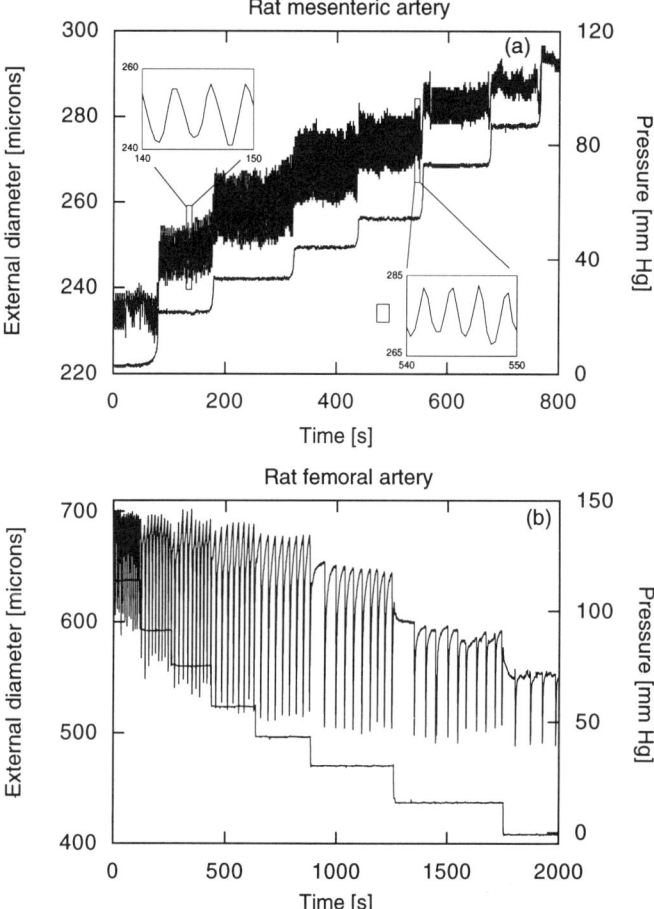

Figure 8. Simultaneous recordings of external diameter and pressure in in vitro preparations of a rat mesenteric (a) and a rat femoral (b) artery. The results show a clear effect of pressure on both frequency and magnitude of vasomotion.

amplitude is normalized to the vessel diameter. Peak vasomotion amplitude is reached at approximately 40 mm Hg for the mesenteric and approximately 55 mm Hg for the femoral arteries. Peak vasomotion amplitude occurs below the usual physiological pressure range which explains why most previous studies reported a monotonic decrease of vasomotion amplitude with pressure (Busse et al., 1982; Osol and Halpern, 1988; Oude Vrielink et al., 1989; Oude Vrielink et al., 1990). Morita-Tsuzuki et al. (1993), however, also reported an inverted U-shape vasomotion amplitude-pressure curve for the cerebral arteries of the cat, although for these arteries peak amplitude was observed for pressures in the range of 60 to 80 mm Hg.

THEORETICAL STUDIES ON ARTERIAL VASOMOTION

Theoretical models of arterial or arteriolar vasomotion can serve two main purposes: first, to propose mechanisms for the development of spontaneous oscillations in muscular

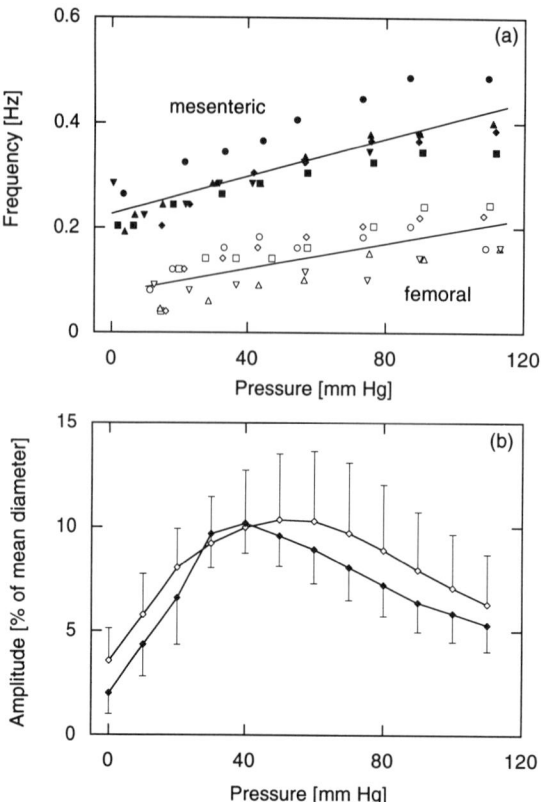

Figure 9. (a) Vasomotion frequency and (b) vasomotion amplitude (mean + standard error) as a function of intraluminal pressure. Results from in vitro experiments in 5 rat mesenteric and 5 rat femoral arteries.

tone, which could be subsequently tested by means of appropriately designed experiments. Second, to facilitate the study of the role of various biophysical or biomechanical parameters on the development and characteristics of vasomotion. The latter constitutes an important advantage of theoretical over experimental models, because it allows the detailed study of certain parameters which are difficult or impossible to manipulate in any in vivo or in vitro preparation.

Several mathematical models have appeared recently in the literature proposing different mechanisms for the origin and dynamics of arterial or arteriolar vasomotion (Achakri et al., 1994; Gonzalez-Fernandez and Ermentrout, 1994; Kireeva and Klochkov, 1982; Ursino and Fabbri, 1992). A common feature of all mathematical models is their basic assumption that spontaneous oscillations in the smooth muscle tone develop as a result of a dynamic system instability. For the purposes of our discussion, the mathematical models of vasomotion will be classified according to the degree of coupling between the vessel's geometry and mechanical properties and the mechanisms controlling smooth muscle tone.

Coupled Models

This category includes theoretical models by Kireeva and Klochkov (1982), Ursino and Fabbri (1992) and Achakri et al. (1994). The origin of vasomotion is claimed to be mechanical instabilities resulting from the interaction between external loads (pressure and flow), the mechanical properties of the artery, and the mechanisms controlling the development of muscular tone. All of the above are lumped-parameter models applied to a segment of an artery. Thus, the geometry and mechanical properties of the arterial segment could, in principle, influence the development and characteristics of vasomotion.

Kireeva and Klochkov (1982) considered a vessel perfused with a viscous fluid from a reservoir under constant pressure through a constant resistance. The vascular material was considered to be a nonlinear modification of a standard viscoelastic solid and it was postulated that the steady state pressure-diameter relationship is not monotonic. The qualitative analysis of the governing equations has shown that self-sustained oscillations may arise. Although the authors have not related the model parameters to specific blood vessels and physiological flow conditions, their analysis concerns rather small resistive blood vessels. On the other hand, the postulated pressure-diameter relationship implies an instantaneous development of the muscular tone as a result of the change of arterial pressure, disregarding the evolution of the active tension generated by the vascular smooth muscle.

A significant contribution towards the analysis of vasomotion, as a phenomenon resulting from mechanical instabilities, is the work of Ursino and Fabbri (1992). Their mathematical model was based on the fact that in microvessels the smooth muscle contractility may be altered by different kinds of external stimuli. Active tension generated by the smooth muscle was assumed to be dependent on state variables, which reproduce the effect of myogenic and metabolic mechanisms. Evolution equations for the state variables have been postulated, reflecting the dependence of the muscular contractility on the instantaneous value and the rate of change of wall tension, as well as on the deviation of the local blood flow from a certain specific value. Computer simulations demonstrated the existence of the oscillations of the vascular caliber, which compared well with the observed vasomotion in microvessels.

Achakri et al. (1994) developed a theoretical model for a single muscular artery, of length L, as a deformable cylinder perfused with a constant inflow Q_i. The artery is ended by a constant resistance Ω representing the hydraulic resistance of all distal vessels. Being a lumped parameter model, all variables were functions of time only. Thus, the continuity equation takes the form:

$$\frac{dV}{dt} = Q_i - Q_o = Q_i - \frac{p}{\Omega} \qquad (2)$$

where $V = \pi r^2 L$ is the intra-arterial volume, p is pressure and Q_o is the outflow. The artery was considered as a membrane made of nonlinear incompressible elastic material and, vasomotion being a quasi-static process, the equilibrium equation was written as:

$$p = \frac{h_o \sigma}{r_o \lambda} \qquad (3)$$

where $\lambda = r/r_0$, is the stretch ratio, r the deformed mid-wall radius, and r_0 and h_0 being the mid-wall radius and thickness, respectively, at zero transmural pressure and in situ length and in the absence of vascular tone. The total Lagrangian circumferential stress was related to the passive and active stress by the relation:

$$\sigma = \sigma_p(\lambda) + f(\lambda)c \qquad (4)$$

The first term on the rhs of Equation 4 gives the passive stress. The second term represents the active stress due to the contraction of the smooth muscle fibers. The function $f(\lambda)$ reflects the dependence of active stress development on λ, as shown by the skewed bell-shape curve of Figure 2b. Furthermore, it was assumed that the active stress is proportional to c defined as the difference between the actual concentration of intracellular Ca^{2+} and a threshold value necessary to initiate the contraction process.

The evolution of the calcium concentration is described by the following balance equation:

$$\frac{dc}{dt} = -\frac{c}{\tau} + \varphi \tag{5}$$

where τ is a time constant that characterizes the decay rate of calcium concentration. To account for myogenic and flow-dependent (EDRF) mechanism, it was postulated that:

$$\varphi = \varphi(p, \tau_w) \tag{6}$$

where τ_w is the wall shear stress. Assuming Poiseuille flow, Equations 2 to 6 can be reduced to a system of two autonomous nonlinear differential equations for the circumferential stretch ratio $\lambda(t)$ and calcium concentration $c(t)$. Achakri et al. proceeded with a dynamic system analysis to derive the conditions for which the system exhibits unstable stationary solutions. It was found that for a range of parameter values the system undergoes a critical Hopf bifurcation which give rise to stable periodic solutions. A necessary condition for instability is the existence of a negative slope in the pressure-diameter relationship. This is shown in Figure 10a. The lower bound of the possible instability zone is point I_1, where the slope of

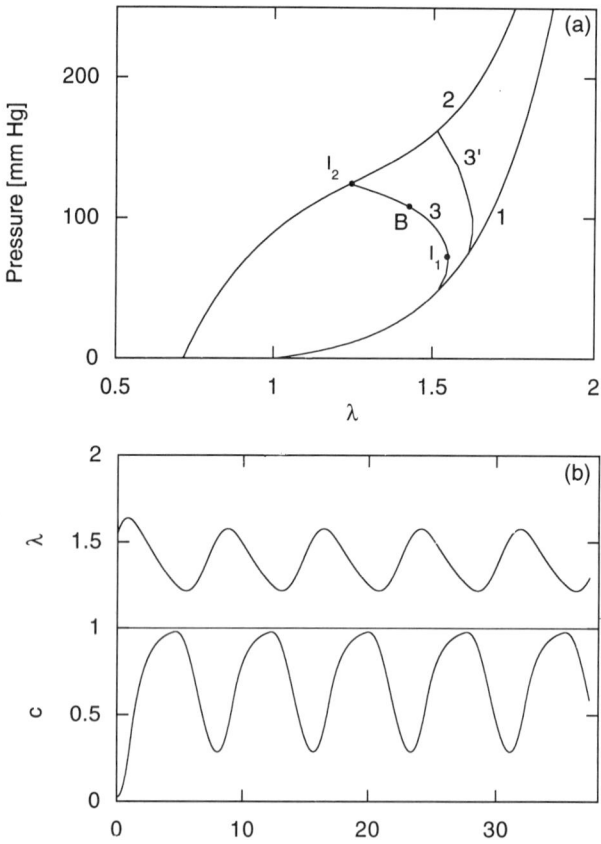

Figure 10. (a) Normalized pressure-stretch ratio curves in an artery under pure myogenic control, as modeled by Achakri et al (1994). Curves 1 and 2 represent the response of a fully relaxed and a fully contracted artery, respectively. Curve 3 defines the state of the artery (stationary solution) at an intermediate contractile state. Pressures p_1 and p_2 define the operational range of the myogenic mechanism. The existence of negative slope indicates a strong feedback regulatory mechanism induced by the myogenic response. (b) Predicted oscillations in the circumferential stretch ratio (diameter oscillations) and the dimensionless calcium concentration for an unstable point in the range B-I_2.

curve 3 becomes infinite. Stability analysis shows that the actual bifurcation point is further up the curve, marked as point B in Figure 10a. The upper limit of the instability zone was point I_2. For purely myogenic control, the position of the bifurcation point B depended on the product of segment length, L, and peripheral resistance, Ω. By varying the parameter $K=L\Omega$, and by tracking the position of the bifurcation point B, a bifurcation diagram was constructed. It was found that a certain critical value, K_{cr}, exists, below which the system is stable. This means that for a given value of the peripheral resistance and for a given inflow, there is a critical segment length below which the artery will not exhibit self-sustained oscillations. Conversely, for an arterial segment of a given length, the model predicts that there is a critical peripheral resistance value below which the segment will be a stable steady state for any inflow.

Achakri et al. attempted also a qualitative assessment of the effect of the shear-stress dependent mechanism on the stability of the system. This was done by letting the smooth muscle cell membrane permeability function, φ, be dependent not only on pressure but also on wall shear stress. This results in a displacement and clockwise shift of curve 3 in Figure 10a, now shown as curve 3'. The result is a reduction in the extent of the negative slope region, which is a prerequisite for the development of spontaneous oscillations. Overall, the shear stress-dependent mechanism was found to have a stabilizing effect.

Uncoupled Models

Recently, Gonzalez-Fernandez and Ermentrout (1994) have developed an elegant theoretical model (abbreviated as GFE model) to explain the origin and mechanisms involved in arterial vasomotion. In the GFE model, vasomotion results from instabilities in the ionic transport mechanisms across the smooth muscle cell membrane. The cause is the interaction of Ca^{2+} and K^+ fluxes, mediated by voltage-gated calcium channels and voltage-calcium-gated potassium channels, respectively. The Ca^{2+}-dependence of the K^+ outflux leads to an inherent time lag between the two ionic transport mechanisms, which in turn gives rise to periodic oscillations in the cytoplasmic calcium concentration.

The basic part of the GFE model, which describes the transport of ions across the smooth muscle cell membrane and which is responsible for the appearance of oscillations, consists of three nonlinear, coupled, first order differential equations defining: 1) the time change in the relative number of open K^+ channels in the cell membrane, 2) the time changes of the membrane potential, and 3) the time changes of the cytoplasmic calcium. The physiological reasoning and specific details in the formulation of the above equations are beyond the scope of this discussion, so the reader is encouraged to consult the original article. Of particular relevance to this discussion, however, is the fact that the three basic equations described above are written for a single smooth muscle cell. Thus, the appearance of vasomotion is not determined or influenced by the geometry and mechanical properties of the vessel (uncoupled model). This constitutes a major difference with the phenomenological models of Ursino and Fabbri and Achakri et al., where vessel geometry and properties are important parameters in the basic equations of the model.

Oscillations in cytoplasmic calcium concentration elicit periodic changes in the phosphorylation levels of myosin light chains which leads to crossbridge formation between overlapping myosin and actin filaments and the development of active stress. Oscillations in the active stress lead to analogous oscillations in the total (hoop) stress and thus, by virtue of Laplace's law, to oscillations in the arterial diameter, therefore vasomotion. Hence, the frequency of vasomotion is set by the frequency of the driving signal (calcium). The magnitude of vasomotion results from an interplay between intracellular events (calcium oscillations, contraction mechanisms), the geometry and mechanical properties of the vessel as well as the external load.

The link between Ca^{2+} oscillations and the development of active force is achieved by means of phenomenological equations describing the relations between the calcium concentration, the level of phosphorylation, and the shortening velocity of the contractile component. Muscle dynamics were represented by a viscoelastic Maxwell model (Figure 2a), with the exception that the stress-velocity of shortening relationship was derived from experimental data (Edman, 1988).

To model the myogenic mechanism, Gonzalez-Fernandez and Ermentrout hypothesized that some of the parameters that control the ionic flux across the membrane are pressure-dependent. Specifically, they assumed that the voltage associated with the opening of half the population of the voltage-dependent calcium channels is a monotonically decreasing function of pressure. This leads to an increase in membrane potential and cytoplasmic calcium concentration, with analogous increase in active stress, as pressure increases. With this assumption, the GFE model was able to reproduce Harder's experimental observations in cat cerebral arteries, showing that an increase in pressure results in an increase in muscle tone and frequency of vasomotion existing only at intermediate pressure and being abolished at low (p << 50 mm Hg) or high (p >> 170 mm Hg) pressures.

Effect of Static and Dynamic Properties of Smooth Muscle Cell on Vasomotion Amplitude: The implications of the GFE model on the existence and pressure-dependence of vasomotion frequency are discussed in detail in the original article. We will extend the discussion into the role of pressure and biomechanical properties on vasomotion amplitude. The variation of vasomotion amplitude with pressure is depicted in Figure 11a (solid line). The amplitude is monotonically decreasing with pressure, which is in agreement with earlier experimental findings (Busse et al., 1982; Harder, 1984; Oude Vrielink et al., 1989). Busse et al. found a 5-fold decrease in relative vasomotion amplitude in the rat tail artery when pressure was increased from 30 to 120 mm Hg, which compares well with the 3-fold decrease predicted by the GFE model.

Direct comparison with the results of Achakri et al. (1995) that showed a bell-shape dependence of vasomotion amplitude on pressure (Figure 9b) cannot be carried out because the theoretical model did not yield oscillations for pressures below 47 mm Hg. To explain the results shown in Figure 9b, Achakri et al. hypothesized that the vasomotion amplitude is correlated with the maximum isometric active stress, $\sigma_{a,max}$, which is also characterized by a bell-shape curve (Figures 2b and 11b). Thus, according to this hypothesis, at each level of pressure (or equivalent stretch ratio), vasomotion amplitude is correlated with the maximum active stress that the vessel can develop.

To examine the above hypothesis, we first altered the static properties of the smooth muscle in the GFE model, by reducing the value of the maximum isometric active stress by approximately a factor of 2 (dashed curve in Figure 11b). This results in a significant decrease in vasomotion amplitude, as shown by the dashed curve in Figure 11a. Thus, there is a clear, global correlation between vasomotion amplitude and the potential of the artery to develop active stress. The correlation of vasomotion amplitude and CC stress development with the maximum isometric stress can be seen in Figure 11c. The three closed loops trace the changes in the length and the active stress developed by the contractile element during the vasomotion cycle, for the pressures of 70, 90, and 110 mm Hg. As pressure increases the loops are displaced to the left due to the shortening of the CC length. We note that, as pressure increases, the mean active stress (average over the vasomotion cycle and shown by the filled circular symbols) increases, which is a manifestation of the myogenic response. However, with an increase in pressure, both the amplitude of the active stress and the amplitude of CC length variations (defined as the height and width of each loop, respectively), are decreasing. This is in direct relation to the corresponding values of $\sigma_{a,max}$. Thus, based on the theoretical

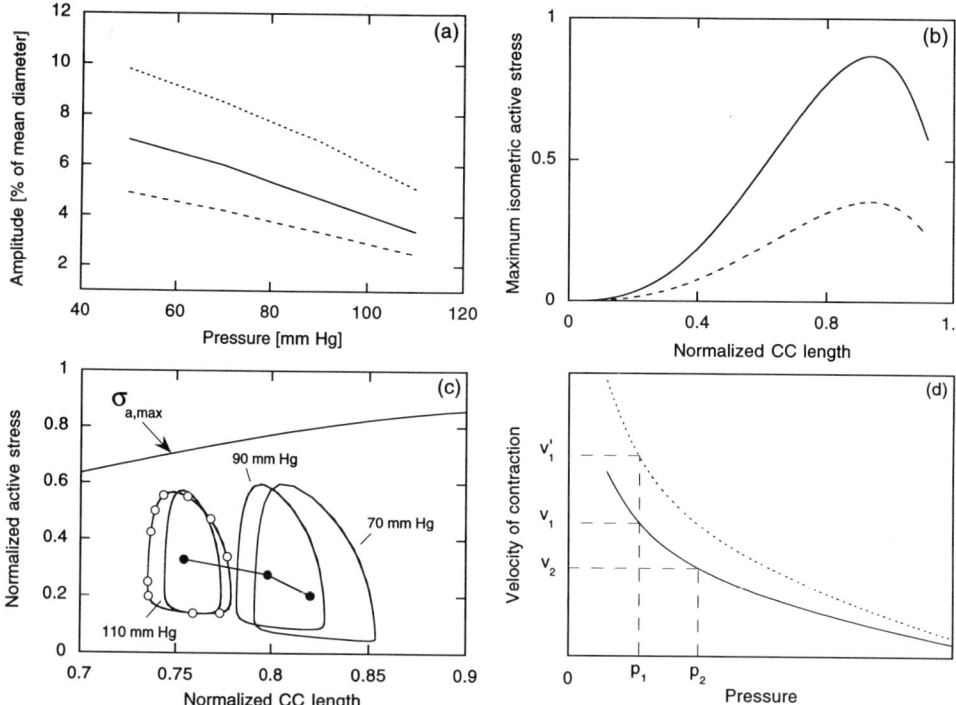

Figure 11. Effect of the active properties of the smooth muscle on vasomotion amplitude. (a) Normalized vasomotion amplitude versus pressure for control (solid line), diminished contraction capacity (dashed line) and augmented velocity of contraction (dotted line). (b) Maximum isometric active stress curves for control (solid line) and diminished contraction capacity (dashed line). (c) Active stress-contractile length loops during vasomotion cycle for different levels of transmural pressure. The second loop at p = 110 mm Hg (line with open circular symbols) represents the active stress-CC length variation for the case of augmented velocity of contraction. Filled circular symbols represent the mean values over the vasomotion cycle. (d) Velocity of contraction-stress curves for control (solid line) and for 50% increase in the velocity of contraction (dotted line).

results shown in Figure 11c, the correlation between vasomotion amplitude and maximum active stress seems to be valid.

We examined also the role of the dynamic properties of the smooth muscle, namely the velocity of shortening-stress curve, on the amplitude of vasomotion. This was done by increasing the velocity of contraction by 50% (Figure 11d). The resulting amplitude of vasomotion-pressure curve is shown in Figure 11a (dotted line). We note a significant increase (in the order of 50%) in vasomotion amplitude for all pressure levels.

Based on the theoretical predictions of the GFE model (Figure 11) and in view of the strong dependence of vasomotion amplitude to the velocity of contraction, we formulated the following hypothesis for the relation between vasomotion amplitude, A_v, frequency, f, and velocity of contraction, v: for a given vasomotion frequency, the velocity of contraction is proportional to the product of vasomotion amplitude and frequency. This heuristic argument results from considering the kinematics of vasomotion as those of a simple harmonic oscillator. Since vasomotion amplitude is the only dependent variable (v and f are functions of pressure but independent of A_v), the above hypothesis is better expressed as:

$$A_v \sim \frac{v}{f} \qquad (7)$$

The implications of the above hypothesis can be easily seen by reference to Figure 11d, where we examine the consequences of an increase in the velocity of contraction and an increase in pressure on the vasomotion amplitude. At a given pressure, p_1, an increase in the velocity of contraction, from the control value of v_1 to the new value of v'_1, means that the amplitude of vasomotion will be increased accordingly, since the frequency of oscillation, $f_1(p_1)$, remains the same. This is in good agreement with the results presented in Figure 11a, which show that proportionality between the increase in contraction velocity (50%) and the increase in amplitude (approximately 50%) is respected. When pressure increases from p_1 to p_2, contraction velocity decreases to v_2 and frequency increases to $f_2(p_2)$ (the GFE model predicts a monotonic increase of frequency with pressure which agrees well with Figure 9a). Therefore, when pressure increases the vasomotion amplitude will decrease. Again, the results presented in Figure 11a support the applicability of our hypothesis.

From a basic mechanics point of view, Equation 7 makes sense. The contractile mechanism is "forced" to follow the oscillations of free intracellular calcium at a rate imposed by the instabilities in the ion exchange subsystem. How far the muscle can swing (amplitude) depends on how fast (velocity) it can catch up with the pace (frequency) of the forcing oscillator. The above hypothesis, however, needs to be viewed with caution because it is evaluated using model predictions and not in real physiological situations.

Higher Order Models, Fractal Dimension and Chaos

Highly irregular patterns of vasomotion (as for example those presented in Figure 5b and 5c) are often recorded in vivo (Intaglietta, 1981; Porret et al., 1995) and in vitro (Griffith and Edwards, 1993). There is substantial evidence that these irregular fluctuations in diameter are chaotic in nature, in the sense that they are governed by a nonlinear deterministic system (Griffith, 1994; Griffith and Edwards, 1994a; Griffith and Edwards, 1994b; Yamashiro et al., 1990). Therefore, the theory for nonlinear dynamical systems or chaos theory is well suited for the analysis of vasomotion. For a more detailed discussion of the possible applications of chaos theory on vasomotion the reader may refer to the review article by Griffith (1994).

Using the nonlinear correlation technique of Grassberger and Procaccia, Griffith and Edwards measured the fractal dimension of histamine-induced vasomotion in isolated rabbit ear resistance arteries (Griffith and Edwards, 1994a; Griffith and Edwards, 1994b). On average this was found to take a value between 2 and 3, although in a small percentage of cases a value between 3 and 4 was obtained. The findings thus suggest that a minimum of 4 independent control variables are needed to define the complexity of the system. The concentration of histamine used to induce tone was found not to affect the fractal dimension even though it markedly influenced the superficial form of the observed responses. Constrictor tone per se can therefore be excluded as a key control variable in this preparation. In contrast, other interventions significantly reduced fractal dimension by inhibiting the two distinct oscillatory subsystems that participate in the genesis of rhythmic activity of this artery type. A fast subsystem (period 5-20 secs) was inhibited by blockade of voltage-dependent Ca^{2+} influx or Ca^{2+}-activated K^+ channels (K_{Ca}). In contrast, the slow subsystem (period 1-5 mins) was inhibited by blockade of Ca^{2+}-induced Ca^{2+}-release from intracellular stores with alkanoid ryanodine (Griffith and Edwards, 1994b). The four dominant control variables that contribute to vasomotion were therefore suggested to be membrane potential, cytosolic $[Ca^{2+}]$, $[Ca^{2+}]$ in the sarcoplasmic reticulum, and the open state probability of K_{Ca} channels. Subsequent nonlinear modeling with 4 coupled differential equations to represent the interactions between these variables have successfully simulated the patterns of vasomotion observed experimentally and also the effects of a variety of pharmacological interventions (Parthimos and Griffith, unpublished).

NEW DIRECTIONS AND CRITICAL QUESTIONS TO BE ANSWERED

In this chapter we have discussed some new experimental findings regarding various physiological and biomechanical aspects of arterial vasomotion. We have also reviewed recent theoretical models of vasomotion and discussed their implications on the origin of the phenomenon. Finally, we have used the model by Gonzalez-Fernandez and Ermentrout to examine the role of active mechanical properties of the smooth muscle on vasomotion amplitude.

Arterial vasomotion affects the apparent mechanical properties of the wall and therefore it has important clinical and diagnostic consequences. The physiological significance of arterial vasomotion is, however, unknown and the underlying mechanisms are still not clear. Further research is needed to understand what causes and controls vasomotion. Pertinent to the topics addressed in this chapter are the following questions:

- Is there a difference between arterial and arteriolar vasomotion? Are the mechanisms implicated in the two types of arteries the same or fundamentally different? The flow dependence of vasomotion presented earlier supports the latter, although this was not true for all cases studied. Neurogenic factors are most influential for the control of smooth muscle tone in large arteries whereas local factors dominate in smallest vessels (Mulvany, 1983). On the other hand, the fractal dimension of arterial vasomotion appears to be of the same order as in arteriolar vasomotion, thus the complexity of the driving mechanisms seems to be the same.
- Which is the preferred model to use? Is vasomotion due only to instabilities in the ion transport and release/uptake mechanisms, as the GFE model suggests? Such models are supported by observations showing spontaneous activity in cell cultures, i.e. in the absence of any coupling with the actual arterial environment (Knot et al., 1991). It is not clear, however, how these models can explain the dependence of vasomotion frequency on arterial size, as clearly demonstrated in Figure 9a. Do models based on instabilities of ion control apply to pacemaker cells or to the average smooth muscle cell of the artery?
- How do models of vasomotion predict the myogenic response? Folkow (1964) and others assumed a tight link between the two phenomena. Does this imply that a model predicting well the characteristics of vasomotion, should also predict well the myogenic response?
- What is the role of flow and the endothelium in vasomotion? Griffith and Edwards have shown that EDRF modulates or suppresses vasomotor activity (Griffith and Edwards, 1993; Griffith and Edwards, 1994a) but does not alter its fractal dimension and thus its intrinsic complexity. This means that EDRF is an external modulating parameter and not a primary control variable that determines the dynamics of the process involved in the genesis of oscillations. Furthermore, the myogenic response appears to be influenced by endothelium function and EDRF activity (Griffith and Edwards, 1990; Harder, 1987), although contradictory results have also been reported in the literature (Falcone et al., 1991).

The above constitute a list, certainly not exhaustive, of questions that need to be addressed in order better to understand the mechanisms of vasomotion. Vasomotion is the manifestation of a series of complex events at the cellular (microscopic) and tissue (macroscopic) level, along which important control mechanisms (myogenic, NO synthase) are implicated. Understanding vasomotion will enhance our basic knowledge of the control mechanisms and functional properties of arteries.

ACKNOWLEDGMENTS

A large part of the material presented in this chapter is the outcome of the Ph.D. work of Dr. Yanik Tardy, the work of graduate students Hassan Achakri and Claude-André Porret, and the diploma thesis work of Alexandre Sauvageot. We thank them all for the help during the preparation of this chapter. We thank also Drs Dimitris Parthimos and Tudor M. Griffith, from the University of Wales, College of Medicine, and Dr. Stephen E. Greenwald, from the London Hospital College of Medicine, Department of Morbid Anatomy, for their critical review of the manuscript.

REFERENCES

Achakri, H., Rachev, A., Stergiopulos, N., and Meister, J.-J., 1994, A Theoretical Investigation of Low Frequency Diameter Oscillations of Muscular Arteries, *Ann. Biomed. Eng.* 22: 253-263.

Achakri, H., Stergiopulos, N., Hoogerwerf, N., Hayoz, D., Brunner, H. R., and Meister, J.-J., 1995, Intraluminal pressure modulates the magnitude and the frequency of induced vasomotion in rat arteries, *J. Vasc. Res.*: 32: 237-246.

Bayliss, W. M., 1902, On the local reactions of the arterial wall to changes of internal pressure, *J. Physiol.* 28: 220-231.

Busse, R., Bauer, R. D., Burger, W., Sturm, K., and Shabert, A., 1982, Correlation between amplitude and frequency of spontaneous rhythmic contractions and the mean circumferential wall stress of a small muscular artery. In: *Cardiovascular System Dynamics,* T. Kenner, R. Busse, and H. Hinghofer-Szalkay, (Eds.), New York: Plenum Press, pp. 363-372.

Colantuoni, A., Bertuglia, S., and Intaglietta, M., 1985, Variations of rhythmic diameter changes at the arterial microvascular bifurcations, *Pfluegers Archiv* 403: 289-295.

Demey, J. G., Boonen, H. C. M., and Strukyer-Boudier, H. A. J., 1988, Rhythmic contractile activity in resistance arteries of spontaneously hypertensive rats. In: *Resistance arteries,* W. Halpern, B. Pegram, J. Brayden, K. Mackey, M. McLaughlin, and G. Osol, (Eds.), NY: Perinatology, pp. 336-341.

Edman, K. A. P., 1988, Double-hyperbolic force-velocity relation in frog muscle fibres, *J. Physiol.* 404: 301-321.

Falcone, J. C., Davis, M. J., and Meininger, G. A., 1991, Endothelial independence of myogenic response in isolated skeletal muscle arterioles, *Am. J. Physiol.* 260: H130-H135.

Folkow, B., 1964, Description of the myogenic hypothesis, *Circ. Res.* 15: 279-287.

Funk, W., and Intaglietta, M., 1983, Spontaneous arteriolar vasomotion, *Prog. Appl. Microcirc.* 3: 66-82.

Gonzalez-Fernandez, J. M., and Ermentrout, B., 1994, On the origin of the vasomotion of small arteries, *Math. Biosci.* 119: 127-167.

Griffith, T. M., and Edwards, D. H., 1990, Myogenic autoregulation of flow may be inversely related to endothelium-derived relaxing factor activity, *Am. J. Physiol.* 258: H1171-H1180.

Griffith, T. M., Hutcheson, I., Randall, M., and Edwards, D. H., 1991, Role of flow in endothelial-mediated responses. In: *Resistance Arteries, Structure and Function,* M. J. Mulvany, C. Aalkjaer, A. M. Heagerty, and N. C. B. Nyborg, (Eds.), Amsterdam: Elsevier, pp. 204-207.

Griffith, T. M., and Edwards, D. H., 1993, Modulation of chaotic pressure oscillations in isolated resistance arteries by EDRF, *Eur. Heart J.* 4(I): 60-67.

Griffith, T. M., 1994, Chaos and fractals in vascular biology, *Vasc. Med. Rev.* 5: 161-182.

Griffith, T. M., and Edwards, D. H., 1994a, EDRF suppresses chaotic pressure oscillations in an isolated resistance artery without influencing their intrinsic complexity, *Am. J. Physiol.* 266: H1786-H1800.

Griffith, T. M., and Edwards, D. H., 1994b, Fractal analysis of the role of smooth muscle Ca^{2+} fluxes in the genesis of chaotic arterial pressure oscillations, *Am. J. Physiol.* 266: H1801-H1811.

Harder, D. R., 1984, Pressure-dependent membrane depolarization in cat middle cerebral artery, *Circ. Res.* 55: 197-202.

Harder, D. R., 1987, Pressure-induced myogenic activation of cat cerebral arteries is dependent on intact endothelium, *Circ. Res.* 60: 102-107.

Hayoz, D., Tardy, Y., Rutschmann, B., Mignot, J. P., Achakri, H., Feihl, F., Meister, J.-J., Waeber, B., and Brunner, H. R., 1993, Spontaneous diameter oscillations of the radial artery in humans, *Am. J. Physiol.* 264: H2080-H2084.

Hayoz, D., Bernardi, L., Noll, G., Weber, R., Porret, C.-A., Passino, C., Wenzel, R., and Stergiopulos, N., 1995, Flow-mediated vasomotion: a potential functional approach to vascular integrity, *J. Hypertens.*: In review.

Hermsmeyer, K., 1973, Multiple pacemaker sites in spontaneously active vascular smooth muscle, *Circ. Res.* 33: 244-251.

Huxley, A. F., 1980, *Reflections on muscle*. Princeton: Princeton University Press.

Intaglietta, M., 1981, Vasomotor activity, time-dependent fluid excange and tissue pressure, *Microvas. Res.* 21: 153-164.

Intaglietta, M., 1991, Arteriolar vasomotion: implications for tissue ischemia, *Blood Vessels* 28: 1-7.

Johansson, B., and Bohr, D. F., 1966, Rhythmic activity in smooth muscle from small subcutaneous arteries, *Am. J. Physiol.* 210: 801-806.

Jones, T. W., 1852, Discovery that veins of bat's wing (which are furnished with valves) are endowed with rhythmical contractility and that onward flow of blood is accelerated by each contraction, *Phil. Trans. R. Soc. Lond.* 142: 131-136.

Khayutin, V. M., 1993, Active arterial function: prompt adaptation of the vascular lumen to the blood flow velocity and viscosity. In: *Contemporary problems of biomechanics*, G. Chernyi and A. Regirer, (Eds.), Moscow: Mir, pp. 142-207.

Kireeva, E. E., and Klochkov, B. N., 1982, Nonlinear model for vascular tone, *Transl. Mech. Compos. Mater. (Russian)* 5: 887-894.

Knot, H. J., de Ree, M. M., Gaehwiler, B. H., and Rueegg, U. T., 1991, Modulation of electrical activity and of intracellular calcium oscillations of smooth muscle cells by calcium antagonists, agonists, and vasopressin., *J. Cardiovasc. Pharmacol.* 18: S7-S14.

Morita-Tsuzuki, Y., Bouskela, E., and Hardebo, J. E., 1993, Effects of nitric oxide synthesis blockade and angiotensin II on blood flow and spontaneous vasomotion in the cat cerebral microcirculation, *Acta Physiol. Scand.* 148: 449-454.

Mulvany, M. J., 1983, Functional characteristics of vascular smooth muscle, *Prog. Appl. Microcirc.* 3: 4-18.

Mulvany, M. J., and Aalkaer, C., 1990, Structure and function of small arteries, *Physiol. Rev.* 70(4): 922-961.

Murphy, R. A., 1980, Mechanics of vascular smooth muscle. In: *Handbook of Physiology*, S. R. Geiger, (Eds.), Bethesda, Maryland: American Physiological Society, pp. 325-351.

Osol, G., and Halpern, W., 1988, Spontaneous vasomotion in pressurized cerebral arteries from genetically hypertensive rats., *Am. J. Physiol.* 254: H28-H33.

Oude Vrielink, H. H. E., Slaaf, D. W., Tangelder, G. J., and Reneman, R. S., 1989, Changes in vasomotion pattern and local arteriolar resistance during stepwise pressure reduction, *Pfluegers Archiv* 414: 571-578.

Oude Vrielink, H. H. E., Slaaf, D. W., Tangelder, G. J., Weijmer-Van Velzen, S., and Reneman, R. S., 1990, Analysis of vasomotion waveform changes during pressure reduction and adenosine application, *Am. J. Physiol.* 258: H29-H37.

Porret, C.-A., Stergiopulos, N., Hayoz, D., Brunner, H. R., and Meister, J.-J., 1995, On the vasomotion of the conduit arteries of the human upper limbs: an in vivo study, *Am. J. Physiol.*: 269: H1852-H1858.

Rembold, C. M., and Murphy, R. A., 1990, Latch-bridge model in smooth muscle: $[Ca^{2+}]_i$ can quantitatively predict stress, *Am. J. Physiol.* 259: C251-C257.

Secomb, T. W., Intaglietta, M., and Gross, J. F., 1989, Effects of vasomotion on micro-circulatory mass transport, *Prog. Appl. Microcirc.* 15: 41-48.

Stergiopulos, N., Meister, J.-J., Achakri, H., Hayoz, D., and Brunner, H. R. (1993) Noninvasive estimation of the properties of the radial artery wall: continuous long-time measurements and the influence of vascular tone. *ASME Summer Bioengineering Meeting*, Breckenridge, Colorado.

Tardy, Y., Meister, J. J., Perret, F., Brunner, H. R., and Arditi, M., 1991, Non-invasive estimate of the mechanical properties of peripheral arteries from ultrasonic and photoplethysmographic measurements, *Clin. Phys. Physiol. Meas.* 12(1): 39-54.

Tardy, Y. 1992 *Non-invasive characterization of the mechanical properties of arteries*. Ph.D. thesis, Swiss Federal Institute of Technology, Lausanne.

Ursino, M., and Fabbri, G., 1992, Role of the myogenic mechanism in the genesis of microvascualr oscillations (vasomotion): analysis with a mathematical model., *Microvasc. Res.* 43: 156-177.

Yamashiro, S. M., Slaaf, D. W., Reneman, R. S., Tangelder, G. J., and Bassingthwaighte, J. B., 1990, Fractal analysis of vasomotion, *Ann. N. Y. Acad. Sci.* 591: 410-416.

9

ARCHITECTURE AND HEMODYNAMICS OF MICROVASCULAR NETWORKS

T. W. Secomb,[1] A. R. Pries,[2] and P. Gaehtgens[2]

[1] Department of Physiology, University of Arizona
Tucson, Arizona 85724
[2] Dept. of Physiology, Freie Universität Berlin
Arnimallee 22, D-14195 Berlin, Germany

INTRODUCTION

The main function of the circulation is to transport materials between different parts of the body. Transport over large distances is accomplished by convection, in blood flowing through large vessels. Exchange of materials between blood and tissues occurs mainly over short distances in the peripheral vascular beds, which consist of numerous very small vessels (the microcirculation). These microvessels provide a large surface area for exchange, and bring blood into close proximity to nearly all parts of most organs. Transport at this microscopic level occurs by diffusion, by active cellular transport, or by convective motion of water through microvessel walls.

The ability of the circulatory system to provide the necessary capacity for exchange of materials, while not placing an excessive mechanical load on the heart, depends crucially both on the architecture of the microcirculation and on the mechanics of blood flow through it. These areas have been studied for many years, but remain incompletely understood. Here, we review basic aspects of microvascular architecture and hemodynamics, and discuss several recent investigations.

NETWORK ARCHITECTURE

Basic Features

The largest conduit vessels (the major arteries and veins) are connected to the smallest microvessels (capillaries) via a hierarchical network of branching vessels. On the arterial side, the blood flow is subdivided among an increasingly large number of successively narrower and shorter segments until it reaches the capillaries; this pattern is reversed on the venous side. Vessels with diameters less than about 300 μm are generally referred to as microvessels and form the microcirculation.

Networks of microvessels exhibit a high degree of topological and geometrical heterogeneity. The number of segments forming each complete pathway through a network varies considerably, and the lengths and diameters of segments at corresponding positions in these pathways also vary. Such architectural heterogeneity is, to some extent, an inevitable consequence of the functions of microvascular beds. The need to supply regions that are more or less distant from the major feeding and draining vessels implies heterogeneity of pathway lengths. The structures of microvascular beds must be capable of continual adaptation in response to growth and changing functional demands. Maintenance of a strictly symmetric architecture would preclude this adaptability, since addition or removal of individual segments would result in loss of symmetry.

The geometrical arrangement of microvessels varies considerably from tissue to tissue, reflecting differences in tissue structure and function (Wiedeman et al., 1981). The mesentery, a thin sheet-like structure, contains a relatively sparse, almost two-dimensional network of microvessels (Zweifach, 1937; Frasher and Wayland 1972; Lipowsky and Zweifach, 1974; Ley et al., 1986). Several skeletal muscles, such as the spinotrapezius and sartorius, are also sheet-like structures, but contain denser networks of capillaries. The capillaries are typically aligned parallel to muscle fibers, with arterioles and venules running across fibers (Engelson et al., 1985; Koller et al., 1987). The networks surrounding alveoli in the lung are extremely dense, almost forming continuous sheets (Fung, 1984). In cardiac muscle and thicker skeletal muscles, capillaries extend throughout the tissue, forming a three-dimensional array (Potter and Groom, 1983). Kidney, spleen and liver microcirculations show specialized features associated with these organs' functions (Wiedeman et al., 1981). Even for a given tissue, network structures may vary significantly with species. For example, arcading microvessel structures are observed in the cat mesentery (Lipowsky and Zweifach, 1974) while they are absent in the rat mesentery (Ley et al., 1986).

As a result of the large number of microvessels, the heterogeneity of their arrangement, and differences between tissues, it is has proved difficult to develop methods for describing microvascular network architecture which are both accurate and generally applicable. Complete, segment-by-segment descriptions of restricted regions are possible, but are specific to the region under investigation. A more generally useful approach is to introduce a classification of the segments based on structural and/or functional criteria, and to describe network architecture with reference to this classification (Popel, 1987). In this way, characteristic features of network architecture can be identified, despite the networks' heterogeneity. Several such classification methods have been used, and they reveal differing aspects of network architecture, as described next.

Methods for Describing Network Architecture

The classic study of vascular anatomy by Mall (1888) was based on a classification of vessels according to morphological characteristics. Mall's widely quoted data (e.g., Burton, 1972) give a broad picture of the typical numbers and dimensions of different types of vessels, and provide a basis for understanding the overall distributions of hemodynamic variables according to vessel type throughout the circulatory system. For instance, they show that the majority of blood volume resides on the venous side of the circulation, while the largest contribution to flow resistance occurs on the arterial side, in the arterioles. The velocity of blood flow decreases with decreasing vessel diameter on both the arterial and venous sides, and is lower on the venous side than in corresponding vessels on the arterial side.

In order to describe the architecture of microvascular beds in more detail, a numerical classification of vessels is desirable. In one frequently used scheme applicable to arteriolar and venular trees, described as centrifugal (Popel, 1987), successive branches of the arteriolar tree are assigned increasing numbers starting from a main feeding vessel, based on a combination of topological and morphological criteria. Although simple in principle, this scheme has the disadvantage of requiring a degree of subjective judgment, making it difficult to apply in a systematic way to large networks.

In an effort to overcome this limitation, Fenton and Zweifach (1981) applied a centripetal scheme, previously used in the description of river systems by Horton (1945) and Strahler (1952). The capillaries are assigned order one, and higher numbers are assigned by proceeding towards the feeding vessel, retaining the higher order when segments of unequal orders converge, and increasing the order by one when equal orders converge.

This so-called "Strahler" scheme provides a basis for quantitative analysis of network architecture. For example, the ratio of numbers of vessels at successive orders is found to be roughly constant through several branching orders, and depends on the degree of asymmetry in the topological structure. Similarly, the mean lengths and diameters of vessel segments typically change by roughly constant ratios from one order to the next (Fenton and Zweifach, 1981). Thus, segment numbers, lengths and diameters are proportional to a quantity of the form R^n where n is the Strahler order, and R is the ratio of the numbers, mean lengths or mean diameters of segments from one order to the next. These three ratios provide a concise characterization of a network's architecture, and can be used as a basis for stochastic generation of simulated networks with similar properties (Dawant et al., 1986; Levin et al., 1986).

However, the very conciseness of the Strahler-based description necessarily implies that it provides only a partial characterization of network anatomy. The number of Strahler orders in a typical network is quite small, and many very disparate networks can have the same distribution of segment numbers with respect to Strahler order (Gaehtgens et al., 1986).

An alternative classification method, using "generation number," was proposed by Gaehtgens et al. (1986) and Ley et al. (1986). In this method, the main feeding vessel is assigned generation one, and the generation number is increased by one at every diverging bifurcation (i.e., generation). The range of generation numbers is typically considerably larger than the range of Strahler order, due to the asymmetric topological structure of microvascular networks which contain pathways consisting of very different numbers of vessel segments. The distribution of segments by generation number contains substantially more information about the topological structure of the network than the distribution with Strahler order (Gaehtgens et al., 1986). Generation numbers of functionally similar vessels, such as capillaries, may vary widely. This variation in generation numbers is functionally significant, since hematocrit is positively correlated with generation number in capillaries, as discussed below (Pries et al., 1992b).

The classifications described above have been developed to describe the topological structures of vessel trees not containing cross-connections between arterioles or between veins, which result in closed loops (arcades). Arcading systems are thought to provide more uniform perfusion of distal microvessels under varying hemodynamic conditions (Mayrowitz, 1986). While arcade systems in different tissues have been qualitatively described for some time (e.g., Lipowsky and Zweifach, 1974), numerical classifications for such systems are not as simple and general as those for tree-like structures (Engelson et al., 1985; Popel et al., 1988).

Network Architecture of the Rat Mesentery

The mesentery is a thin sheet-like tissue with a relatively sparse microvascular bed, allowing its vasculature to be visualized relatively easily, with a complete segment-by-segment description of its structure (Figure 1). For this reason, it has been used in several studies

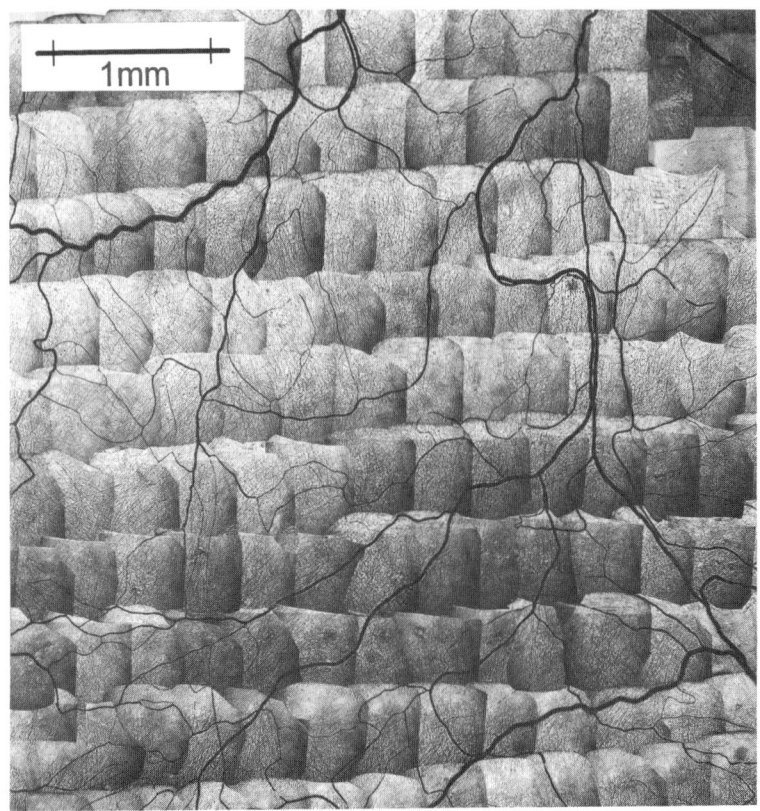

Figure 1. Photomontage of a microvascular network in the rat mesentery, obtained as described by Ley et al. (1986).

of network architecture, and is probably the best characterized microvascular bed in this respect.

Ley et al. (1986) described the architecture of complete microvascular networks in the rat mesentery. Vessel segments connecting two diverging bifurcations were described as "arteriolar," segments linking converging bifurcations were termed "venular," and segments connecting diverging to diverging bifurcations were termed "av-segments." In the following, such segments will be described as "arterioles", "venules" and "capillaries" respectively for convenience, although it must be remembered that these vessel types do not necessarily correspond to a classification based on morphology. Arterioles and venules were classified numerically according to the generation method described above. For capillaries, generation numbers with respect to both the arteriolar and the venular trees were determined. An example of the distribution of vessels with generation number is shown in Figure 2a.

Geometric properties of segments show a high degree of variability. The standard deviation of arteriolar segment length almost equals the mean length of 0.35 mm, and similar results were found for venules and capillaries (Ley et al., 1986). Diameters range widely in accordance with results obtained in other tissues (Potter and Groom, 1983); for instance, capillaries were observed with diameters ranging from 5 to 22 µm. An example of the variation of mean diameter with generation number is shown in Figure 2b. At each generation, venules are consistently larger in diameter than arterioles, which are in turn larger than capillaries. As might be expected, arteriolar and venular diameters tend to decrease with increasing generation, while mean capillary diameters even exhibit a slight increase with generation number.

The generation numbers of capillaries vary widely, from two to twenty, with a mean of about twelve (Ley et al., 1986). Arteriolar and venular generation numbers sometimes differ substantially, but are positively correlated. The distribution of capillaries according to generation number provides a measure of the asymmetry of the arteriolar (or venular) tree. In a completely symmetric tree, all capillaries would have the same generation number, while in the most extreme case of asymmetry, all capillaries would branch off a single parent vessel, giving exactly one capillary segment with each generation number up to the maximum for the network. Hypothetical trees with intermediate degrees of asymmetry may be generated by a stochastic process (Ley et al., 1986) in which additional branches are successively added to a network at randomly selected existing terminal segments ("random-terminal branching") or at randomly selected segments throughout the tree ("random-segment branching"). The latter method produces a more asymmetric tree. Actual distributions of capillaries are found to be intermediate between the expected distributions for random-terminal and random-segment trees. The lower generation numbers follow the random-terminal model, while higher generations correspond more closely to the random-segment model. These findings presumably reflect the biological processes governing growth and adaptation of microvascular networks, but their significance in this regard has not been established.

These descriptions of microvascular architecture provide a basis for investigations of network hemodynamics, as described below.

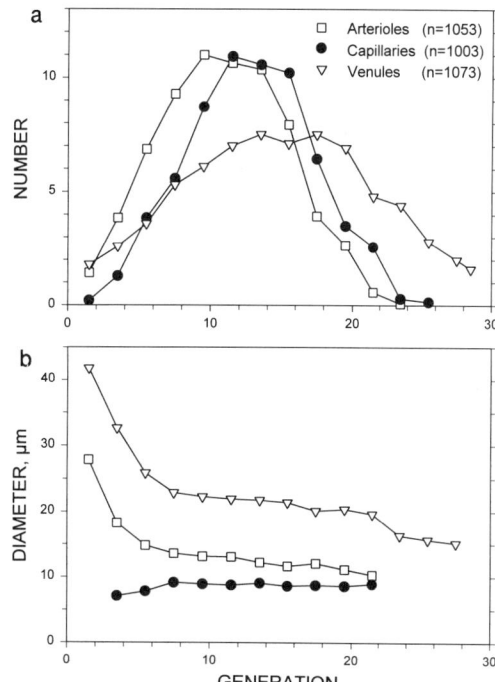

Figure 2. (a) Distribution of numbers of arteriolar, capillary and venular vessel segments with generation number. (b) Mean diameter of segments at each generation. For both parameters, mean values were determined for all segments fed by the main arteriolar inputs in seven microvascular networks in rat mesenteric preparations. For details of the procedures used to obtain these data, see Ley et al. (1986) and Pries et al. (1990).

NETWORK HEMODYNAMICS

Factors Influencing Blood Flow in Microvascular Networks

When a microvascular network is perfused with blood at a given driving pressure, the total blood flow and the distribution of flow within the network are determined by its architecture and by the flow behavior of blood within it. Blood is not a homogeneous fluid, and its flow properties are strongly influenced by the cells it contains in suspension, particularly red blood cells. To understand the dynamics of blood flow in microvascular networks, it is necessary to consider first the composition of blood and its rheological behavior in narrow tubes and in their branch points. Data on these aspects, together with information on network architecture, provide a basis for interpretation and quantitative analysis of the dynamics of blood flow in networks, as discussed in the following sections.

Blood Flow in Microvessels

Mechanics of Blood's Constituents: Blood is a concentrated suspension of cells. The suspending medium is plasma, a solution of proteins, electrolytes and other substances, which is an incompressible, virtually newtonian fluid. Red blood cells (erythrocytes), white blood cells (leukocytes) of several different types, and platelets are the main suspended elements. Normal human blood has a hematocrit (volume fraction of red cells) of about 45%, and so red cells strongly influence blood's flow properties. Blood cell dimensions are comparable to microvessel diameters, and a continuum description of blood's rheological properties, while appropriate for large vessels, is not adequate to describe blood flow in microvessels.

In the absence of external stresses, a human red cell is a biconcave disk, approximately 8 μm in diameter and 2 μm thick, consisting of a thin viscoelastic membrane surrounding a viscous incompressible cytoplasm (Skalak, 1976; Hochmuth and Waugh, 1987). The membrane consists of a lipid bilayer and a cytoskeleton (network of protein molecules). Its elastic shear modulus is very much less than its modulus of isotropic dilation and so the membrane shears readily but resists area changes. Also, its bending resistance is small unless very small radii of curvature are involved. Consequently, red cells are highly deformable, as long as changes in surface area or volume are not required. This allows them to pass through capillaries with diameters much less than 8 μm.

White blood cells are much stiffer than red blood cells. They are typically smaller in diameter than red cells, but have volumes up to twice as large. Despite the fact that they are much less numerous than red cells, white cells may contribute significantly to microvascular flow resistance (Schmid-Schönbein et al., 1981). Their role in network hemodynamics has been analyzed by Fenton et al. (1985) and Warnke and Skalak (1990). Platelets are much smaller than red cells, and do not contribute significantly to flow resistance.

Blood Flow in Narrow Tubes. From a hemodynamic point of view, a crucial property of a microvessel is the resistance it presents to blood flow, defined as the ratio of the driving pressure Δp to the volume flow rate Q. For steady laminar flow of a Newtonian fluid in a cylindrical tube, Poiseuille's Law gives:

$$Q = \frac{\pi}{128} \frac{\Delta p\, D^4}{\mu L} \qquad (1)$$

where L is the tube length, D is the diameter and μ is the fluid viscosity. In microvessels, however, μ is not a known constant, because blood does not behave as a continuum. Resistance to blood flow in microvessels is conveniently described in terms of the *apparent viscosity*, obtained by rearranging equation (1):

$$\mu_{app} = \frac{\pi}{128} \frac{\Delta p\, D^4}{Q L} \quad (2)$$

The *relative apparent viscosity* is $\mu_{rel} = \mu_{app}/\mu_p$, where μ_p is the viscosity of the suspending fluid (plasma).

For glass tubes with diameters below about 500 µm, apparent viscosity is found to decline to levels substantially lower than the bulk viscosity. This is known as the Fåhraeus-Lindqvist effect (Martini et al., 1930; Fåhraeus and Lindqvist, 1931; Pries et al., 1992a). In the diameter range 5 µm to 10 µm, μ_{rel} of human blood is below 1.3, i.e., the apparent viscosity is less than 30% above that of plasma. Below this diameter range, μ_{rel} increases rapidly, due to constraints on the passage of red cells resulting from their incompressibility and conservation of their membrane area (Halpern and Secomb, 1989). Apparent viscosity depends on hematocrit, increasing with increasing hematocrit, particularly strongly at hematocrits above 45% (Pries et al., 1992a). At very low flow velocities, μ_{rel} is dependent on velocity, both in very narrow tubes in which red cells flow in single file (Secomb, 1987), and in slightly larger tubes in which aggregation and sedimentation of red cells occur (Reinke et al., 1987). However, this dependence is weak at moderate and high flow velocities.

The reduced apparent viscosity in narrow tubes results primarily from the tendency of red cells to migrate away from the tube wall in flow, creating a layer of zero or low hematocrit adjacent to the wall. Removal of red cells from this region, where fluid shear rates are highest, results in reduced flow resistance. A further consequence is a dynamic reduction of hematocrit, known as the Fåhraeus effect (Fåhraeus, 1928). Because the slowly moving fluid near the wall consists mainly of plasma, mean red cell velocity exceeds mean blood velocity. Consequently, mean transit time of red cells is less than overall mean transit time, and so the concentration of red cells within the tube (tube hematocrit) is less than the concentration in the blood entering or leaving the tube (discharge hematocrit).

The low apparent viscosity of blood in narrow tubes is generally considered to imply a substantial reduction in resistance to blood flow in the peripheral circulation, compared to the resistance to flow of a homogeneous fluid with the same bulk viscosity. It should be emphasized, however, that the experimental data referred to above were obtained using uniform glass tubes. Due to technical difficulties, few direct measurements of apparent viscosity of blood in microvessels in living tissues have been carried out. Lipowsky et al. (1978) performed such measurements and obtained apparent viscosity values substantially higher than those observed in glass tubes with corresponding diameters. This discrepancy will be discussed later.

Models for Blood Flow in Narrow Tubes: The small size of microvessels implies that the Reynolds number is small, and inertial effects can generally be neglected. However, the particulate nature of blood has to be taken into account, and the known mechanical properties of individual red blood cells provide a basis for quantitative models.

Most progress has been made in the case of single-file flow, which typically occurs in capillaries with diameters of 6 µm or less. Red cells deform to fit into the capillary, and are surrounded by a sleeve of plasma. The cell shapes and the width of the plasma layer depend on the mechanics of cell deformation and plasma flow. Theoretical analyses were developed by Lighthill (1968) and Barnard et al. (1968), assuming axisymmetric shapes and using lubrication theory to describe the plasma flow in the gap between the cell and the wall.

Further developments, with more realistic representations of red cell mechanics, were made by Zarda et al. (1977), Secomb et al. (1986) and Halpern and Secomb (1989), and are reviewed by Secomb (1991, 1995). Predictions of μ_{rel} for tube diameters ranging from 3 μm to 8 μm agree well with available experimental data from glass tubes. Actual red cell shapes in capillaries are not axisymmetric, and exhibit continuous "tank-treading" motion of the membrane relative to the cell (Gaehtgens and Schmid-Schönbein, 1982). However, the effects of asymmetry and tank-treading on apparent viscosity seem to be slight (Hsu and Secomb, 1989).

In larger microvessels, red cells do not usually flow in single file, and their motion is more difficult to model. A simple two-layer model, consisting of a central core with viscosity equal to bulk blood viscosity and a peripheral plasma layer of width 1.8 μm, predicts μ_{rel} values close to those observed, for diameters from 30 μm to 1000 μm (Secomb, 1995). However, the width of the cell-free layer is a fitted parameter in this model. In reality, the layer is generally cell-depleted rather than cell-free. Adequate theories for the variation of hematocrit near the wall, taking into account the deformability and high concentration of red cells, are not available. For diameters below about 30 μm, even this two-layer model is unlikely to be realistic, since the core cannot be well approximated as a continuum.

Blood Flow in Bifurcations

In a microvascular network, vessel segments meet at bifurcation points. Each bifurcation generally connects three segments. At "diverging" bifurcations, flow from a parent vessel is divided between two daughter vessels. Because of non-continuum behavior, the partition of red cells between the daughter vessels is generally not proportional to the partition of total flow, resulting in unequal hematocrits in the daughter vessels ("phase separation"). Consequently, microvessel hematocrits can vary within a network, with significant consequences both for its hemodynamics and for its transport functions.

Two mechanisms of phase separation may be distinguished. In the parent vessel upstream of a bifurcation, the concentration of red cells is generally reduced near the vessel wall, as mentioned previously. At the bifurcation, the daughter vessel receiving the lower share of total flow may be fed predominantly from this low hematocrit region of the parent vessel and therefore receive a lower hematocrit. This has been called 'plasma skimming' (Krogh, 1921). In addition, due to the particulate nature of blood, fluid forces at the bifurcation may cause red cells to depart from fluid streamlines, leading to uneven hematocrits in the daughter vessels ('red cell screening') (Pries et al., 1989).

The phase separation behavior of a bifurcation may conveniently be described by expressing the fraction (ϕ) of red cells from the parent vessel entering one daughter vessel as a function of the total flow fraction (ψ) entering that vessel (Schmid-Schönbein et al., 1980). Since zero flow implies zero red cell flux, $\phi = 0$ when $\psi = 0$ and $\phi = 1$ when $\psi = 1$. In relatively large vessels, ϕ is approximately equal to ψ, i.e., phase separation is insignificant. In smaller microvessels, however, the relationship is markedly nonlinear, and generally sigmoidal. If the flow fraction ψ in one daughter vessel is small, that vessel may receive only plasma from the parent vessel, so that $\phi = 0$. At higher ψ, ϕ starts to rise. The relationship between ψ and ϕ depends on the distribution of red cells in the cross-section of the parent vessel (Schmid-Schönbein et al., 1980).

Pries et al. (1989) observed the phase separation behavior of 65 arteriolar bifurcations in the rat mesentery. Based on these observations, they developed parametric representations of the ψ - ϕ relationship, including its dependence on the diameter and hematocrit of the parent vessel, and on the relative diameters of the daughter vessels (Figure 3). Several theoretical analyses of phase separation have also been developed, and are reviewed by Secomb (1995).

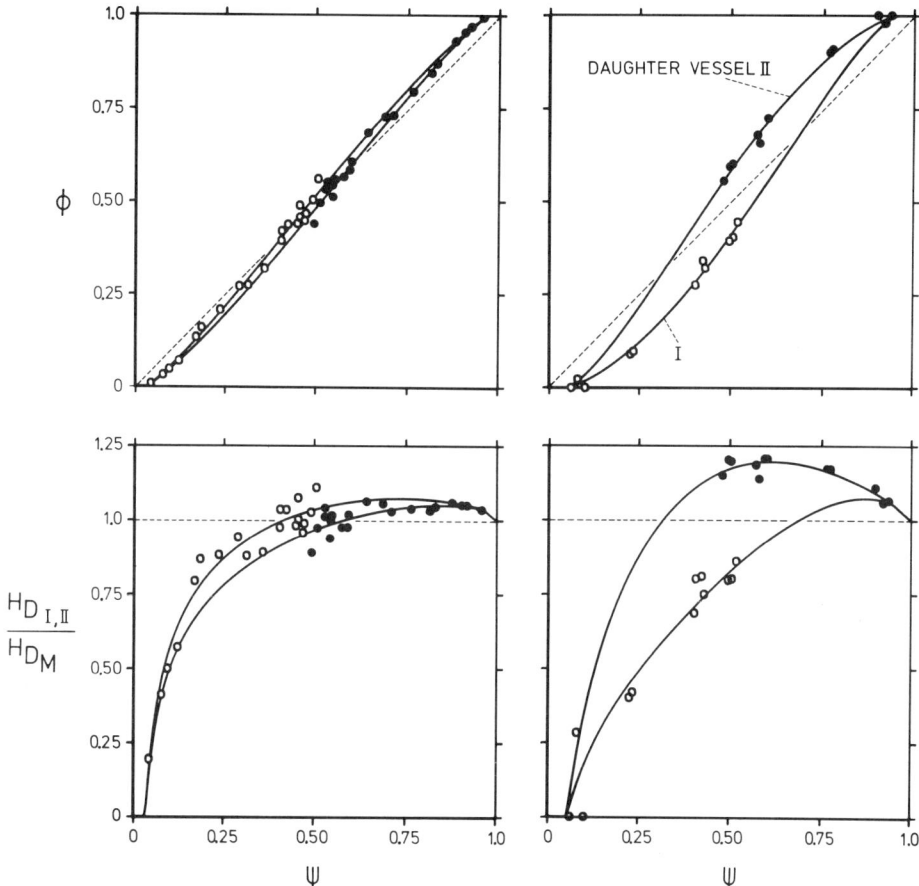

Figure 3. Phase separation in diverging bifurcations. Upper panels: Red cell fraction (ϕ) entering daughter vessels I and II. Lower panels: Ratios of hematocrit H_D in daughtervessels (I and II) to hematocrit in parent vessel (M). All quantities are shown as functions of the total flow fraction (ψ) entering the same daughter vessel. The experimental data (open and solid circles) in the left panels were obtained in a bifurcation with a parent vessel diameter of 20 µm; in the right panels the parent vessel diameter was 8 µm. Solid curves show fitted empirical relationships. Dashed lines show behavior without phase separation. The asymmetry of cell distribution and daughter vessel hematocrit is greater in the smaller bifurcation. From Pries et al. (1989).

Blood Flow in Microvascular Networks

Observations of Network Hemodynamics: Quantitative studies of network hemodynamics are made difficult by the complex and heterogeneous architecture of microvascular networks. Network heterogeneity leads to large segment-to-segment variability in hemodynamic variables, such as velocity, pressure and hematocrit. Furthermore, the complexity of network architecture makes it difficult to interpret the significance of the measurements in terms of overall network hemodynamics. Such interpretation is aided by the use of a method to classify vessel segments according to position within the network, so that observed hemodynamic variables can be presented in terms of the chosen classification.

Until recently, vessel diameter has been most often used for this classification, with vessels on the arterial side distinguished from those on the venous side. This is based on the

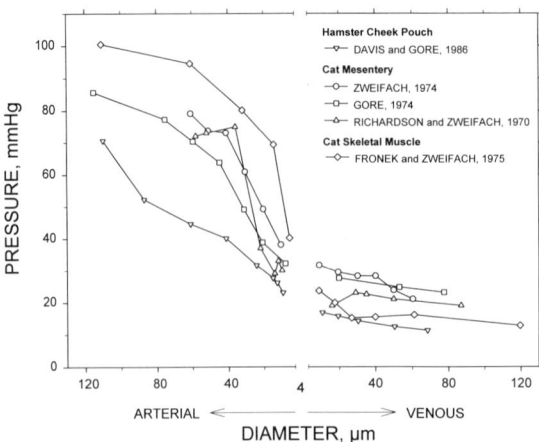

Figure 4. Variation of intravascular pressure with vessel diameter for arterioles and veins of several different tissues and species, based on direct measurements with micropipets.

idea that blood passing through a network encounters a sequence of increasingly narrow arterioles, until it reaches the capillaries, and then passes through increasingly wide venules. It has long been recognized (Zweifach, 1974) that this approach is not entirely adequate, since the topological organization of the network is not incorporated in such presentations of hemodynamic data.

Nonetheless, hemodynamic data from individual segments, presented as functions of vessel diameter, demonstrate several important features of microvascular network hemodynamics. For example, the velocity of blood flow is positively correlated with vessel diameter (Zweifach and Lipowsky, 1977). When a wide range of diameters is considered, an approximately linear dependence is seen, particularly within the arteriolar side of the network. Consequently, the volumetric blood flow rate is nearly proportional to the cube of vessel diameter (Mayrowitz and Roy, 1983). A further implication is that the wall shear rate is approximately uniform within the arteriolar tree. On the venous side, velocities are substantially lower than in arteriolar vessels with corresponding diameters, and wall shear rates are therefore lower than in arterioles.

The existence of higher wall shear rates in arterioles implies that pressure gradients are higher on the arteriolar side than on the venular side, for comparable vessel diameters. Therefore, the pressure drop through the microcirculation might be expected to occur mainly on the arteriolar side. This has been demonstrated in several studies (Figure 4).

As the principal site of flow regulation, the arterioles are capable of substantial active diameter changes, modulating their flow resistance. The distribution of pressure through the microvasculature is clearly sensitive to such changes. In general, arteriolar constriction will result in an increase in the fraction of the total pressure drop occurring on the arteriolar side and therefore in a decrease of mean capillary pressure, and vice versa (Mellander and Bjornberg, 1992).

Further characteristics of microvascular network hemodynamics will be described below. First, however, theoretical simulations of blood flow through networks are briefly discussed, since such simulations have been used in conjunction with experimental observations in several recent studies of network hemodynamics.

Simulations of Network Hemodynamics: Because of the complexity and heterogeneity of network architecture, many studies of network hemodynamics have used theoretical simulations. Such simulations, in combination with experimental data, can provide insight into aspects of network hemodynamics that result from the complicated network architecture and would not be present in an idealized network consisting of symmetric arrays of identical vessels in parallel. Simulations can provide more complete data on network hemodynamics than can be obtained by direct measurements, and represent a detailed quantitative framework within which experimental data can be analyzed. Comparisons between simulations and observations provide critical tests of the assumptions underlying the simulations.

The principal determinants of network hemodynamics are network architecture, flow resistance in vessel segments, and red cell partition in diverging bifurcations. Information in each of these areas, as described above, provides a basis for simulating the distribution of flow and hematocrit throughout a network. The basic method of such simulations (Lipowsky and Zweifach, 1974; Fenton and Zweifach, 1981; Popel, 1987; Pries et al., 1990, 1994) is as follows. Given an initial estimate of the flow resistance of each segment, the flow in each segment can be expressed in terms of the pressures at the nodes (bifurcations). The flows entering or leaving each node must add to zero, giving a system of linear equations to be solved for the nodal pressures, subject to specified pressures or flows in vessels feeding or draining the network. The hematocrit in each segment can then be computed based on information about phase separation in diverging bifurcations. Since apparent viscosity depends on hematocrit, this procedure leads to revised estimates of the flow resistance in each segment. This procedure is continued iteratively, leading eventually to estimates of the flows and hematocrits satisfying all the imposed conditions.

In these simulations, two different approaches have been used to represent the distribution of hematocrit within the network. One approach is to track the simulated motion of individual red blood cells through the network (Schmid-Schönbein et al., 1980). This approach allows simulations of transient effects resulting from fluctuations in the number of red cells in each segment, but requires much computation for large networks (Kiani et al., 1993). Alternatively, a continuum approach can be used, in which an average hematocrit is assigned to each segment based on the phase separation behavior in bifurcations described above (Pries et al., 1990, 1994). In studies of the effects of white blood cells on network hemodynamics, Fenton et al. (1985) and Warnke and Skalak (1990) combined these approaches, tracking the motion of individual white cells, but treating red cells as a continuum.

Resistance to Blood Flow in Microvessels In Vivo: Pries et al. (1990, 1994) used experiments and simulations to study network hemodynamics in the rat mesentery, in which virtually all vessels within selected regions could be visualized. Seven such regions were studied, containing 383 to 913 segments, and segment lengths, diameters, interconnections, hematocrits and flow velocities were determined. The observed network architectures formed the basis for simulations of hemodynamics, using the method outlined above. Hematocrit distribution was simulated using the continuum approach. Predicted segment hematocrits and flow velocities were compared with measured values on a segment-by-segment basis.

In initial simulations, flow resistance in each segment was estimated from *in vitro* data. The frequency distributions of predicted hematocrits and velocities agreed well with measured values, but correlations between predicted and measured values in individual segments were poor. Known sources of error in measurement and in model assumptions could account for only part of this discrepancy. To account for the remainder of the discrepancy, it was necessary to assume a different dependence of apparent viscosity on microvessel diameter, so that apparent viscosities in the smallest vessels are substantially higher than observed in glass tubes (Figure 5). This conclusion was confirmed by further direct measurements of flow resistance in similar networks. The Fåhraeus-Lindqvist effect is still present in the assumed *in vivo* relationship, but its reversal occurs at a larger diameter than in glass tubes (40 μm instead of 6 μm).

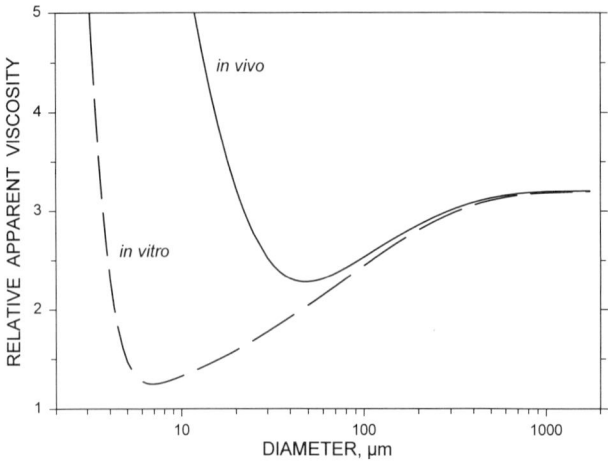

Figure 5. Relative apparent viscosity of blood as a function of diameter at a feed hematocrit of 45%. Dashed curve: *in vitro* result (Pries et al., 1992). Solid curve: *in vivo* result based on experiments and simulations in rat mesentery (Pries et al., 1994).

Possible causes of this discrepancy between *in vivo* and *in vitro* resistance to blood flow in narrow tubes include (Pries et al., 1990): flow obstruction by white cells, which are generally removed for glass-tube experiments; possible reduction of the cross-section available for flow, caused by macromolecules lining the interior of microvessels (Desjardins and Duling, 1987); asymmetric radial distributions of red cells within microvessels *in vivo*; and irregularity of the internal cross-sections of microvessels *in vivo*. Further work is required to determine the importance of these or other effects, and whether a similar phenomenon exists in other tissues. Nonetheless, the "viscosity law" (dependence of apparent viscosity on diameter and hematocrit) deduced from these network simulations has been used in further investigations into network hemodynamics in the rat mesentery.

Distribution of Hematocrit in Microvascular Networks: The unequal partition of hematocrit in diverging bifurcations leads to substantial variations in hematocrit among the microvessels in a network. At each bifurcation, the higher-flow pathway tends to receive a higher hematocrit than the pathway with lower flow, as already discussed. This leads to wide variations in the oxygen-carrying capacity of blood from one microvessel to another.

The distribution of hematocrit in networks was studied by Pries et al. (1986), who introduced the concept of a complete flow cross-section, i.e., a set of segments which together carry the total flow of the network, such that each fluid element of the blood traverses only one of the segments. The set of capillaries, as defined earlier, is an example. Among the segments of each complete flow cross-section, flow velocity and hematocrit are positively correlated, as a consequence of unequal hematocrit partition at bifurcations. Therefore, the total red cell flux through the flow cross-section exceeds the product of the total flow and the mean tube hematocrit (red cell concentration) in the vessels forming the cross-section (Vicaut, 1986). In other words, high-flow, high-hematocrit pathways more than compensate for low-flow, low-hematocrit pathways in their contribution to red cell flux. This condition is indicated schematically in Figure 6.

The reduction in tube hematocrit by the Fåhraeus effect, as already described, is a consequence of the correlation between local velocity and local hematocrit with individual vessels. The phenomenon described in the previous paragraph is thus analogous to the Fåhraeus effect, and causes a further lowering of mean tube hematocrit within a microvascular network, relative to that entering or leaving the network. Therefore, it is called the network Fåhraeus effect. Its relationship to the covariance of velocity and hematocrit is given by Secomb et al. (1989). In studies of the rat mesentery, Pries et al. (1986) showed that this

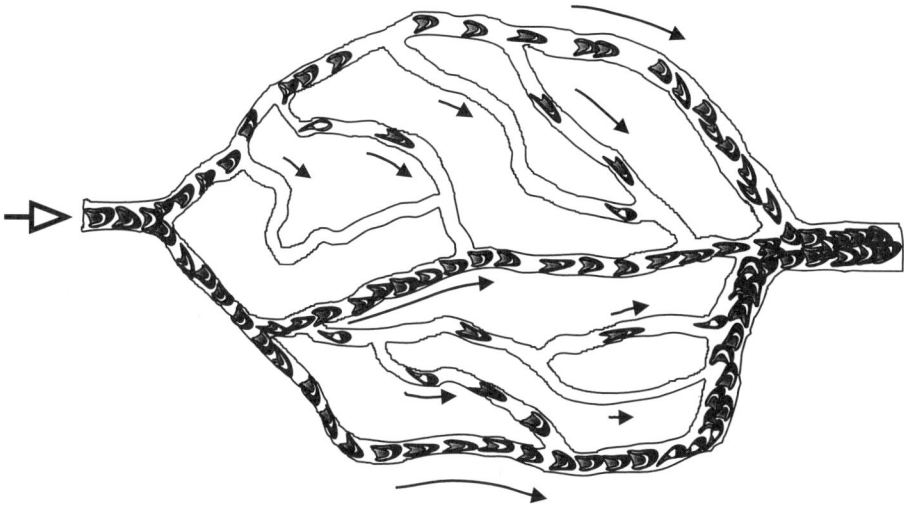

Figure 6. Schematic diagram of conditions leading to the network Fåhraeus effect. If vessels which exhibit high flow velocities also exhibit high hematocrits, i.e., if flow velocity and hematocrit are positively correlated, the mean hematocrit on a complete flow cross-section is lower than that in the feeding or draining segments of the network. Flow velocities are indicated by the lengths of the solid arrows beside each of the capillaries representing one complete flow cross-sections of the network.

effect resulted in a reduction of tube hematocrit in the capillaries by about 10% relative to its expected value in the absence of the network Fåhraeus effect.

Further insight into the distribution of hematocrit in networks can be obtained by analyzing variations in capillary hematocrit with generation number (Pries et al., 1992). Capillaries with low generation numbers branch off relatively large, high-flow vessels, and consequently receive relatively low hematocrits, due to the effects of plasma skimming and red cell screening. Conversely, high-generation capillaries are at the ends of these high-flow pathways and receive relatively high hematocrits. This "pathway effect" has the interesting consequence that the relatively low flow rates expected in longer pathways through the network may be compensated, in terms of oxygen delivery, by increased hematocrit.

Pries et al. (1992) also examined the effects of hemodilution on hematocrit distribution in networks of the rat mesentery. The "pathway effect" was accentuated when overall hematocrit was reduced, with red cells increasingly going to distal regions at the expense of proximal regions. This results from the dependence of red cell partition in individual bifurcations on the hematocrit in the parent vessel: at lower hematocrits, the partition becomes more unequal. The overall effect is a more homogeneous distribution of red cell fluxes between proximal and distal regions in hemodilution, a possibly beneficial result.

The above results were obtained by a combination of experimental observations and theoretical simulations using the model described earlier, for same networks, with and without hemodilution (Pries et al., 1992). The predicted distributions of hematocrit with generation agreed closely with observations, indicating that the observed "pathway effect" and its sensitivity to hemodilution could be accounted for by known rheological phenomena, particularly the phase separation at bifurcations.

Heterogeneity of Network Hemodynamics and Network Architecture: The unequal distribution of hematocrits in networks exemplifies the overall heterogeneity of network hemodynamics. As a result of geometric heterogeneity (in lengths and diameters of segments), different pathways through a network differ in flow resistance. Topological heterogeneity of network structure implies that different pathways through a single network consist of varying numbers of vessel segments. This also contributes to variations in pathway flow resistance, and consequently to heterogeneous flow rates. At diverging bifurcations, different flow rates produce heterogeneity of microvessel hematocrit, as already discussed.

A high degree of heterogeneity in network architecture and hemodynamics is a consistent feature of the microcirculation, and has significant functional consequences (Duling and Damon, 1987). Such heterogeneity implies that mass transport processes occurring in individual microvessels may not be well described in terms of a "typical" vessel to which is assigned mean parameter values for its class. It thus complicates the estimation of mass transport properties of microvascular networks. For instance, if two hemodynamic parameters, e.g. flow rate and hematocrit, are positively correlated, then the product of their means is an underestimate of the mean of their product, e.g. mean capillary red cell flux (Vicaut, 1986). This fact underlies the network Fåhraeus effect, described above. Similarly, if a transport parameter depends nonlinearly on a hemodynamic parameter, it cannot be estimated reliably based on the mean value of that parameter. An example is the clearance of a diffusive solute from blood, which depends nonlinearly on blood flow (Renkin, 1985). It may be speculated that heterogeneity of flow confers a physiological advantage by allowing smoother modulation of solute exchange over a wide range of flow rates than would occur in a very regular network. However, this idea remains to be examined in a systematic way.

Since hemodynamic heterogeneity is a direct consequence of structural heterogeneity, it is possible in principle to establish quantitative relationships between structural and hemodynamic heterogeneity in microvascular networks. In particular, theoretical models can be used to predict the effects of topological and geometric heterogeneity of network architecture on the dispersion in observed flow velocities and hematocrits among vessels of the same class, and in transit time of blood through the network.

This approach was taken by Dawant et al. (1986) and Levin et al. (1986), using computer-generated "stochastic" networks. To introduce topological heterogeneity, they used data from the conjunctival microcirculation (Fenton and Zweifach, 1981) to estimate the distribution of vessel numbers according to the Strahler ordering system. Geometric heterogeneity (in segment lengths and diameters) was introduced by choosing values at random from appropriate statistical distributions. Both the average values and the distributions of hemodynamic variables were shown to depend on the heterogeneity of network topology and geometry. The dispersions in capillary flow and pressure were found to depend mainly on the geometric heterogeneity of the network, with less sensitive dependence on topological heterogeneity (Dawant et al., 1986). Similarly, the reduction in mean hematocrit was found to depend mainly on geometric heterogeneity and to be only slightly influenced by topological heterogeneity (Levin et al., 1986).

As pointed out earlier, the Strahler scheme provides only a partial characterization of network architecture. Subsequent to the studies referred to above, more complete anatomical information has become available, for instance in the rat mesentery as described above. This provides a basis for re-examining the relationship between network architecture and hemodynamic heterogeneity, using more realistic network structures. Such studies are currently in progress. On the basis of results obtained so far, it can be stated that the high degree of heterogeneity of hemodynamic variables typically observed in microvascular networks is consistent with the observed structural heterogeneity. Topological heterogeneity appears to strongly influence overall flow resistance and pressure distribution with vessel generation in the network, while geometric heterogeneity has less effect on these factors. Conversely, network Fåhraeus effect appears to depend mainly on the geometric heterogeneity of the network, as found by Levin et al. (1986). Both geometric and topological heterogeneity appear to be important in determining the distribution of transit times through the network.

CONCLUSIONS

The architecture of microvascular networks is typically irregular and asymmetric, leading to a high degree of heterogeneity in hemodynamic variables, such as velocity and hematocrit. Blood is a suspension of cells with dimensions comparable to microvessel diameters. These features of the microcirculation have impeded efforts to understand microvascular hemodynamics at a quantitative level, although several qualitative features have long been known, e.g., the fact that the pressure drop in the microcirculation occurs predominantly in the arterioles.

Recently, experiments have been made in which entire microvascular networks have been visualized and mapped, and their hemodynamics observed and compared with theoretical simulations. These studies have led to the following conclusions:

- Resistance to blood flow in microvessels is larger than expected from experiments using glass tubes with corresponding diameters. The reasons for this discrepancy may include irregularity of microvessel lumens and bifurcations, and effects of vessels' macromolecular linings.
- Uneven partition of hematocrit in bifurcations leads to a high degree of heterogeneity in microvascular hematocrit. Furthermore, it contributes to a reduction of mean microvascular hematocrit relative to the feed hematocrit.
- Geometric and hemodynamic parameters exhibit broad variations within microvascular networks, even among vessels of the same functional type, e.g., capillaries. The observed heterogeneity of microvascular hemodynamic parameters can be accounted for based on observed variability of microvessel dimensions and asymmetry of network topology.

However, several aspects require more investigation:

- Most of the results referred to above are for mesenteric preparations, and other tissues have been studied in less detail.
- A mechanistic basis for the higher apparent viscosity of blood observed *in vivo* has not been established.
- Although several different systems for describing network architecture have been developed, it has not been established that any of them, other than a complete segment-by-segment description, provides an adequate characterization of network architecture, in the sense that all relevant features of network hemodynamics can be deduced from the descriptions they provide.
- The processes governing the growth and adaptation of network architectures are not well understood, and are the subject of current research.
- The physiological consequences of the findings described above require further exploration, particularly with regard to the mass transport functions of the circulatory system.

Thus, despite more than a hundred years of study, the architecture and hemodynamics of microvascular networks still present many unanswered questions.

REFERENCES

Barnard, A.C.L., Lopez, L. and Hellums, J.D. (1968) Basic theory of blood flow in capillaries. *Microvasc. Res.* 1, 23-34.
Burton, A.C. (1972) *Physiology and Biophysics of the Circulation.* 2nd ed. Chicago: Year Book Medical Publishers.
Dawant, B., Levin, M. and Popel, A.S. (1986) Effect of dispersion of vessel diameters and lengths in stochastic networks. I. Modeling of microcirculatory flow. *Microvasc. Res.* 31, 203-222.
Davis, M.J. and Gore, R.W. (1986) Pressure distribution in the microvascular network of the hamster cheek pouch. In *Microvascular Networks: Experimental and Theoretical Studies*, ed. A.S. Popel and P.C. Johnson, pp. 142-154. Basel: Karger.
Desjardins, C. and Duling, B.R. (1987) Microvessel hematocrit: measurement and implications for capillary oxygen transport. *Am. J. Physiol.* 252, H494-H503.
Duling, B.R. and Damon, D.H. (1987) An examination of the measurement of flow heterogeneity in striated muscle. *Circ. Res.* 60, 1-13.
Engelson, E.T., Skalak, T.C. and Schmid-Schönbein, G.W. (1985) The microvasculature in skeletal muscle. I. Arteriolar network in the rat spinotrapezius muscle. *Microvasc. Res.* 30, 29-44.
Fåhraeus, R. (1928) Die Strömungsverhältnisse und die Verteilung der Blutzellen im Gefässsystem. *Klin. Wschr.* 7, 100-106.
Fåhraeus, R. and Lindqvist, T. (1931) The viscosity of the blood in narrow capillary tubes. *Am. J. Physiol.* 96, 562-568.
Fenton, B.M. and Zweifach, B.W. (1981) Microcirculatory model relating geometrical variation to changes in pressure and flow rate. *Ann. Biomed. Eng.* 9, 303-321.
Fenton, B.M., Wilson, D.W. and Cokelet, G.R. (1985) Analysis of the effects of measured white blood cell entrance times on hemodynamics in a computer model of a microvascular bed. *Pflügers Arch.* 403 396-401.
Frasher, W.G. and Wayland, H. (1972) A repeating modular organization of the microcirculation of cat mesentery. *Microvasc. Res.* 4, 62-76.
Fronek, K. and Zweifach, B.W. (1975) Microvascular pressure distribution in skeletal muscle and the effect of vasodilation. *Am. J. Physiol.* 228, 791-796.
Fung, Y.C. (1984) *Biodynamics: Circulation.* New York: Springer.
Gaehtgens, P. and Schmid-Schönbein, H. (1982) Mechanisms of dynamic flow adaptation of mammalian erythrocytes. *Naturwissenschaften* 69, 294-296.
Gaehtgens, P., Ley, K., and Pries, A.R. (1986) Topological approach to the analysis of microvessel structure and hematocrit distribution. In *Microvascular Networks: Experimental and Theoretical Studies*, ed. A.S. Popel and P.C. Johnson, pp. 52-60. Basel: Karger.
Gore, R.W. (1974) Pressures in cat mesenteric arterioles and capillaries during changes in systemic blood pressure. *Circ. Res.* 34, 581-591.
Halpern, D. and Secomb, T.W. (1989) The squeezing of red blood cells through capillaries with near-minimal diameters. *J. Fluid Mech.* 203, 381-400.
Hochmuth, R.M. and Waugh, R.E. (1987) Erythrocyte membrane elasticity and viscosity. *Ann. Rev. Physiology* 49, 209-219.
Horton, R.E. (1945) Erosional development of streams and their drainage basins; hydrophysical approach to quantitative morphology. *Geol. Soc. Amer. Bull.* 56, 275-370.
Hsu, R. and Secomb, T.W. (1989) Motion of non-axisymmetric red blood cells in cylindrical capillaries. *J. Biomech. Eng.* 111, 147-151.
Kiani, M.F., Cokelet, G.R. and Sarelius, I.H. (1993) Effect of diameter variability along a microvessel segment on pressure drop. *Microvasc. Res.* 45, 219-232.
Koller, A., Dawant, B, Liu, A., Popel, A.S., and Johnson, P.C. (1987) Quantitative analysis of arteriolar network architecture in cat sartorius muscle. *Am. J. Physiol.* 253, H154-H164.
Krogh, A. (1921) Studies on the physiology of capillaries. II. The reactions to local stimuli of blood vessels in the skin and web of the frog. *J. Physiol. (London)* 55, 412-422.
Levin, M., Dawant, B. and Popel, A.S. (1986) Effect of dispersion of vessel diameters and lengths in stochastic networks. I. Modeling of microvascular hematocrit distribution. *Microvasc. Res.* 31, 223-234.
Ley, K., Pries, A.R. and Gaehtgens, P. (1986) Topological structure of rat mesenteric microvessel networks. *Microvasc. Res.* 32, 315-332.
Lighthill, M.J., (1968) Pressure-forcing of tightly fitting pellets along fluid-filled elastic tubes. *J. Fluid Mech.* 34, 113-143.

Lipowsky, H.H. and Zweifach, B.W. (1974) Network analysis of microcirculation in rat mesentery. *Microvasc. Res.* 7, 73-83.
Lipowsky, H.H., Kovalcheck, S. and Zweifach, B.W. (1978) The distribution of blood rheological parameters in the microvasculature of cat mesentery. *Circ. Res.* 43, 738-749.
Mall, J.P. (1888) Die Blut- und Lymphwege im Dünndarm des Hundes. *Königl. Sächs. Gesellsch. der Wissensch., Abhandlung der Math. Physikal. Klasse*, Leipzig, Vol. XIV, 153-161.
Martini, P., Pierach, A., and Schreyer, E. (1930) Die Strömung des blutes in engen Gefässen. Eine Abweichung vom Poiseuille'schen Gesetz. *Dtsch. Arch. Klin. Med.* 169, 212-222.
Mayrovitz, H.N. (1986) Hemodynamic significance of microvascular arteriolar anastamosing. In *Microvascular Networks: Experimental and Theoretical Studies*, ed. A.S. Popel and P.C. Johnson, pp. 197-209. Basel: Karger.
Mayrovitz, H.N. and Roy, J. (1983) Microvascular blood flow: evidence indicating a cubic dependence on arteriolar diameter. *Am. J. Physiol.* 245: H1031-H1038.
Mellander, S. and Bjornberg, J. (1992) Regulation of vascular smooth muscle tone and capillary pressure. *News in Physiol. Sci.* 7, 113-119.
Popel, A.S. (1987) Network models of peripheral circulation. In *Handbook of Bioengineering*, ed. R. Skalak and S. Chien, pp. 20.1-20.24. New York: McGraw-Hill.
Popel, A.S., Torres Filho, I.P., Johnson, P.C. and Bouskela, E. (1988) A new scheme for hierarchical classification of anastomosing vessels. *Int. J. Microcirc. Clin. Exp.* 7, 131-138.
Potter, R.F., and Groom, A.C. (1983) Capillary diameter and geometry in cardiac and skeletal muscle studied by means of corrosion casts. *Microvasc. Res.* 25, 68-84.
Pries, A.R., Ley, K. and Gaehtgens, P. (1986) Generalization of the Fåhraeus principle for microvessel networks. *Am. J. Physiol.* 251, H1324-H1332.
Pries, A.R., Ley, K., Claasen, M. and Gaehtgens, P. (1989) Red cell distribution at microvascular bifurcations. *Microvasc. Res.* 38, 81-101.
Pries, A.R., Secomb, T.W., Gaehtgens, P. and Gross, J.F. (1990) Blood flow in microvascular networks - Experiments and simulation. *Circ. Res.* 67: 826-834.
Pries, A.R., Neuhaus, D. and Gaehtgens, P. (1992a) Blood viscosity in tube flow: dependence on diameter and hematocrit. *Am. J. Physiol.* 263, H1770-1778.
Pries, A.R., Fritzsche, A., Ley, K., and Gaehtgens, P. (1992b) Redistribution of red blood cell flow in microcirculatory networks by hemodilution. *Circ. Res.* 70, 1113-1121.
Pries, A.R., Secomb, T.W., Gessner, T., Sperandio, M.B., Gross, J.F. and Gaehtgens, P. (1994) Resistance to blood flow in microvessels *in vivo*. *Circ. Res.* 75: 904-915.
Reinke, W., Gaehtgens, P. and Johnson, P.C. (1987) Blood viscosity in small tubes: effect of shear rate, aggregation and sedimentation. *Am. J. Physiol.* 253, H540-H547.
Renkin, E.M. (1985) Regulation of the microcirculation. *Microvasc. Res.* 30, 251-263.
Richardson, D.R. and Zweifach B.W. (1970) Pressure relationships in the macro- and microcirculation of the mesentery. *Microvasc. Res.* 2, 474-488.
Schmid-Schönbein, G.W., Skalak, R., Usami, S. and Chien, S. (1980) Cell distribution in capillary networks. *Microvasc. Res.* 19, 18-44.
Schmid-Schönbein, G.W., Sung, K.-P., Tözeren, H., Skalak, R. and Chien, S. (1981) Passive mechanical properties of human leukocytes. *Biophys. J.* 36, 243-256.
Secomb, T.W. (1987) Flow-dependent rheological properties of blood in capillaries. *Microvasc. Res.* 34, 46-58.
Secomb, T.W. (1991) Red blood cell mechanics and capillary blood rheology. *Cell Biophysics* 18: 231-251.
Secomb, T.W. (1995) Mechanics of blood flow in the microcirculation. To appear in *Biological Fluid Dynamics*, ed. C.P. Ellington and T.J. Pedley. Cambridge, Company of Biologists.
Secomb, T.W., Skalak, R., Özkaya, N. and Gross, J.F. (1986) Flow of axisymmetric red blood cells in narrow capillaries. *J. Fluid Mech.* 163, 405-423.
Secomb, T.W., Pries, A.R., Gaehtgens, P. and Gross, J.F. Theoretical and experimental analysis of hematocrit distribution in microcirculatory networks. In *Microvascular Mechanics*, ed. J.S. Lee and T.C. Skalak, Springer, New York, 1989, pp. 40-49.
Skalak, R. (1976) Rheology of red blood cell membrane. In *Microcirculation, Vol. I*, ed. J. Grayson and W. Zingg, pp. 53-70. New York: Plenum.
Strahler, A.N. (1952) Hypsometric (area-altitude) analysis of erosional topography. *Geol. Soc. Amer. Bull.* 63, 1117-1142.
Vicaut, E. (1986) Statistical estimation of parameters in microcirculation. *Microvasc. Res.* 32, 244-247.
Warnke, K.C. and Skalak, T.C. (1990) The effects of leukocytes on blood flow in a model skeletal muscle capillary network. *Microvasc. Res.* 40, 118-136.

Wiedeman, M.P., Tuma, R.F. and Mayrovitz, H.N. (1981) *An Introduction to Microcirculation*. New York: Academic Press.

Zarda, P.R., Chien, S. and Skalak, R. (1977) Interaction of viscous incompressible fluid with an elastic body. In *Computational Methods for Fluid-Solid Interaction Problems*, ed. T. Belytschko and T.L. Geers, pp. 65-82. New York: American Society of Mechanical Engineers.

Zweifach, B.W. (1937) The structure and reactions of the small blood vessels in Amphibia. *Am. J. Anat.* 60, 473-514.

Zweifach, B.W. (1974) Quantitative studies of microcirculatory structure and function. I. Analysis of pressure distribution in the terminal vasculature in cat mesentery. *Circ. Res.* 34, 843-857.

Zweifach, B.W. and Lipowsky, H.H. (1977) Quantitative studies of microcirculatory structure and function. III. Microvascular hemodynamics of cat mesentery and rabbit omentum. *Circ. Res.* 41, 380-390.

10

MASS TRANSPORT THROUGH THE WALLS OF ARTERIES AND VEINS

M. John. Lever

Centre for Biological & Medical Systems
Imperial College of Science, Technology & Medicine
London SW7 2BY, United Kingdom

INTRODUCTION

In the microcirculation, the blood vessel wall is the major barrier which limits the transport of materials between the blood and tissues. Mass transport studies in the microcirculation are, therefore, primarily concerned with the mechanisms by which metabolites and catabolites are exchanged across the capillary wall. In contrast, in large vessels, there is more interest in the processes by which materials move into and out of the wall tissue itself. Somewhat paradoxically, the walls of large arteries are relatively avascular, having a slab of intimal and medial tissue which may be several hundred microns thick between the blood in the vessel lumen and that in the capillaries of the vasa vasorum. As thick walled vessels grow, angiogenesis occurs and capillaries penetrate the media from the adventitial side, but with increased thickening of the intima on ageing, the tissue may again become more avascular. For the tissue to undergo efficient metabolism, respiratory gases, nutrients and catabolites must be rapidly exchanged. In the field of large vessel mass transport though, the major interest has been in the movement of larger materials such as plasma proteins because of their putative role in atherosclerosis. Because this condition normally affects only arteries, transport processes in these vessels have been studied far more extensively than in veins or pulmonary vessels.

Large vessel walls are heterogeneous tissues comprising intimal, medial and adventitial layers, each of which is composed of cells embedded in an interstitial space containing fluid, fibres and proteoglycans. Movement of molecules into and through the walls may involve their passage through multiple pathways which can be both in parallel and in series with each other. This causes great complexity, both in interpreting the results of experimental studies and in devising models of the transport processes which are able to give an adequate description of the physiological processes.

This chapter is not intended to be a full review of the extensive work performed in the area, but briefly discusses some of the important factors which are involved in large vessel mass transport.

THERMODYNAMICS OF TRANSPORT

The rate of transport of any molecular species is dependent on its chemical potential gradient, and the transport 'resistance' of the medium through which it is moving. For this reason any modification of a material, whether by binding to the tissue matrix or by chemical reaction will alter the gradient and, in the latter case, may create new gradients. For example the conversion of low density lipoproteins within the intimal layer of vessel wall tissue to oxidized forms (which may be a significant process in the initiation of an atherosclerotic lesion) will increase the gradient for transport of the unmodified moiety between the luminal blood and the intima, decrease the gradient for its efflux across the medial layer to the adventitia but create new gradients for the outflow of the oxidized moiety in both directions. Similarly complications arise when fluorescently-labelled or radiolabelled molecules are used for studying the transport processes of the native forms. Introduction of these modified molecules into the circulation creates an unnaturally high gradient between the blood and the tissues and so their initial rates of transport can be very much greater than those of the materials which they are designed to model. The native material is likely to have a fairly constant steady state distribution across all tissue and any chemical potential gradients are likely to be much smaller. Considerable errors can therefore arise if the assumption is made that the fluxes of native proteins are identical to those of labelled tracers.

The chemical potential of a molecular species, is dependent on its chemical activity (which in turn is related to its concentration), the local hydrostatic pressure and on the properties of the medium in which it is dissolved. Rigorous mathematical expressions for transport processes based on chemical potentials can be derived (Silberberg 1989). More commonly, somewhat simpler approaches partition the transport between diffusive processes driven by the random motion of solute molecules imposed by the bombardment by surrounding solvent molecules, and convective processes driven by hydrostatic pressure gradients. Because of the high pressures within arteries, fluid flow rates across the wall can be appreciable and convection may be more important than in other tissues.

Diffusive and Convective Transport

An index of the relative importance of diffusion and convection in the transport of a molecule is the Peclet number Pe, given by:

$$Pe = \frac{U \chi x}{D}$$

Where U is the fluid velocity through the medium x, is the slip coefficient (the velocity of the molecule relative to that of the fluid), x is the distance across which transport is occurring and D is the diffusion coefficient of the molecule in the medium. Small molecules have high diffusion coefficients, giving rise to low Peclet numbers and their transport rates can be approximated by the Fick equations for steady state and unsteady diffusion, since fluid velocities in the tissues are usually higher than the convective velocities (in contrast to the velocity of blood within the vessels themselves). The steady state flux J_s of a solute across an area A is given by:

$$\frac{J_s}{A} = D \frac{dC}{dx}$$

where dC/dx is the concentratio n gradient through the medium.

Proteins, having much lower diffusion coefficients will have higher Peclet numbers and are therefore more likely to be convected (although if they are moving through tissue spaces of molecular dimensions, then x, and hence Pe will be reduced). Solute flux occurring under the combined influence of convection and diffusion can be approximated by the expression:

$$J_s = \frac{D}{x}(C_1 - C_2)\left(1 - \frac{Pe}{2}\right) + U\chi C_1$$

where C_1 and C_2 are the concentrations of the solute at the upstream and downstream boundaries of the tissue.

An excellent review of mass transport phenomena derived from fundamental thermodynamics and applied to capillary transport processes is given by Curry (1984). Exactly the same principles are applicable when considering transport in large blood vessels, but their application becomes more difficult because of the much greater complexity of the structure of the walls.

Transport Resistance of the Tissue

Among factors which determine the transport resistance of the wall tissue is the distribution volume or partition coefficient (this is the inverse of the degree of exclusion). The partition coefficient is the ratio of the concentration of a material within the tissue to that in free solution under equilibrium conditions. It cannot be determined precisely from *in vivo* concentrations, since the pressure and concentration gradients which exist across tissues may allow the development of a steady state, but will prevent the attainment of equilibrium. Measurements can be made *in vitro* but values may be different between unpressurized tissues and those maintained under normal physiological stresses (Tedgui & Lever 1987). For materials which are excluded from cells the partition coefficients will be dependent on the size of the extracellular space. Those of larger molecules are invariably smaller than those of smaller ones because of their geometric exclusion from the proteoglycans, elastin and collagen fibres which constitute the extracellular tissue matrix.

TRANSPORT OF LIPIDS INTO THE VESSEL WALL

The motivation for much of the research on transport processes in arteries has been to gain an understanding about the initiation and development of atheromatous plaques. The pathogenesis of atheroma appears to be a multi-factorial process and there is still considerable uncertainty about the most significant mechanisms (Ross 1993). It is possible that the initial lesion involves a failure of normal homeostatic transport processes. There may also be chemical modification of plasma components during their passage through the wall tissue such as oxidation, polymerization or degradation processes. These products may then precipitate the migratory and proliferative phases involving intimal accumulation of smooth muscle cells and matrix. Alternatively it may be these, or other processes, which potentiate the intimal accumulation of the plasma components.

It is evident, however, that transport processes have an important role to play in atherogenesis. Although the endothelium lining large blood vessels has a very low permeability to macromolecules (10^{-9} - 10^{-7} cm.s^{-1}) a very large number of studies using labelled proteins have indicated that they can pass between the blood and vessel wall tissue (Fry 1987, Stender & Hjelms 1988). Furthermore, lipoproteins and fibrinogen, and the various products of these proteins, which are present even in the earliest atherosclerotic lesions originate from the blood plasma (Smith et al. 1979). Some of the major components of the lesions, including cholesterol in its free and esterified forms, phospholipids and triglycerides are transported into the wall tissue in association with lipoproteins (Stender & Christensen 1977). In fat fed animals, focal increases in wall LDL levels precede the appearance of atherosclerotic lesions (Schwenke & Carew 1989).

ROUTES INTO AND ACROSS THE WALLS OF LARGE BLOOD VESSELS

Depending on the gradients of chemical potential, materials can move from the blood into wall tissue either from the lumen of the vessel or from the vasa vasorum, in both cases crossing a layer of endothelial cells. Once through the endothelial layer, the pathway from the luminal side further into the tissue is anatomically quite different from that on the adventitial side. In addition, the effects of convection on transport will be quite different between the luminal and adventitial surfaces, because the hydrostatic pressure gradient is in the outward direction.

Role of the Endothelium

Large vessels have a complete layer of endothelial cells covering the luminal surface. Earlier studies suggested that regions of cell denudation were present even in normal vessels and these observations served to underlie the 'response to injury' models of atherosclerosis. However, it now appears that these findings probably resulted from damage to the fragile endothelial cells during tissue preparation. Molecules entering the wall across the intimal surface must therefore penetrate an intact layer. The route taken is not the same for all

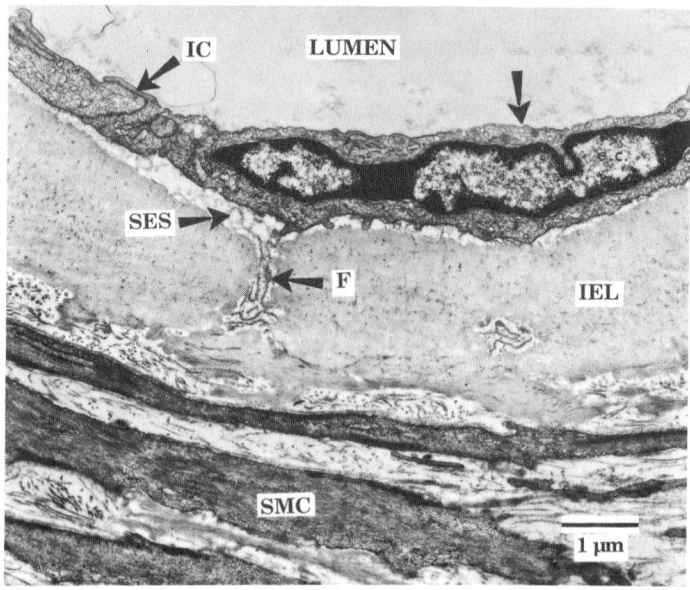

Figure 1. Transverse electron micrograph through an arterial endothelial cell showing the potential routes for the transport of materials between the blood and vessel wall. IC = intercell cleft, SES = sub-endothelial space, F = fenestra, IEL = internal elastic lamella, SMC = smooth muscle cell.

materials; lipid soluble materials are able to pass through the cell membrane, and small water soluble molecules can travel through the intercellular clefts. The path taken by larger molecules is less clear and is still an area of great controversy. Endothelial cells have large numbers of vesicles, many of which open onto either the luminal or abluminal surface. It has been thought that they may provide a route for transendothelial transport either by their own movement between opposing faces of the cell, or by the formation of transcellular channels (Vasile et al. 1983). Serial sectioning, however, suggests that such channels are exceptionally rare (Frøkjær-Jensen 1980). Although the intercell cleft appears to have tight regions which are too narrow for the passage of the larger plasma proteins, there may be routes for them around these junctional complexes (Bundgaard 1984). The occurrence of regions of the endothelium with increased permeability is discussed later.

Many studies have shown that the endothelium offers the primary barrier for the entry of proteins into the wall, and damage to it can cause very large increases in the rates of transport into the tissue (Chobanian et al. 1983, Ramirez et al. 1984). It has a lesser role however, in determining the rate of transport of smaller molecules because the resistance it offers to their transport is so much lower, and apparently not much different from that of other vessel wall components. The permeability of the endothelium to small water soluble molecules is in the range 10^{-5} to 10^{-4} cm s^{-1} and can be much higher for lipid soluble molecules.

Role of Endothelial Cell Receptors

Endothelial cell membranes contain lipoprotein receptors allowing at least two routes for transport into and through the intimal surface. The relative importance of the receptor mediated and receptor independent routes has been assessed by simultaneous uptake studies using both native and modified forms of LDL which are not bound by the receptor (Wiklund et al. 1985, Curmi et al. 1989). These studies appear to indicate that the receptor mediated route represents a significant but minor pathway. Studies with cultured endothelial cells have suggested that both native and modified LDL may be transported by the receptor mediated path much faster from the intimal side of the endothelium to the luminal side than in the opposite direction (Kim et al. 1994). This could have very important implications in determining the rate at which materials accumulate within the wall.

Transport from the Luminal Surface

Below the luminal endothelium, the intimal layer can vary markedly in structure, composition and thickness from one site to another. In most young arteries and in all veins, it is extremely thin, comprising a collagen IV basement membrane and an interstitial space rich in fibronectin and laminin. In young arteries near branching sites (Stary 1989) and in older arteries, the intimal layer is thickened and can exceed the thickness of each of the other layers. The thickened intima contains smooth muscle cells and a fibrous interstitium, but it is probably a relatively porous structure. The protein distribution volume is at least twice that of the medial layer (Cary et al.1987) and can accommodate very high concentrations of plasma proteins (Smith & Staples 1980, Frank & Fogelman 1989).

Transport through the Medial Layer

Even very large proteins such as fibrinogen (Bell et al. 1974) and low density lipoprotein (LDL) (Bratzler et al. 1977b) can penetrate the very dense interstitium of the medial layer although they are excluded from this tissue to a much greater extent than in that of the thickened intima. Transport into the layer occurs relatively rapidly, measurable

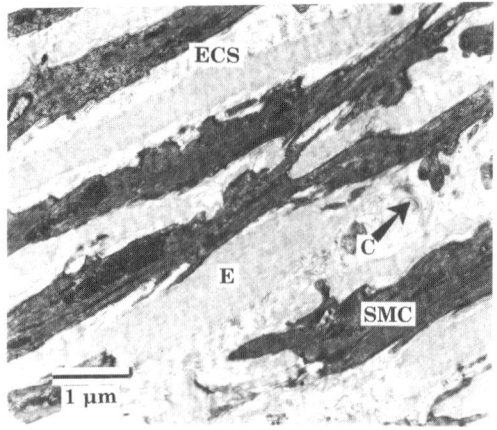

Figure 2. Transverse electron micrograph through the media of an artery showing potential transport routes. ECS = extracllular space, C = collagen fibres, E = elastin, SMC = smooth muscle cell.

quantities of labelled proteins being present in all parts of even relatively thick vessels within about 10 minutes. However, the steady state distributions of the large proteins across the thickness of the media are different from those of smaller proteins such as albumin or HDL (Bratzler 1977a, Bratzler et al.1977b, Lever and Coleman 1994). Indeed in innermost regions below the intima large molecules appear to be preferentially retained, being ultrafiltered by the tissue matrix, while in outer medial regions, closer to the adventitial layer they may have a lower concentration, partly because of increased convective flux (higher Peclet numbers because of lower diffusion coefficients) and partly because of their retention upstream. The degree of ultrafiltration of the larger proteins by the matrix appears to be greater in thicker walled arteries than in thin walled vessels. There are also large differences between the transport properties of the medial layers of systemic arteries and those of veins and pulmonary vessels; these will be discussed later.

Exchange Processes Involving the Vasa Vasorum

Large blood vessels develop an extensive circulation in the outer layers of their walls which continues to develop as the vessels grow. In thin walled vessels, the vasa vasorum is confined to the adventitial layer but if the media grows to more than 1 mm in thickness, capillaries will penetrate from the adventitia into the media. The vasa vasorum appears to be essential in supplying nutrients to, and maintaining the metabolism of the wall tissue; oxygen tensions across the thickness of the vessel wall fall from both the luminal and adventitial side suggesting inward diffusional transport to replace that being utilized by the medial smooth muscle cells (Klinowski et al.1982). If large molecules are being metabolized or otherwise transformed within the wall they could also enter from both sides, but the transport resistance may not be the same for the endothelial cells lining the lumen and those lining the vasa vasoral capillaries (Baldwin & Chien 1984).

The transport resistance of the adventitial tissue appears to be much lower than that of the medial layer of systemic arteries (though not that of pulmonary vessels or veins) reflecting the less dense and less fibrous matrix structure (Lever & Jay 1990)

In experimental studies on transport between blood and vessel wall tissue using labelled tracers, entry of material from the vasa vasorum can be inferred from a fall in concentration from the adventitia through the outer layers of the media. As mentioned above, however, the introduction of the tracer into the circulation introduces an unphysiological stepwise change in the chemical potential gradients between blood and tissue and may not

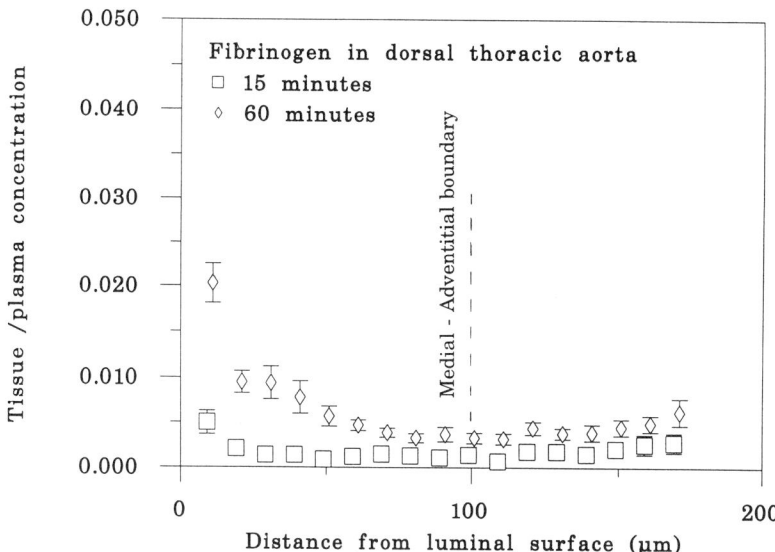

Figure 3. Diagram showing the concentration of fibrinogen at different sites across the wall of the rabbit aorta after the labelled protein has been present in the circulation for 15 min. Note that the concentration falls from both the luminal and the adventitial surfaces, suggesting that the protein is entering the wall both from the blood in the lumen of the vessel and that in the vasa vasorum.

be representative of the processes occurring in the more nearly steady state conditions which prevail *in vivo*.

METHODS FOR STUDYING TRANSPORT PROCESSES

Most experimental methods for studying vascular transport have utilized tracers, materials which have been modified to permit their detection, usually by addition of a dye or a radio-isotope. Some of the earliest studies employed Evans blue dye which enters the vessel wall in association with plasma albumin to which it strongly binds and they demonstrated considerable regional variability of the endothelial surface (McGill et al. 1957). The amount of dye entering the tissue can be measured using reflectometric techniques (Fry 1977), but until recently, most quantitative studies have been performed using radiolabelled materials. Spatial variation of the uptake of the labelled molecules has been determined by point sampling (Bell et al.1974, Stemerman et al 1986), and the distribution of radioactivity through the thickness of the wall has been measured using *en face* sectioning techniques (Bratzler et al.1977a,b) and by autoradiography (Tompkins et al.1989). The latter measurements allow a distinction to be made between material which enters the wall across the luminal surface and that which enters via the vasa vasorum. Quantitative assays can now be performed on histological sections of tissue containing molecules tagged with fluorescent markers (Weinberg 1988). These methods have a major advantage in that they allow measurements with spatial resolution down to a few micrometres, enabling localization of substances within the smallest compartments in the tissue.

Tracer techniques suffer from a variety of problems. These include alteration of the properties of a molecule on addition of the label, so that transport rates are not representative of those of the original, and lack of chemical stability. If the quantifiable moiety (the radio-isotope or dye) becomes dissociated from the molecule, then very large errors can arise if its transport rate is different from that of the molecule to which it was attached. Dissociation can, however, be an advantage in some circumstances; in longer term uptake studies the quantity of a labelled protein in the wall tissue will not be a measure of how much has entered from the blood stream, since some will almost certainly have been lost by normal efflux processes or by metabolism. Labelling the protein with materials such as tyramine-cellobiose which is stripped from the protein and bound intracellularly allows the estimation of the total quantity that has entered the wall throughout the whole duration of the experiment (Schwenke & St Clair 1993).

To overcome the problems associated with labelled materials, attempts have been made to measure the concentration of certain naturally occurring materials within the wall tissue (eg Smith & Staples 1982). Such measurements are invaluable for quantifying the steady state concentrations of proteins within different parts of the tissue, but yield less information about the kinetics of the transport processes.

A useful measure of the rate at which materials can pass across the endothelial layer is the permeability coefficient which is the quantity transported per unit time per unit concentration difference across it. This index derives from basic flux equations and its determination is difficult in tracer uptake experiments unless their duration is sufficiently short that the concentration in the wall tissue is negligible compared with that in the plasma. For transport processes into the wall that are largely convective the concentration which determines the quantity entering the wall is only that which is in the plasma.

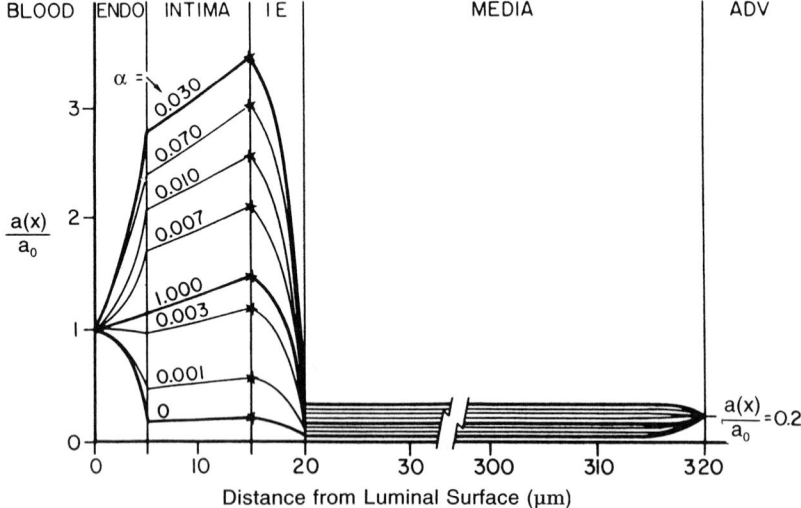

Figure 4. Predictions of the distribution of proteins across the thickness of the different layers of an artery. (Endo = endothelium, IE = internal elastic lamella, adv = adventitia. The concentration at different distances (x) from the luminal surface relative to those in the plasma are defined as $a(x)/a_0$. The curves correspond to endothelial layers with different fractional areas of leaky regions, denoted by α. (For a more detailed explanation, see Fry, 1987, Arterio-sclerosis,7:88-100)

MODELS OF ARTERIAL TRANSPORT

Mathematical modelling has offered many insights into the processes controlling the movement of materials in and out of blood vessel walls. Early models (eg Truskey et al.1981) assumed that the wall was a homogeneous slab of tissue. Different transport coefficients were ascribed to the endothelial layer which formed a boundary between the blood and the wall, and to the compartment which represented the remainder of the wall. Combined diffusion and convection equations were used to test the validity of the chosen coefficients against existing experimental data. Further refinement involved a sub-division of the wall into multiple layers including endothelium, intima, internal elastic lamella and media, allowing the prediction of the relative concentrations of the transported species throughout the wall thickness (eg. Fry 1987).

The models above are all one-dimensional, since it was assumed that the molecules follow a linear path, in the radial direction through the wall. More recently, two-dimensional models have been employed to take account of spatial non-uniformity of the sites of entry of materials through the endothelium and inhomogeneity within individual layers such as the internal elastic lamella (Yuan et al.1991).

MECHANICAL FORCES AND VESSEL WALL TRANSPORT

Blood vessel walls are subjected to relatively large forces associated with the blood pressure and haemodynamic shearing stresses. These forces vary temporally, not only within the cardiac cycle, but also with a large variety of time-constants, in response to many physiological stimuli. They can have a direct effect on transport, by determining, for example, the driving force for fluid transport through the wall. In addition, because of the very compliant nature of the vascular wall, the forces can induce changes in tissue structure, which can also modify transport.

Effects of Altered Transmural Pressure

A) Increase in Driving Force for Convective Fluxes: The pressure gradient across the walls of blood vessels sets up a continuous movement of fluid. Although the pressure difference is normally smaller in veins than in arteries, the total flux may be greater in the former, presumably because of their thinner walls (and greater endothelial permeability, which will be discussed later). Water movement through the tissue occurs at relatively low velocities, usually less than 10^{-5} cm s^{-1}, though they are probably greater through restricted spaces such as endothelial cell junctions. Small molecules will therefore have Peclet numbers which are too low for convective processes to contribute to their transport, but this will not necessarily be the case for proteins (Tedgui & Lever 1985). Alterations in water flux caused by changes in transmural pressure may then have considerable effects on the rate of transport of the proteins into and through the vessel wall.

B) Effects of Tissue Compliance: Stretching the vessel wall on increasing the transmural pressure would appear to increase the surface area available for transport into the wall. In view of the complexity of the wall structure, however, account must be taken of transport routes which may be affected in different ways from those expected from gross changes in morphology. For example, changes in the total blood/wall interface may occur mainly by deconvolution of the endothelial cells, with little change in the geometry of the intercell clefts. Similarly, although the thickness of the medial layer will decrease on raising the

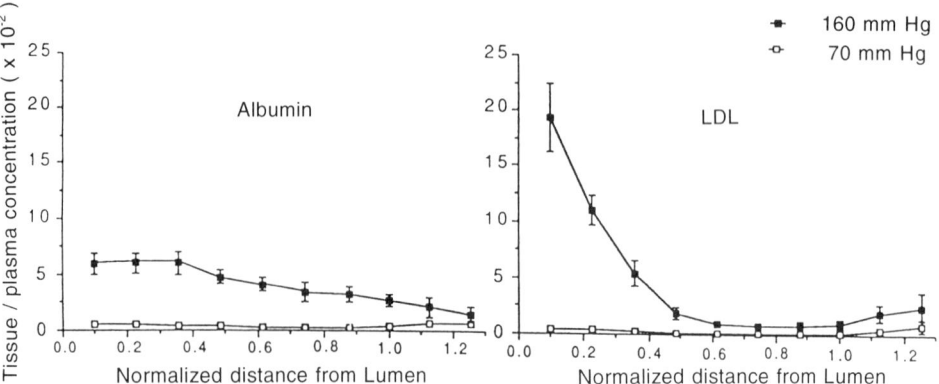

Figure 5. Graphs showing the relative concentrations of LDL (molecular radius approx 11nm) and albumin (molecular radius approx 3.5nm) in the wall of the rabbit aorta when the transmural pressures were 70 and 160mm Hg. Note that at the higher pressure, the concentration of LDL in the inner part of the wall is much higher than that of albumin. (From Curmi et al 1990, Circulation Research, 66: 1692-1702).

pressure, pathways through the matrix may become more tortuous. There have been several reports that the total quantity of protein in the wall is relatively unaffected by transmural pressure (Duncan et al. 1965, Fry et al. 1992, Tedgui & Lever 1987). Care is required in interpreting these results, however, since uptake measurements will under-estimate the total flux through the tissue, if the increasing pressure accelerates the rate of efflux.

Blood vessel wall tissue may undergo compaction, as well as circumferential distension as a consequence of steepening of the pressure gradient through it (Kenyon 1979, Klanchar & Tarbell 1987). This phenomenon might explain the complex relationship which has been observed between pressure and the hydraulic conductivity of the wall (Tedgui & Lever 1984, Baldwin et al.1990). Since tissue compaction will inevitably increase the degree of exclusion of larger molecules, it might also be responsible for increased trapping of lipoproteins, relative to smaller molecules, which is observed at elevated pressures (Curmi et al.1990).

Effects of Altered Blood Flow

The non-uniform uptake of Evans blue dye by the vessel wall (Packham et al.1967), gave rise to the idea that transport through the intimal surface was sensitive to the shear forces associated with blood flow over the surface. Mechanisms proposed for this dependence have included shear-induced damage to the endothelial cells (Fry 1968) and the formation of unstirred, mass transport boundary layers (Caro et al.1971). Various attempts have been made to investigate the effects of shear, either by perfusing vessels under carefully controlled flow conditions *in vitro*, or by the modification of blood flow patterns *in vivo* by the surgical formation of stenoses or fistulae. Some of these studies have been criticized because the techniques used may have induced endothelial damage. Experiments which were designed to overcome these problems have demonstrated a small dependence of the transport of proteins into the wall (Berceli et al.1990) and of transmural fluid transport (Lever et al. 1992b). This is not surprising, in view of the numerous responses of the vascular endothelium to shear, such as altered cell morphology (Davies et al.1983), and the release of agents known to affect permeability including histamine (Hollis & Rosen 1972) and nitric oxide (Yuan et al.1992).

Chronic Changes in Mechanical Forces

Long term alteration of pressure, for example in hypertension, causes extensive tissue remodelling which can alter the transport resistance of the tissue. Different changes may occur in the permeability of the endothelium and media (Tedgui et al. 1992). The media is invariably thickened, with the incorporation of dense fibrous matrix which may hinder the outward efflux of proteins. It is particularly interesting to note, therefore, that when pressure is elevated in veins, following their use as arterial grafts, or in pulmonary vessels in pulmonary hypertension, these previously unsusceptible vessels become rapidly prone to the development of atherosclerosis. Changes in wall structure also result from modification of normal blood flow patterns and these are also expected to alter transport processes.

REGIONAL VARIATION IN THE TRANSPORT OF MATERIALS ACROSS THE WALL

As early as the 1950's, it had been observed that the intimal surface of the aorta showed a spatially non-uniform uptake of Evans blue dye (McGill et al.1957), which when injected into the blood binds to albumin and enters the vessel wall attached to the protein. Different authors have described blue and white patches located at similar sites in the large arteries of dogs, rabbits, monkeys, pigs and pigeons (Fry 1977).

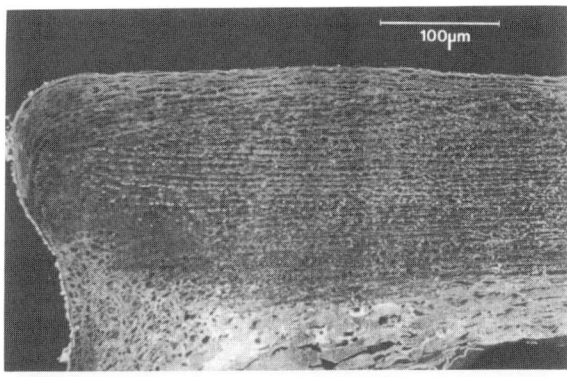

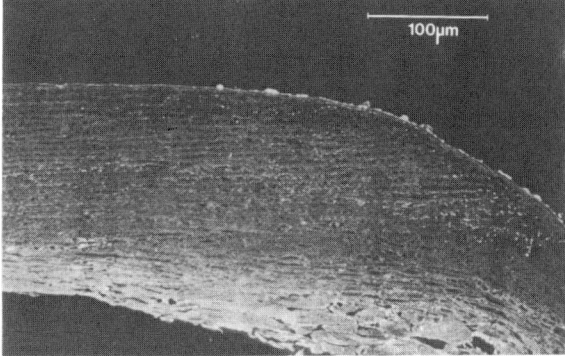

Figure 6. Fluorescence of lissamine-rhodamine labelled albumin in regions both upstream and downstream of the intercostal ostium of the rabbit aorta. Note how the albumin fluorescence reflects different features of the wall structure, with particularly high levels in the adventitia and no fluorescence from the medial smooth muscle cells. (From Weinberg, Atherosclerosis 1988, 74:139-148).

Variation of Transport Within Large Arteries

Many studies have been performed, using a wide variety of different tracers and experimental animals, of protein uptake at different sites along the length of the aorta. The results invariably confirm the earlier observations with Evans blue that uptake is greatest in the ascending aorta and the aortic arch and then progressively falls peripherally (Duncan et al 1963). Some authors have reported that the decrease is not uniform and that there is a slight rise around the diaphragm (Nielsen et al. 1992). These changes have been related to many factors, including the morphology of the endothelial cells (Gerrity et al.1977), and the structure of the endothelial glycocalyx (Haldenby et al. 1994).

Variation around the Ostia of Branches: One of the more obvious features of Evans blue uptake studies has been the non-uniformity of uptake around sites of branching from the aorta. The commonest finding has been that uptake is greatest in regions distal to the flow dividers of the ostia (Caro et al 1971, Bell et al. 1972). At bifurcations, there appears to be a good correlation between uptake and the properties of the glycocalyx (Gorog & Born 1983). When examining uptake at such sites, quantitative fluorescence microscopy has been particularly valuable, because of the high spatial resolution of the technique.

Differences Between Systemic Arteries, Veins and Pulmonary Vessels

Compared with numerous studies which have been performed on arteries, there have been far fewer experiments on the transport properties of other large vessels within the circulation. However, over recent years it has emerged that the flux of several materials occurs more rapidly across the intimal surfaces of veins and pulmonary vessels than those of arteries. Higher rates of entry have been observed for lipids and several different plasma proteins (Christensen et al. 1982, Tompkins 1989, Chuang et al. 1990, Jay & Lever 1992). The differences seen between the large vessels may reflect those which are observed in the microcirculation between arterioles and venules (Michel & Levick 1977). The pulmonary artery has also been shown to be much more permeable to small molecules (Glatz and Massaro 1976), probably reflecting differences in glycosaminoglycan composition and content between the vessels.

The transmural pressures are normally much smaller in veins and pulmonary vessels than in arteries and so less convective transport of larger molecules might be expected.

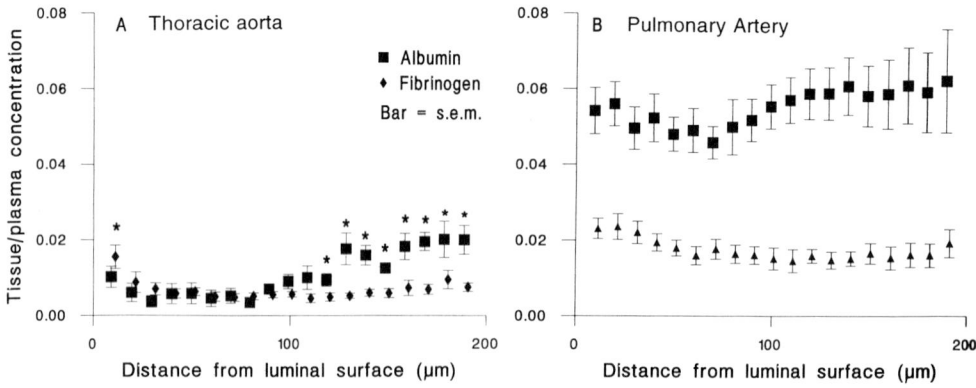

Figure 7. The distribution of fibrinogen and albumin after simultaneous uptake into the walls of the rabbit aorta and pulmonary artery. (from Lever and Coleman, 1994, J. Physiol. 475P:115).

However, the hydraulic conductivity may be lower in the latter and the pressure gradients may also be smaller because of the greater thickness of their walls (Lever & Jay 1993).

Differences are also seen in the medial layers of large blood vessels, the pulmonary artery and veins having much higher partition coefficients for proteins than arteries (Lever and Jay 1990). This behaviour probably is a reflection of structural differences arising from the different transmural pressures that are borne by these vessels. A consequence of these different partition coefficients is that thicker walled arteries appear to retain more of the larger atherogenic proteins such as LDL and fibrinogen than smaller proteins such as HDL. In contrast, the medial layers of veins and pulmonary vessels do not appear to preferentially trap the larger proteins (Lever and Coleman 1994).

Intimal "Hot Spots"

When visible tracers, such as Evans blue dye or horse-radish peroxidase are introduced into the circulation for short periods of time, they are found to be localized in the wall in spots, varying in dimensions from 50 to a few hundred microns (Stemerman et al. 1986, Lin et al 1988). The spots are observed in veins and pulmonary vessels as well as in arteries (Jay & Lever 1992) and are found in regions where there is enhanced uptake of radio-labelled proteins. Evans blue is particularly suitable for demonstrating these sites because once the albumin-dye complex enters the wall, the dye is preferentially bound to elastin and it remains localized close to the site of entry (Adams 1981). With appropriate illumination, the dye fluoresces, and so its distribution can be studied not only by *en face* observation, but also in transverse sections of the vessel wall. These also show that protein enters the wall not only through the 'hot spots' but also through the whole of the luminal surface (Lever & Jay 1994).

The spots are not uniformly distributed varying along the length of the aorta and around the ostia of branches (Barakat et al. 1992). In rabbits, the highest concentration of spots was in the ascending aorta (0.2 - 3 spots per mm^2 of endothelial surface), progressively declined in the thoracic aorta (0 - 1 spot per mm^2), and then increased again in the abdominal aorta, particularly around the ostia of the main visceral arteries. This distribution may be consistent with a role of haemodynamic shearing stresses in determining the formation of these spots, but it appears that either the relationship is not a simple one or that the patterns

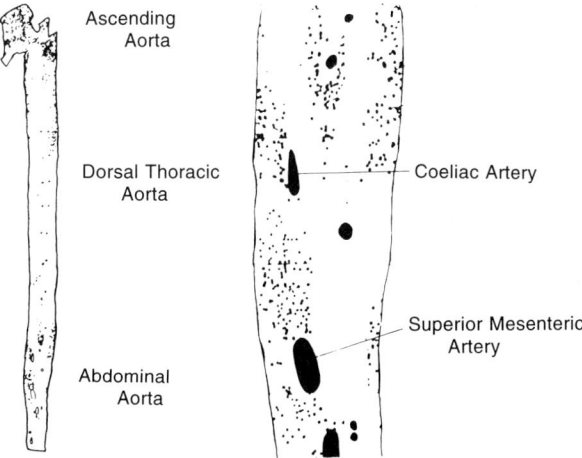

Figure 8. Schematic diagram showing the sites of hot spots along the whole length and in the abdominal aorta of one rabbit aorta (From Barakat et al, J Biomechanical Engineering 1992, 114: 283-292).

of blood flow are extremely complex. Many, but not all of the spots appear at sites where endothelial cells are undergoing mitosis (Lin et al 1990). Around ostia, proteins such as IgG have been found to be taken up into endothelial cells, implying that they may be abnormal in these areas (Hansson et al. 1980).

VARIATIONS IN VESSEL WALL TRANSPORT IN RELATION TO REGIONAL DIFFERENCES IN SUSCEPTIBILITY TO ATHEROSCLEROSIS

It has commonly been believed that those regions where blood components can enter the wall most rapidly, are those where lesion development is most likely to occur. This has been found to be the case in many animal experiments in which blood lipid levels have been dramatically elevated above normal levels (Nielsen et al. 1992), but evidence is mounting that this is not the case in more 'normal' pathophysiological conditions. Many studies have now shown that pulmonary blood vessels and veins exhibit higher transport rates than most arteries, though the former are invariably spared from disease. Even in naturally susceptible animals such as the white carneau pigeon, LDL has been shown to enter the upper part of the aorta, which is spared from the disease, more rapidly than the more distal regions, where the lesions are more commonly found (Schwenke and St Clair 1993).

Role of Transformation or Efflux in the Atherosclerotic Process

The observations described above suggest that the rate of influx of protein or lipid cannot account on its own for the accumulation of intimal material during the development of atherosclerotic lesions. An additional process which must be considered is the rate of efflux, which will depend on the permeability of the luminal endothelium and on the ease with which materials can leave the wall across the outer surface via the medial layer (Nordestgaard et al. 1990). When the entry of material into the wall is grossly elevated, for example in lipid-fed animals, the rate of efflux may be unable to keep pace with the rate of influx and accumulation will inevitably follow. Such a mechanism may account for the different pattern of localization of lesions observed in patients with familial hyperlipidaemia; their lesions are much more prevalent in the more highly permeable regions such as the aortic arch, than is the case in normal subjects (Sugrue et al. 1981).

Chemical transformation of plasma components within the wall tissue will alter chemical potential gradients both for the incoming materials and also for the reaction products. This can simultaneously increase the rate of influx but also increase the rate of efflux of the transformed material. An important example is an alteration of the blood-tissue gradients resulting from the interconversion of cholesterol esters and free cholesterol within the wall. Indeed, it has been proposed that oestrogens may exert an anti-atherogenic effect, by modifying the activity of cholesterol ester hydrolase and synthetase and thereby freeing cholesterol which can be removed from the tissue by HDL (Hough & Zilversmit 1986).

In atherogenesis, considerations other than the steady state wall concentrations of atherogenic proteins are likely to be relevant. The products of chemical transformations may cause an adverse reaction in the tissue such as a chemotactic response for monocytes, as is the case when lipoproteins become oxidised (Chatterjee 1992). Fibrinogen and its degradation products are known to stimulate smooth muscle cell proliferation and migration (Thompson et al. 1987, Naito et al. 1990). In these cases the controlling factors may be both the local concentration of the reactants and their residence time within in the tissue. Any conditions which bring about either a reduction in tissue concentration or an increase in residence time may be palliatives.

ULTRAFILTRATION OF PROTEINS BY THE VASCULAR WALL

Most tissues are very much more permeable to water than to macromolecules and when there is an appreciable convective flux, there may be a sieving of the larger molecules by the tissue matrix. When the solvent drag induced movement of molecules towards a potentially sieving layer is greater than diffusion in the opposite direction, then a region of elevated concentration can arise, a so-called concentration polarization zone, or an unstirred layer. If there is stirring, or a component of fluid flow parallel to the surface, the polarized layers will tend to disperse. Within blood vessels there is normally a vigorous flow of blood over the endothelial surface which is much greater than the convective flow of fluid driven by the pressure gradient between the lumen and the exterior of the vessel. Concentration polarization is only likely to occur at sites where there are stagnation zones within the blood or regions of exceptionally low shear stresses (Colton et al.1982), and may occur within the intercellular clefts below the blood-wall interface (Lever et al.1992). It is also possible that concentration polarization can occur more deeply within the wall tissue because of the heterogeneity of the wall structure and particularly because of the presence of the internal elastic lamella between the intima and the media, and the numerous lamellae throughout the medial layer.

Barrier Properties of the Internal Elastic Lamella and the Medial Layer

It has been suggested that the internal elastic lamella, a layer of dense elastin, which forms a boundary between the intimal and medial layers of the wall, can act as a transport barrier, thereby causing the accumulation of plasma components within the intima (Sims 1989, Svendsen et al. 1990). This could explain why the protein concentration in the interstitium of the intimal layer can rise to very high levels (Smith & Staples 1982, Frank & Fogelman 1989). Following intimal damage, atherosclerotic lesions tend to form under regenerating endothelial cells rather than in denuded areas (Minick et al. 1977); plasma components would only be able to accumulate to excessively high levels upstream of the intimal-medial barrier, if their back-diffusion into the blood is impeded by an endothelial layer.

However, it is observed that even the largest labelled proteins can enter the medial layer from the intimal surface suggesting that they penetrate this layer relatively easily. Numerous fenestrae are present within the internal elastic lamella, and these could offer a route between the intima and medial layers (Song & Roach 1983).

TRANSMURAL TRANSPORT

It has been proposed that many materials can pass across the whole thickness of the wall from the luminal surface to the adventitia (Adams et al.1964, Walton 1975). As indicated above, the chemical potential gradients between the blood and the tissue may be exceptionally high for tracer materials introduced into the circulation. Under natural conditions, gradients may be very much less because the proteins will be more uniformly distributed across all tissue compartments. However, because of the less hindered movement of water relative to the native macromolecules, there will always be a concentration gradient between blood and lymph. There is an extensive network of lymphatic vessels in the adventitial layers of arteries, veins and pulmonary vessels (Johnson 1969, Sacchi et al.1990) and these can act as sinks not only for material originating from the adventitial vasa vasorum but also for material crossing the wall.

Direct evidence for transmural transport has been provided by *in vitro* experiments, in which labelled proteins have been shown to move from the lumen of arteries and veins into a surrounding bath, in the absence of any perforating branch vessels (Jay & Lever 1988, Lever & Jay 1993). Absence of Evans blue staining in these vessels indicated that there was no endothelial damage (Bjokerud & Bondjers 1972). Both albumin and fibrinogen were found to penetrate the whole thickness of the wall very rapidly - within a few minutes, and at a rate much greater than predicted from their rates of accumulation within the medial tissue. This could indicate the presence of channels through the tissue (presumably less dense regions of the extracellular space) which can act as rapid transit pathways. The steady state rates of transport across the rabbit carotid artery with functionally intact endothelial cells were approximately 5×10^{-8} cm.s^{-1} for albumin and 4×10^{-9} cm.s^{-1} for fibrinogen. Fluxes across the inferior vena cava were 3×10^{-7} cm.s^{-1} for albumin and 0.9×10^{-7} cm.s^{-1} for fibrinogen.

Attempts have also been made to study transmural flux in vivo using collars placed around blood vessels to collect fluid and solutes passing through the vessels wall (Allen et al.1988, Coleman et al.1993). These studies have shown that proteins as large as fibrinogen and LDL can pass readily across the whole thickness of the wall, but at rates which are lower than those of albumin or HDL. The magnitudes of the fluxes may be greatly in excess of those normally prevailing *in vivo*, because of inflammatory changes induced by the collars. As a result, fibrinogen and LDL concentrations within the wall tissue of the collared vessels are elevated up to 50 times those of normal vessels and to a proportionately greater extent than HDL or albumin concentrations (Coleman et al.1994). These observations are consistent with ultrafiltration mechanisms described above causing fractionation of the differently sized proteins with preferential retention of the larger ones in the tissue. The collared vessels undergo extremely accelerated intimal thickening, a process which may be related to the augmented tissue concentrations of the atherogenic proteins.

With a continuous flow of proteins across the wall, the local concentration of each at different sites within the tissue will be a reflection of the porosity of the layers through which they move. In this context, it is interesting to note that there is marked variation not only in the endothelial, but also the medial permeability among different large vessels of the circulation (Lever & Jay 1990). Thus in the pulmonary vessels which appear to have both high permeability of the luminal endothelial layer and a high medial porosity, there will be little tendency for proteins to accumulate within the tissue. In peripheral systemic arteries, proteins will cross the endothelial surface more slowly but they will be retarded by a medial layer of low porosity and may then accumulate to a greater degree.

EFFECTS OF DRUGS AND HORMONES ON WALL UPTAKE

There is great potential importance in findings ways of modulating transport processes so as to delay the development of atherosclerosis and to cause its regression. Many naturally occurring and synthetic compounds have been shown to alter both rates of transport across the endothelial surface and also retention within the wall. Sometimes, the effects of agents will be adverse. For example, long-term exposure of animals to nicotine increases the number of 'hot spots' on the intimal surface of the aorta (Lin et al.1992) and thereby increases the rate of influx of proteins. In some cases changes may be due, in part, to alteration of other factors such as blood pressure, particularly by agents such as catecholamines (Cardona Sanclemente & Born 1992). Thus Angiotensin II appears to alter luminal flux both by changing the permeability of the vessel wall (Robertson & Khairallah 1972) and by altering the driving pressure for convective flux (Feig et al. 1982). It has more marked effects on transport of atherogenic proteins into normal than hypertensive animals (Cardona Sanclemente et al. 1994)

The porosity of the medial layer appears to depend on vascular smooth muscle tone which can be altered by both vasodilators and vasoconstrictors (Caro & Lever 1983, Lever et al. 1992a). Altered porosity is likely to change both the turnover rate of potentially atherogenic proteins, and their accumulation.

EFFECTS OF LESION DEVELOPMENT ON TRANSPORT INTO THE WALL

Transport processes do not occur at a uniform rate once an atherosclerotic lesion starts to develop. In general the permeability of the endothelial layer tends to increase (Nicoll et al. 1981) giving rise to a potential positive feedback accumulation within the intima, with an elevation of the rate of accumulation of LDL, particularly when aggravated by high levels of blood cholesterol (Schwenke & St Clair 1993). Lesions may not be uniformly permeable, however, and in uptake studies with Evans blue, it was noted that transport occurred at an elevated rate around their margins (Packham et al 1967).

ACKNOWLEDGMENT

Much of the work on large vessel mass transport in the author's group has been supported by the British Heart Foundation.

REFERENCES

Adams, C.W.M., Bayliss, O.B., Davison, A.N. and Ibrahim, M.Z.M., 1964, Autoradiographic evidence for the outward transport of ^{3}H-cholesterol through rat and rabbit aortic wall, *J. Path. Bact.* 87:297-304.
Adams, C.W.M., 1981, Permeability in atherosclerosis - an artefact with Evan's blue staining, *Atherosclerosis*, 39: 131-132.
Allen, D.R., Browse, N.L., Rutt, D.L., Butler, L. and Fletcher, C., 1988, The effect of cigarette smoke, nicotine and carbon monoxide on the permeability of the arterial wall, *J. Vasc. Surg.* 7:139-152.
Baldwin, A.L., Chien, S., 1984, Endothelial transport of anionized and cationized ferritin in the rabbit aorta and vasa vasorum, *Arteriosclerosis* 4:372-382.
Baldwin, A.L., Wilson L.M. and Simon, B.R., 1992, Effect of pressure on aortic hydraulic conductance, *Arteriosclerosis & Thrombosis,* 12:163-171.
Barakat, A.I., Uhthoff, P.A.F. and Colton, C.K., 1992, Topographical mapping of sites of enhanced HRP permeability in the normal rabbit aorta, *J. Biomechanical Eng.* 114: 283-292.
Bell, F.P., Somer, J.B., Craig, I.H. and Schwartz, C.J., 1972, Patterns of aortic Evans blue uptake *in vivo* and *in vitro*. *Atherosclerosis,*16: 369-375.
Bell, F.P., Gallus, A.S. and Schwartz, C.J., 1974, Focal and regional patterns of uptake and the transmural distribution of ^{131}I-fibrinogen in the pig aorta in vivo. *Exp. Mol. Pathol.* 20: 281-292.
Berceli S.A., Warty V.S., Sheppeck R.A., Mandarino W.A., Tanksale S.K. and Borovetz H.S., 1990, Hemodynamics and very low density lipoprotein metabolism. *Arteriosclerosis.* 10: 688-694.
Björkerud, S. and Bondjers, G., 1972, Endothelial integrity and viability in the aorta of the normal rabbit and rat as evaluated with dye exclusion tests and interference contrast microscopy, *Atherosclerosis* 15: 285-300.
Bratzler, R.L., Chisolm, G.M., Colton, C.K., Smith, K.A., Zilversmit, D.B. and Lees, R.S., 1977a The distribution of labelled albumin across the rabbit thoracic aorta in vivo, *Circulat. Res.* 40: 182-190.

Bratzler, R.L., Chisolm, G.M., Colton, C.K., Smith, K.A. and Lees, R.S., 1977b, The distribution of low density lipoproteins across the rabbit thoracic aorta in vivo, *Atherosclerosis* 28:289-307.

Bundgaard, M., 1984, The three dimensional organization of tight junctions in a capillary endothelium revealed by serial scanning electron microscopy, *J. Ultrastruct. Res.* 88:1-17.

Cardona Sanclemente, L.E., and Born G.V.R., 1992, Adrenaline increases the uptake of low density lipoprotein by carotid arteries of rabbits, *Atherosclerosis* 91: 215-218.

Cardona Sanclemente, L.E., Medina, A. and Born G.V.R.,, 1994, Effect of increasing doses of angiotensin II infused into normal and hypertensive wistar rats on low density lipoprotein and fibrinogen uptake by artery walls, *Proc. Nat. Acad. Sci.* 91: 3285-3288.

Caro C.G., Fitzgerald J.M., and Schroter R.C. 1971, Atheroma and arterial wall shear - Observation, correlation and proposal of a shear dependent mass transfer mechanism for atherogenesis. *Proc. Roy Soc (London) B.*177:109-159.

Caro C.G. and Lever M.J., 1983, Effect of vasoactive agents and applied stress on the albumin space of excised rabbit carotid arteries, *Atherosclerosis* 46:137-146.

Cary, N, Jay, M.T. and Lever, M.J., 1987, Distribution volumes of albumin in the intima, media and adventitia of human mesenteric arteries, *J. Physiol.* 388: 26P.

Chatterjee, S., 1992, Role of oxidized human plasma low density lipoproteins in atherosclerosis - effects on smooth muscle cell proliferation, *Mol. Cell. Biochem.* 111:143-147.

Chobanian, A.V., Menzoian, J.O., Shipman, J., Heath, K. and Haudenschild, C.C., 1983, Effects of endothelial denudation and cholesterol feeding on in vivo transport of albumin, glucose, and water across rabbit carotid artery. *Circulat. Res.* 53:805-814.

Christensen,S., Stender, S., Nyvad, O. and Bagger, H., 1982, In vivo fluxes of plasma cholesterol, phosphatidylcholine and protein into mini-pig aortic and pulmonary segments. *Atherosclerosis* 42: 309-319.

Chuang, P-T., Cheng, H-J., Lin S-J., Jan K-M., Lee, M.M.L. and Chien, S., 1990, Macromolecular transport across arterial and venous endothelium in rats. *Arteriosclerosis* 10: 188-197.

Coleman P.J., Lever M.J. and Martin J.F., 1993, Transarterial flux in the anaesthetized rabbit, *J. Physiol.* 467:41P.

Coleman P.J., Lever M.J. and Martin J.F., 1994, Fibrinogen retention in rabbit carotid arteries *in vivo* following application of a silastic collar, *J. Physiol.* 475P: 115-116.

Colton, C.K., Friedman, S., Wilson, D.E. and Lees, R.A., 1972, Ultrafiltration of lipoproteins through a synthetic membrane, *J. Clin. Invest.* 51: 2472-2481.

Curmi, P.A., Renaud, G., Juan, L., Chiron B. and Tedgui, A., 1989, Role of LDL receptors in the *in vitro* uptake and degradation of LDL in the media of rabbit aorta, *Circulat. Res* . 64:957-966.

Curmi, P.A., Juan, L. and Tedgui, A. 1990, Effect of transmural pressure on low density lipoprotein and albumin transport and distribution across the intact arterial wall, *Circulation Res.* 66: 1692-1702.

Curry F.E. 1984, Mechanics and thermodynamics of transcapillary exchange. In Handbook of Physiology. Section 2: The cardiovascular system. Volume IV. Microcirculation, Part 1. Ed E.M. Renkin & C.C. Michel. American Physiological Society, Bethesda, Maryland.

Davies, P.F., Dewey, C.F., Bussolari, S.R., Gordon, E.J. and Gimbrone, M.A., 1983, Influence of haemodynamic forces on endothelial cell function. *In vitro* studies of shear stress and pinocytosis in bovine aortic endothelial cells, *J. Clin. Invest.*, 73:1121-1129.

Duncan, L.E., Buck, K. and Lynch, A. 1963, Lipoprotein movement through canine aortic wall, *Science,* 142: 972-973.

Duncan, L.E., Buck, K. and Lynch, A., 1965, The effect of pressure and stretching on the passage of labelled albumin into canine aortic wall. *J. Atherosclerosis Res.* 5: 69-79.

Feig, L.A., Pappas, N.A., Colton, C.K., Smith, K.A. and Lees, R.S., 1982, The effect of angiotensin II on *in vivo* albumin transport in the rabbit, *Atherosclerosis,* 44: 307-318.

Frank, J.S. and Fogelman, A.M., 1989, Ultrastructure of the intima in WHHL and cholesterol-fed rabbit aortas prepared by ultra-rapid freezing and freeze-etching, *J. Lipid Res.* 30:967-978.

Frøkjær-Jensen, J., 1980, Three-dimensional organization of plasmalemmal vesicles in endothelial cells. An analysis by serial sectioning of frog mesenteric capillaries, *J. Ultrastruct. Res.* 73:9-20.

Fry, D.L., 1968, Acute vascular endothelial changes associated with increased blood velocity gradients, *Circulation Research.* 22: 165-197.

Fry, D.L., 1977, Aortic Evans blue accumulation: its measurement and interpretation, *Am. J. Physiol.* 232:H204-H222.

Fry, D.L., 1987, Mass transport, atherogenesis and risk, *Arteriosclerosis.* 7:88-100.

Fry, D.L., Haupt, M.W. and Pap, J.M. 1992, Effect of endothelial integrity, transmural pressure, and time on the intimal-medial uptake of serum ^{125}I-albumin and ^{125}I-LDL in an in vitro porcine arterial organ-support system, *Arteriosclerosis and Thrombosis,* 12: 1313-1328.

Gerrity R.G., Richardson M., Somer J.B., Bell F.P. and Schwartz C.J., 1977, Endothelial cell morphology in areas of *in vivo* Evans blue uptake in the aorta of young pigs. II Ultrastructure of the intima in areas of differing permeability to plasma proteins, *Am. J. Pathol.* 89:313-334.

Glatz, C.E. and Massaro, T.A., 1975, Influence of glycosaminoglycan content on mass transfer behaviour of porcine aortic wall, Part 1. Diffusive transport of $^{45}Ca^{2+}$ and ^{3}HHO, *Atherosclerosis,* 25:153-163.

Gorog P. and Born G.V.R., 1983, Uneven distribution of sialic acids on the luminal surface of aortic endothelium, *Brit. J. Exp. Pathol.* 64:418-424.

Haldenby K.A., Chappell, Winlove C.P., Parker K.H. and Firth J.A., 1994, Focal and regional variations in the composition of the glycocalyx of large vessel endothelium, *J. Vasc. Res.* 31: 2-9.

Hansson. G.K., Bondjers., G., Bylock A. and Hjalmarsson L., 1980, Ultrastructural studies on the localization of IgG in the aortic endothelium and the subendothelial intima of atherosclerotic and nonatherosclerotic rabbits, *Exp. Mol. Pathol.* 33:302-315.

Hollis T.M. and Rosen L.A., 1972, Histidine decarboxylase activity of bovine aortic endothelium and intima media, *Proc. Soc. Exp. Biol.* 141: 978-981.

Hough, J.L. and Zilversmit, D.R., 1986, Effects of 17- estradiol on aortic cholesterol content and metabolism in cholesterol-fed rabbits, *Arteriosclerosis* 6: 57-63.

Jay M.T. and Lever M.J., 1988, Transmural pressure dependence of EDTA and albumin flux across the rabbit isolated carotid artery wall, *J. Physiol.* 407:26P.

Jay, M.T. and Lever, M.J., 1992, Spatial variations in the permeability of the luminal surfaces of large blood vessels of the rabbit. *J. Physiol.* 452: 5P.

Johnson, R.A., 1969, Lymphatics of blood vessels, *Lymphology*, 2:44-56.

Kenyon, D.E., 1979, A mathematical model of water flux through aortic tissue, *Bull. Math. Biol.* 41:79-90.

Kim, M-J., Dawes, J. and Jessup, W., 1994, Transendothelial transport of modified low-density lipoproteins. *Atherosclerosis.* 108: 5-17.

Klanchar, M. and Tarbell J.M., 1987, Modelling water flow through arterial tissue, *Bull. Math. Biol.* 49:651-669.

Klinowski, J., Korsner, S.E. and Winlove, C.P., 1982, Problems associated with the micropolarographic measurement of the arterial wall pO_2. *Cardiovascular Research* .16: 448-456.

Lever, M.J. and Coleman, P.J., 1994, Fractionation of ^{125}I-fibrinogen and ^{131}I-albumin by rabbit vascular tissue in vivo, *J. Physiol.* 475P: 115.

Lever, M.J. and Jay, M.T., 1990, Albumin and Cr-EDTA uptake by systemic arteries, veins and pulmonary artery of rabbit, *Arteriosclerosis* 10:551-558.

Lever, M.J. and Jay, M.T., 1991, The endothelial permeability to albumin of various large blood vessels in the anaesthetized rabbit., *J. Physiol.* 435: 5P.

Lever M.J. and Jay M.T., 1993, Convective and diffusive transport of plasma proteins across the walls of large blood vessels, *Frontiers in Medical and Biological Engineering*, 5: 67-72.

Lever M.J. and Jay M.T., 1994, The role of the endothelial and medial layers in the transport of plasma proteins in the walls of large blood vessels, In Recent Progress in Cardiovascular Mechanics, Ed. Hosoda, S., et al Harwood Academic Publishers, Chur, Switzerland.

Lever, M.J., Jay M.T., Debaer F. and Thome G., 1992a, Effect of vasomotor tone on the hydraulic conductivity and albumin reflection coefficient of the medial layers of isolated rabbit aorta and carotid artery, *J. Physiol.* 452: 18P.

Lever, M.J., Tarbell, J.M. and Caro, C.G., 1992b, Effect of luminal flow in rabbit carotid artery on transmural fluid transport, *Experimental Physiol.* 77:553-563.

Lin, S-J. Jan, K-M., Schuessler, G., Weinbaum, S. and Chien, S., 1988, Enhanced macromolecular permeability of aortic endothelial cells in association with mitosis, *Atherosclerosis* 73: 223-232.

Lin, S-J., Hong, C-Y., Chang, M-S., Chiang, N.B. and Chien, S., 1992, Long-term nicotine exposure increases aortic endothelial cell death and enhances transendothelial macromolecular transport in rats, *Arteriosclerosis & Thrombosis,* 12:1305-1312.

McGill, H.C., Geer, J.C., Holman, R.L., 1957, Sites of vascular vulnerability in dogs demonstrated by Evan's blue, *Arch. Pathol.* 64: 303-311.

Michel, C.C., and Levick, J.R., 1977, Variation in permeability along individually perfused capillaries of the frog mesentery, *Quart. J. Exp. Physiol.* 62: 1-10.

Minick, C.R., Stemerman, M.B. and Insull, W. Jr., 1977, Effect of regenerating endothelium on lipid accumulation in the arterial wall, *Proc. Nat. Acad. Sci.* 74:1724-1728.

Naito, M., Hayashi, T., Kuzuya, M., Funaki, C., Asai, K. and Kuzuya, F., 1990, Effect of fibrinogen and fibrin on the migration of vascular smooth muscle cell in vitro, *Atherosclerosis*, 83: 9-14.

Nicoll A., Duffield R. and Lewis B., 1981, Flux of lipoproteins into human arterial intima, *Atherosclerosis* 39:229-242.

Nielsen, L.B., Nordestgaard, B.G., Stender, S. and Kjeldsen, K., 1992, Aortic permeability to LDL as a predictor of aortic cholesterol accumulation in cholesterol-fed rabbits, *Arteriosclerosis and Thrombosis,* 12: 1402-1409.

Nordestgaard, B.G., Hjelms, E, Stender, S. and Kjeldsen, K., 1990, Different efflux pathways for high and low density lipoproteins from porcine aortic intima. *Arteriosclerosis.* 10:477-485.

Packham, M.A., Rowsell, H.C., Jorgensen, L. and Mustard, J.F. 1967 Localized protein accumulation in the wall of the aorta, *Exp. Mol. Pathol.* 7:214-232.

Ramirez, C.A., Colton, C.K., Smith, K.A., Stemerman, M.B. and Lees, R.S., 1984, Transport of ^{125}-albumin across normal and deendothelialized rabbit thoracic aorta in vivo, *Arteriosclerosis* 4:283-291.

Robertson, A.L., and Khairallah P.A., Effects of Angiotensin II and some analogues on vascular permeability in the rabbit, *Circulat. Res.* 31:923-931.

Ross, R, 1993, The pathogenesis of atherosclerosis: a perspective for the 1990's, *Nature,* 362: 801-809.

Sacchi G., Weber E., and Comparini L., 1990, Histological framework of lymphatic vasa vasorum of major arteries - an experimental study, *Lymphology*, 23:135-139.

Schwenke, D. and Carew, T.E, 1989, Initiation of atherosclerotic lesions in cholesterol-fed rabbits: I. Focal increases in arterial LDL concentration precede development of fatty streak lesions, *Arteriosclerosis* 9: 895-907.

Schwenke, D.C. and St Clair, R.W. 1993, Influx, efflux and accumulation of LDL in normal arterial areas and atherosclerotic lesions of white carneau pigeons with naturally occurring and cholesterol aggravated aortic atherosclerosis, *Arteriosclerosis and Thrombosis* 13: 1368-1381.

Silberberg A., 1989, Transport through deformable materials. *Biorheology* 26:291-313.

Sims F.H., 1989, The internal elastic lamina in normal and abnormal human arteries. A barrier to the diffusion of macromolecules from the lumen, *Artery* 16: 159-173.

Smith E.B., Staples E.M., Dietz H.S. and Smith R.H., 1979, Role of endothelium in sequestration of lipoprotein and fibrinogen in aortic lesions, thrombi, and graft pseudo-intimas, *Lancet* 2:812-816.

Smith, E.B. and Staples E.M., 1980, Distribution of plasma proteins across the human aortic wall: Barrier function of endothelium and internal elastic lamina, *Atherosclerosis* 37: 579-590.

Smith, E.B. and Staples E.M., 1982, Plasma protein concentrations in interstitial fluids from human aortas. *Proc. Roy. Soc. Lond. (B)* 217: 59-75.

Song, S.H. and Roach, M.R., 1983, Quantitative changes in the size of fenestrations of the elastic laminae of sheep thoracic aorta studies with SEM, *Blood Vessels* 20: 145-153.

Stary, H.C., 1989, Evolution and progression of atherosclerotic lesions in coronary arteries of children and young adults, *Arteriosclerosis,* 9 19-32.

Stemerman, M.B., Morrel, E.M., Burke, K.R., Colton, C.K., Smith K.A. and Lees, R.S., 1986, Local variation in arterial wall permeability to low density lipoprotein in normal rabbit aorta, *Arteriosclerosis.* 6: 64-69.

Stender, S., and Christensen S., 1977, The concomitantly measured transfer of free cholesterol, esterified cholesterol, phospholipids and phosphoprotein from plasma into the aortic wall of stilboestrol-treated cockerels, *Atherosclerosis,* 28:15-28.

Stender S. and Hjelms E., 1988, In vivo transfer of cholesterol ester from high and low density plasma lipoproteins into human aortic tissue, *Arteriosclerosis* 8:252-262.

Svendsen E., Dregelid E. and Eide G.E., 1990 Internal elastic membrane in the internal mammary artery and left descending coronary arteries and its relationship to intimal thickening, *Atherosclerosis* 83:239-249.

Sugrue D.D., Thompson G.R., Oakley C.M., Trayner I.M. and Steiner R.E., 1981, Contrasting patterns of coronary atherosclerosis in normo-cholesterolaemic smokers and patients with familial hypercholesterolaemia, *Brit. Med. J.* 283:1358-1360.

Tedgui A., and Lever M.J., 1984, Filtration through damaged and undamaged rabbit aorta, *Am. J. Physiol.* 247: H784-H791.

Tedgui A., and Lever M.J., 1987, Effect of pressure and intimal damage on ^{131}I-albumin and ^{14}C-sucrose spaces in aorta, *Am. J. Physiol.* 253: H1530-1539.

Tedgui, A. and Lever, M.J., 1985, The interaction of convection and diffusion in the transport of ^{131}I-albumin within the media of the rabbit thoracic aorta, *Circulation Research* 57: 856-863.

Tedgui, A., Merval, R. and Esposito, B., 1992, Albumin transport characteristics of rat aorta in early phase of hypertension. *Circulation Research.* 71: 932-942.

Thompson, W.D., McGuigan, C.J., Snyder, C., Keen, G.A. and Smith E.B., 1987, Mitogenic activity in human atherosclerotic lesions, *Atherosclerosis* 66:85-93.

Tompkins, R.G., Yarmush. M.L., Schnitzer, J.J., Colton, C.K., Smith, K.A. and Stemerman, M.B., 1989, Low-density lipoprotein transport in blood vessel walls of squirrel monkeys, *Am. J. Physiol.* 257: H452-H464.
Truskey, G.A., Colton, C.K., Smith, K.A., 1981, Quantitative analysis of protein transport in the arterial wall. In Structure and function of the circulation, C.J. Schwartz et al. (Eds),289-355, Plenum Press, New York.
Vasile, E., Simionescu, M. and Simionescu, N., 1973, Visualization of the binding, endocytosis and transcytosis of low density lipoproteins in the arterial endothelium *in situ, J. Cell. Biol.* 96:1677-1689.
Walton K.W., 1975, Pathogenetic Mechanisms in Atherosclerosis, *Am J. Cardiol.* 35:542-558.
Weinberg, P.D., 1988, Application of fluorescence densitometry to the study of albumin uptake by the rabbit aortic wall up- and downstream of intercostal ostia, *Atherosclerosis.* 74: 139-148.
Wiklund, O., Carew, T.E. and Steinberg D., 1985, Role of the low density lipoprotein receptor in penetration of low density lipoprotein into rabbit aortic wall, *Arteriosclerosis* 5:135-141.
Yuan, Y., Granger, H.J., Zawieja, D.C., Chilian, W.M., 1992, Flow modulates coronary venule permeability by a nitric oxide - related mechanism, *Am. J. Physiol.* 263: H641-H646.
Yuan, F., Chien, S. and Weinbaum, S., 1991, A new view of convective and diffusive transport in the arterial intima, *J. Biomech. Eng.* 113: 314-329.

11

BLOOD CROSS FLOW FILTRATION THROUGH ARTIFICIAL MEMBRANES

M. Y. Jaffrin

URA CNRS 858, Dept of Biological Engineering
Technological University of Compiègne
BP 529, 60205 Compiegne Cedex
France

INTRODUCTION

Membrane separation is used in various stages of blood purification. The best known application is hemodialysis used for the treatment of end-stage renal failure (1, 2). In this process, uremic toxins are removed by diffusion through a semi-permeable membrane which separates the circulating blood from an ionic solution, the dialysate.

The membrane must retain plasma proteins and the formed elements while the dialysate contains ions and solutes which must not be removed from the blood. At the same time, excess water (from 1 to 4 l) is removed from the plasma by ultrafiltration resulting from a transmembrane pressure gradient caused by partial clamping of the retentate line or by placing a pump on the permeate. Ultrafiltration also contributes to toxin removal since the filtration flux carries along all solutes smaller than the membrane pore diameter. In hemofiltration proposed in the U.S. by Henderson et al. (3) and in France by Man and Funck Brentano (4) no dialysate is used and the entire toxin removal is achieved by the ultrafiltration flux which must be relatively large (1.5 - 2.0 ml/s during a 3 to 4 h period) and necessitates reinjection to the patient of a substitution solution containing the ions which have been removed with the permeate.

Another application proposed initially by Solomon et al. (5) is the separation of plasma from whole blood by cross flow microfiltration instead of centrifugation. The membrane pore diameters must be between 0.2 and 0.8 μm in order to transmit the largest immunoglobulins while retaining platelets and red cells. This technique can be used for plasma collection from donors. Blood is withdrawn from the donor's vein by a peristaltic pump through a needle. It circulates along the membrane and plasma filtrates under the action of transmembrane pressure. This technique yields a cell free plasma which avoids for the recipient the immunological hazards of contamination by platelets and cellular fragments and is relatively atraumatic for the red cells, if precautions are taken to avoid hemolysis. But the relatively high cost of the disposable filter makes it more expensive than traditionnal centrifugation. Therefore,

Biological Flow, Edited by Michel Y. Jaffrin and Colin Caro
Plenum Press, New York, 1995

it is important to optimize the filtration process in order to maximize the filtration flux and minimize both membrane area and collection time.

The second application of plasma separation is the treatment of autoimmune diseases and of familial hypercholesterolemia by plasma exchange (6, 7) or by plasmafractionation (8, 9). These diseases are due to pathological immuno-globulins which must be removed. Thus two techniques are possible. Either the pathological plasma is discarded after separation from the red cells and replaced by albumin solution, or it is filtrated through a secondary filter equipped with an ultrafiltration membrane which will retain the pathological immunoglobulins and transmit albumin. However, since the size difference between albumin and immunoglobulins is not very large, actual membranes are not selective enough to transmit albumin completely while rejecting all the pathological immunoglobulins and a compromise has to be found. The plasma separation step is similar to the case of plasma collection from donors except that blood flows and extracted plasma volumes are larger (3 l instead of 0.6 l), which puts more severe constraints on the membrane capacity to avoid fouling.

The secondary filtration step is generally carried out in quasi dead end mode with a retentate flow rate exiting from the filter which amounts to less than 15% of the inlet flow in order to minimize albumin losses in the discarded retentate. This technique permits recovery of the patient's albumin and avoids the need of reinfusing a costly albumin solution. But the process needs to be optimized in order to be as selective as adsorption methods such as sulfate dextran columns for LDL cholesterol removal (10).

Flux and Mass Transfer Limitations through Membranes

When the membrane is free from deposited particles or macromolecules, which is the case at the very onset of filtration or when the solution is very dilute, the membrane behaves as a porous medium and the filtration flux J is governed by Darcy's law:

$$J = \frac{1}{\mu R_m} (p_{tm} - \Delta\pi) \tag{1}$$

where p_{tm} is the transmembrane pressure, $\Delta\pi$ the osmotic pressure difference, μ the permeate viscosity and R_m the membrane resistance to filtration.

Concentration Polarization: Since the filtrate flux drags towards the membrane particles and macromolecules which are too large to pass through the membrane pores, a high concentration layer forms near the membrane and grows with time as long as the incident mass flux exceeds the returning flux to the bulk solution. During this short period, the permeate flux falls rapidly since the membrane resistance R_m in eq (1) replaced by $R_m + R_c(t)$ where $R_c(t)$ is the additionnal resistance of a deposited layer of macromolecules or particles. However, it is observed (11) that, after this initial phase, the permeate flux becomes constant or decays more slowly, meaning that the resistance of the deposited layer has reached an equilibrium. This is due to the combination of two phenomena. The incident mass flux to the membrane decreases as the deposited layer thickens and its resistance increases. At the same time, the diffusive mass flux of proteins and red cells to the bulk solution increases due to increasing concentration. When the incident flux is balanced by the returning diffusive mass flux and the flux transmitted through the membrane (in case of partial rejection) a steady state is reached and the local concentration c in the boundary layer is governed by, according to Blatt et al. (12)

$$Jc = D \frac{dc}{dy} + Jc_F \tag{2}$$

where c_F is the concentration in the permeate. For constant solute diffusivity D, eq. 2 may be integrated across the concentration boundary layer of thickness δ (Fig. 1) to give:

$$J = k \ln \frac{c_w - c_F}{c_b - c_F} \quad (3)$$

where $k = D/\delta$ is the mass transfer coefficient c_w and c_b are the membrane and bulk concentration respectively.

For laminar flow in a tube or a hollow fiber the mass transfer coefficient averaged along the tube may be expressed by (12):

$$k = 0.81 \left(\frac{\gamma_w D^2}{L} \right)^{1/3} \quad (4)$$

where γ_w is the wall shear rate and L the tube length. Thus if the permeate flux becomes independent of p_{tm} it means that c_w must have reached a limiting value independent of pressure. For plane flow, the mass transfer coefficient is also given by Eq. 4 but with a numerical coefficient equal to 0.96.

Osmotic Pressure: Another cause of flux limitation in ultrafiltration is the increase in osmotic pressure due to the high membrane concentration. In blood, the osmotic pressure is normally 30 mmHg for a protein concentration of 70 g/l. But since protein concentration at the membrane may exceed 300 g/l (13), the osmotic pressure at the membrane can reach 250 mmHg which causes the permeate flux to drop according to Eq. 1.

Cake Formation: The cake which forms in cross flow microfiltration is a stagnant layer of particles compacted by the transmembrane pressure. According to Romero and Davies (14) the stagnant cake forms on the membrane when the shear stress acting on the cake is not high enough to cause all the particles in the layer to flow. This will occur necessarily at some distance from the inlet since the shear stress decreases as the concentration boundary layer thickens. This cake represents an additionnal resistance to filtration in series with the membrane resistance and decreases the permeate flux. Since the cake layer grows slowly with time and becomes less and less permeable due to its compressibility, it induces a slow decay of permeate flux with time which is usually irreversible without chemical cleaning.

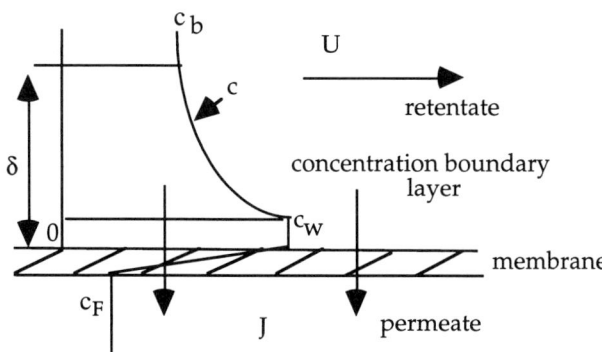

Figure 1. Schematic of concentration polarization

Table 1. Available plasma filters and dimensions

Manufacturer	Model	Membrane	Pores μm	S m^2	L cm
Hollow fiber					
Asahi	Plasmaflo	Cellulose Acetate	0.2	0.55	18.3
Kuraray		Polyvinyl Alcool	0.4	0.5	
Nipro	pF	Cellulose triacet.	0.55	0.5	22.5
Dideco	Hemaplex	Polypropylene	0.55	0.2	15
Dideco	BT 900	Polypropylene	0.55	0.1	12
Fresenius	Plasmaflux	Polypropylene	0.5	0.5/0.25	26
Travenol	CPS 10	Polypropylene	0.55	0.17	21.4
Organon Technica	Curesis	Polypropylene	0.65	0.12	14
Organon Technica	Plasmapur	Polypropylene	0.65	0.07	14.2
Press type					
Cobe	TPE	P V C	0.6	0.13	20

PLASMA SEPARATION FROM WHOLE BLOOD BY STEADY FLOW MICROFILTRATION

The objectives are to obtain a filtrate with a protein concentration as close as possible to the donor's plasma entirely cell free and which represents an important fraction of the incoming blood flow (20 - 30%). This plasma should be obtained with minimal hemolysis and with filters of the least membrane area. Several biocompatible membranes exist for these applications, mostly in hollow fiber form (Table 1).

It is interesting to note the large variation is size of these filters, the recent filters being generally the smallest.

The principle is illustrated on Figure 2. Blood flows tangentially along the membrane and plasma is filtrated through the pores.

Since plasma proteins pass freely through the membrane, concentration polarization concerns only the formed elements namely the platelets and red blood cells. The platelets are smaller (2 μm in diameter) and are less subject to the wall exclusion phenomenon than red cells which are 8 μm discs. However red cells are much more numerous than platelets (5 10^6/mm^3 versus 2 10^5/mm^3). Therefore, it is reasonable to assume that the concentration polarization layer is made up mostly from red cells although platelet concentration seems to play a significant role as we shall see later.

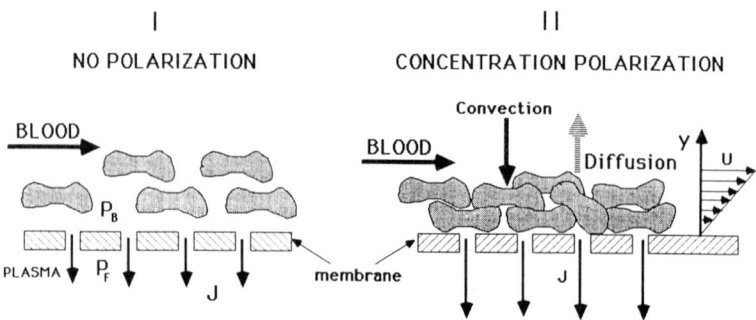

Figure 2. Principle of plasma separation from blood by cross flow microfiltration.

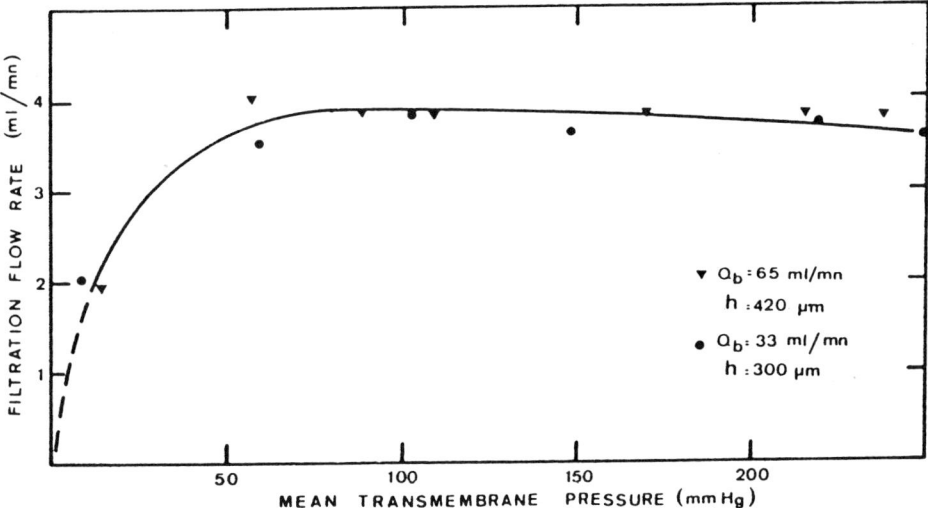

Figure 3. Variation of plasmafiltration flow with transmembrane pressure (from Ref. 15) for a plane membrane.

In Vitro Filtration Data

Variation of Plasma Filtration Flux with Transmembrane Pressure and Wall Shear Rate: The variation of filtration rate with mean transmembrane pressure defined as the arithmetic average between inlet and outlet pressure minus permeate pressure follows a classical pattern (15), increasing almost linearly with pressure at low p_{tm} before attaining a plateau (Fig. 3). The plateau is reached at a relatively low pressure of the order of 70-100 mmHg which increases with increasing shear rate. The plateau level also increases

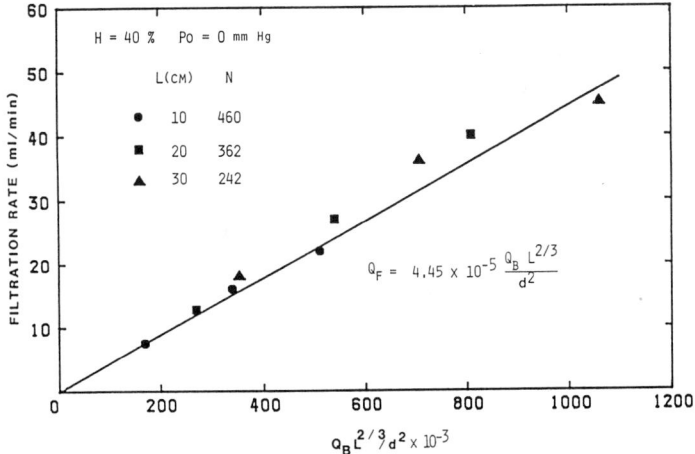

Figure 4. Variation of maximum plasmafiltration flow with inlet wall shear rate for hollow fiber polypropylene membranes (from Ref. 16).

with increasing wall shear rate but with an exponent larger than the value 1/3 predicted by Eq. 4 (Fig. 4).

Variation of Filtration Flux with Membrane Length: Gupta et al. (16) have measured the filtration rate with fresh bovine blood in filters made of the same polypropylene fibers but of various lengths (10, 20 and 30 cm). The permeate flux was found to vary proportionaly to $L^{-0.34}$ in accordance with Eq. 4. A similar dependence on L was also found by Raff et al. (17) using filters of 11.5, 14 and 18 cm. This dependence with L is, of course, due to the flux decrease along the length. Using bovine blood at 40% hematocrit and filters of various lengths and numbers of fibers, the following correlation was found for the maximum flux at plateau (16):

$$J_m = 0.304 \ 10^{-8} \frac{\gamma_w}{L^{1/3}} \quad r^2 = 0.996 \tag{5}$$

with S.I. units.

Eq. 5 has been shown to remain valid for different fiber inner diameters. An interesting consequence of Eq. 5 is that the filtration flow rate for a fixed inlet blood flow is independent of the number of fibers in parallel. This explains why the plasma flow rate is not very sensitive to the membrane area. Since the flow inside the fibers is laminar and the length-to-diameter ratio is large, the wall shear is given by Poiseuille's law:

$$\gamma_w = 32 \ Q_B /(\pi d^3 \ N). \tag{6}$$

Thus the filtration rate at plateau in S.I. units becomes:

$$Q_{FM} = 9.73 \ 10^{-6} \ Q_B \ L^{2/3} /d^2. \tag{7}$$

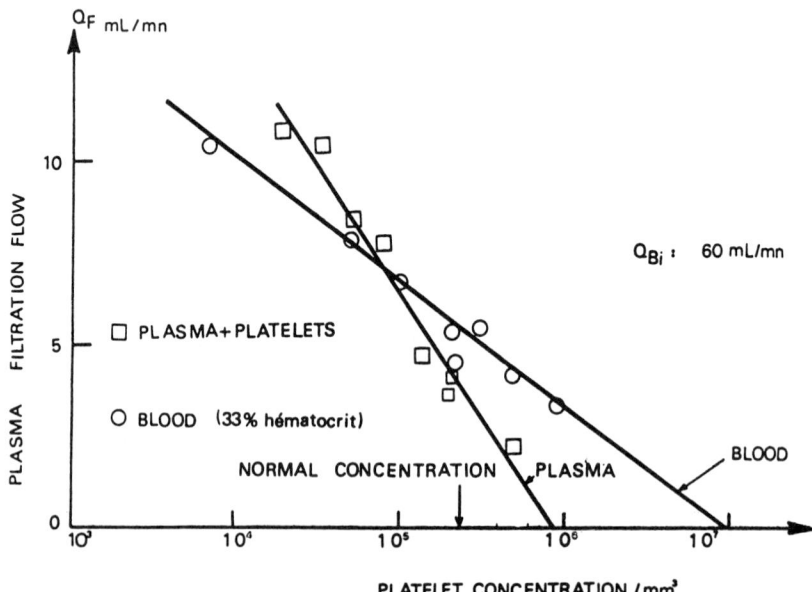

Figure 5. Variation of maximum plasmafiltration rate with the logarithm of platelet concentration for blood and platelet rich plasma (from Ref. 18).

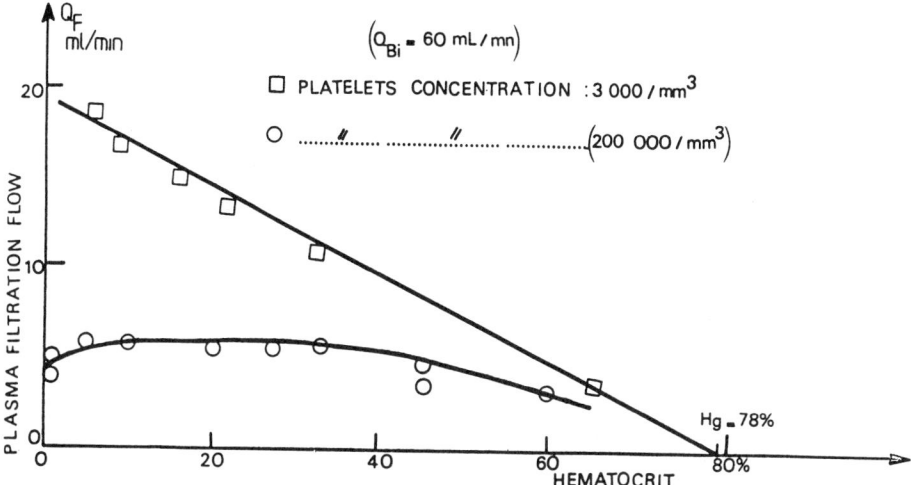

Figure 6. Variation of maximum plasmafiltration rate with hematocrit for normal and low platelet concentrations (from Ref. 18).

Variation of Filtration Flux with Blood Composition: Jaffrin et al. (18) have investigated the effect of platelet concentration and hematocrit on the plasma filtration flux using a single channel filterpress device equipped with a polyacrylonitrile membrane of 170 cm² area with a mean pore size of 0.5 μm. The platelet and red cell concentration were varied by mixing 0⁺ type blood cells and platelet concentrates with human plasma.

They found (see Fig. 5) that both for blood and plasma + platelets solutions, the maximum filtration rate (at plateau) decreased with increasing bulk platelet concentration c_p according to equation:

$$J_m = k \ln \frac{c_w}{c_p} \tag{8}$$

with $k = 1.4 \; 10^{-6}$ m/s and $c_w = 10^7$/mm³ for blood
       $k = 2.6 \; 10^{-6}$ m/s and $c_w = 9 \; 10^5$/mm³ for plasma

Eq. 8 seems to indicate that the plasmafiltration rate is governed by platelet concentration polarization since it has the same form as Eq. 3 with $c_F = 0$ because the platelets do not cross the membrane. However, the limiting platelet concentration c_w is much higher in blood than in plasma. A possible interpretation may be that, in the absence of red cells, the platelets get easily deposited on the membrane where they form a high resistance layer, while in blood, red cell motion and rotation prevent a large fraction of the platelets from reaching the membrane and a much higher bulk platelet concentration is needed to stop the filtration.

The variation of filtration flux with hematocrit is represented on Fig. 6 for two platelet concentrations (2 10⁵/mm³ (normal blood) and 5 10³/mm³).

Since platelets are very fragile, the lower platelet concentration is representative of blood from blood banks which is used by many investigators. Fig. 6 reveals a striking difference between the filtrations at these two platelet concentrations. For platelet poor blood the flux decreases almost logarithmically with increasing hematocrit which is characteristic of red cell concentration polarization. Similar data have been reported by Zydney and Colton

(19) using packed red cells obtained from the American Red Cross. The experiments were performed 4 to 5 days after blood collection so the platelet concentration was lower than normal. On the other hand, when platelet concentration was at its normal level, the filtration flux remains roughly independent of hematocrit until 40% and drops thereafter. Other data have been published by Sueoka et al. (20) using heparinized bovine blood. They give the following correlation for the maximum filtrate flux in S.I. units:

$$J_m = 0.83 \; 10^{-8} \ln\left(\frac{H_w}{H}\right) \bar{\gamma}_\omega^{0.9} \tag{9}$$

with a wall hematocrit $H_w = 112\%$ which, of course, should be regarded as a mathematical parameter and not a physical one. Here $\bar{\gamma}_w$ is the mean wall shear rate along the filter. Unfortunately, the platelet concentration was not given but ranged from 0 to $5 \; 10^5$ /mm^3. They also found that the maximum permeate flux dropped linearly with increasing protein concentration. Their conclusion was that plasmafiltration in the presence of concentration polarization was mainly governed by shear rate and *blood* viscosity μ rather than by blood composition. They finally suggest the following correlation for the maximum plasmafiltration rate in S.I. units valid for bloods with hematocrits ranging from 10 to 50% and variable protein concentrations:

$$J_m = 4.06 \; 10^{-8} \bar{\gamma}_\omega^{0.9} \left(\frac{1000}{\mu}\right)^{1.1}. \tag{10}$$

Although Eq. 10 appears to fit the data, it is hard to interpret since the filtration flux should be expected to be a function of plasma (filtrate) viscosity rather than blood (retentate) viscosity.

Models of Plasma Microfiltration

Most models assume that the filtration of plasma is limited by the formation of a high concentration layer of formed elements deposited on the membrane but disagree over the back transport phenomena governing the growth and therefore the filtration resistance of this layer. The back transport models can be divided into 3 categories : those based on concentration polarization (21, 22), those based on lift force (23, 24) and those using axial convection (25).

Lift force models determine the steady state filtration rate by balancing the drag force from the filtration with the hydrodynamic lift force acting on the cells (26). But these models predict plasmafiltration fluxes one order of magnitude smaller than those observed and will not be discussed here. We have seen that the concentration polarization model of Eqs. 3 and 4 fitted the observed dependence of J_m with L but was in contradiction with the observed dependence on wall shear rate. Zydney and Colton (21) resolved this discrepancy by using the concept of enhanced self diffusion of red cells. Their basic assumption was that the plasmafiltration flux was limited by red blood cell concentration polarization with a red cell effective diffusion coefficient proportional to the wall shear rate. It is well known (27) that the effective diffusivity is larger in concentrated flowing suspensions than in the same medium at rest and that it increases with shear rate. The reason is that migration and rotation of large particles create additional random transport which can completely overtake diffusion. Eckstein et al. (28) have investigated the self diffusion of large spherical particles in a cylindrical Couette flow device. They have found that, when particle concentration exceeded 20% in volume, the effective self diffusivity was correlated by:

$$D_{eff} = 0.025 \, a^2 \, \gamma w \qquad (11)$$

where a is the particle radius. For application to blood, a was taken by Zydney and Colton equal to 2.75 µm which is the radius of a sphere with same volume as a red cell. It is clear that the substitution of Eq. 11 into Eq. 4 yields a linear relation for k and therefore for the permeate flux J_m with wall shear rate as observed experimentally. Taking for the hematocrit at the membrane $H_w = 95\%$, Zydney and Colton obtain the following equation, in S.I. units:

$$J_m = 0.269 \, 10^{-8} \frac{\gamma_w}{L^{1/3}} \frac{\ln 0.95}{H} \qquad (12)$$

$$= 0.2326 \, 10^{-8} \, \gamma_w / L^{1/3} \text{ for } H = 40\%$$

which is about 25% lower than the experimental correlation of Eq. 5 of Gupta et al. (16).

Independently, Malbrancq et al. (22) have devised a similar model but assuming that concentration polarization concerned mainly the platelets, in accordance with experimental data of Jaffrin et al. (18), and using Wang and Keller (27) experimental data for platelet effective diffusivity in flowing blood:

$$D_{eff} = 0.332 \, 10^{-12} \, \gamma_w^{0.886}. \qquad (13)$$

It must be noted that Eq. 11 gives the same effective diffusivity for platelet and red cells equal to $9.5 \, 10^{-11}$ m²/s for $\gamma_w = 500$ s^{-1} and that the platelet diffusivity from Eq. 13 at the same shear rate is also very close ($8.1 \, 10^{-11}$ m²/s). The main quantitative difference

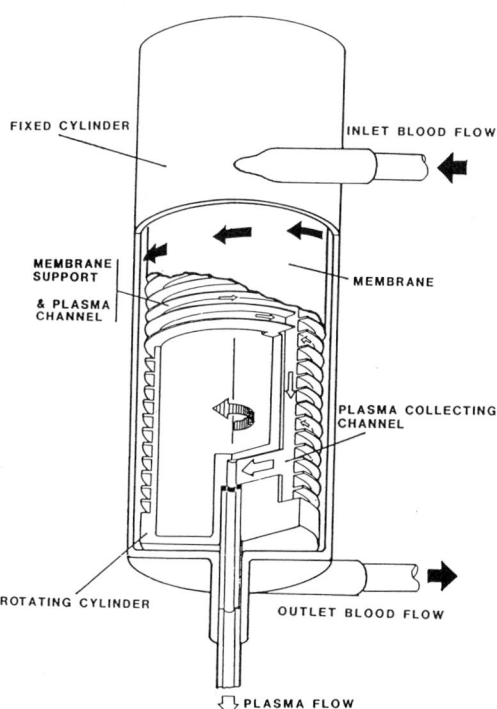

Figure 7. Schematic of the Hemascience Couette flow device for plasma collection from donors.

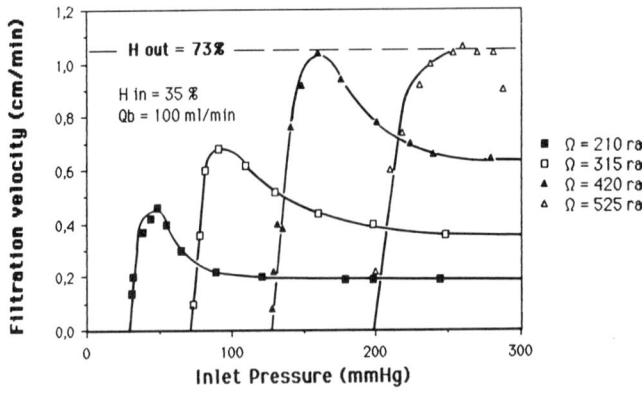

Figure 8. Variation of filtration velocity with inlet pressure at an inlet blood flow of 100 ml/min and various speeds of rotation (rad/s • $\Omega = 210$; $\Omega = 315$; Δ $\Omega = 420$; Δ $\Omega = 525$) (from Ref. 30)

between the two models lies in the polarization factor c_w/c_p which can be higher for the platelets than for the red cells. Substitution of Eq. (13) into Eq. (4) yields, in S.I. units:

$$J_m = 0.388 \cdot 10^{-8} \frac{\gamma_w^{0.92}}{L^{1/3}} \ln \frac{c_w}{c_p}. \tag{14}$$

For $\gamma_w = 10^3 \text{ s}^{-1}$ Eq. 14 becomes identical to the experimental correlation of Eq. 5 if $c_w/c_p = 3.90$ which is similar to the values found for blood protein concentration polarization (13).

Plasmafiltration by a Rotating Membrane

It was shown in the previous section that the plasma flux is approximately proportional to the wall shear rate. But the wall shear rate is limited by the flow available from the patient blood access and by the admissible pressure drop in the filter. If the number of fibers in parallel N in the filter is decreased in order to increase the shear rate, the fiber length must increase in order to maintain the membrane area and the pressure drop along the fibers increases as N^{-2}, which will in turn causes hemolysis. Thus, the only way to increase the shear rate without increasing the pressure drop is to use a rotating membrane, either in a cylindrical Couette device or as a flat disc.

In 1985, an American Company (Hemascience, Santa Ana, California) introduced a disposable rotating membrane device for plasma collection from donors (29). A polycarbonate membrane of 58 cm² area with 0.8 µm pores was mounted on the inner cylinder rotating at 3600 RPM inside a concentric cylinder with a 0.09 cm gap (Fig. 7).

The inner cylinder radius was 1.34 cm. Blood inlet and outlet were mounted tangentially respectively at the top and the bottom of the outer cylinder and the plasma was collected through groves molded on the inner cylinder and left the device through a duct located on the rotation axis and exiting at the unit bottom.

Experimental Results: Operation of this unit was tested extensively in vitro with fresh bovine blood by Beaudoin and Jaffrin (30) at various speeds of rotation, pressure and inlet blood flows. Since the filtrated plasma rotates with the outer surface of the inner cylinder before being collected, the permeate pressure at the membrane is proportional to the square of the angular speed of rotation Ω. Thus, the inlet pressure must exceed a threshold value to overcome the permeate counter pressure before filtration starts. The variation of filtration velocity (permeate flux) with inlet pressure is plotted on Fig. 8.

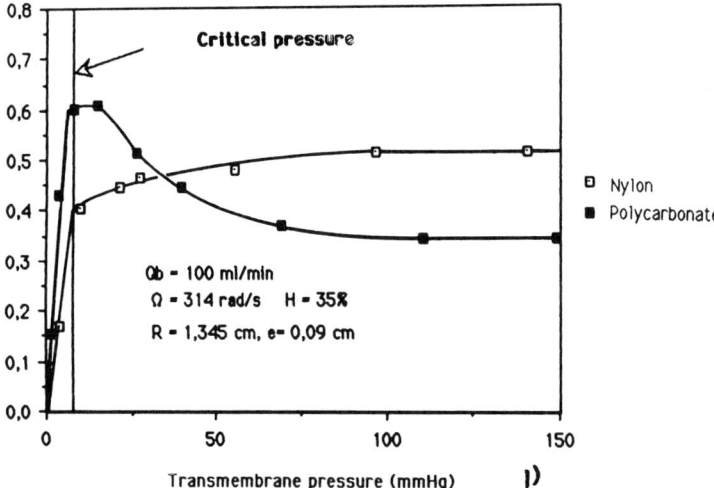

Figure 9. Variation of filtration velocity with transmembrane pressure for a polycarbonate and polyamide membranes (from Ref. 30).

The effective transmembrane pressure can be estimated as the difference between the inlet pressure and the pressure at the onset of filtration for each curve. It is seen that filtration does not increase at speeds above 420 rad/s and an unusual feature not seen in microfiltration with stationary membrane is the presence of a sharp peak of filtration at a small transmembrane pressure of the order of 25 mmHg before a decay to a plateau at higher pressure. This peak has also been observed for the same device equipped with a polycarbonate membrane by Ohashi et al. (31). It is also important to note that the permeate fluxes both at the peak and at the plateau, are at least one order of magnitude higher than those obtained in hollow fibers. The reason is that, at low p_{tm}, the rotating membrane appears to be free from red cells and the permeate flux rises very steeply when the transmembrane pressure is increased from 0 to 20 mmHg.

The variation of permeate flux with p_{tm} is sensitive to the membrane material. In 1986, the 0.8 μm polycarbonate membrane was replaced by the manufacturer by a polyamide one with 0.5 μm pore and the filtration peak at low pressure disappeared (Fig. 9) while the height of the plateau increased.

The high filtration velocities observed with rotating membranes are not a direct consequence of the centrifugal forces which keep the cells away from the membrane but rather an indirect one. This was demonstrated by Beaudoin and Jaffrin (30) who modified the device by mounting a membrane inside the fixed outer cylinder. The filtration fluxes were identical at the same p_{tm} for the two membranes. The explanation for the high performance, as suggested by other authors (32, 33) who have built rotating membranes for non medical applications, is due to the generation of Taylor vortices in the circular gap which increase the shear rate according to the equation (Ref. 34):

$$\gamma_w = 0.23 \, T_a^{1/2} \, \Omega R/e \tag{15}$$

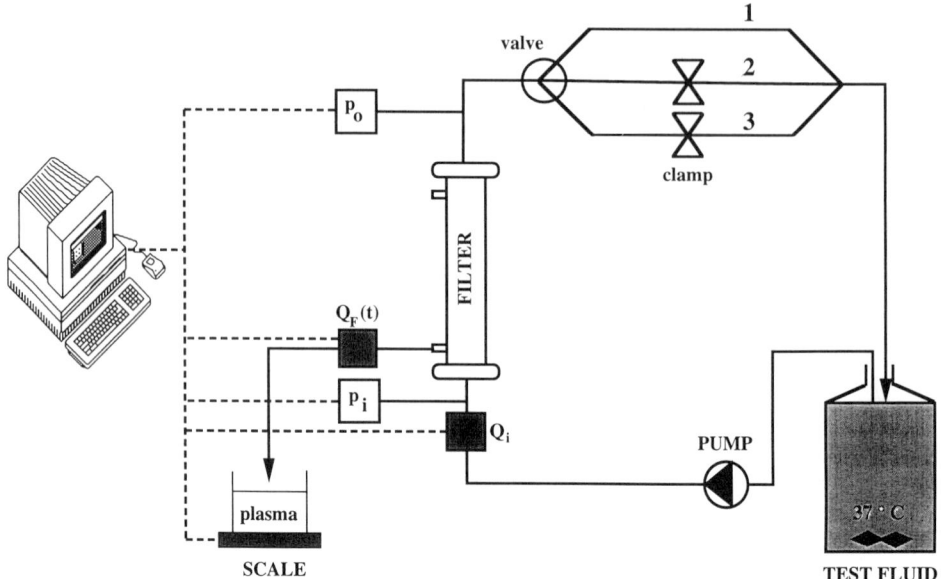

Figure 10. Schematic of experimental set-up for creating sudden transmembrane pressure changes (from Ref. 11).

where R is the inner radius cylinder, e the gap and $T_a = \Omega R^{1/2} e^{3/2} / \nu$ is the Taylor number when $e \ll R$, ν is the blood kinematic viscosity. This was also confirmed by experiments with blood at various Taylor numbers (35) which showed that the filtration flux increased sharply with T_a above a critical Taylor number of about 50, close to the classical value of 42 given by the theory of hydrodynamic stability (36). According to eq. (15), the wall shear rate in the Hemascience device is about 20 000 s^{-1} at 3600 RPM, while it is only 500 - 1000 s^{-1} in hollow fibers filters. This alone explains why the filtration fluxes were one order of magnitude higher in rotating devices than in conventional plasmafilters since they are proportional to shear rates.

UNSTEADY PLASMAFILTRATION

Transient Effects

The investigations of plasma separations from whole blood reported so far were conducted in the steady state regime when the layer of formed elements deposited on the membrane was established and steady since the mass flux of incident particles was in equilibrium with the mass flux returning to the bulk solution. We now investigate the kinetics of formation of the concentration polarization layer when the incident mass flux is larger than the return flux. Similarly, we also investigate the reversibility of concentration polarization•when the return flux becomes larger than the incident flux.

Experiments to this effect were performed by Jaffrin et al. (11) which consisted in raising suddenly the transmembrane pressure by switching the filter outlet from a low resistance circuit (circuit 1) to one of two high resistance circuits (circuits 2 and 3) using a three-way electronic valve (Fig. 10). These high resistances were created by clamping the flexible tubes.

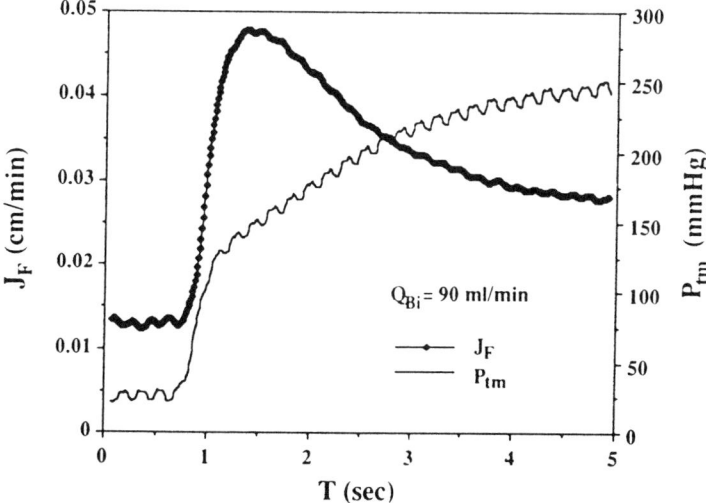

Figure 11. Variation of transmembrane pressure and permeate flux with time following a sudden pressure increase, using a 0.1 m² polypropylene plasmafilter (from Ref. 11).

Similarly, the reversibility of the polarization layer can be investigated by switching from a high resistance circuit to circuit 1. The instantaneous transmembrane pressure and permeate flow rates are monitored by pressure transducers and a medical electromagnetic flowmeter respectively. The blood was fed into the filter by a volumetric peristaltic pump which kept the flow constant when the pressure changed. Fresh bovine blood at 37° and 36% hematocrit was used.

Results: The variation of transmembrane pressure and permeate flux with time when switching the outlet from circuit 1 to circuit 2 at an inlet blood flow of 90 ml/min is plotted on Fig. 11. Oscillations on the signals are due to the peristaltic pump.

The initial pressure, at 30 mmHg, was below the onset of concentration polarization while the final pressure was 230 mmHg, well above the plateau threshold. The permeate flux

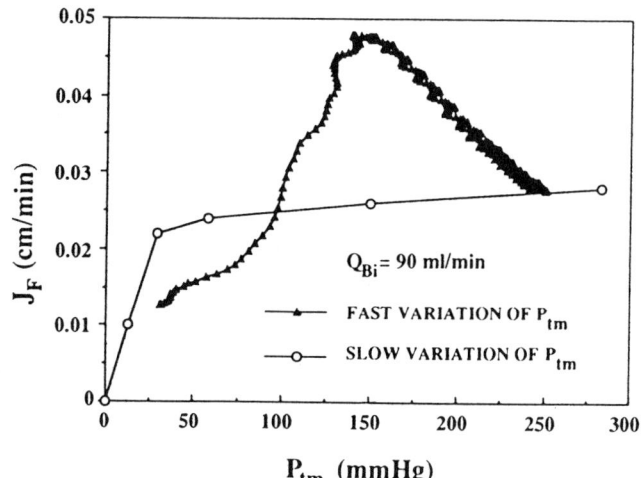

Figure 12. Representation of data of Fig. 11 on a flux - p_{tm} plot for fast and slow variation of pressure (from Ref. 11).

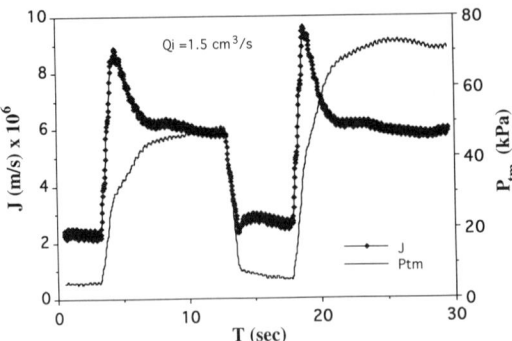

Figure 13. Variation of transmembrane pressure and permeate flux with time when switching from circuit 1 to circuit 2, returning to circuit 1 and switching to circuit 3 (from Ref. 11).

increases sharply to a peak in less than one sec. and decreases approximately exponentially with time to a plateau even though the pressure continues to increase. The height of the plateau corresponds to the concentration polarization equilibrium value. In fact, the high peak of filtration is responsible for the progressive pressure rise, since the filter outlet pressure which is equal to the pressure drop across the clamp increases when the permeate flow decreases.

Another representation of the same experiment is plotted on Fig. 12 in a flux versus pressure plot, together with the corresponding equilibrium curve when the transmembrane pressure is varied slowly and the return mass flux balances the incident flux.

Fig. 12 demonstrates that a fast rise in transmembrane pressure permits to avoid temporarily the phenomenon of concentration polarization and the permeate flux exceeds the plateau level during a few seconds but by a wide margin.

The reversibility and reproducibility of the membrane response to successive pressure increments are illustrated on Fig. 13 which shows the permeate flux produced by a first pressure rise from 80 to 400 mmHg, a return to the basal level followed 5 s later by a second pressure rise from 80 to 600 mmHg (Fig. 13).

It is seen that, when the pressure returns to the basal value, the filtration flux returns instantly to its initial level. Even though the second pressure rise is higher than the first, the peak and plateau plasma fluxes are nearly the same. However, the flux decay after the peak is faster at higher pressure. It can be concluded that the polarization layer disappears in less than 5 s when the pressure is lowered and that the plateau reached a few seconds after the peak is indeed the pressure independent plateau due to concentration polarization in steady state experiments.

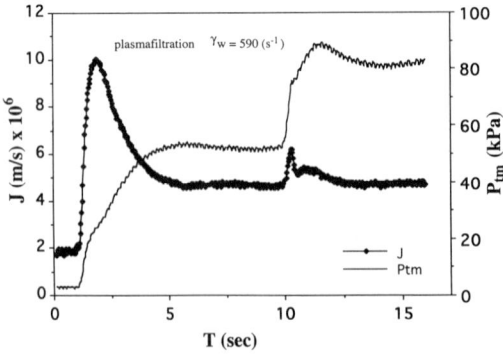

Figure 14. Same as Fig. 13 but without returning to circuit 1 before switching to circuit 3 (from Ref. 11).

If the pressure is increased to 600 mmHg in two steps without returning to the basal level between increments (Fig. 14), the second pressure increment only generates a small spike for the flux which returns almost instantaneously to the plateau.

This observation can be explained by the packing of the red cell layer deposited on the membrane rather than by a thickening of this layer. If the permeate flux was controlled by a thickening of the layer, this thickening would take place over a finite time. Since the pressure is raised by 50%, the layer thickness should also increase by 50% to maintain the flux at the same limit and the amplitude of the second peak over the plateau should be 50% of that of the first peak. So the data of Fig. 14 suggest that the increase in resistance is rather due to packing of the cell layer which is greatly facilitated by the high red cell deformability.

The thickness of the red cell layer in membrane plasmapheresis has been estimated to range from 4 to 9 μm (37) when the plasma flux is of the order of $5 \cdot 10^{-4}$ cm/s. A similar estimation (less than 13 μm) was made by Jaffrin et al. (11) from the data of Fig. 11 by computing the number of red cells brought by the filtration to the membrane before equilibrium is reached.

Hydrodynamic Plasma Flux Enhancement

Introduction: As said in the beginning of this chapter, it is important from an economical point of view to reduce the membrane area while extracting a plasma flow rate representing about 30% of the inlet blood flow. This goal cannot be reached with conventional steady flow filtration because the shear rate is limited by available blood flow and the admissible pressure drop.

However, the transient effects described in the preceding section give a clue on how to "cheat" with concentration polarization. If the transmembrane pressure is suddenly raised, the membrane transmits a permeate flux higher than the polarization-limited plateau during a brief instant before concentration polarization gets established in 2 or 3 s. Therefore, it can be hoped that, by repeating successive pressure pulses at a frequency of 1 Hz, an increase in permeate can be obtained if the membrane returns to its low pressure state between pulses.

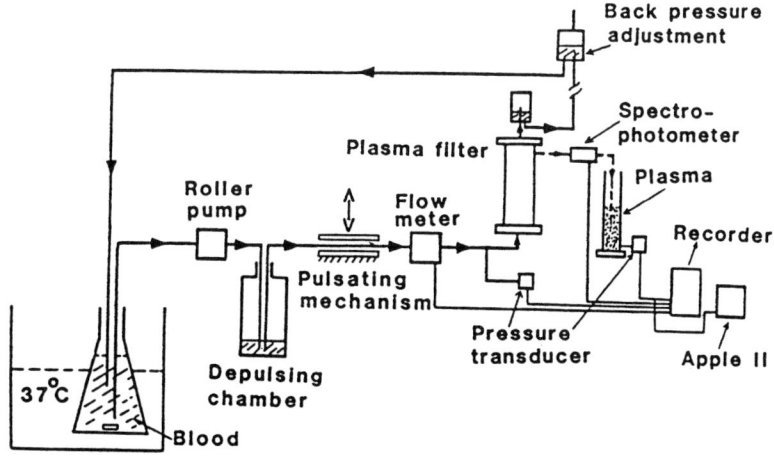

Figure 15. Experimental set-up with pulsation generator (from Ref. 39).

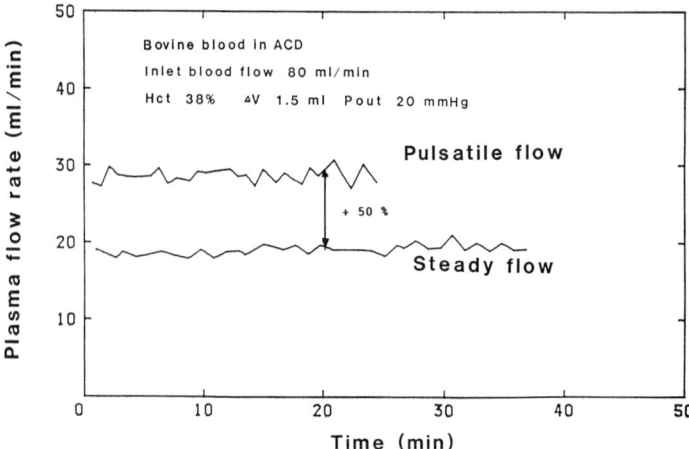

Figure 16. Variation of instantaneous plasmafiltration rate with time for steady and pulsatile inlet blood flow with a 0.1 m² filter (from Ref. 40).

This could be achieved by creating a fluid acceleration which sweeps the polarization layer before it gets compacted by the pressure.

Flux Enhancement by Pulsatile Flow: Enhancement of plasmafiltration by pulsatile flow was first reported by Galletti et al. (38) who used a press type filtering device equipped with a 74 cm² flat membrane. Increases of 50 to 100% over the filtration rate at the same constant blood flow rate were reported but, due to the small membrane area in relation to the blood flow and inefficient design of the cell, the plasma flow rate did not exceed 1,5% of the time-mean inlet blood flow. Our group (39-41), using commercially available polypropylene hollow fiber filters of 0.1 m², showed that appropriate pulsations of blood flow could significantly increase the plasma permeate flux, even when it represented already 25% of the inlet blood flow at steady flow. In our set-up, pulsations were created by periodic

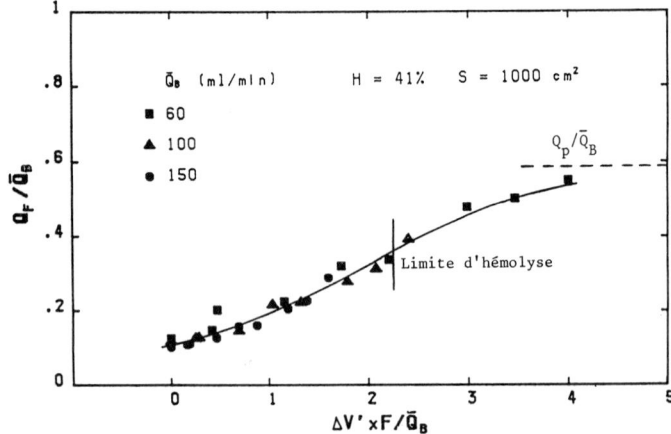

Figure 17. Variation of the ratio of plasmafiltration rate to mean inlet blood flow as a function of the dimensionless parameters (from Ref. 41).

squeezing of a silicone rubber tube placed upstream of the filter between a fixed plate and a moving plate driven by a motor (Fig. 15).

When the moving plate releases the tube, it causes a flow reversal in the filter. A closed vessel containing air is placed between the roller pump in order to damp out both oscillations from the roller pump and pressure waves produced upstream by the moving plate. The pulsation frequency was adjusted between 1 and 4 Hz by changing the speed of the motor driving the plate. The volume amplitude of the pulsations could be varied by inserting shims between the tube and the fixed plate.

RESULTS

This system typically permits to increase the filtration rate by 50% from 20 to 30 ml/min (40) for an inlet blood flow of 80 ml/min at a frequency F = 2 Hz and a pulsatile volume ΔV = 1.5 ml (Fig. 16).

In fact, it can be shown (41), that the dimensionless parameter governing the enhancement is $\Delta V.F/\overline{Q}_B$ where $\overline{Q}_B$ is the time-mean inlet blood flow. This parameter represents in fact the ratio of the pulsatile flow amplitude over the steady flow. It is possible to extract almost all the plasma from the blood by increasing this ratio to 4 (Fig. 17). But hemolysis sets in before this limit is reached.

Jaffrin et al. (41) also observed that, when velocity reversal is prevented by a one-way valve, the filtration rate is only slightly increased by pulsations. They suggested two possible explanations to the enhancement mechanism :

a) by analogy with Rayleigh oscillatory boundary layers (36), it can be inferred that periodic flow reversal limits the growth of the concentration boundary layer thickness δ. Since the mass transfer coefficient is inversely proportional to δ, it will be higher than for steady flow and the concentration polarization limited flux will also be higher.
b) the return mass flux of cells and platelets to the bulk solution is increased by flow reversal, which creates flow separation and vortices, increases the effective diffusivity and accelerates the disappearance of concentration polarization between successive pulses.

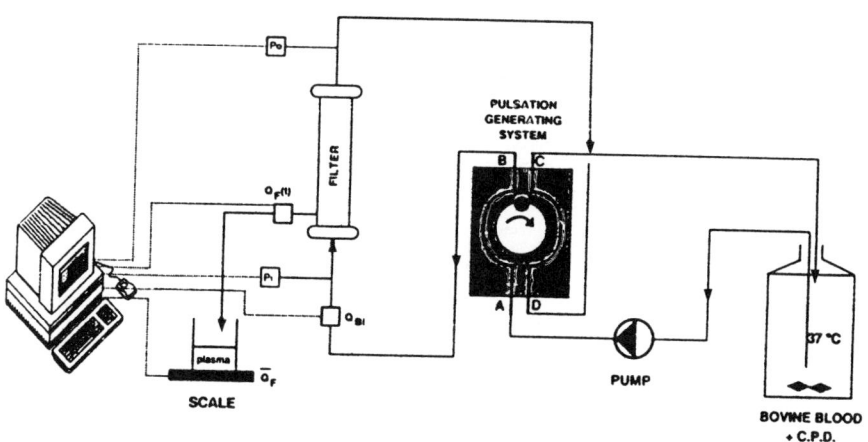

Figure 18. Experimental set-up with single roller peristaltic pump as pulsation generator (from Ref. 43).

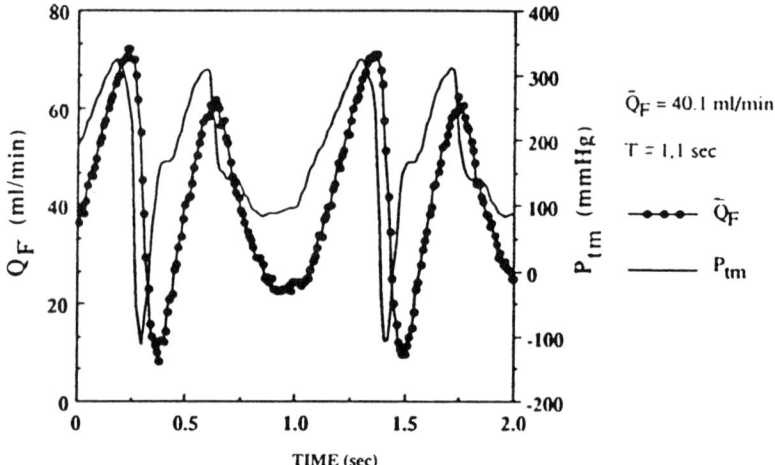

Figure 19. Plasma flow and transmembrane pressure signals when both tubes are squeezed (from Ref. 43).

This pulsatile flow plasmapheresis technique was tested clinically in 40 volunteer donors in Nancy Transfusion Center in France, by Schooneman et al. (42) using a Dideco Filtra plasmapheresis monitor and 0.1 m² polypropylene hollow fiber filters. In a first series of tests the average time for collecting 600 ml of plasma was 52 min without pulsations and 44 min with pulsations, i.e. a 15% improvement. In a second series of tests, the gain was of the order of 20%. Composition of collected plasma under pulsatile filtration was found to be identical to that in the blood plasma before filtration except for a slightly more elevated plasma hemoglobin.

A more effective way of generating pulsations was later reported by Ding et al. (43). This pulsating mechanism consisted in a modified peristaltic pump in which one of the two rollers was removed and two holes were drilled in B and C through the circular back plate to permit passage of the blood lines (Fig. 18).

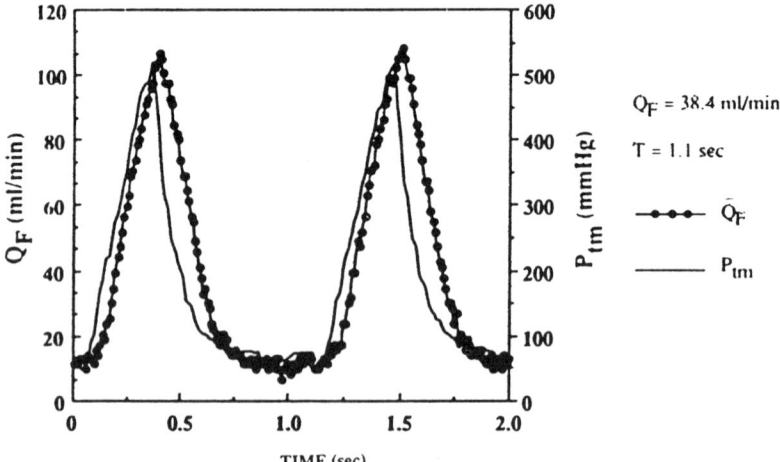

Figure 20. Same as Fig. 19 but with outlet tube squeezed (from Ref. 43).

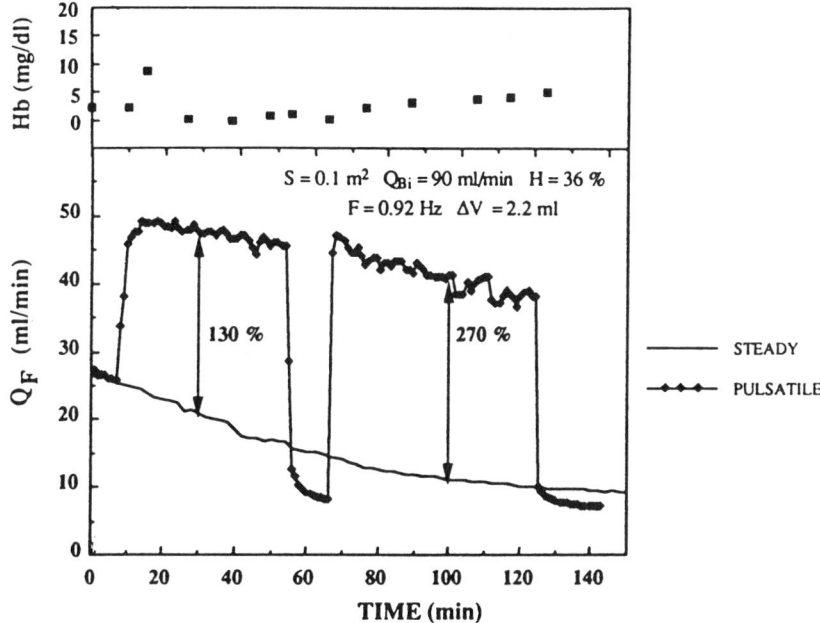

Figure 21. Comparison of filtration rate and hemoglobin concentration with time for steady and pulsatile flows (both lines squeezed).

During its rotation, the single roller alternatively squeezes the two silicone tubes, over segments AB and CD which are located respectively upstream and downstream of the plasmafilter. The squeezing of segment AB produces a forward wave which accelerates blood through the filter and raises the inlet blood pressure p_i. The squeezing of line CD produces a retrograde wave towards the feed pump which reverses the flow direction in the filter and raises the outlet pressure. These waves form two peaks of transmembrane pressure per period (Fig. 19). It is seen on this figure that the instantaneous variation of the permeate follows closely the p_{tm}. The slight shift between the two signals is an artefact due probably to air in the permeate compartment or to permeate inertia.

But there are two other modes of operation. In the 2nd mode, only the outlet tube CD is introduced in the modified roller pump and squeezed by the roller, while the filter inlet is connected directly to the feed pump. In this mode, there is only one maximum for p_{tm} per cycle caused by the retrograde wave which pushes blood against the steady flow from the

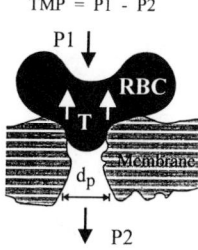

Figure 22. Schematic representation of a red cell deformed into a pore of the membrane.

pump. But the pressure and permeate flow peaks are higher than in the first mode and the time-mean plasma filtration rate is only slightly less (Fig. 20).

In the 3rd mode, only the inlet line AB is squeezed by the roller and the filter outlet is directly connected to the downstream reservoir. In this case, the acceleration of the blood by the combined action of the roller and the feed pump produces a depression inside the filter causing a backfiltration which reduces the time-mean permeate flow rate. Thus, the increase over the steady flow value is smaller than for the other modes (20% instead of 100%).

This technique can be very effective both for enhancing filtration rate and reducing membrane plugging as illustrated on Fig. 21. When pulsations are started after 10 min of steady flow, the plasma flux increases by 130%.

When stopping the pulsations 45 min later, the permeate flux drops below the corresponding steady flow value after the same time of filtration. When pulsations are resumed, the plasma flux returns instantaneously to its former level when pulsations were discontinued. But, because fouling of the membrane under steady flow condition is more severe than under pulsatile flow, the plasma flow rate, in presence of pulsations, exceeds the steady flow value by 270% after 100 min of filtration. This spectacular performance was obtained without significant hemolysis.

DISCUSSION

The simultaneous recordings of instantaneous transmembrane pressure and permeate flux when pulsations are created by the single roller show that the increase in time-mean filtration is due to the transient phenomena described in the preceding section, namely the finite time necessary for establishment of concentration polarization. On Figs. 19 and 20, it can be seen that the transmembrane pressure reaches its peak in about 0.3 s which is too short for establishment of concentration polarization. Thus the membrane responds to the pressure rise by a simultaneous increase in flux which decays at the same rate as the pressure when it drops. Even though the low pressure phase only lasts 0.5 s, it is enough to remove or to loosen the deposited layer of formed elements and the pressure rise is reproducible. This reversibility of the polarization is facilitated by a sudden acceleration of the blood in reverse direction when the roller leaves the tube.

This dynamic interpretation backed by detailed pressure and flux recordings is probably more realistic than the explanation given earlier, when instantaneous flux recordings were not available.

Flux Enhancement by Microvortices: Another method for enhancing plasmafiltration was proposed by Bellhouse et al. (44). This method, directly inspired by a technique used previously by the same group for increasing gas transfer in oxygenators (45), consists in circulating the blood over a furrowed or dimpled membrane. In addition to the main flow, the blood is submitted to an oscillatory motion by two flexible diaphragms placed at each of the channel and operated synchroneously by pistons linked to a motor. As the flow decelerates, microvortices grow in the hollows of the membrane until they bulge into the central channel. When the flow reverses, these vortices direct the reversed flow around the channel wall, increasing the wall shear and mixing. The unit tested had a 200 cm^2 polysulfone membrane with 0.2 µm pores and half circle dimples 0.5 mm deep which were shaped by heat forming the membrane into a brass mold. The membranes were supported by molded rigid plates with hollows matching the dimples in the membranes. The diaphragms had a stroke volume of 2.8 ml and were operated at frequencies from 3 to 6 Hz. Tests were run with citrated bovine blood at 35% hematocrit and an inlet flow of 85 ml/min. Extremely high plasmafiltration rates, in view of the small membrane area, from 22 ml/min at 3 Hz to 36 ml/min at 6 Hz were reported. This correspond to a maximum plasma flux of nearly 3 10^{-5} m/s which is five times higher than those obtained in hollow fibers or press type filters at steady blood flow. However, it is probable that, due to the weak dependence of permeate flow rate on membrane area in plasma separation, the permeate fluxes in the presence of microvortices would drop in larger units.

Clinical tests were carried out in 24 healthy volunteers. The mean plasma flow rate was 18 ml/min for a mean inlet blood flow of 48 ml/min and a frequency of 3 Hz. The mean values of hematocrit were 39.6% at inlet and 61.7% at outlet. Negligible hemolysis and no platelet activation were reported. A similar system consisting of a dimpled membrane unit with built in bellows and an appropriate monitor is presently marketed by Stryker Company as an autotransfusion device which washes and concentrates blood from the surgical field.

HEMOLYSIS IN PLASMAPHERESIS

Hemolysis is a serious constraint in membrane plasmapheresis since free hemoglobin is toxic above a certain concentration. It was observed in early works on membrane plasmapheresis (46) that hemolysis occurred when transmembrane pressure exceeded a certain threshold of the order of 70 to 90 90 mmHg. This observation prompted Travenol Company (now known as Baxter) to equip its plasmafilters with a transmembrane pressure limiting device consisting in a rigid chamber containing a diaphragm. Blood retentate circulates on one side of the diaphragm while the filtrated plasma circulates on the other side. Thus, an increase in blood pressure displaces the diaphragm, causing an obstruction on the permeate outflow which raises the permeate pressure and regulates the transmembrane pressure to its previous level.

Another device, the Cobe TPE press type with 6 compartments, was constructed with limited compliance and was operated being compressed by an hydraulic press. Thus the blood film thickness could be adjusted according to available blood flow to limit pressure drop and therefore transmembrane pressure to a predetermined value in order to avoid hemolysis.

However another observation that hemolysis actually decreased at high shear rates was overlooked by filter manufacturers since early commercial plasma filters had large membrane areas, above 0.4 m^2 and were operated at low shear rates of the order of 500 s^{-1}.

Zydney and Colton (21, 47) were the first to propose a model of mechanisms leading to hemolysis which fitted the observations. Their basic assumption is that a red cell deposited on the membrane will hemolyze if it is forced into a pore where it deforms and if the strain on the membrane caused by the deformation is sufficient to cause its rupture (Fig. 22). This is supported by the observation that most of the released hemoglobin is found in the permeate side. Since according to Rand (48) the red cell membrane is viscoelastic, its strain increases with time and hemolysis will occur if the red cell is trapped by a pore over a sufficient time. If one assumes that the portion of the red cell inside the pore is hemispherical, the tension σ in the cell membrane is related to the pore radius R_p and transmembrane pressure by:

$$\sigma = p_{tm} R_p/2. \qquad (16)$$

Investigating the lysing of red cells sucked into a micropipette, Blackshear and Anderson (49) found that the critical membrane tension in N/m required to lyse a cell was a function of exposure time to the stress which could be correlated by:

$$\sigma = 33\frac{10^{-3}}{1+0.2t} \quad t<4s \tag{17}$$

$$\sigma = 2 \cdot 10^{-3} + \frac{0.06}{t} \quad t>4s \tag{18}$$

In addition, Zydney and Colton assume that the residence time t_R of a cell over a pore is inversily proportional to the shear rate:

$$t_R = \alpha/\gamma_w \tag{19}$$

since at large shear, red cells have a higher probability of being dislodged. Application of eq. (19) to the data of Rand for tension at onset of hemolysis yielded $\alpha = 7.2 \cdot 10^5$ and to the data of Blackshear, $\alpha = 2.4 \cdot 10^4$. Both give residence time much larger than 4s for shear rates normally encountered (less than 2000 s^{-1}). The elimination of time and α between eqs. (16), (18) and (19) yields the following condition for avoiding hemolysis, in S.I. units:

$$p_{tm} \leq 4\frac{10^{-3}}{R_p} + \frac{0.12}{R_p\alpha}\gamma_w. \tag{20}$$

Eq. 20 exhibits the correct qualitative features that hemolysis at a given transmembrane pressure is reduced if shear rate increases or if the pore radius decreases. It is found also reasonnably accurate if the lower value of α is used.

Taking a pore radius of 0.25 µm as in polypropylene membranes and the value of $\alpha = 2.4 \cdot 10^4$, we obtain for the hemolysis boundary:

$$p_{tm} = 16 \cdot 10^3 + 20\gamma_w. \tag{21}$$

For instance, at a shear of 500 s^{-1}, the threshold pressure at onset of hemolysis is 26 10^3 Pa or 19 5 mmHg, which is about 50% higher than experimental observations (50).

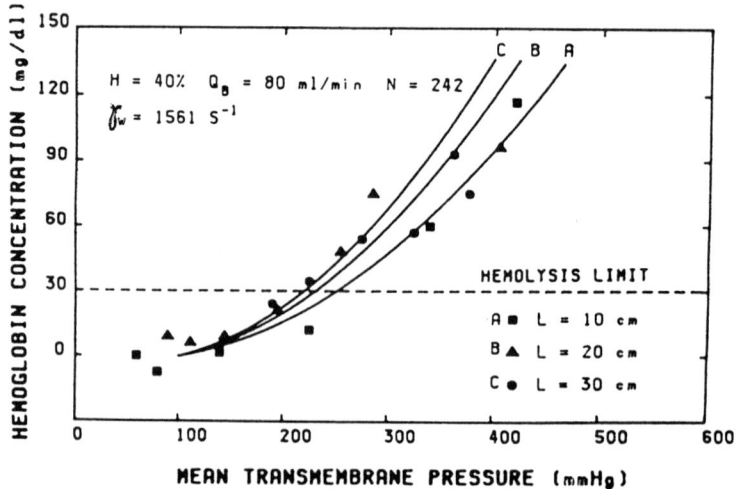

Figure 23. Variation of plasma hemoglobin concentration with transmembrane pressure and fiber length. Solid lines one calculated from Eq. 25 with K = 0.2 (from Ref. 50).

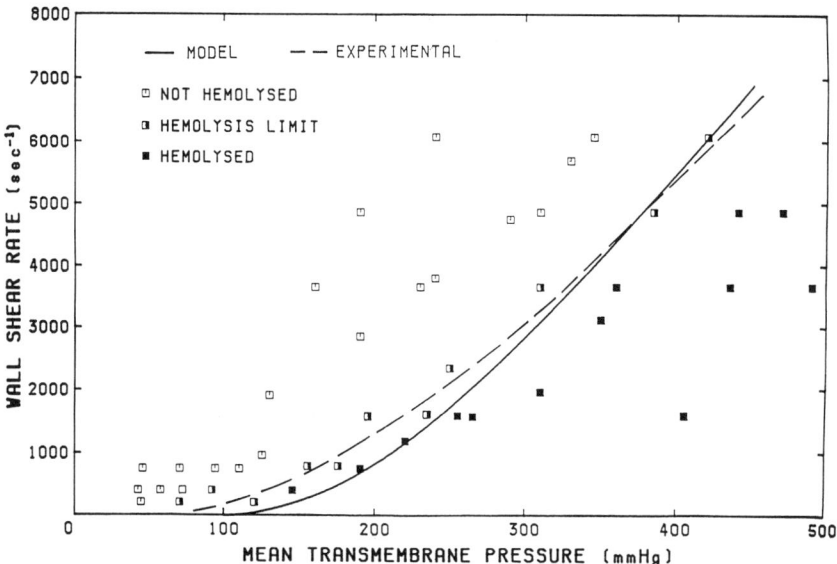

Figure 24. Comparison of experimental and theoretical hemolysis boundaries in the wall shear rate and mean transmembrane pressure plane (From Ref. 50 with permission).

A similar but slightly more elaborate model of hemolysis caused by microporous membrane was set up by Ding et al. (50) which, in addition, permits to predict the variation of plasma free hemoglobin with geometrical and hemodynamical parameters. The basic assumption of this model is that statistically the number of red cells N_t trapped by the membrane is proportional to the membrane area S and to the transmembrane pressure which holds down the cell into a pore:

$$N_t = K_i \, p_{tm} \, S \quad (22)$$

If a cell gets hemolysed after a certain residence time t_R, the average number of lysed cells per unit time will be:

$$N_H = N_t / t_R \quad (23)$$

and if m_H is the hemoglobin mass released by a lyzed cell, the plasma free hemoglobin concentration will be:

$$C_H = m_H N_H / Q_F \quad (24)$$

On substituting in Eq. 24, Q_F from Eq. 5, t_R from Eq. 18, and σ from Eq. 16, we obtain:

$$C_H = \frac{K L^{1/3}}{\gamma_w} p_{tm} (R_p \, p_{tm} - 4 \cdot 10^{-3}) \quad (25)$$

with $K = 2.74 \cdot 10^9 \, m_H \, K_i$ in S.I. units.

A comparison of Eq. 24 with in vitro data obtained using fresh bovine blood and polypropylene membranes of 0.5 μm pores is shown on Fig. 23. The agreement is found to be good at different lengths

Eq. 25 predicts that hemolysis starts, during filtration when the transmembrane pressure exceeds $4 \cdot 10^{-3}/R_p$, i.e. 120 mmHg for a pore diameter of 0.5 μm. However, the clinical definition of hemolysis is different. Since hemolysis occurs naturally with blood due to ageing cells, the blood is said to be hemolyzed when the plasma free hemoglobin exceeds a certain threshold taken generally to be 30 mg/dl as shown on Fig. 23. Thus, the hemolysis boundary in the shear-transmembrane pressure plane is obtained by setting $C_H = 30$ mg/dl in Eq. 25 and has a parabolic shape instead of a straight line as the boundary of Eq. 20. This hemolysis boundary is represented on Fig. 24 together with in vitro data from Ding et al. (50). The model overestimates slightly the transmembrane pressure at which hemolysis occurs but the difference is not large. It is important to note that, in contrast to the common belief before 1984, high transmembrane pressure (above 300 mmHg) can be sustained without hemolysis if the wall shear rate is high enough.

Using filters of the same membrane material but with different lengths and area, Philp et al. (51) observed that the concentration of plasma free hemoglobin was proportional to membrane area. Thus, in order to compare the performance of filters of different areas, they proposed as hemolysis index the concentration of plasma free hemoglobin divided by the membrane area, expressed in mg/dl m². They also introduced a new criterion for defining hemolysis in membrane plasmafiltration. The filtration was considered as hemolytic if the hemolysis index due to filtration, after deduction of natural hemolysis, was above 30 mg/dl m². This criterion corresponds to lower hemolysis thresholds for small filters than these used clinically since it corresponds to 3 mg/dl for a 0.1 m² filter. Using fresh citrated bovine blood, they obtained a linear hemolysis boundary in the γ_w - p_{tm} plane which was almost independent of membrane area and could be expressed as:

$$p_{tm} = 8 \cdot 10^3 + 46 \gamma_w \qquad (26)$$

Eq. 26 indicates a stronger effect of the shear rate for preventing hemolysis than Eq. 21 but gives the same pressure threshold at a shear rate of 300 s⁻¹.

It is also legitimate to check whether the onset condition for hemolysis of Eq. 26 is modified when pulsations are used to enhance plasmafiltration.

Since pressure and shear rates vary during the pulsation cycle, it seems logical to use the time mean transmembrane pressure and wall shear rate but the instantaneous transmembrane pressure exceeds the mean one during a part of the pulsation cycle and even if the point representing time mean condition lies in the non hemolyzed zone, the point representing instantaneous shear and pressure may intrude in the hemolyzed region during the cycle. Not surprisingly, Philp et al. found that under pulsatile flow, the region in the γ_w - p_{tm} plane exempt from hemolysis was reduced when compared to the region in steady state filtration. In addition, for the same time mean wall shear rate, hemolysis occurred at a lower transmembrane pressure for longer filters. This may be explained by the fact that, in the presence of pulsations, the peak flow induces a large pressure drop along the membrane and therefore a peak in transmembrane pressure and this effect is more pronounced in long filters which have a larger pressure drop.

CONCLUSION AND PERSPECTIVES

Spurred by the competition with centrifugation, plasma separation from whole blood by microfiltration has been the subject of extensive studies over a relatively brief period mainly from 1980 to 1987 and considerable progress has been made in the understanding of the hemodynamic and fluid dynamical factors governing the plasma flux and the occurrence of hemolysis. As witness to this progress, it is interesting to recall the continuous trend of decrease in membrane area of commercially available plasmafilters which for the most part have been evaluated by Gurland et al. (52). The first commercially available plasma filter was the ASAHI Plasmaflo AP 06 with a cellulose diacetate membrane of 0.65 m^2 and a cost in France of 2.900 FF (560 US $) in 1978. The next generation of filters with polypropylene, polycarbonate and polymethyl-metacrylate membranes had areas ranging from 0.2 to 0.5 m^2 without noticeable loss in performance since the permeate flux proved to depend little upon membrane area (53). But fear of hemolysis prevented for a while further membrane area reduction since a filter with fewer fibers and a larger pressure drop would have to operate at a larger transmembrane pressure. However, the discovery that high shear rates permitted plasma filtration at relatively high transmembrane pressure without hemolysis encouraged a further reduction in size : 1000 cm^2 for the Dideco Hemaplex BT 900 at 180 FF and 700 cm^2 for the Plasmapur from Organon Technica.

Both used polypropylene hollowfibers made by AKZO Company with 330 µm inner diameter. Further area reduction without loss of performance would have been possible according to Eqs. 6 and 14 by reducing the fiber inner diameter. But the limited size of the market did not motivate AKZO Company to modify its product. Lysaght et al. (53) estimated at 600 cm^2 the lower practical limit for donor's plasma filters. A 600 cm^2 two channel press type unit was indeed developed by Rhône-Poulenc Company in 1982 for plasma collection from donors and successfully tested. But its manufacturing cost was judged to be too high to compete with centrifugation.

Of course the models presented earlier become unrealistic if the membrane area gets too small since the permeate flux declines rapidly due to protein adsorption on the membrane and in the pores (51, 54, 55). The highest plasma fluxes (0.5 cm/min) were obtained with the rotating membrane device of Hemascience, the Autopheresis C (now distributed by Fenwall Division from Baxter Health Care). Even though this device was conceived for collecting 600 ml of plasma from donors, it has also been successfully applied to therapeutic plasma exchange by Kaplan et al. (56) who, after increasing the inlet pressure setting, were able to filtrate 3 l of plasma in 70 to 90 min depending upon the site of blood access.

An ingenious modification to this device consists in separating first the blood into a platelet rich plasma and red cells by a combination of Taylor vortices and centrifugation using a similar Couette device without a membrane. Then the platelets are separated from the plasma by the Autopheresis C. In this way, 600.000 platelets/µl can be collected with minimal contamination by leukocytes and red cells (57).

An interesting clinical technique called spontaneous plasmapheresis was proposed by Lysaght and Schmidt in Münich (58, 59). Taking advantage of the observation that plasma can be extracted under low transmembrane pressure, they connected the plasma filter to the femoral artery as an arterio venous shunt and blood circulated through the filter at a flow rate from 110 to 140 ml/min depending upon the filter size without a pump. Adequate transmembrane pressure is insured by placing the plasma collecting bag below the filter. High plasma flow rates ranging from 37 to 80 ml/min can be obtained without any other apparatus.

Another recent application of blood microfiltration concerns blood salvage during surgery which permits to avoid risks associated with transfusion. Usually, the blood is collected by suction from the operative field mixed with an anticoagulant and prefiltered in order to remove microaggregates and debris. During cardiac surgery, the blood gets diluted with cardioplegic fluid and has to be reconcentrated by a factor from 2 to 3, to an hematocrit of at least 50%. Orthopaedic surgery hemolyses the blood and the level of plasma free hemoglobin has to be decreased by a factor up to 6 by dilution with saline. Automatic blood cell processors for autotransfusion based on centrifugation are presently available but they are too costly (around $ 50,000) for small surgical units. In addition, the loss of platelets and the possibility of future hemolysis caused by centrifugation have to be considered. A membrane based auto-transfusion device using an ultrafiltration membrane was proposed by Solem et al. (60). But the choice of an ultrafiltration membrane to permit protein recovery precluded removal of plasma free hemoglobin. Legallais and Jaffrin (61) investigated a system containing successive hollow fibers plasma filters with saline introduction between filters. By proper optimization of the system, 500 ml of recovered blood could be washed with 750 ml of saline and a total membrane area of 0.5 m^2. The hematocrit was raised from 20% to 50% while the plasma free hemoglobin was lowered by a factor of five. Another membrane based system, the Haemocell 350, which uses the Bellhouse (44) concept of micro vortices generated along a dimpled microporous PVDF membrane was tested clinically for autotransfusion by Kalra et al. (62). This unit permitted a satisfactory recovery of all formed elements and a decrease of plasma hemoglobin. But, due to the very small membrane area used, 450 cm^2, the final hematocrits reached after washing the cells were not very high.

An original approach to plasma filtration from blood as been pioneered by Sakai's group in Tokyo (63, 64) who used 0.2 µm ceramic tubular and flat sheet filters to filtrate bovine blood. The advantages of mineral membranes over polymeric ones is that they can be completely regenerated after use by heating in an electric furnace or by rinsing with sodium hypochlorite. They are also easy to sterilize and a have a narrower pore size distribution than polymeric membranes. Due probably to this narrow pore distribution no hemolysis was observed even at high transmembrane pressures. The filtration characteristics were comparable to those of polymeric membranes and regeneration was repeated up to 40 times without loss of performance.

ACKNOWLEDGMENTS

Many of the models and of the data presented in this review were the results of a collaborative effort with my colleagues, L.H. Ding, B.B. Gupta and our doctoral students. The experimental research was made possible by the support of Fresenius and Akzo-Nobel companies.

REFERENCES

1 Kolff, W.J., 1947, New ways of treating uremia, *Churchill Ltd, London.*

2 Frost, T.H., Ed. 1978, Technical aspects of renal dialysis, *Pitman Medical.*

3 Henderson, L.W., Colton, C.K. and Ford, C.A., 1975, Kinetics of hemodiafiltration. II. Clinical characterization of a new blood cleansing modality, *J. Lab. Clin. Med.* 85:372-380.

4 Man, N.K. and Funck-Brentano, J.L., 1977, L'hémofiltration. *Actualités néphrologiques de l'Hôpital Necker. Flammarion Paris.* 1: 387-395.

5 Solomon, B.A., Castino, F., Lysaght, M.J., Colton, C.K., and Friedman, L.I., 1978, Continuous flow membrane filtration of plasma from whole blood, *Trans. Am. Soc. Artif. Int. Organs* 24:21-26.

6 Nose, Y., and Malchesky, P.S., 1981, Therapeutic membrane plasmapheresis, *Therapeutic plasmapheresis, Ota et ed., Schattauer, Stuttgart,* 3-14.

7 Gurland, H.J., Samtleben, W., and Blumenstein, M., 1983, Therapeutic membrane plasmapheresis, Present state and future aspects. *Life Support Systems,* 1:61-70.

8 Agishi, T., Kaneko, I., Hasuo, Y. et al., 1980, Double filtration plasmapheresis. *Trans. Am. Soc. Artif. Int. Organs,* 26:406-410.

9 Lysaght, M.J., Samtleben, W., Schmidt, B., Gurland, H.J., 1984, Analytical comparison of single pass and dead-end operation in cascade filtration plasmapheresis. *Artificial Organs,* 8:481-488.

10 Homma, Y., Mikami, Y., Tamachi, H. et al., 1986, Comparison of selectivity of LDL removal by double filtration and dextran sulfate cellulose column plasmapheresis, *Atherosclerosis*, 60:23-27.
11 Jaffrin, M.Y., Ding, L.H., Laurent, J.M., 1992, Kinetics of concentration-polarization formation in cross flow filtration of plasma from blood : experimental results, *J. Memb. Sci.*, 72:267-275.
12 Blatt, W.F., Dravid, A., Michaels, A.S., Nilsen, L., 1970, Solute polarization and cake formation in membrane ultrafiltration : causes, consequences and control techniques, *In Flinn F.E. (ed) Membrane Science and Technology, New-York, Plenum*, 47:97.
13 Colton, C.K., Henderson, L.W., Ford, C.A., Lysaght, M.J., 1975, Kinetics of hemodiafiltration : I. In vitro characteristics of a hollow fiber blood ultrafilter, *J. Lab. Clin. Med.*, 85:355-371.
14 Romero, C.A., Davis, R.H., 1988, Global model of cross flow microfiltration based on hydrodynamic particle diffusion, *J. Memb. Sci.*, 39:157-185.
15 Malbrancq, J.M., Jaffrin, M.Y., Bouveret, E., Anglerand, R., Vantard, G., 1984, Plasma filtration through a microporous membrane, *ASAIO J.*, 7:16-24.
16 Gupta, B.B., Jaffrin, M.Y., Ding, L.H., Dohi, T., 1986, Membrane plasma filtration through small-area hollow fiber filters, *Artificial Organs*, 10:45-51.
17 Raff, M., Ohmayer, M., Göhl, H., Samuelson, G., 1984, Influence of geometric parameters on filtration flux in plasmafilters, *In Therapeutic Apheresis : A critical look, Ed. by Nose,Y, Malchesky, P.S., Smith, J.W., ISAO Press, Cleveland*, 115-121.
18 Jaffrin, M.Y., Malbrancq, J.M., Vantard, G., Martin, T., Faure, A., 1985, Rheological aspects of plasma separation by membrane, *Clinical Hemorheology*, 5:231:240.
19 Zydney, A.L., Colton, C.K., 1987, Fundamental studies and design analyses for cross flow membrane plasmapheresis in artificial organs, *Ed. by Andrade, J.D., et al., VCH Publishers, Florida, USA*, 343-358.
20 Sueoka, A., Malchesky, P.S., Nose, Y., 1983, Effects of blood composition on membrane plasma separation, *Plasmapheresis, Ed. by Nose, Y., Malchesky, P.S., Smith, J.W., ISAO Press, Cleveland*, 93-103.
21 Zydney, A.L., Colton, C.K., 1982, Continuous flow membrane plasmapheresis : theoretical models for flux and hemolysis prediction, *Trans. Am. Soc. Artif. Intern. Organs*, 28:408-412.
22 Malbrancq, J.M., Jaffrin, M.Y., Bouveret, E., Angleraud, R., Vantard, G., 1982, Factors governing plasma filtration rate in plasmapheresis by plane microporous membranes, *Proc. 9th ESAO, Life Support Systems*, 11:46-50.
23 Forstrom, R.J., Bartlett, K., Blackshear, P.L., Wood, T., 1975, Formed element deposition onto filtering walls, *Trans. Am. Soc. Artif. Intern. Organs*, 21:602-607.
24 Malchesky, P.S., Wojcicki, J., Moorman, M., Pentermann, E.J., Nose, Y., 1984, Blood cell effects in membrane plasma separation, *Trans. Am. Soc. Artif. Intern. Organs*, 30:313-317.
25 Vassilief, C.S., Leonard, E.F., Stepner, T.A., 1985, The mechanism of cell rejection in membrane plasmapheresis, *Clinical Hemorheology*, 5:7-15.
26 Cox, R.G., Mason, S.G., 1971, Suspended particles in fluid flow through tubes, *Annual Review of Fluid Mechanics*, 3:291-316.
27 Wang, N.M.L., Keller, K.H., 1979, Solute transport induced by erythrocyte motions in shear flow, *Trans. Am. Soc. Artif. Intern. Organs*, 25:14-17.
28 Eckstein, E.C., Bailey, D.G., Shapiro, A.H., 1977, Self diffusion of particles in shear flow of a suspension, *J. of Fluid Mech.*, 79:191-208.
29 Rock, G. Titley, P., Mc Combie, N., 1986, Plasma collection using an automated membrane device, *Transfusion*, 26:269-271.
30 Beaudoin, G., Jaffrin, M.Y., 1989, Plasma filtration in Couette flow membrane devices, *Artificial Organs*, 13:43-51.
31 Ohashi, K., Tashiro, K., Kushiya, F. et al., 1988, Rotation-induced Taylor vortex enhances filtrate flux in plasma separation, *Trans. Am. Soc. Artif. Intern. Organs*, 34:300-307.
32 Vigo, F., Uliana, C., Lupino, P., 1985, The performance of a rotating module in oily emulsions ultrafiltration, *Separation Sci. and Technology*, 20:213-230.
33 Kroner, H.K., Nissinen, V., 1988, Dynamic filtration of microbial suspensions using an axially rotating filter, *J. of Memb. Sci.*, 36:85-100.
34 Taylor, G.I., 1936, Fluid friction between rotating cylinders, 1. Torque measurements, *Proc. Royal. Soc.*, 157:546-564.
35 Jaffrin, M.Y., Beaudoin, G., Ding, L.H., Djennaoui, N., 1989, Effect of membrane characteristics on the performance of Couette rotating plasma separation devices, *Proc. Am. Soc. Artif. Intern. Organs*, 35:690-693.
36 Schlichting, H., 1968, Boundary layer theory, *Mac Graw Hill, New-York*.
37 Zydney, A.L., Saltzman, W.M., Colton, C.K., 1989, Hydraulic resistance of red cell beds in an unstirred filtration cell, *Chem. Eng. Sci.*, 44:147-155.

38 Galletti, P.M., Richardson, P.D., Trudell, L.A., 1983, Oscillating blood flow enhances membrane plasmapheresis, *Trans. Am. Soc. Artif. Intern. Organs*, 23:279-282.
39 Jaffrin, M.Y., Gupta, B.B., Cannon, R.I., Ding, L.H., 1984, Enhancement of plasma filtration in hollow fiber filters by pulsatile blood flow, *Proc. XI ESAO, Life Support Systems*, Vol. 2, Suppl. 1:207-210.
40 Gupta, B.B., Ding, L.H., Jaffrin, M.Y., 1985, High efficiency small membrane area plasmapheresis using pulsatile blood flow, *Progress in artificial organs, Ed. by Nose Y., ISAO Press*, 891:895.
41 Jaffrin, M.Y., Ding, L.H., Gupta, B.B., 1987, Rationale of filtration enhancement in membrane plasmapheresis by pulsatile blood flow, *Life Support Systems*, 5:267-271.
42 Schooneman, F., Stoltz, J.F., Streif, F. et al., 1986, Technical study and biological results obtained with a pulsatile flow plasmapheresis system, *Life Support Systems*, 4:362-365.
43 Ding, L.H., Laurent, J.M., Jaffrin, M.Y., 1991, Dynamic filtration of blood : a new concept for enhancing plasma filtration, *The Int. J. of Artif. Organs*, 14:365-370.
44 Bellhouse, B.J., Lewis, W.H., 1988, A high efficiency membrane separator for donor plasmapheresis, *Trans. Am. Soc. Artif. Intern. Organs*, 34:747-754.
45 Bellhouse, B.J., Bellhouse, F.N., Curl, C.M. et al., 1973, A high efficiency membrane oxygenator and pulsatile pumping system and its application to animal trials, *Trans. Am. Soc. Artif. Intern. Organs*, 19:72-79.
46 Solomon, B.A., 1981, Membrane separation : technological principles and issues, *Trans. Am. Soc. Artif. Intern. Organs*, 27:345-350.
47 Zydney, A.L., Colton, C.K., 1984, A red cell deformation model for hemolysis in cross flow membrane plasmapheresis, *Chem. Eng. Commun*, 30:191-207.
48 Rand, R.P., 1964, Mechanical properties of the red cell membrane. II. Viscoelastic break down of the membranes, *Biophysical J.*, 4:303-316.
49 Blackshear, P.L. Anderson, R.J., 1977, Hemolysis threshold in microporous techniques, *Blood Cells*, 3:377-390.
50 Ding, L.H., Jaffrin, M.Y., Gupta, B.B., 1986, A model of hemolysis in membrane plasmapheresis, *Trans. Am. Soc. Artif. Int. Organs*, 32:330-333.
51 Philp, J.L., Jaffrin, M.Y., Ding, L.H., 1994, Hemolysis during membrane plasma separation with pulsed flow filtration enhancement, *J. of Biomechanical Eng., Trans. ASME*, 116:514-520.
52 Gurland, H.J., Lysaght, M.J., Samtleben, W., Schmidt, B., 1984, Comparative evaluation of filters used in membrane plasmapheresis, *Nephron*, 36:173-182.
53 Lysaght, M.J., Samtleben, W., Schmidt, B., Gurland, H.J., 1983, Contemporary technical issues in membrane plasmapheresis : controversies and reconciliation, *in Plasma Separation and Plasma Fractionation, Karger, Baser*, 315-328.
54 Roberts, C.G., Schindhelm, K., Farrell, P.C., 1983, Protein membrane interactions in membrane plasma separation, *in Plasmapheresis, ed. by Nose Y., Malchesky P.S., Smith J.W., ISAO Press, Cleveland*, 81-91.
55 Young, B.R., Pitt, W.G., Cooper, S.L., 1988, Protein adsorption on polymeric biomaterials. II - Adsorption kinetics, *J. Colloïd Interface Science*, 125:246-253.
56 Kaplan, A.E., Halley, S.E., 1988, Evaluation of a rotating filter for use with therapeutic plasma exchange, *Trans. Am. Soc. Artif. Intern. Organs*, 34:274-276.
57 Stromberg, R.R., Friedman, L.I., Schorr, J.B., 1991, Donor plasmapheresis technology, *in Advances in Hemapheresis, ed. by C.Th. Smit Sibinga, L. Katers, Klewer Academic Publish., Dordrecht*, 29-38.
58 Lysaght, M.J., Samtleben, W., Schmidt, B., Stoffner, D., Gurland, H.J., 1983, Spontaneous membrane plasmapheresis, *Trans. Am. Soc. Artif. Intern. Organs*, 29:506-510.
59 Schmidt, B., Lysaght, M.J., Samtleben, W., Gurland, H.J., 1983, Plasmapheresis without pumps for therapeutic and donor purposes, *in Plasma separation and plasma fractionation, ed. by Lysaght M.J., Gurland H.J., Kerger, Basil*, 188-196.
60 Solem, J.O., Steen, S., Olin, C.A., 1986, A new method for autotransfusion of shed blood, *Acta Chirurg. Scand.*, 152:421-425.
61 Legallais, C., Jaffrin, M.Y., 1993, A feasibility study of a filtration type autotransfusion device, *J. Biomed. Eng.*, 15:143-147.
62 Kalra, M., Beech, M.J., Al Khaffaf, H., Charlesworth, D., 1993, Autotransfusion in aortic surgery, the haemocell 350 cell saver system, *Br. J. Surg.*, 80:32-35.
63 Sakurai, H., Ozawa, K., Takesawa, S., Sakow, K., 1986, Design of a plasma separation using ceramic membranes, *Trans. Am. Soc. Artif. Intern. Organs*, 32:410-413.
64 Ozawa, K., Kin, H.B., Sakurai, H., Takesawa, S., Sakai, K., 1986, Novel utilization of ceramic membranes in plasma treatment, *in Progress in Artificial Organs, ed. by Nosé Y. et al., ISAO Press*, 913-920.

12

CARDIO-VASCULAR INTERACTION DETERMINES PRESSURE AND FLOW

N. Westerhof

Laboratory for Physiology
Institute for Cardiovascular Research (ICaR-VU)
Free University of Amsterdam
Van der Boechorststraat 7
1081 BT Amsterdam, The Netherlands

INTRODUCTION

The arterial system as part of the circulation has been a subject of study since Harvey's monograph of 1628. In 1735 Stephen Hales was the first to measure arterial blood pressure and noticed the oscillatory aspects of it. He also suggested that reduction of the arterial pressure oscillations resulted from arterial compliance. By the end of the last century the Windkessel model was proposed by Frank as a lumped model of the systemic arterial tree (Frank, 1899). Although wave travel was neglected in this model, the importance of arterial compliance as part of the load on the heart was clearly brought forward. With modern techniques simultaneous high quality pressure and flow data became available. These data could be analyzed with the newly developed computers and led to the derivation of input impedance, a comprehensive description of the arterial system. Knowledge of the input impedance provided a great step forward in the understanding of arterial function.

The heart as a pump was studied by Frank (1895). He used the isolated frog heart and found for isovolumic contractions that with an increase in diastolic cardiac volume developed pressure increased. He was one of the first to present the cardiac pump in terms of the pressure-volume diagram. A few years later with the work of Langendorff (1899) it became possible to study isolated mammalian hearts where the coronary perfusion was provided for. Somewhat later Starling and his associates studied the dog heart in the heart-lung preparation (Patterson & Starling, 1914). In this preparation the heart was loaded with a Starling resistor implying that 'aortic' pressure was kept constant and an increase in cardiac output with an increase in ventricular filling was found.

In the middle seventies the pressure-volume relation became a subject of study again in the mammalian heart (Suga et al., 1973). This provided new information on pump function and also led to better understanding of how cardiac oxygen consumption is related to mechanics. About the same time the heart as a pump in terms of output and pressure generation was studied by Elzinga and Westerhof (1973, 1979). This approach made it possible to study the interaction of arterial system and heart because both systems were characterized by pressure-flow relations.

The heart and arterial system form a matched pair in the sense that heart rate is related to the RC-time of the arterial system resulting in similar systolic and diastolic aortic pressures in most mammals (Elzinga & Westerhof, 1991).

In the present review we will discuss the arterial system and the heart in terms of pressure and flow, compare the descriptions with other characterizations and finally combine our knowledge to describe cardiac-arterial matching.

THE ARTERIAL TREE

The arterial system can be considered as the load on the heart. We will discuss the arterial system in terms of pressure-flow relations. The relations between pressure and flow here derived apply to all vascular beds and parts thereof, but we will here concentrate on the systemic arterial tree except when clearly stated.

Resistance and Impedance

Resistance and impedance are descriptions of the entire arterial system derived from the measured pressure and flow in the ascending aorta. It is feasible to measure these quantities, together with left ventricular pressure, simultaneously in the human since the sensors can be mounted on a single catheter (Murgo et al., 1980). An example of these measurements is given in Fig. 1.

A straightforward and simple characterization of the arterial system is given by its resistance on the basis of Ohm's law: mean pressure drop (arterio-venous pressure difference) divided by mean flow. This characterization is in principle independent of the pump, the heart. However, the relation between pressure and flow is usually not a proportional relation (Sagawa and Eisner, 1975) and the relation is subject to change due to humoral and nervous control. Thus although the arterial system is not linear and the properties may vary in time, peripheral resistance is often fruitfully used to characterize the arterial tree in the

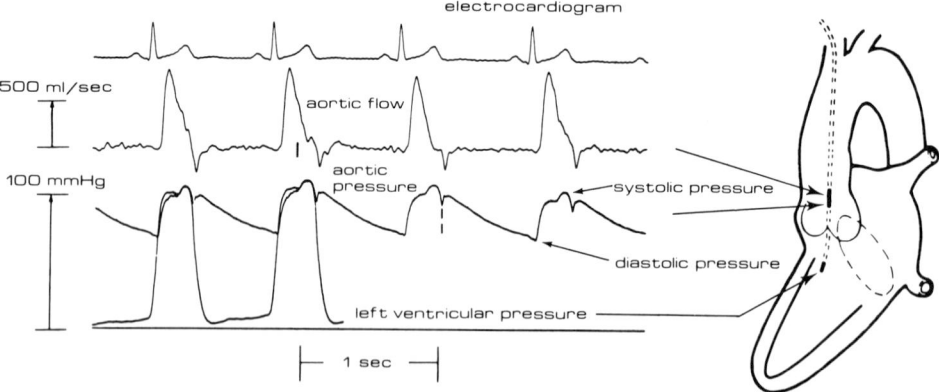

Figure 1. The electrocardiogram, instantaneous aortic flow and ventricular and aortic pressure measured simultaneously in the conscious human. Pressure and flow velocity sensors are located on a single catheter. Aortic pressure and flow are used to characterize the arterial system and ventricular pressure and aortic flow are used to characterize the heart as a pump by means of the pump function graph. Data from Murgo et al. (1980).

steady state. Resistance is mainly determined by the resistance in the arterioles and we therefore call it peripheral resistance. Because in the systemic circulation venous pressure is small with respect to arterial pressure, the pressure drop in the calculations is often replaced by arterial pressure alone. Obviously, this approximation does not allow for the pulmonary arterial tree where arterial pressure is 20 mmHg and venous pressure 5 mmHg. The pressure difference also avoids possible misconceptions. When, for instance, pressure and flow in the femoral artery are measured use of Ohm's law leads to the calculation of resistance of the femoral bed alone, not the combination of the femoral bed and the other beds, because it is the pressure drop from femoral artery to femoral vein which determines the bed.

A comprehensive description of an arterial tree or part thereof can be given by its input impedance. The derivation and its limitations have been discussed earlier by Westerhof et al. (1979, 1993). When input impedance is calculated the wave shapes of pressure and flow are taken into account, not only their mean values. Input impedance is also obtained on the basis of Ohm's law. To derive impedance, both the pressure and flow are written as a Fourier series. The Fourier series can be calculated for signals that are repetitive, as are pressure and flow in the steady state, and contain a limited number of discontinuities so that the sharp incisura of the aortic pressure wave and the backflow in the flow wave form no problem. Pressure and flow are analyzed similarly, therefore we will restrict ourselves to the pressure wave here.

The Fourier series reads:

$$P(t) = P_m + A_1 \cos \omega t + A_2 \cos 2\omega t + \ldots + B_1 \sin \omega t + B_2 \sin 2\omega t + \ldots \qquad (1)$$

with P_m mean pressure and $\omega = 2\pi f$ circular frequency and f frequency. The n is the so-called harmonic number with the mean term often designated as the zeroth harmonic. The Fourier coefficients A_n and B_n can be derived straightforwardly from integration of the signal as a function of time:

$$A_n = \frac{2}{T} \int_0^T P(t) \cos n\omega t \, dt \ll\_\gg n = 1,2,3,\ldots \qquad (2a)$$

$$B_n = \frac{2}{T} \int_0^T P(t) \sin n\omega t \, dt \ll\_\gg n = 1,2,3,\ldots \qquad (2b)$$

Integration is to be carried out over a single or an integer number of cardiac cycles. In the application of Ohm's law pressure and flow are to be divided. Therefore it is preferred to write the Fourier series as:

$$P_n(\omega) = P_n \exp j(n\omega t + \phi_{p,n}). \qquad (3)$$

with $P_n^2 = A_n^2 + B_n^2$ and $\tan \phi_n = B_n / A_n$.

The P_n and ϕ_n are called the modulus and phase of the harmonic. Impedance is now calculated on basis of Ohm's law from division of pressure and flow harmonics, i.e. impedance is calculated as:

$$|Z_{in}| = P_n / F_n \qquad (4a)$$

and:

$$\angle Z_{in} = \phi_{p,n} - \phi_{f,n}. \qquad (4b)$$

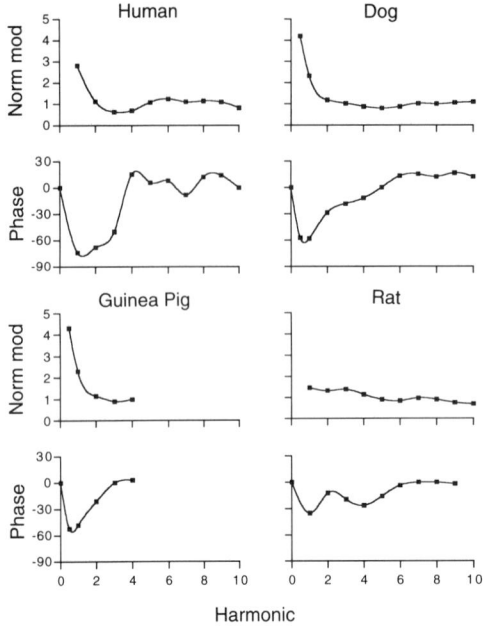

Figure 2. Averaged input impedance measured in the human (n=7), dog (n=6), guinea pig (n=16), and rat (n=31). All frequencies are relative to heart rate (harmonics) and the impedance modulus is normalized to characteristic impedance. The peripheral resistance and characteristic impedance values (in $g \cdot cm^{-4} \cdot s^{-1}$) are respectively 1,400 and 82 for the human, 5,100 and 240 for the dog, 102,000 and 4,300 for the guinea pig and 140,000 and 8,370 for the rat (Westerhof & Elzinga, 1991). Body mass and heart rate are 75 kg and 69 beats per minute (bpm) in the human, 20-43 kg and 94 bpm in the dog, 0.73 kg and 270 bpm in the guinea pig and 0.28-0.75 kg and 420 bpm in the rat. These data have been presented by a large number of authors such as Murgo et al. (1980), Van den Bos et al. (1982), Avolio et al. (1976) and Mitchell et al. (1994), and are summarized by Westerhof and Elzinga (1991).

The modulus and phase of the input impedance are a function of frequency and therefore impedance is called a frequency domain characterization of the arterial system.

As with the calculation of resistance the basic assumption in the application of Ohm's law is that the system under study is linear. In terms of mean values of pressure and flow we have indicated above that this implies a proportional relationship. In the context of sine waves this means that when pressure is a sine wave, flow should be a sine wave as well. We know that the arterial system is nonlinear (Anliker et al., 1968) but several tests have suggested that this nonlinearity is not leading to large errors (Dick et al., 1968; Reuderink et al., 1989). More investigations are necessary to decide if and how nonlinearity contributes to errors in the input impedance. Stergiopulos et al. (1995b) showed that the scatter found in the impedance spectrum may result from nonlinearities (See Fig. 7 below). If Fig. 7 gives an indication of errors due to nonlinear characteristics of the arterial system then the important overall features of impedance seem correct.

Calculation of input impedance, in analogy to the derivation of resistance, should take into account the pressure drop instead of arterial pressure alone. However, the oscillations of pressure in the veins are small and usually disregarded.

The phase of the harmonics of pressure and flow ($\phi_{p,n}$ and $\phi_{f,n}$) depends on the starting point of the integration, usually the R-wave in the electrocardiogram. When the phase of the impedance is calculated and the phase angles of pressure and flow subtracted this arbitrariness disappears.

We see that input impedance can be derived from the pressure and flow waves that are generated by the combination of heart and arterial system without the need to introduce pumps. Thus impedance is by its derivation a characterization of the arterial system per se and can be obtained without great intrusion on the system. Direct division of pressure and flow in the time domain does not lead to useful results (Westerhof et al., 1979). We can see this from the simplest wave shape, the sine wave. When pressure and flow would be sine waves that are out of phase instantaneous division would lead to a ratio that varies in time between minus and plus infinity.

It can be shown that the amplitude of the harmonics of the Fourier series of pressure and flow decreases with harmonic number (Westerhof et al., 1979). This implies that information at high harmonics is subject to greater error and that only information is obtained for a limited number of harmonics. For aortic pressure and flow this number is about 15 while for pulmonary signals and peripheral systemic arterial pressures and flows this number is lower. Thus data on high frequencies are difficult to obtain (Newman et al., 1986). Also impedance data are obtained at multiples of the frequency of the heart beat only. By pacing the heart at different rates the impedance can be obtained at other frequencies. In this way it is possible to improve the frequency resolution (see Fig. 7).

In Fig. 2 the input impedance of the systemic arterial tree of four species is shown. To compare animals impedance was normalized, i.e. the modulus of the impedance is presented relative to aortic characteristic impedance (Westerhof et al., 1979). Normalization of the input impedance modulus with respect to peripheral resistance instead of aortic characteristic impedance leads to similar results, because the ratio of characteristic impedance and peripheral resistance is the same in most mammals (Westerhof & Elzinga, 1991). Instead of plotting impedance as a function of frequency it is presented as a function of harmonics, i.e. multiples of heart rate.

The value of the input impedance modulus at zero Hz equals peripheral resistance. The impedance modulus decreases rapidly with frequency and then hovers about a constant value, we call this value aortic characteristic impedance (see below). The phase of the impedance is zero at zero Hz, negative for low frequencies and returns to about zero at high frequencies. We will later explain the features of the input impedance in relation to arterial function.

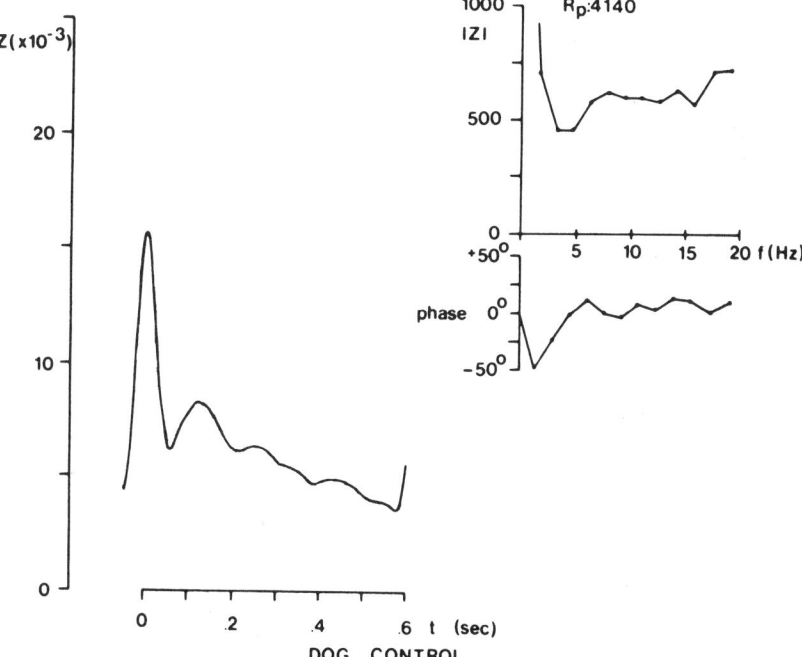

Figure 3. Impulse response function of the systemic arterial tree of the anaesthetized dog. The input impedance is presented in the inset. Peripheral resistance and impedance modulus in $g.cm^{-4}s^{-1}$; impulse response in $g.cm^{-4}s^{-1}$. Modified from Laxminarayan et al. (1978).

Impulse Response Function

The arterial system can also be characterized in the time domain using the impulse response (Sipkema et al., 1980). The impulse response is the pressure resulting from a (infinitely) short impulse in flow, and is therefore a time domain characterization of the arterial system. The area under the flow impulse is unity and the response is thus given in pressure per volume. The impulse response function can be derived from the input impedance (Laxminarayan et al., 1978). In other words it can be calculated from the measurement of pressure and flow without the need of a pump. An example of the input impedance and the impulse response function of the systemic arterial tree of the dog is given in Fig. 3.

The impulse response and the input impedance present the same information in different manners. The two characterizations form a Fourier pair. Through Fourier analysis of the impulse response or inverse Fourier transformation of the impedance they can be converted into each other. The calculations, however, may be involved (Laxminarayan et al., 1978).

In certain cases it may be useful to physically apply an impulse with a generator and measure the impulse response function to characterize the arterial system. Newman et al. (1986) applied impulses to the systemic arterial tree of the dog to increase the high frequency content of the pressure and flow signal. In this way they were able to obtain input impedance information at very high frequencies and concluded that the impedance remains at a constant level for these high frequencies. Van Huis et al. (1987) used an impulse generator to obtain the impulse response of the coronary arterial tree in the beating heart. Impulses were given in systole and diastole and the impulse responses measured. These impulse responses were then transformed to impedances. It turned out that the input impedance of the left coronary circulation did not vary between diastole and systole, except for the zero Hz term. In a time-varying system Fourier analysis of the pressure and flow signals in the steady state of oscillation is permitted but the calculation of impedance is not useful. The impulse response lasted only a fraction of the heart beat and it was therefore assumed that during systole and diastole the coronary arterial system was invariant. The unexpectedly small difference in coronary arterial impedance between systole and diastole is explained as follows (also see explanation of impedance below). The zero Hz term is determined by the arterioles, while the impedance at the higher the frequencies is determined by the more proximal arterial properties. Apparently during contraction only the very small vessels, the resistance vessels, are affected.

Reflections

Pressure and flow in the arterial system can also be viewed as consisting of separate waves travelling from the heart towards the periphery (forward or initial waves) and waves travelling towards the heart (backward or reflected waves) (Westerhof et al., 1972). The relation between the forward and backward waves is given by the reflection coefficient. Simple division of the time functions of the backward and forward pressures does not lead to useful information. Therefore the reflection coefficient should be expressed in terms of harmonics and thus is a function in the frequency domain. The reflection coefficient is related to impedance which is also a frequency domain function.

The reflection coefficient can best be explained on the basis of a single uniform blood vessel with an arterial system as its load impedance (Z_l). The single uniform vessel can be characterized by its characteristic impedance (Z_c), which is the input impedance of the vessel when no reflections are present or when the vessel is infinitely long (Westerhof et al., 1969). The characteristic impedance depends on the basic properties of the vessel such as diameter,

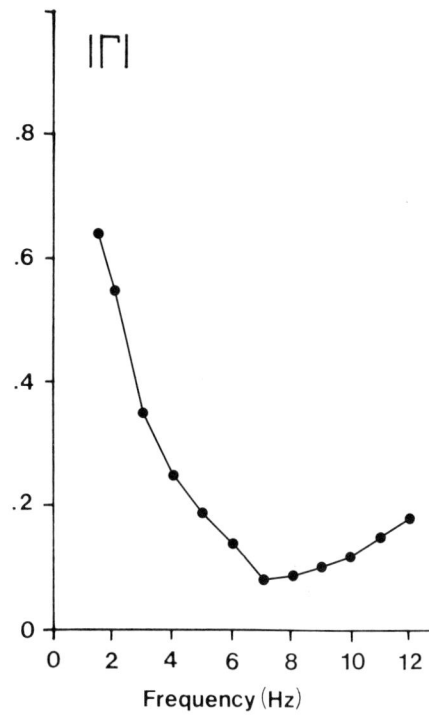

Figure 4. Modulus of the global reflection coefficient in the ascending aorta of the anaesthetized dog. Modified from Westerhof et al. (1972).

wall thickness and wall elasticity and on the properties of the blood, i.e. viscosity and density (Jager et al., 1965). The reflection coefficient is:

$$\Gamma = (Z_l - Z_c) / (Z_l + Z_c). \tag{5}$$

On basis of this formula it is clear why the reflection coefficient is a function in the frequency domain. Using this formula two reflection coefficients have been introduced: the local and global reflection coefficient (Westerhof, 1968).

Again if we consider a single tube with a load we can calculate a local reflection coefficient at any location. In the uniform section of the tube the local reflection coefficient is zero and of little interest. However, at the end of the tube, at the junction with the load, the local reflection coefficient is a measure of the mismatch at that location. Also at bifurcations the local reflection coefficient gives information on the impedance mismatch. It was generally observed that the local reflection coefficient at arterial bifurcations is low with in general small phase shifts (Westerhof, 1968).

The global reflection coefficient gives information about the amount of reflected waves present at the location of interest. The global reflection coefficient at any location of the tube is different from zero and depends on the location. The global reflection coefficient at the root of the aorta was first calculated by Westerhof et al. (1972). The input impedance of the entire arterial tree was taken as the load impedance and the characteristic impedance of the ascending aorta as Z_c, and inserted in equation 5. This reflection coefficient gives information about the amount of reflected waves present in the ascending aorta.

The modulus of the global reflection coefficient in the ascending aorta of the dog is given in Fig. 4.

Figure 4 shows that reflection is strongest for low frequencies and decreases strongly with frequency.

The relation between global reflection coefficient and input impedance at the root of the aorta is obvious. For low frequencies input impedance deviates strongly from aortic characteristic impedance and the reflection coefficient is large. For high frequencies the input impedance approaches aortic characteristic impedance and the reflection coefficient is small.

We conclude that three characterizations of the arterial system are possible: input impedance, impulse response and global reflection coefficient.

Forward and Backward Travelling Waves

When the reflection coefficient is known the pressure and flow waves in the arterial system can be separated into their forward and reflected components. We should stress that pressure and flow and therefore their forward and backward components as well are the result of the interaction of heart and arterial load. This means that the separation is of interest and can teach us much but characterization of the arterial system or the heart is not presented in this form.

Separation of pressure and flow waves is performed as follows (Westerhof et al., 1972). Fourier analysis produces the individual harmonics of pressure and flow. Input impedance is derived and the characteristic impedance of the aorta is measured or estimated. This can be done from the high frequencies of the input impedance (Westerhof et al., 1972), from the initial rise of the pressure and flow wave as time functions (Li et al., 1986) or from the measurement of pressure and diameter. When for each harmonic the reflection coefficient is known the relation between the forward and backward harmonics of pressure (P_f, P_b) and flow (F_b, F_f) is known as well:

$$P_b(n\omega) = \Gamma(n\omega) \cdot P_f(n\omega) \quad \text{and} \quad F_b(n\omega) = -\Gamma(n\omega) \cdot F_f(n\omega) \tag{6}$$

It should be noted that the reflection coefficient of the pressure and flow waves is the same in magnitude. However, a positive reflection of the pressure relates to a negative reflection for flow, i.e. the reflection coefficients are 180 degrees out of phase. Since the measured pressure and flow harmonics (P_m and F_m) are the sum of their forward and backward components it also holds that:

$$P_m(n\omega) = P_f(n\omega) + P_b(n\omega) \quad \text{and} \quad F_m(n\omega) = F_f(n\omega) + F_b(n\omega) \tag{7}$$

From equations 6 and 7 we calculate that:

$$P_f(n\omega) = P_m(n\omega)/(1 + \Gamma(n\omega)) \quad \text{and} \quad F_f(n\omega) = F_m(n\omega)/(1 - \Gamma(n\omega)) \tag{8}$$

and with P_f and F_f known the harmonics of the backward waves of pressure and flow are calculated from equation (6).

After performing this calculation for all harmonics of pressure and flow and subsequent addition of the harmonics (Westerhof et al., 1979) the measured pressure and flow together with their forward and backward components are derived as a function of time (Westerhof et al., 1972).

This calculation can be simplified when the characteristic impedance of the aorta (Z_c) is a real rather than complex (in the mathematical sense) quantity (Murgo et al., 1981). Assuming a real quantity is acceptable on basis of Womersley's theory (Jager et al., 1965), we can write the equations in the time domain:

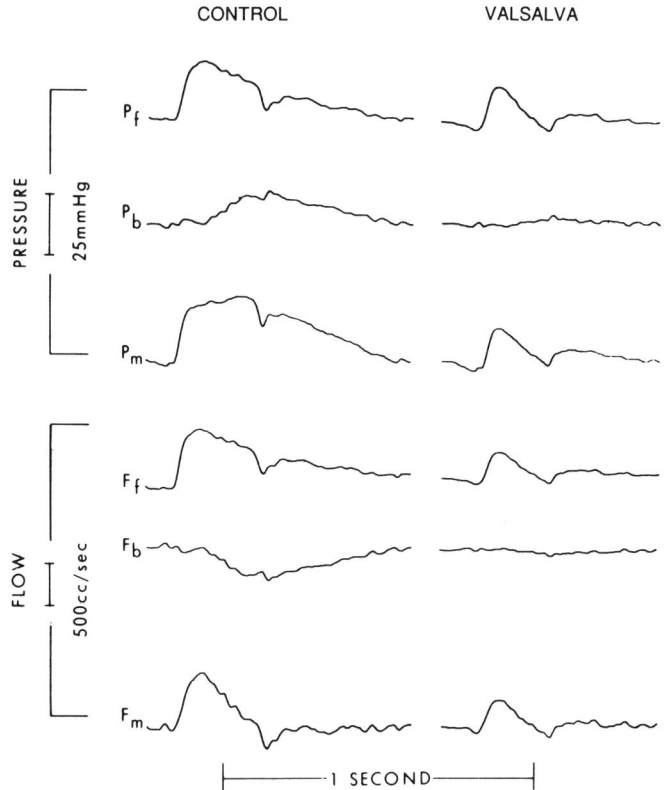

Figure 5. Forward and backward waves of pressure and flow in the conscious human during control and Valsalva maneuver. Modified from Murgo et al. (1981).

$$P_m(t) = P_f(t) + P_b(t) \quad \text{and} \quad F_m(t) = F_f(t) - F_b(t). \tag{9}$$

With a real characteristic impedance the forward pressure and flow are of the same shape. The backward waves of pressure and flow are also of the same shape, but due to the negative reflection of flow, inverted. Thus we can write:

$$P_f(t) = Z_c \cdot F_f(t) \quad \text{and} \quad P_b(t) = -Z_c \cdot F_b(t). \tag{10}$$

Combining equations 9 and 10 leads to:

$$P_f(t) = (P_m(t) + Z_c \cdot F_m(t))/2 \quad \text{and} \quad P_b(t) = (P_m(t) - Z_c \cdot F_m(t))/2 \tag{11}$$

and:

$$F_f(t) = (P_m(t) + Z_c \cdot F_m(t))/2Z_c \quad \text{and} \quad F_b(t) = -(P_m(t) - Z_c \cdot F_m(t))/2Z_c. \tag{12}$$

This calculation is relatively easy to perform and only needs information on aortic characteristic impedance and leads directly to the separation of pressure and flow waves without the necessity to perform Fourier analysis. Recently a method for separation of waves into their forward and backward components in a nonlinear arterial system was presented (Stergiopulos et al., 1993).

In Fig. 5 the pressure and flow with their separation into forward and backward components is shown for the control situation and during the Valsalva maneuver (Murgo et al., 1981).

It may be seen that during the Valsalva maneuver the backward wave is negligible in amplitude. This implies that reflections in the system are small during the Valsalva maneuver and aortic input impedance is very close to aortic characteristic impedance. This was indeed found to be the case (Murgo et al., 1981).

Arterial Function Explained

The characteristic findings on input impedance and reflection in the arterial tree are explained as follows.

For the mean pressure and flow the ratio is peripheral resistance, i.e. the impedance modulus at zero Hz equals peripheral resistance. At low frequencies the impedance modulus decreases strongly with increasing frequency and the phase angle is negative. This is characteristic for a compliance. For high frequencies the input impedance approaches the characteristic impedance of the aorta and the modulus reaches constant values while the phase angle returns to zero (real characteristic impedance). We can explain these findings from wave theory. Waves of pressure and flow are reflected at all bifurcations and discontinuities. The reflection coefficients may be small but there are many discontinuities. The further the waves travel into the periphery the shorter the distances between discontinuities become so that more reflection is found the further we are in the periphery. The reflected waves return to the aorta. Since for low frequencies the wave length is not short with respect to the distance of the reflection sites from the aorta and local refection coefficients exhibit small phase shifts (Westerhof, 1968; Westerhof et al., 1969) most waves return in phase and thus add to a considerable magnitude. For high frequencies the phase angles of the local reflection coefficients increase (Westerhof, 1968) and the wave length decreases so that waves return with random phases and cancel. This effect of reflection was first explained by Taylor (1966a; 1966b) using a randomly branching model of the systemic arterial tree. When reflections are small the input impedance approaches the characteristic impedance of the aorta, i.e. the reflectionless tube or infinitely long tube situation is approximated. Thus the input impedance decreases from the value of the peripheral resistance at zero Hz to the aortic characteristic impedance at high frequencies.

From this explanation we can also see the following. At zero Hz the resistance vessels in the periphery are mainly determining the impedance. For the intermediate frequencies the large, elastic arteries contribute in terms of compliance, while for very high frequencies the proximal aorta determines the input impedance. In other words, with increasing frequency input impedance is determined by more proximal properties of the arterial system. This also explains the small difference in coronary arterial input impedance in systole and diastole (see above).

Arterial Modelling

Several models of the arterial system have been described. We can separate these models into lumped models (Frank, 1899; Cope 1960; Cope, 1961; Westerhof, 1968; Westerhof et al., 1971; Burattini et al., 1994), models based on simplified geometry, such as the single uniform tube model with a single windkessel load (Sipkema and Westerhof, 1975; Berger et al., 1994) or with a number of windkessel loads for head, kidneys and lower body (Sipkema et al., 1990) and the T-tube models (O'Rourke & Cartmill, 1971; Burattini et al.,

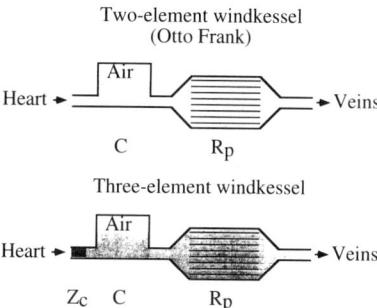

Figure 6. A. The two-element windkessel consisting of total arterial compliance (C) and peripheral resistance (R_p). B. The three-element windkessel. The Z_c is the characteristic impedance of the proximal ascending aorta. The characteristic impedance is determined by aortic compliance and blood mass: $Z^2_c = (dP/dA)\cdot r/A$, with A cross-sectional area, P pressure and ρ blood density.

1989) and the transmission line models based on actual vascular anatomy (Westerhof et al., 1969; O'Rourke et al., 1980; Stergiopulos et al., 1992) and models based on random geometry in the sense of lengths (Taylor, 1966a; 1966b). The windkessel models are conceptually simple and easy to program on a digital computer. Their input impedance may show a remarkable likeness to the actual arterial impedance. These models will be discussed in detail below. The T-tube models, representing the circulation to the upper and lower body are also relatively simple to program. They show that with two parallel tubes of different lengths the impedance is close to the physiological one. Obviously these models include wave travel (O'Rourke & Cartmill, 1971). Especially after the introduction of windkessel loads at the ends of the two tubes the model became very realistic (Burattini, 1989). The transmission line models have proven that the arterial system characteristics are determined by the geometry of the system together with the properties of the wall and blood. These models also are fit to describe the travelling waves in the arterial system, and allow calculation of reflections, allow studies of the effect of local changes in the arterial system on input impedance, etc. The newest of these models even includes nonlinear aspects of convective flow and nonlinear wall elasticity (Stergiopulos et al., 1992). However, these models are more difficult to program.

The lumped models are also called windkessel models. The two-element windkessel, proposed and worked out by Frank (1899), consists of a peripheral resistance and an arterial compliance determined by the elasticity of the large arteries (see Fig. 6, top). The model was deduced from the pressure in the aorta only and explained the diastolic aortic pressure wave form well. The diastolic decay of aortic pressure is exponential in this model with a time constant $\tau = R_p \cdot C$ with R_p peripheral resistance and C total arterial compliance. However, when flow measurements became available and input impedance could be calculated it soon became clear that the input impedance of the two-element windkessel deviated strongly from the actual input impedance at high frequencies. On basis of the observation that input impedance approaches the characteristic impedance of the aorta a third element was introduced: the characteristic impedance of the aorta (Fig. 6, bottom).

This three-element windkessel has an input impedance very close to the actual impedance as shown for the dog in Fig. 7 (Westerhof & Elzinga, 1988).

This model was used extensively as a load in isolated heart studies (Elzinga & Westerhof, 1973; Elzinga & Westerhof, 1979; Suga et al., 1973). Up till now it is still the best artificial load for isolated hearts.

The three-element windkessel was also used to estimate arterial parameters (Toorop et al., 1987), mainly total arterial compliance. By applying pressure measured in the biological preparation to the three-element windkessel model and comparing the model calculated flow with measured flow the parameters of the model were estimated. It was shown that the fit of the model flow was very close to the measured flow. Recently,

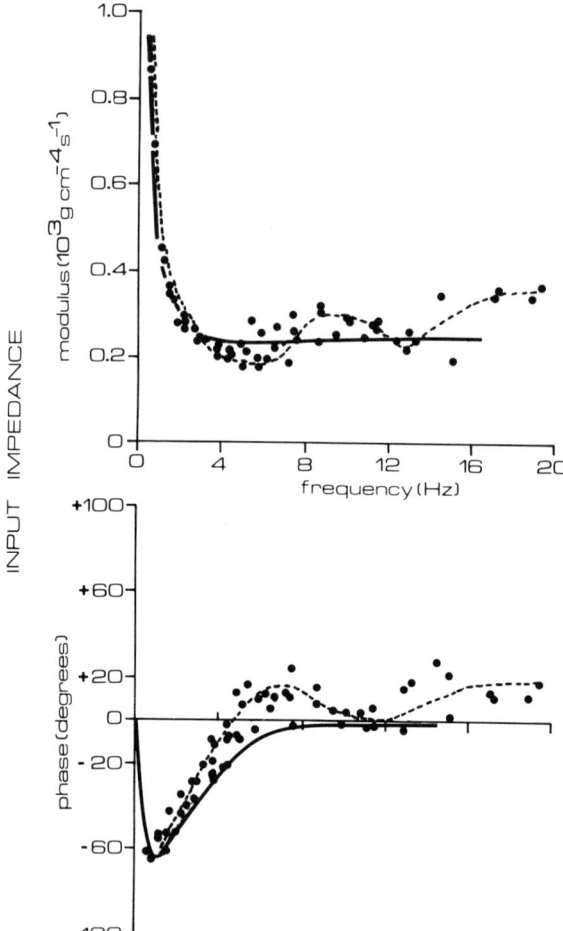

Figure 7. Systemic arterial input impedance of the anaesthetized dog. High frequency resolution is obtained by pacing the heart at different rates. The dashed line is drawn by hand. The fully drawn line is the input impedance of the three-element windkessel model. Adapted form Westerhof and Elzinga (1988).

Stergiopulos et al. (1994) showed that the fit of model derived flow may be very good but the parameter estimates of compliance were consistently deviating from the real values. The explanation is found in the fact that the three-element windkessel model uses a resistor as characteristic impedance. This resistor contributes to mean pressure and flow while a characteristic impedance only plays a role in the oscillatory pressure and flow. Toorop et al. (1987) studied the possibility of introducing nonlinear elasticity to the three-element windkessel model but found limited improvement only. Recently, Vrettos and Gross (1994) however, suggested advantageous effects of changes in arterial compliance during ejection in terms of energy transfer.

We conclude that the three-element windkessel model is a very good one to use as a load on the heart but that it should not be used to estimate arterial compliance.

Several methods exist to estimate arterial compliance and they have recently been compared (Stergiopulos et al., 1995a). The methods based on the two-element windkessel were, in general more accurate than those based on the three-element windkessel. The systolic and diastolic pressure obtained by feeding the measured aortic flow into the two-element windkessel and fitting them to their measured counterparts turned out to be an accurate and practical method to estimate (total) arterial compliance at all locations in the arterial tree (Stergiopulos et al., 1994).

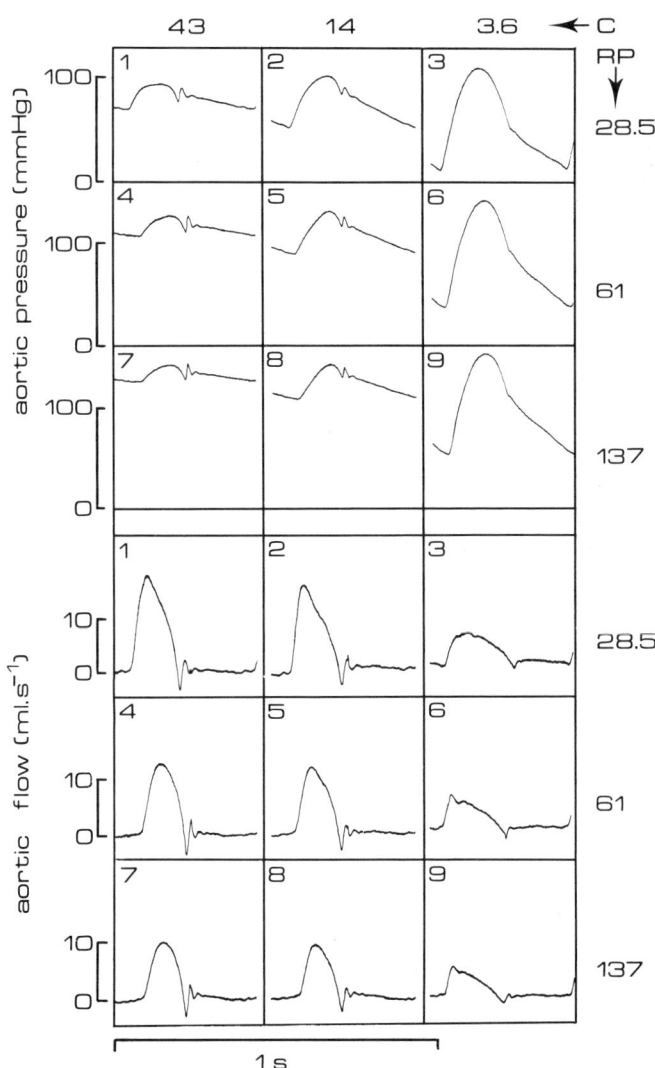

Figure 8. Aortic pressure and flow generated by the isolated cat heart loaded with the three-element windkessel. The different panels present different values of total arterial compliance (in $10^{-6} \cdot cm^4 \cdot s^2 \cdot g^{-1}$) and peripheral resistance (in $10^3 \cdot g \cdot cm^{-4} \cdot s^{-1}$). Modified from Elzinga and Westerhof (1973).

Peripheral Resistance and Arterial Compliance Form Load on the Heart

When the isolated heart is loaded with the three-element windkessel it is easy to change total arterial compliance and peripheral resistance without affecting heart rate, ventricular filling and cardiac contractility (Elzinga & Westerhof, 1973; Elzinga & Westerhof, 1979). In this manner one can study what the effects of changes in compliance and resistance are on pressure and flow. Results are given in Fig. 8.

It may be noted that pressure and flow in the control setting of the windkessel model (top left panel) are both in magnitude and shape very close to pressure and flow found in the intact animal, the cat in this example. In Fig. 8 the effects of changes in total arterial

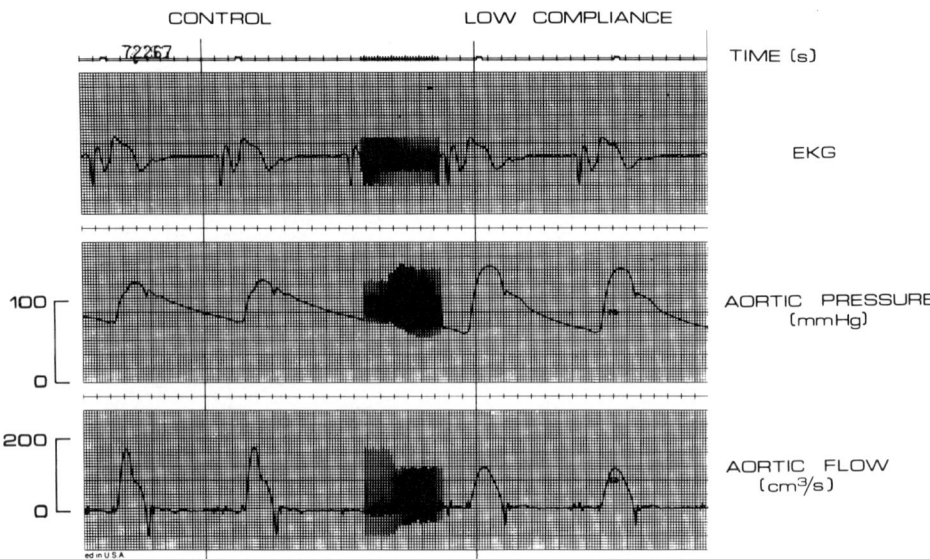

Figure 9. Recording of the electrocardiogram, aortic pressure, and aortic flow measured simultaneously in the anaesthetized dog before (left) and after sudden reduction in arterial compliance. The heart is paced after the production of A-V block.

compliance and peripheral resistance are shown. A decrease in compliance results in a decrease in flow and a widening of the pulse pressure. Systolic pressure is little affected but diastolic pressure decreases (top row panels of pressure and flow in Fig. 8). An increase in resistance results in an increase in mean pressure and a decrease in mean flow. Pulse pressure decreases somewhat (left hand columns of pressure and flow in Fig. 8).

When the total arterial compliance is decreased in the intact closed thorax dog by replacing the aorta by a stiff tube (Randall et al., 1984) we find Fig. 9. We see that with the decrease in compliance pulse pressure increases and flow decreases. However, here systolic pressure increases while diastolic pressure decreases with the decrease in compliance. In this intact situation peripheral resistance tended to increase somewhat but this was not significant (Randall et al., 1984). Regulatory mechanisms, such as the baroreflex and ventricular filling are differences between the intact animal and the isolated heart. Detailed information about decreased arterial compliance on blood pressure has been published by Randall et al. (1984).

THE HEART

The heart as a pump will be described in two ways: the pressure-volume diagram and the pump function graph. After discussion of both these descriptions their interrelationship will be given. We will concentrate on the left heart.

The Pressure-Volume Diagram

The heart as a pump can be presented by the pressure-volume diagram of the (left) ventricle. The first attempt to use this diagram was by Frank (1895). Later Wiggers (1921) and others used this approach. However, with the studies on the isolated blood perfused dog

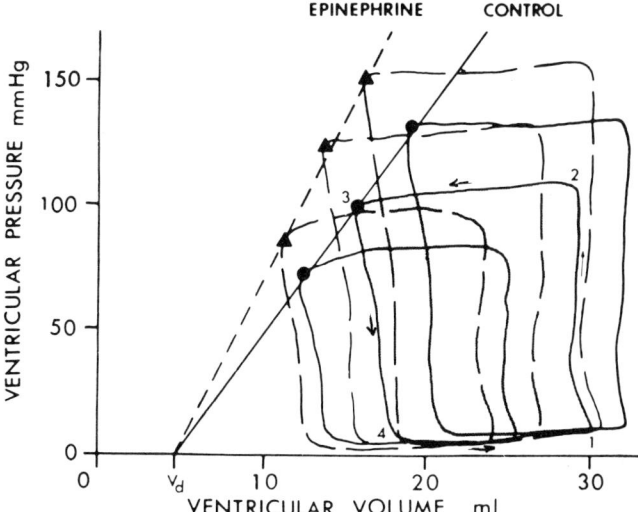

Figure 10. Pressure-volume relations of the isolated dog heart during control and increased contractile state (Suga et al., 1973, with permission).

heart loaded with the three-element windkessel by Suga et al. (1973) real progress in the understanding of cardiac pump function was achieved.

We start the description with the approximation of straight relations as originally suggested by Suga et al. (1973) and Sagawa (1978) and as shown in Fig. 10.

The relations can be characterized by their intercept (V_d) and their slope (Elastance, E). The pressure-volume relation of the ventricular cavity turns out to be a time-dependent relation with constant intercept, i.e. elastance is a function of time over the cardiac cycle: E(t), the time varying elastance. The relation in diastole is characterized by a low elastance (large ventricular compliance) and increases over systole to a maximum value (E_{max}, also

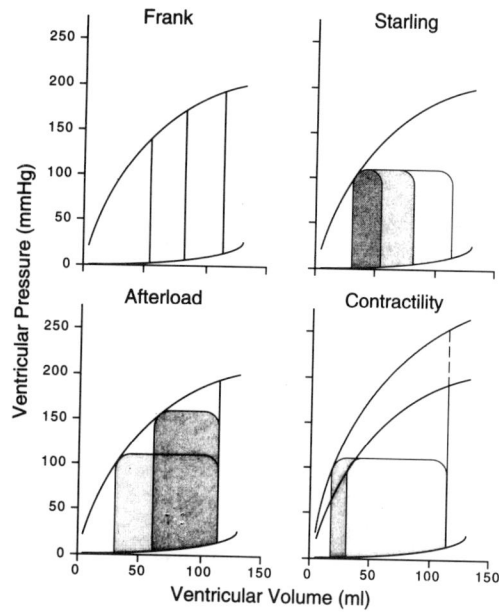

Figure 11. Schematic presentation of ventricular pressure-volume relations. The diastolic and end-systolic pressure-volume relations are plotted together with pressure-volume loops. The top two panels show the experiments by Frank (1895) and Patterson and Starling (1914) where the effects of ventricular filling were studied. Frank mainly studied isovolumic contractions while Starling made the heart pump against a constant pressure by means of a Starling resistor. The bottom two panels show the effect of afterload (left) and contractility (right) changes for ejecting beats.

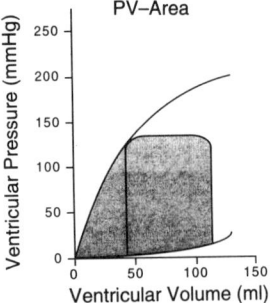

Figure 12. Schematic presentation of ventricular diastolic and end-systolic pressure-volume relations. The area under the pressure-volume loop plus the potential energy area (lower left area, in principle energy available for generation of external energy) form the 'pressure-volume area' (PVA) which is linearly related to cardiac oxygen consumption.

called end-systolic pressure-volume relation) and then decreases again to the diastolic relationship. It therefore holds that:

$$E(t) = P(t)/(V(t) - V_o). \tag{13}$$

The two extremes of the family of curves, the diastolic and systolic, E_{max}, relations are the major lines in the diagram. It is here implicitly assumed that the $E(t)$ is independent of preload and afterload so that it is a characteristic of the heart alone. Pressure and volume change depending on the cardiac load but the slope of the relation is, at any moment in time, a given quantity. In this approach increased E_{max} implies increased contractile state (see Fig. 10). With the straight line approximation the increase in contractility is simply the increase in slope of the end-systolic pressure-volume relation.

In practice, and as was later verified, the relations are not straight but curved, probably with a varying intercept so that a single slope or a single $E(t)$ value is difficult to define. It was also found that a single E_{max} curve probably does not exist. This means that in some ejecting conditions the E_{max} line may not be reached ('deactivation of shortening') or an overshoot may be present (Hunter, 1989). These complications of the simple concept

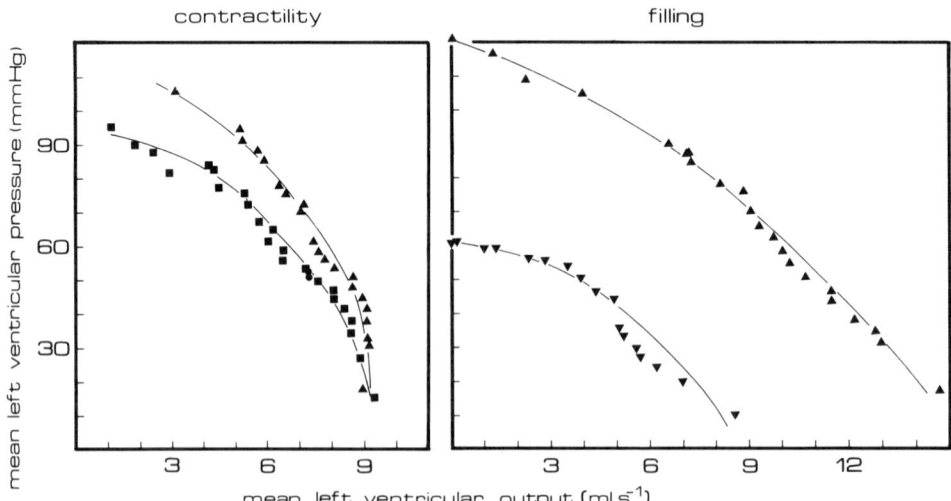

Figure 13. The effects of changes in contractility and ventricular filling on the pump function graph determined in the isolated cat heart loaded with the three-element windkessel.

of straight relations, however, do not reduce the importance of the concept. The effects of filling and contractility on the end-systolic pressure-volume relation are schematically presented in Fig. 11.

In the same figure the experiments of Frank and Starling are schematically presented. In the isovolumically contracting heart ventricular pressure increases with increased volume. This is an aspect of the basic property of muscle where increased volume corresponds to increased overlap of thick and thin filaments leading to greater force and thus greater pressure. In the Starling experiment cardiac output increases when volume is increased while the heart ejects against a constant pressure. In Starling's experiment the increase in cardiac output simply results from the end-diastolic volume increase, it does not represent the typical muscle property that with increased volume, i.e. increased overlap of thick and thin filaments, the force of contraction is greater.

The varying elastance concept of the ventricles was carried further to form the basis of a relation between cardiac oxygen consumption and cardiac mechanics, the PVA-concept. In this concept the area under the pressure-volume relation, related to external work plus the potential energy (see Fig. 12) is linearly related to oxygen consumption per beat (Suga, 1979; Suga et al., 1983; Suga, 1990).

The area indicated in Fig. 12 is called the pressure-volume area (or PVA for short). Thus it is not the external work that relates well with oxygen consumption but the external work plus potential energy. The relation between oxygen consumption and PVA is a linear one with an intercept with the oxygen consumption axis. The intercept is found when the heart ejects isobarically, i.e without the development of pressure. The intercept can be divided into two parts. One part is related the basic metabolic processes such as ionic pumps and the other is related to excitation-contraction. This implies that for increased contractility this part of the intercept increases. In mathematical form this can be stated as (Suga, 1979; Suga et al., 1983; Suga, 1990):

$$VO_2 = \alpha \, PVA + \beta \, E_{max} + \text{constant} . \tag{14}$$

For a thorough review and basic information see the comprehensive book by Sagawa et al. (1988). The pressure-volume relation is a qualitative image of the basic properties of cardiac muscle. The force-length relations of muscle in diastole and systole form the basis of the pressure-volume relations. However, the quantitative relations are difficult to work out since factors such as ventricular geometry, muscle fiber directions in the wall, connections between muscle fibers and other structures such as vessels all play a role in the pressure-volume relationships (Schouten et al., 1992; Allaart et al., 1995). It should be realized that since the force-length relations of heart muscle in diastole and systole are not straight it is highly unlikely that pressure-volume relations of a ventricle would be straight.

The Pump Function Graph

The pump function graph is a straightforward characterization of the heart as a pump in terms of pressure and flow. When the left ventricle is contracting against a closed aortic valve, left ventricular pressure will be high and flow negligible. When, on the other hand, the aorta would be opened to the atmosphere pressure development would be negligible and flow would be large. For other, less extreme loads, pressure increases and flow decreases with increased impedance of the arterial system. As for the arterial system, where peripheral resistance gives, in first approximation, a characterization of the system, mean values of pressure and flow are considered here only. Since the aortic valves form a strongly nonlinear element the pump function graph was taken as the relation between mean ventricular pressure and mean flow. Also on basis of physical analogy left ventricular pressure should be used in

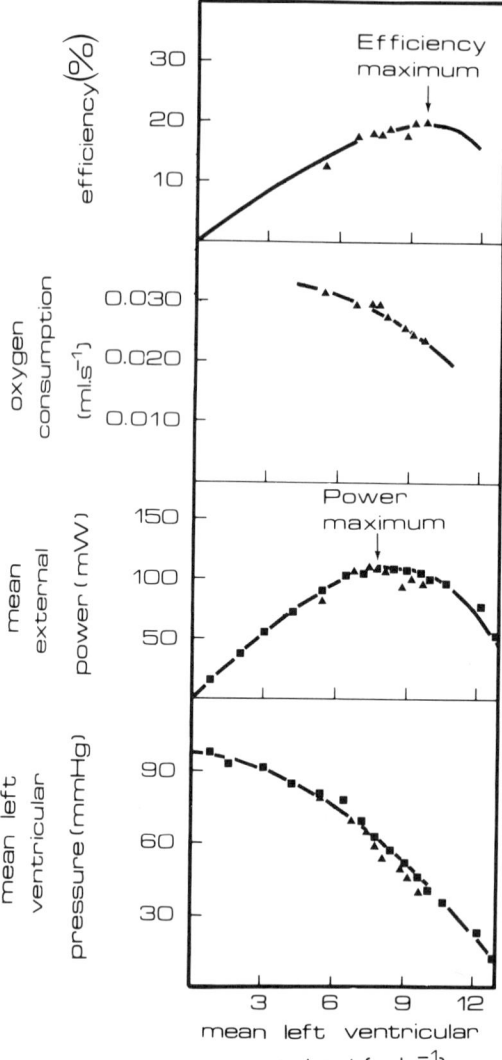

Figure 14. The cardiac efficiency, oxygen consumption, mean external power and mean left ventricular pressure presented as a function of mean flow. The bottom panel is the pump function graph. Power exhibits a maximum and so does efficiency. Oxygen consumption decreases with flow, i.e. increases with pressure. Adapted form Elzinga and Westerhof (1980).

the pump function graph (Elzinga & Westerhof, 1979). When the heart as a pump is approximated by a pressure generator with a resistor in series (the source resistance) an inverse straight line relationship between mean ventricular pressure and mean flow is obtained with the slope equal to the source resistance. Using this approach the heart was characterized by changing its arterial load and constructing the pump function graph. It was found that in the isolated cat heart loaded with the three-element windkessel the relation can be obtained from changes in peripheral resistance and changes in arterial compliance (see Fig. 7, Elzinga & Westerhof, 1973; Elzinga & Westerhof, 1979). It was also worked out what the effects of changes in contractility, ventricular filling, and heart rate are on the pump function graph. This is presented in Fig. 13.

It was shown that the pump function graph is a characterization arising from basic properties of cardiac muscle (Elzinga & Westerhof, 1981; Elzinga & Westerhof, 1982). In these studies the isolated papillary muscle was made to contract as if in the wall of the ventricle. The calculations were carried out assuming a cylindrical geometry for the left ventricle with circumferentially arranged muscle fibres. Pressure was calculated from muscle stress and flow was derived from the rate of strain. Their mean values were used to construct the pump function graph. The pump function graph of the muscle qualitatively exhibited the same characteristics as that of the ventricle.

The relation between the pump function graph and cardiac external power and cardiac efficiency at constant filling, heart rate and contractility, is shown in Fig. 14.

It may be seen that external power, the product of pressure and flow (Milnor, 1989) when plotted as a function of mean flow, exhibits a maximum value. This maximum can be understood as follows. For isovolumic contractions when flow is zero, external power is zero as well. For isobaric contractions, when no pressure development is possible, flow is large but the product of pressure and flow is negligible. Thus the end-points of the relation between power and mean flow are zero. Therefore a maximum must be present for intermediate values of flow and pressure. It may also be seen in Fig. 14 that oxygen consumption decreases with increasing flow. This observation is in line with well known rate-pressure-product (Katz et al., 1989), i.e. for constant heart rate oxygen consumption increases with pressure. When we refer to the pump function graph we see that a decrease in pressure and thus in oxygen consumption, relates to an increase in flow. It may also be seen from the figure that the efficiency of the heart is about 25%. This implies that about 75% of the oxygen consumption is converted into heat. It has been shown that heat is transported by diffusion to ventricular lumen and thorax and by convection by the coronary circulation. In the order of 50-70% of the heat is removed from the heart by coronary perfusion depending on flow (Ten Velden et al., 1982).

It was shown in the in situ heart that there is a proportional relationship between mean left ventricular pressure and mean aortic pressure (Van den Horn et al., 1985). This is so because of the fact that changes in load in the intact animal are not as large as in the isolated heart with an artificial load. Thus, for all practical purposes, and for not too extreme values in load, the pump function graph may also be presented as a relation between mean aortic pressure and mean aortic flow. The pump function graph may be approximated with a parabola of the form (Van den Horn et al., 1984):

$$P = P_{max} (1 - F^2/F^2_{max}). \tag{15}$$

From the mathematical formulation of the pump function graph (eq. 15) it can be derived that the product of pressure and flow exhibits its maximum value at $F_{max}/\sqrt{3} = 0.58 \cdot F_m$ (Van den Horn et al., 1984).

Later it was found that in the intact animal the power maximum was found for the pressure and flow in the control situation (Toorop et al., 1988). In other words in the control conditions the heart appears to work at the maximum of its external power. What the possible explanation might be will be discussed below.

From equation 15 and from Fig. 13 is may be seen that the heart is neither a pressure source nor a flow source. A pressure source is a pump that generates the same pressure for all loading conditions and thus corresponds with a horizontal line in the pump function graph. A flow source produces the same flow irrespective of its loading conditions and thus corresponds to a vertical pump function graph. It may thus be seen from Fig. 13 that in the working range the heart is neither a pressure source nor a flow source but that for large loads (low flows) a pressure source is approximated and for small loads (large flows) a flow source is approached.

In terms of the experiments of Frank and Starling the pump function graph plays the following role. Increases in ventricular volume result in a parallel shift of the pump function graph (Fig. 13). Frank studied isovolumic contractions which implies that the intercepts of

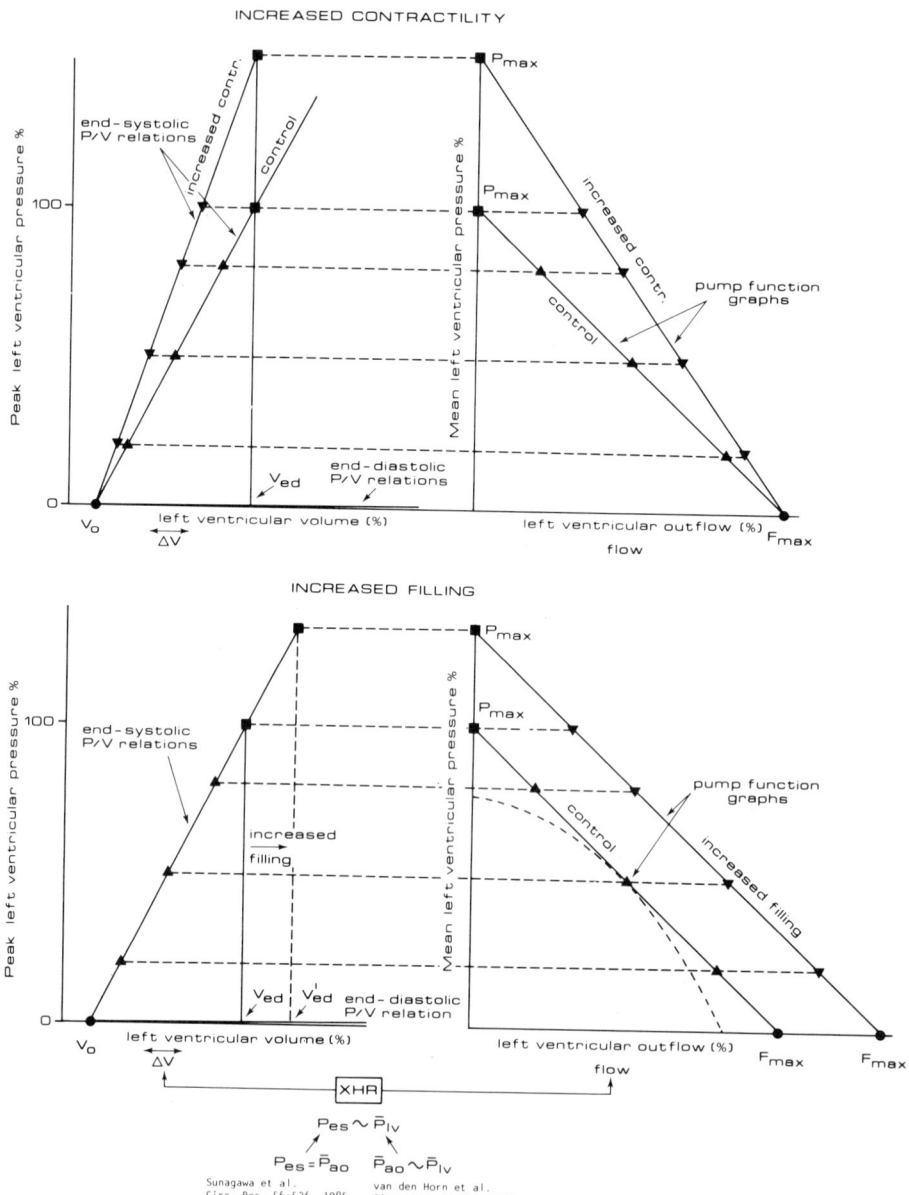

Figure 15. Schematic presentation of pressure-volume relations and the effects of contractility (top) and filling (bottom) are given on the left side. The related effects on the pump function graph are shown on the right side. For simplicity all relations are taken as straight lines (the dashed line on bottom left reminds the reader that all relations are curved). The pressure-volume relations can be connected to the pump function graphs because mean ventricular pressure is related to end-systolic ventricular pressure and mean flow is related to stroke volume when heart rate is constant.

the pump function graph with the vertical axis (no flow) are compared. Increased volume means increased mean pressure and also increased systolic ventricular pressure. Starling loaded the heart with a constant pressure. This means that the intersection of the family of pump function graphs and a horizontal line (constant pressure) is found for increases in volume. In other words flow increases with filling (Westerhof & Elzinga, 1993).

Relation between Pressure-Volume Relation and Pump Function Graph

The pressure-volume relations and the pump function graph are both representations of cardiac muscle mechanics; they present the same information in different form. Westerhof and Elzinga (1978) published a pump model based on the varying elastance concept with the three-element windkessel as its load. From that model the pump function graph was obtained by changing the windkessel parameters and computing mean pressure and flow. From this model it was therefore shown that on the basis of the varying elastance concept the pump function graph could be obtained. This finding was later corroborated in other studies (Elzinga & Westerhof, 1980; Elzinga & Westerhof, 1984). In Fig. 15 the relations between the two characterizations of the heart as a pump are qualitatively worked out.

On the left hand side the pressure-volume relations are given and on the right hand side the pump function graphs. End-systolic ventricular pressure, the quantity of interest for the end-systolic pressure-volume relation is related to mean pressure as follows. According to the work of Sunagawa et al. (1983), at constant heart rate end-systolic pressure is proportional to mean aortic pressure, and Van den Horn et al. (1985) showed that mean aortic pressure and mean ventricular pressure are proportional. Thus end-systolic ventricular pressure and mean ventricular pressure are proportional. Similarly a ventricular volume change (stroke volume) is proportional to mean flow for constant heart rate. If we now find in the pressure-volume diagram, at a chosen pressure, a certain volume change then mean flow and mean pressure follow. For an increased end-systolic pressure the volume change is less so that the increased mean pressure corresponds with a smaller mean flow. An increase in contractility implies an increase in isovolumic pressure, i.e. an increase in end-systolic and thus in mean pressure. For low loads the volume change is maximal but not different with changes in contractility and mean flow is also maximal but not dependent on contractility (Fig. 15, top panel). A similar reasoning can be held for increased ventricular filling. The end-systolic pressure-volume relation is not changed but all volumes and also the volume changes are increased by the same amount. In the pump function graph this corresponds to a shift of the relation with a constant amount (Fig. 15, bottom panel). Heart rate changes do not affect the end-systolic pressure-volume relation, except for possibly induced alterations in contractility, while the pump function graph is affected. It turns out that a change in heart rate mainly affects the duration of diastole, and because diastolic values of ventricular pressure and aortic flow are close to zero, the mean values of these quantities are proportional to heart rate. This implies that the pump function graphs for different heart rates can be superimposed when proportionality with heart rate is accounted for (Elzinga & Westerhof, 1980).

From Fig. 15 it may also be seen that an increase in contractility has only a limited effect on stroke volume and mean flow especially for low pressures.

Advantages and Disadvantages of the Characterizations of the Pump

The pressure-volume relation obviously requires the measurement of pressure in the left ventricle and ventricular volume. Volume measurements are rather difficult but the new developments with the so-called volume catheter (Baan et al., 1984) and Magnetic Resonance Imaging make measurements possible. The changes required may be alterations in preload and in afterload, with changes in preload being easy to perform by means of a vena cava balloon.

The pump function graph requires the determination of ventricular pressure and aortic flow. These measurements are feasible since the early eighties with a single arterial catheter equipped with pressure and flow sensors (Murgo et al., 1980; Murgo et al., 1981). The changes needed to obtain the pump function graph are those in the afterload while keeping filling constant. Afterload changes without alterations in ventricular filling are difficult to perform in intact animal and man.

We conclude that in practice, with the new techniques available to obtain ventricular cavity volume, the pressure-volume approach is the preferred one.

MATCHING OF HEART AND ARTERIAL SYSTEM

As mentioned above the heart appears to work at optimum power output (Fig. 14), a finding first reported by Wilcken et al. (1964) and later more extensively studied by Toorop et al. (1988). We consider this an aspect of matching between heart and arterial tree. A possible explanation will now be discussed.

Several matching criteria between heart and arterial system have been suggested, an important candidate being the relation between heart period (T, inverse of heart rate) and arterial time constant (τ). This idea of matching was based on experiments by Broemser (1935). He used a pump and model of the arterial system and found that power could be *minimized* when the pump frequency was related to a characteristic time of the arterial model. However, his artificial pump was a flow source and above we have shown that the heart is neither a flow source nor a pressure source. Later O'Rourke (1965) and Milnor (1979) suggested that heart rate was related to the frequency of the minimum in the modulus of the input impedance, leading to minimisation of power. Westerhof reasoned that this matching is unlikely because oscillatory power is small with respect to cardiac oxygen consumption and power is not related to the modulus of the impedance but to the real part of the impedance (Westerhof, 1994a).

We suggested heart period, a cardiac parameter, to be related to the RC-time i.e. time constant, τ, of the arterial system (Elzinga & Westerhof, 1991). The time constant of the arterial system is $\tau = R_p \cdot C$ with R_p peripheral resistance and C total arterial compliance (Elzinga & Westerhof, 1991; Westerhof & Elzinga, 1991). The diastolic aortic pressure decay is an exponential one and diastolic pressure is determined by mean pressure, the arterial

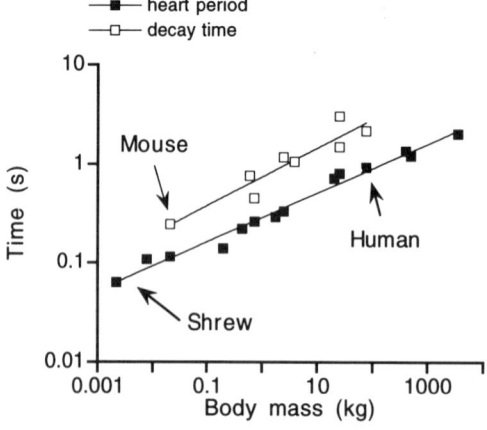

Figure 16. Double logarithmic plot (allometric plot) of heart period (cardiac parameter) and decay time of the systemic arterial tree (product of peripheral resistance and total arterial compliance) plotted as a function of body mass for mammals.

Cardio-Vascular Interaction Determines Pressure and Flow

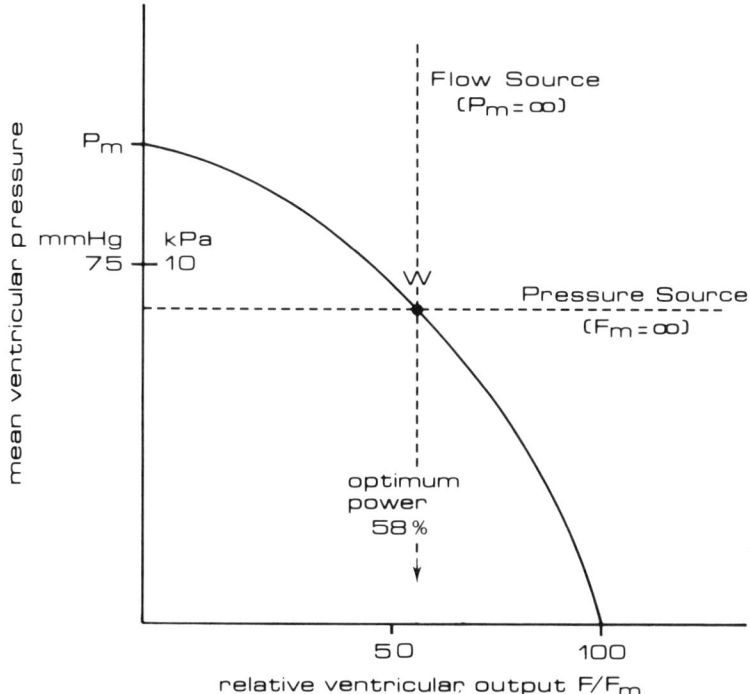

Figure 17. The pump function graph with normalized flow on the horizontal axis. The working point W is given because mean pressure (same in mammals) and flow (determined by body weight) are given. Many pump function graphs through W are possible the two extremes are the pressure source and the flow source. If the heart were a pressure source the pump function graph would be the horizontal dashed line (with F_m infinite), if a flow source the vertical dashed line would be found with P_m infinite.

decay time (τ), and the duration of diastole (T_d). The latter two parameters were determined in different mammals and it was found that their ratio was the same (Elzinga & Westerhof, 1991; Westerhof & Elzinga, 1991). Mean pressures (at brain level) in mammals are similar. With a similar value of τ/T_d this means that pulse pressure and thus diastolic and systolic pressure in all mammals are the same. This prediction is in agreement with the observations in most mammals including the human. It turns out that the duration of diastole is in all mammals a constant fraction of the heart period (Westerhof & Elzinga, 1991). Since coronary perfusion virtually only takes place in diastole the fraction of time that the coronary system is perfused is similar in mamals. Heart period and arterial decay time as functions of body mass for a large range of mamals is given in Fig. 16.

We interpret the findings as follows. Mean pressure (at brain level) is the same in mammals to guarantee brain perfusion. Similar diastolic pressures (coronary perfusion pressure) and similar fraction of time for coronary flow guarantee similar conditions for coronary perfusion.

This important matching criterion of heart period and decay time also results in the explanation why input impedance, when plotted as function of harmonic number rather than frequency, and normalized with respect to characteristic impedance or peripheral resistance, is the same for all mammals studied thus far as (shown in Fig. 2, Westerhof & Elzinga, 1991). It can be shown that the RC-time of the arterial system, and thus also the heart period, is proportional to body length ((Westerhof & Elzinga, 1993; Westerhof, 1994).

Why does the heart pump at maximum power? Cardiac output is determined by body mass and mean pressure is the same in mammals. This implies that the normal working point on the pump function graph for an animal is given. Through the working point many pump function graphs can be drawn. Each of these pump function graphs has intercepts with the pressure and flow axes. When the pump function graph is going towards a flow source (Fig. 17) the P_m increases. This means that for isovolumic contractions very high pressures can be generated. Assuming given muscle properties this results, on basis of the law of Laplace, in a wall thickness that has to be large. Inversely, when a pressure source is approached (Fig. 17) the F_m must increase. It means that mean flow can be very large for low loads, requiring a large lumen volume of the ventricle since heart rate is given. A large lumen volume implies a thick wall as well on basis of the law of Laplace. Thus when the pump function graph rotates around the fixed working point total cardiac volume, i.e. lumen plus wall volume, changes. This approach was formulated in mathematical terms for a simple spherical model of the left ventricle (Elzinga & Westerhof, 1991). It could be shown on basis of this simple model that total ventricular volume, i.e. lumen volume plus wall volume, is minimal when the maximal flow value is 1.55 times the flow in the working point (see Fig. 18).

Inversely, flow in the working point is then about 65% of maximal flow. Optimum power was found at 58% of maximal flow. We conclude that the heart works at optimum power because its total volume is minimized.

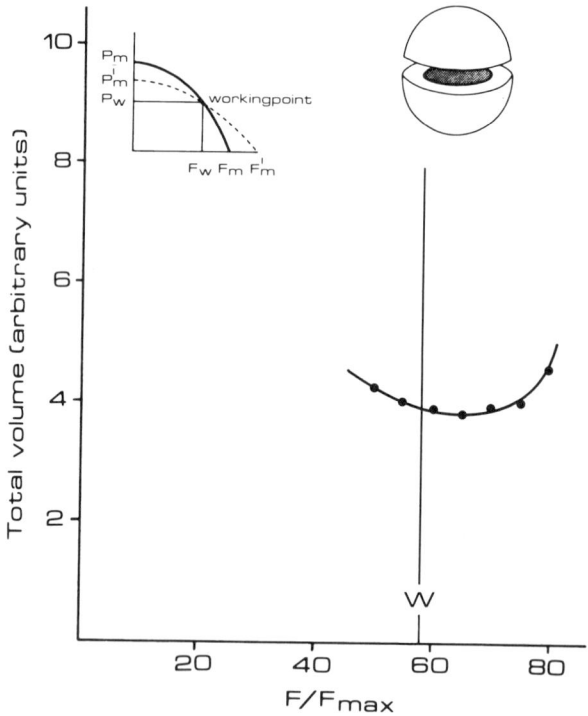

Figure 18. The working point W on the pump function graph is given and all pump function graphs through the working point correspond to P_m and F_m values (inset). For each of these pump function graphs total ventricular volume (lumen plus wall volume) can be calculated. It is found that total ventricular volume is minimal when F/F_m is about 0.65 a value close to the F/F_m at optimum power (0.58).

REFERENCES

Allaart, C.P., Sipkema, P., and Westerhof, N., 1995, Effect of perfusion pressure on diastolic stress strain relations of isolated rat papillary muscle. *Am. J. Physiol.* 268:H945-H954.

Anliker, M., Histand, M.B., and Ogden, E., 1968, Dispersion and attenuation of small artificial pressure waves in the canine aorta. *Circ. Res.* 23:539-551.

Avolio, A.P., O'Rourke, M.F., Mang, K., Bason, P.T., and Gow, B.S., 1976, A comparative study of pulsatile arterial hemodynamics in rabbits and guinea pigs. *Am. J. Physiol.* 230:868-875.

Baan, J., Van der Velde, E.T., De Bruin, H.G., Smeenk, G.J., Koops, J., Van Dijk, A.D., Temmerman, D., Senden, P.J., and Buis, B., 1984, Continuous measurement of left ventricular volume in animals and humans by conductance catheter. *Circulation* 70:812-823.

Berger, D.S., Li, J. K.-J., and Noordergraaf, A., 1994, Differential effects of wave reflections and peripheral resistance on aortic blood pressure: a model based study. *Am. J. Physiol.* 266:H1626-H1642.

Broemser, Ph., 1935, Über die optimalen Beziehungen zwischen Hertztätigkeit und physikalischen Konstantes des Gefäss-systems. *Zeitschr. f. Biol.* 96:1-10.

Burattini, R., and Campbell, K.B., 1989, Modified asymmetric T-tube model to infer arterial wave reflection at the aortic root. *IEEE Trans. Biomed. Eng.* 36:805-814.

Burattini, R., Fogliardi, R., and Campbell, K.B., 1994, Lumped model of terminal aorta impedance in the dog. *Ann. Biomed. Eng.* 22:381-391.

Cope, F.W., 1960, An elastic reservoir theory of the human systemic arterial system using current data on aortic elasticity. *Bull. Math. Biophys.* 22:19-26.

Cope, F.W., 1961, A modified windkessel (elastic reservoir) theory of the human systemic arterial system using modern data on aortic elasticity so as to yield computational accuracy sufficient for clinical usefulness. In: *Circulatory Analog Computers*. A. Noordergraaf, G.N. Jager, and N. Westerhof (eds.). Amsterdam, North Holland Publ. Company, pp. 44-55.

Dick, D.E., Kendrick, J.E., Matsom, G.L., and Rideout, V.C, 1968, Measurement of nonlinearity in the arterial system of the dog by a new method. *Circ. Res.* 22:101-111.

Elzinga, G., and Westerhof, N., 1973, Pressure and flow generated by the left ventricle against different impedances. *Circ. Res.* 32:178-186.

Elzinga, G., and Westerhof, N., 1979, How to quantify pump function of the heart. The value of variables derived from measurements on isolated muscle. *Circ. Res.* 44:303-308.

Elzinga, G., and Westerhof, N., 1980, Pump function of the feline left heart: changes with heart rate and its bearing on the energy balance. *Cardiovasc. Res.* 14:81-92.

Elzinga, G., and Westerhof, N., 1981, "Pressure-volume" relations in isolated cat trabeculae. *Circ. Res.* 49:388-394.

Elzinga, G., and Westerhof, N., 1982, Isolated cat trabeculae in a simulated feline heart and arterial system: contractile basis of cardiac pump function. *Circ. Res.* 51:430-438.

Elzinga, G., and Westerhof, N., 1984, Does the history of contraction affect the pressure-volume relationship? *Fed. Proc.* 43:2402-2407.

Elzinga, G., and Westerhof, N., 1991, Matching between ventricle and arterial load follows from evolution. *Circ. Res.* 68:1495-1500.

Frank, O., 1895, Zur Dynamik des Herzmuskels. *Zeitschr. f. Biol.* 32:370-447.

Frank, O., 1899, Die Grundform des Arteriellen Puls. *Zeitschr. f. Biol.* 37:483-526.

Hunter, W.C., 1989, End-systolic pressure as a balance between opposing effects of ejection. *Circ. Res.* 64:265-275.

Jager, G.N., Westerhof, N., and Noordergraaf, A., 1965, Oscillatory flow impedance in electrical analog of arterial system. *Circ. Res.* 16:121-133.

Katz, L.A., Swain, J.A., Portman M.A., and Balaban, R.S., 1989, Relation between phosphate metabolites and oxygen consumption in heart in vivo. *Am. J. Physiol.* 256:H265-H274.

Langendorff, O., 1899, Zur Kenntnisse des Blutlaufs in den Kranz-Gefässen des Herzens. *Pflügers Arch.* 78:423-440.

Laxminarayan, S., Sipkema, P., and Westerhof, N., 1978, Characterization of the arterial system in the time domain. *IEEE Trans. BME* 25:177-184.

Li, J.K.-J., (1986), Time resolution of forward and reflected waves in the aorta. *IEEE Trans. BME* 33:783-785.

Milnor, W.R., 1979, Aortic wave length as a determinant of the relation between heart rate and body size. *Am. J. Physiol.* 237:R3-R6.

Milnor, W.R., 1989, *Hemodynamics*. Baltimore, Williams and Wilkins, pp. 282-284).

Mitchell, G.F., Pfeffer, M.A., Westerhof, N., and Pfeffer, J.M., 1994, Measurement of aortic input impedance in rats. *Am. J. Physiol.* 267:H1907-H1915.

Murgo, J.P., Westerhof, N., Giolma, J.P., and Altobelli, S.A., 1980, Aortic input impedance in normal man: relationship to pressure wave forms. *Circulation* 62:105-116.

Murgo, J.P., Westerhof, N., Giolma, J.P., and Altobelli, S.A., 1981, Manipulation of ascending aortic pressure and flow wave reflections with the Valsalva maneuver: relationship to input impedance. *Circulation* 63:122-132.

Newman, D.L., Sipkema, P., Greenwald, S.E., and Westerhof, N., 1986, High frequency characteristics of the arterial system. *J. Biomech.* 19:817-824.

O'Rourke, M.F., 1965, Pressure and flow in arteries. *Ph.D. Dissertation.* University of Sydney, Sydney, Australia.

O'Rourke, M.F., and Cartmill, T.B., 1971, Influence of aortic coarctation on pulsatile hemodynamics in the proximal aorta. *Circulation* 44:281-292.

O'Rourke, M.F., and Avolio, A.P., 1980, Pulsatile flow and pressure in human systemic arteries: studies in man and in a multi-branched model of the human systemic arterial tree. *Circ. Res.* 46:363-372.

Patterson, S.W., and Starling, E.H., 1914, On the mechanical factors which determine the output of the ventricles. *J. Physiol.* (London) 48:357:379.

Randall, O.S., van den Bos, G.C., and Westerhof, N., 1984, Systemic compliance: does it play a role in the genesis of essential hypertension? *Cardiovasc. Res.* 18:455-462.

Reuderink, P., Hoogstraten, H.W., Sipkema, P., Hillen, B., and Westerhof N., 1989, Linear and non-linear one-dimensional models of pulse wave transmission at high Womersley numbers. *J. Biomech.* 22:819-827.

Sagawa, K., 1978, The ventricular pressure-volume diagram revisited. *Circ. Res.* 43:677-687.

Sagawa, K., and Eisner, A., 1975, Static pressure-flow relation in the total systemic vascular bed of the dog and its modification by the baroreceptor reflex. *Circ. Res.* 36:406-413.

Sagawa, K., Maughan L., Suga, H., and Sunagawa, K., 1988, Cardiac contraction and the pressure-volume relationship. New York, Oxford University Press.

Schouten, V.J.A., Allaart, C.P., and Westerhof, N., 1992, Effect of perfusion pressure on force of contraction in thin papillary muscles and trabeculae from rat hart. *J. Physiol.* 451:585-604.

Sipkema, P., and Westerhof, N., 1975, Effective length of the arterial system. *Ann. Biomed. Engng.* 3:296-307.

Sipkema, P., Westerhof, N., and Randall, O.S., 1980, The arterial system characterised in the time domain. *Cardiovasc. Res.* 14:270-279.

Sipkema, P., Latham, R.D., Westerhof, N., Rubal, B.J., and Slife, D.M., 1990, Isolated aorta set up for hemodynamic studies. *Ann. Biomed. Engng.* 18:491-503.

Stergiopulos, N., Young, D.F., and Rogge, T.R., 1992, Computer simulation of arterial flow with applications to arterial and aortic stenosis. *J. Biomech.* 25:1477-1488.

Stergiopulos, N., Tardy, Y., and Meister, J.-J., 1993, Nonlinear seperation of forward and backward running waves in arteries. *J. Biomech.* 26:201-209.

Stergiopulos, N., Meister, J.-J., and N. Westerhof, N., 1994, Simple and accurate way for estimating total and segmental arterial compliance: The pulse pressure method. *Ann. Biom. Engng.* 22:392-397.

Stergiopulos, N., Meister, J.-J., and Westerhof, N., 1995a, Evaluation of methods for estimation of total arterial compliance. *Am. J. Physiol.* 268: H1540-H1548.

Stergiopulos, N., Meister, J.-J., and Westerhof, N., 1995b, Scottes in input inpedance. *Am. J. Physiol.* 269: H1490-H1495.

Suga, H., Sagawa, K., and Shoukas, A.A., 1973, Load dependence of the instantaneous pressure-volume relation of the canine left ventricle and the effect of epinephrine and heart rate on the ratio. *Circ. Res.* 32:314-322.

Suga, H., 1979, Total mechanical energy of a ventricle model and cardiac oxygen consumption. *Am. J. Physiol.* 236:H499-H505.

Suga, H., Hisano, R., Goto, Y., Yamada, O., and Ogarshi, Y., 1983, Effect of positive inotropic agents on the relation between oxygen consumption and systolic pressure volume area in the canine left ventricle. *Circ. Res.* 53:306-318.

Suga, H, 1990, Ventricular energetics. *Physiol. Review* 70:247-275.

Sunagawa, K., Maughan, W.L., Burkhoff, D., and Sagawa, K., 1983, Left ventricular interaction with arterial load, studied in isolated canine ventricle. *Am. J. Physiol.* 245:H773-H780.

Taylor, M.G., 1966a, The input impedance of an assembly of randomly branching tubes. *Biophys. J.* 6:29-51.

Taylor, M.G., 1966b, Wave transmission through an assembly of randomly branching tubes. *Biophys. J.* 6:697-716.

Ten Velden, G.H.M., Elzinga, G., and Westerhof, N., 1982, Left ventricular energetics: heat loss and temperature distribution of canine myocardium. *Circ. Res.* 50:63-73.

Toorop, G.P., Westerhof, N., and Elzinga, G., 1987, Beat-to-beat estimation of peripheral resistance and total arterial compliance during pressure transients. *Am. J. Physiol.* 252:H1275-H1283.

Toorop, G.P., Van den Horn, G.J., Elzinga, G., and Westerhof, N., 1988, Matching between feline left ventricle and arterial load: optimal external power or efficiency. *Am. J. Physiol.* 254:H279-H285.

Van den Bos, G.C., Westerhof, N., and Randall, O.S., 1982, Pulse wave reflection: can it explain the differences between systemic and pulmonary pressure and flow waves? *Circ. Res.* 51:479-485.

Van den Bos, G.C., Westerhof, N., and Randall, O.S., 1982, Pulse wave reflection: can it explain the differences between systemic and pulmonary pressure and flow waves? A study in dogs. *Circ. Res.* 51:479-485.

Van den Horn, G.J., Westerhof, N., and Elzinga, G., 1984, Interaction of heart and arterial system. *Ann. Biomed. Eng.* 12:151-162.

Van den Horn, G.J., Westerhof, N., and Elzinga, G., 1985, Optimal power generation by the left ventricle: a study in the anesthetized open-thorax cat. *Circ. Res.* 56:252-261.

Van Huis, G.A., Sipkema. P., and Westerhof, N., 1987, Coronary input impedance during the cardiac cycle as determined by impulse response method. *Am. J. Physiol.* 253:H317-H324.

Vrettos, A.M., and Gross, D.R., 1994, Instantaneous changes in arterial compliance reduce energetic load on left ventricle during systole. *Am. J. Physiol.* 267:H24-H32.

Westerhof, N., 1968, Analog studies of human systemic arterial hemodynamics. *Ph.D. Dissertation*. University of Pennsylvania, Philadelphia, Pa.

Westerhof, N., Bosman, F., de Vries G.J., and Noordergraaf, A., 1969, Analog studies of the human systemic arterial tree. *J. Biomechanics* 2:121-143.

Westerhof, N., Elzinga, G., and Sipkema, P., 1971, An artificial arterial system for pumping hearts. *J. Appl. Physiol.* 31:776-781.

Westerhof, N., P. Sipkema, P., van den Bos, G.C., and Elzinga, G., 1972, Forward and backward waves in the arterial system. *Cardiovasc. Res.* 6:648-656.

Westerhof, N., and Elzinga, G., 1978, The apparent source resistance of heart and muscle. *Ann. Biomed. Engng.* 6:16-32.

Westerhof, N., Murgo, J.P., Sipkema, P., Giolma, J.P., and Elzinga, G., 1979, Arterial impedance. In: *Quantitative Cardiovascular Studies*. N.H.C. Hwang, D.R. Gross, and D.J. Patel (eds.). University Park Press, Baltimore, pp. 111-150.

Westerhof, N., and Elzinga, G., 1988, Hemodynamics. In: *Encyclopedia of Medical Devices and Instruments*. J.G. Webster (ed.). John Wiley and Sons, pp. 1493-1509.

Westerhof, N., and Elzinga, G., 1991, Normalized input impedance and arterial decay time over heart period are independent of body size. *Am. J. Physiol.* 261:R126-R133.

Westerhof, N., and Elzinga, G., 1993, Why Smaller Animals have Higher Heart Rates. In: Interaction Phenomena in the Cardiovascular System. *Advances Exp. Med. and Biol.* S. Sideman and R. Beyar (eds.), 346:319-329.

Westerhof, N., 1993, Arterial Hemodynamics. In: *The Physics of Heart and Circulation*. J. Strackee, and N. Westerhof (eds.). IOP Press, pp. 355-382.

Westerhof, N., 1994a, Heart period is related to time constant of arterial system and not to minimum of impedance modulus. In: *Recent Progress in Cardiovascular Mechanics*. S. Hosoda, T. Yaginuma, M. Sugawara, M.F. Taylor, and C.G. Caro (eds.). Harwood Acad. Publ., Chur Switzerland, pp. 115-127.

Westerhof, N., 1994b, Heart period is proportional to body length. *Cardioscience* 5: 283-285.

Wiggers, G.J., and Katz, L.N., 1921, Contour of ventricular volume curves under different conditions. *Am. J. Physiol.* 58:439-475.

Wilcken, D.E.L., Charlier, A.A., Hoffman. J.I.E., and Guz, A., 1964, Effects of alterations in aortic impedance on the performance of the ventricles. *Circ. Res.* 14:283-293.

13

MECHANICS OF INTRAMURAL BLOOD VESSELS OF THE BEATING HEART

F. Kajiya, M. Goto, T. Yada, Y. Ogasawara, and K. Tsujioka

Kawasaki Medical School
Matsushima 577
Kurashiki, Japan

BRIEF HISTORICAL PERSPECTIVE OF CORONARY CIRCULATION

The blood flow of the coronary circulation is unique in that it supplies blood to the myocardium which pumps blood into the systemic and pulmonary circulations. In 1628, Harvey (1) found that the heart is a pump which expels blood through two circulations in series and also that channels exist in the walls of the heart for its own nourishment (2). In 1695, Scaramucci (3), who is frequently called the father of coronary circulation, postulated that the deeper coronary vessels are squeezed empty by cardiac contraction and they are refilled from the aorta during diastole.

Since ever after that "age of discovery," many uncertainties regarding 1) the effect of cardiac contraction on the coronary blood flow, 2) the force causing an instantaneous blood flow pattern, and 3) the linkage of the myocardial blood flow to the cardiac muscles remained mainly due to technological problems. From the pathophysiological view point, the effect of cardiac contraction on coronary blood flow has been investigated in relation to the mechanisms underlying the predilection for subendocardial underperfusion and ischemia (4).

Recently, our understanding of the interaction between the coronary circulation and heart beating has improved through sophisticated methodologies for measuring detailed coronary blood flow and obtaining fine vascular images of the coronary arteries and veins (5).

METHODS TO MEASURE INSTANTANEOUS MYOCARDIAL ARTERIAL AND VENOUS BLOOD FLOWS, AND CHARACTERISTICS OF THEIR VELOCITY WAVEFORMS

Coronary blood flow is pulsatile due to the direct mechanical effects of cardiac contraction and relaxation on the coronary vessels. In coronary circulation research, thus,

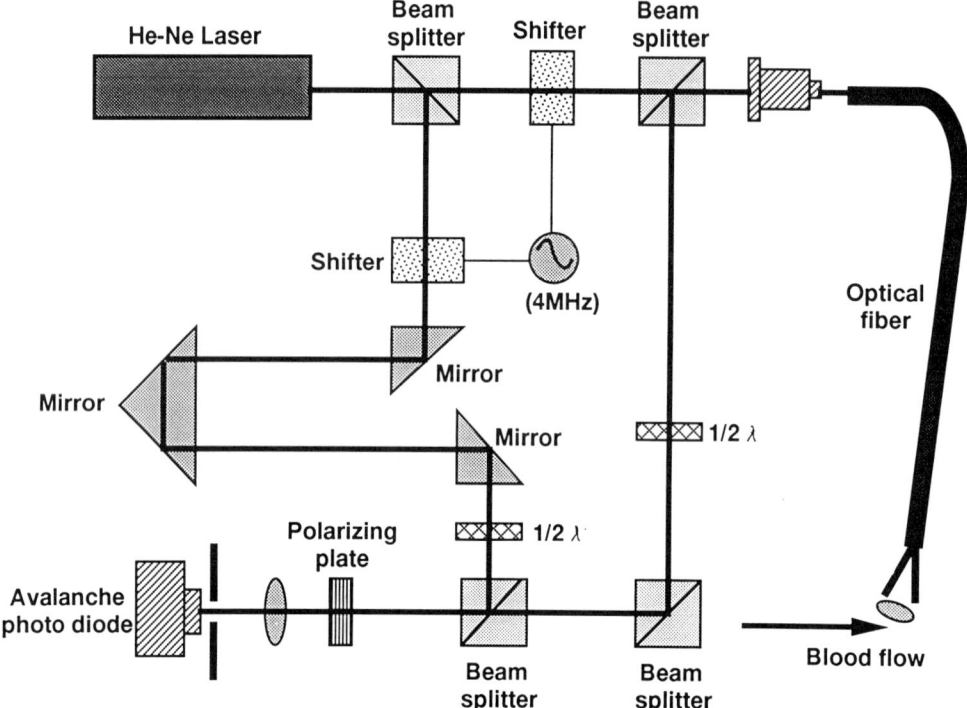

Figure 1. Schematic diagram of the laser Doppler velocimeter with an optical fiber. Mirrors were used to get an optimal path-length of laser beam by changing mirror position. (From Kajiya et al. [10]).

there has been a great need to develop approaches for studying the instantaneous coronary blood flow with enough temporal resolution to analyze the effect of cardiac contraction on the flow. Earlier investigators have evaluated the phasic blood flow pattern of the coronary artery using the electromagnetic flowmeter (6,7). However, the electromagnetic flowmeter is only applicable to the relatively proximal isolated coronary arteries. In the past few decades, thus, new approaches to study the pulsatile coronary artery and venous flows have been introduced to minimize the capacitance effect of relatively large epicardial arteries (8-13). Among recent advances in modern technologies, laser and ultrasound flowmetries have made it possible for us to investigate the effect of cardiac contraction on the phasic pattern of the coronary inflow to the myocardium even in intramyocardial arteries; e.g., the septal artery. From the studies using the laser and ultrasound flowmeters, now we know that arterial blood inflow to the myocardium is almost exclusively limited to diastole. On the other hand, the venous outflow from the myocardium is predominant in systole (13,14).

The principles of the laser and ultrasound flowmetries and their applications for the myocardial vessels are given below. Fig. 1 shows the system of a fiber optic laser Doppler velocimeter (LDV, 15). The He-Ne laser beam (632.8 nm, 5 mW) is divided into reference and sampling beams by a beam splitter (BS). The greater part of the initial light passed through the BS is focused onto the entrance of a graded-index multimode fiber (crad diameter = 62.5 μm, core diameter = 50 μm), and transmitted through the fiber into the blood stream. Part of the light reflected by flowing erythrocytes is re-entered into the same fiber, and transmitted back to the entrance. The other part of the initial light divided by the BS is used as a reference beam. Two frequency shifters (82 and 78 MHz) are interposed on the path of

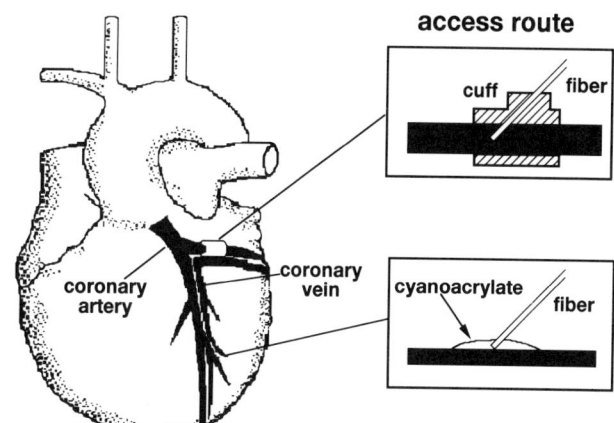

Figure 2. Two different routes of access of the fiber probe to coronary vessels. Route 1 (upper) was applied for velocity measurements in large and middle-sized epicardial coronary vessels. Route 2 (lower) was for velocity measurements in small epicardial vessels.

the incidence and reference beams to differentiate the forward flow from the reverse. Thus, the difference between the shifter frequencies (82 - 78 = 4 MHz) corresponds to zero flow velocity. When Doppler-shift frequency is greater than 4 MHz the direction of blood flow is toward the fiber tip, and when it is less than 4 MHz the direction is away from the fiber tip. The optical heterodyning is obtained by mixing the Doppler-shift signal from the moving erythrocytes with the reference beam. The photocurrent from the photodetector (APD) is fed into a spectrum analyzer to measure the Doppler-shift frequency.

The reflected light signal has Doppler-shift frequency:

$$\Delta f = 2n \, V\cos\theta/\lambda$$

where V is the blood flow velocity; n is the refractive index of blood, approximately 1.33; θ is the angle between the fiber axis and axis of the blood flow stream; and λ is the laser wavelength of 632.8 nm in a free space. When θ is equal to 60°, the Doppler-shift frequency of 1 MHz corresponds approximately to a blood velocity of 48 cm/s. The sample volume of our system is approximately $\pi \times 0.05^2 \times 0.1$ mm^3, and the temporal resolution is 8 msec.

Two different approaches are available for the LDV to evaluate the blood flow velocity in coronary arteries and veins of the beating heart (Fig. 2). First, to measure the blood flow velocity in large and middle-sized epicardial coronary arteries and veins, we placed a cuff around the vessel and inserted the optical fiber into the vascular lumen through a small hole in the cuff. The fiber tip was moved stepwise across the vascular lumen to obtain a profile of the blood flow velocity across the vascular lumen. Second, to measure the blood flow velocity in small epicardial arteries and veins, whose walls were thin enough to be transparent to laser light, we placed the fiber tip on the outer surface of the vessel and fixed it with a drop of cyanoacrylate. The second method made it possible to evaluate the instantaneous blood velocity pattern of the small artery and vein even of the atria (16). Fig. 3 shows a representative tracing of blood flow velocities in left atrial small artery and vein under control conditions. The blood velocity waveform of the arterial flow resembled the aortic pressure pattern, except during the atrial contraction. The atrial contraction caused a transient sharp decrease in arterial flow velocity. The blood velocity waveform of the atrial veins was characterized by a prominent atrial systolic velocity wave. These findings show that myocardial contraction of the atria, like the ventricles, impedes arterial inflow into the atrial capacitance vessels, and promotes venous outflow from them. The extravascular force of atrial myocardial contraction may be due to shortening and thickening of adjacent muscle fibers with increase in their elastance.

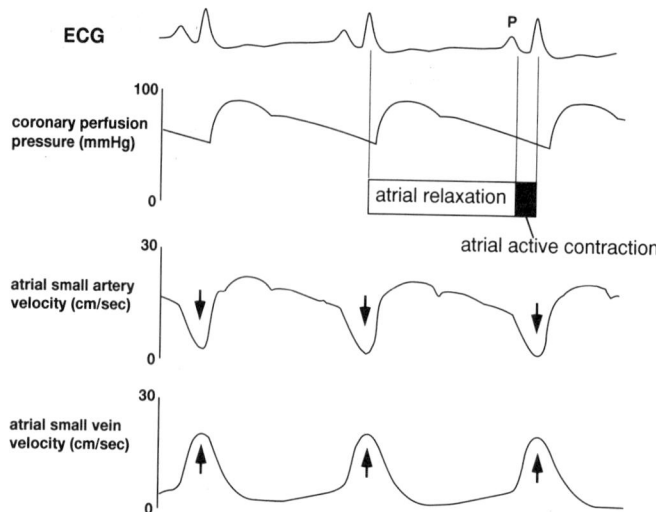

Figure 3. Schematic presentation of the phasic blood velocity pattern in a left atrial small artery and vein. The arterial blood flow corresponded with the pattern of coronary perfusion pressure, but flow was impeded by atrial contraction. In contrast, the venous blood velocity increases during atrial contraction. There is a 180 degree phase shift in these blood flows. (From Kajiya et al.[16]).

Fig. 4 shows the principle of the 80 channel 20 MHz ultrasound pulsed Doppler velocimeter which is applicable to the myocardial arteries and veins of both humans and animals (11,17,18). The transducer consists of a $\pi \times 0.5^2$ mm² piezoelectric crystal with a 20 MHz carrier frequency. Since the depth resolution is 0.2 mm, the sample volume for each sampling point is approximately $\pi \times 0.5^2 \times 0.2$ mm³. This system has 80 sampling gates. Doppler signals from the multicircuits are analyzed by a zero-cross method, and the signal from an optional channel is analyzed by a fast Fourier transform (FFT) method, both in real time. The unit data length of the Fourier analysis is 128 data points. The frequency resolution, temporal resolution, and maximum detectable frequency of our system are 290 Hz, 256 msec, and 25 KHz, respectively. This velocimeter made it possible to measure the blood velocity profile not only of the extramural coronary artery, but also of the intramural septal artery

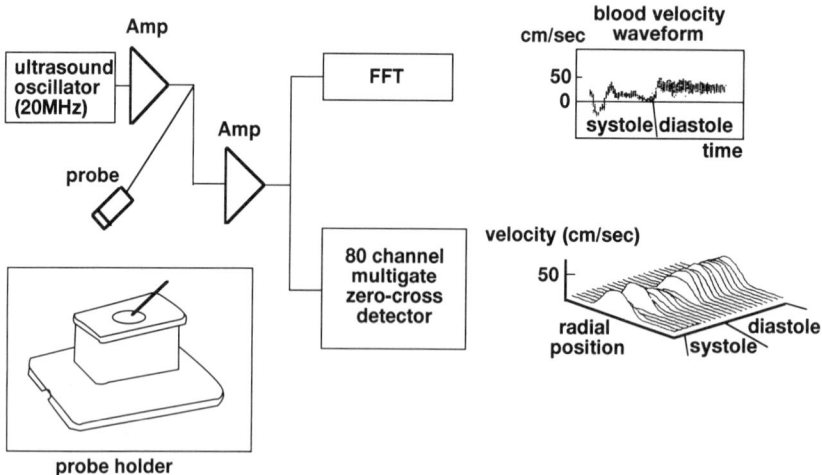

Figure 4. Principle of the 80 channel 20 MHz pulsed Doppler velocimeter. Doppler signals from 80 channels are detected by a multigated zero-cross method and a Doppler signal from one optional channel is analyzed by a fast-Fourier transform method (FFT). (From Kajiya et al.[11]).

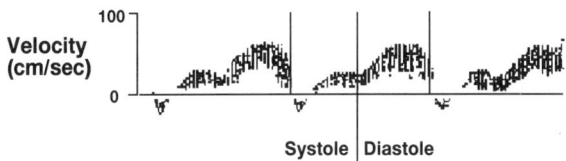

Figure 5. The septal arterial blood velocity throughout the cardiac cycle. Septal arterial blood velocity was measured by our 20 MHz pulsed Doppler velocimeter with a probe. The phasic septal arterial blood velocity waveform was analyzed by a FFT method during a cardiac cycle. (From Kimura et al.[20]).

with relative ease. The right bottom panel shows blood velocity profiles across the left anterior descending coronary artery of humans measured during cardiac surgery. The profiles exhibit a parabolic pattern throughout a cardiac cycle in this case. The right upper panel indicates the velocity waveform by the FFT which is similar to the waveform shown in Fig. 5. To measure the septal artery blood velocity in dogs, the probe of the Doppler system was placed on the artery, where less than half of the circumference of the vessel was isolated for the ultrasound beam. The angle between the ultrasound beam and blood column was almost 40 degrees. The probe was carefully placed on the artery without compression of the vessel, and the holder was manipulated so that the maximum flow diameter was obtained by 80 channel zero-cross detector, indicating that the ultrasound beam passed on or near the midpoint of the vessels. Then, the sampling point was fixed on or near the central axial region and the blood velocity was analyzed by the FFT.

Fig. 5 shows the septal arterial blood velocity throughout the cardiac cycle. The septal arterial blood velocity was predominant in diastole. This is a unique characteristic of normal left coronary artery flow. Beside the diastolic predominancy of the blood flow velocity pattern, it should be noted that retrograde blood flow velocity component can be seen always in the septal artery. With this method, the effects of cardiac contraction on the myocardial perfusion can be evaluated without an influence of compliance of extramural coronary arteries. It is clearly shown that the instantaneous blood flow patterns are different between extra- and intra-mural coronary arteries, because of the compliance of the coronary arteries (19). When the phasic pattern of the coronary inflow to the myocardium was evaluated at the peripheral portion of the artery or in the septal artery to minimize the influence of compliance of extramural coronary arteries, it was confirmed that inflow to the myocardium is almost exclusively limited to diastole and cardiac contraction even causes retrograde flow in the coronary artery (9,15).

CORONARY "SLOSH" PHENOMENON AND CORONARY ARTERY STENOSIS

The presence of a systolic retrograde blood flow velocity component indicates that a substantial amount of the blood which enters the myocardium during diastole is returned to the proximal coronary arteries during the next systole. Thus, the blood moving backward in systole does not contribute to the perfusion of the myocardial bed. This backward movement of the intramyocardial blood affects the systolic "impeding effects" which reduce the time-averaged forward flow to the myocardium. According to the nature of the blood flow, i.e., the to-and-fro movements of the blood between epicardial coronary arteries and intramyocardial arteries, we call it "coronary slosh phenomenon"(20). This phenomenon,

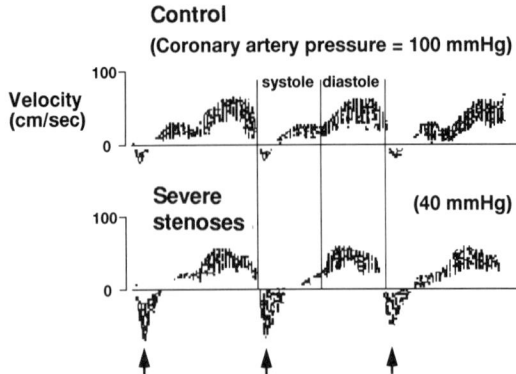

Figure 6. Typical recordings of septal arterial blood flow velocities for no stenosis (upper) and stenosis (lower). With stenosis, the diastolic antegrade blood velocity component decreased, but the systolic retrograde blood velocity component increased. (From Kimura et al.[20]).

when enhanced, may disturb the nutritional blood from the epicardial coronary arteries to the myocardial vascular bed, especially the endomyocardial bed.

To obtain insights into transmural myocardial perfusion during coronary artery stenosis, we evaluated the characteristics of septal arterial blood flow velocity using a 20 MHz multichannel pulsed Doppler velocimeter (20). Fig. 6 shows the septal arterial blood velocity throughout the cardiac cycle in the absence and presence of coronary artery stenosis. With stenosis, the diastolic antegrade blood velocity component decreased, but the systolic retrograde blood velocity component increased. Pharmacological coronary vasodilation further increased the systolic retrograde blood flow. Fig. 7 shows typical septal arterial blood velocities before (upper panel) and after intracoronary nitroglycerin administration (lower panel) during proximal coronary artery stenosis (21). Although nitroglycerin is effective in relieving myocardial ischemia, intracoronary nitroglycerin often fails to relieve angina and has been reported to have deleterious effects on subendocardial blood flow. In the study mentioned above, we have documented that intracoronary administration of nitroglycerin increased diastolic forward flow, but augmented systolic reverse flow markedly. As a result, nitroglycerin increased subepicardial flow, but it failed to increase the subendocardial flow. The subendocardial to subepicardial flow ratio decreased significantly. The experimental results following adenosine administration (20) were almost similar to those of nitroglycerin. Thus, the increased systolic reverse flow following intracoronary administration of vasodilators may not improve subendocardial blood flow during coronary artery stenosis without

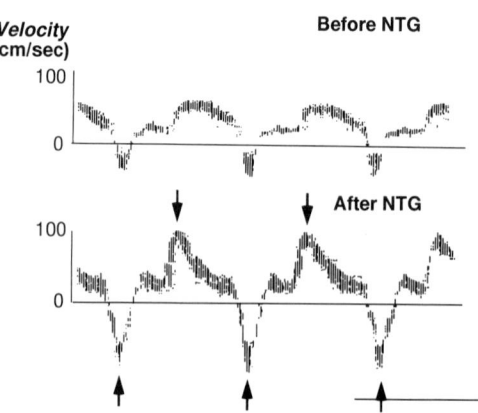

Figure 7. Typical septal arterial blood velocity before (upper) and after intracoronary nitroglycerin administration (lower). After administration of nitroglycerin, the diastolic antegrade blood velocity component increased but the systolic retrograde blood velocity component was augmented more markedly. (From Goto et al.[21]).

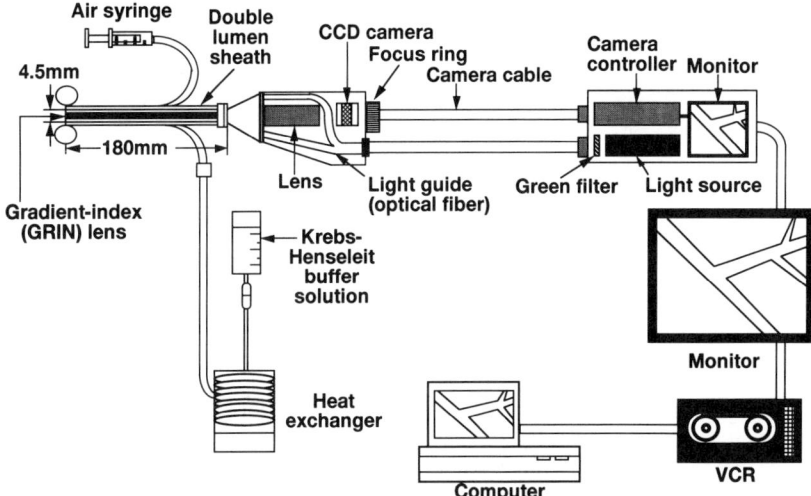

Figure 8. Illustration of the needle-probe video-microscope with a CCD camera.

sufficient development of the collateral vessels, if the vasodilator does not increase the diameter of the stenotic segment.

Further studies are required to understand the mechanisms and the pathophysiological importance of the coronary slosh phenomenon in various forms of cardiac and noncardiac diseases which affects the dynamic nature of the intramyocardial blood flow. In-vivo evaluation of myocardial vascular behavior is necessary to analyze the slosh phenomenon in relation to intramyocardial vascular behavior.

IMAGING OF THE INTRAMYOCARDIAL VESSELS AND THEIR BEHAVIOR DURING THE CARDIAC CYCLE

To understand the effects of cardiac contraction on myocardial blood flow, it is necessary to assess the degree of systolic narrowing of the intramyocardial vessels in the beating heart. So far, the effects of cardiac contraction on the myocardial microvessels have been evaluated by observing the myocardial microvessels from the epicardial surfaces with an intravital floating microscopic system (22). Subepicardial arterial diameter changed very little during a cardiac cycle and subepicardial venular diameter increased during systole (23, 24). Recently, observation of the subendocardial microvessels from the endocardial surface in the beating heart has been possible by introducing a novel needle-probe videomicroscope (25).

The microscope system consists of a needle probe, a camera body containing a CCD camera, a lens and light guide, a control unit, a light source, a monitor, and videocassette recorder. A needle-probe containing a gradient-index (GRIN) lens with a length of 180 mm is used to obtain the images of the subendocardial microcirculation of the left ventricle. (From Yada et al. [25]).

Fig. 8 shows our system of the needle-probe videomicroscope with a CCD camera (VMS 1210, Nihon Kohden, Tokyo, Japan). The system consists of a needle-probe, a camera body containing a CCD camera, a lens and light guides, a control unit, a light source, a monitor and a videocassette recorder. The needle-probe (diameter: 4.5 mm, length: 65 mm and 180 mm) contains a gradient index (GRIN) lens surrounded

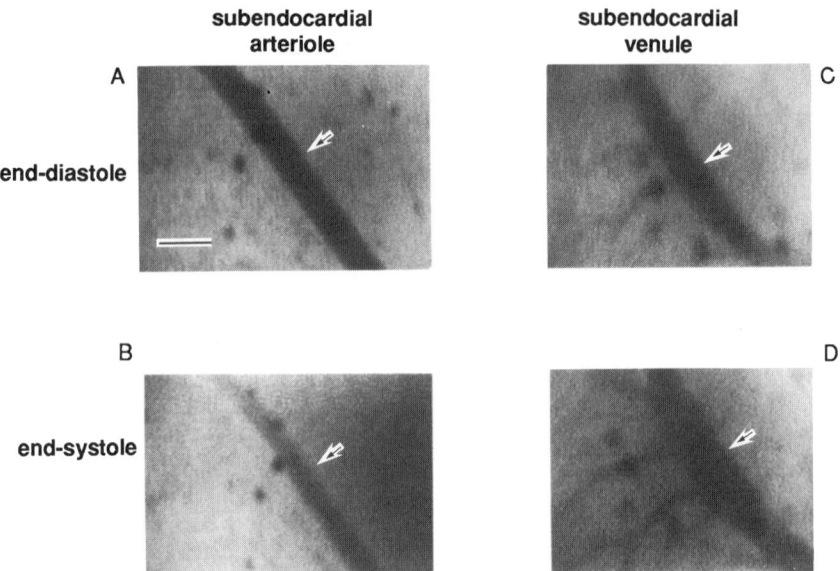

Figure 9. Images of a subendocardial arteriole (left) and venule (right) in end-diastole and end-systole. A green filter was used to contrast the images of the vessels against surrounding tissue. A; image of a subendocardial arteriole in end-diastole; arrowhead = 120 µm, B; image of a subendocardial arteriole in end-systole; arrowhead = 90 µm, C; image of a subendocardial venule in end-diastole; arrowhead = 195 µm, D; image of a subendocardial venule in end-systole; arrowhead = 135 µm. Bar = 200 µm.

by an annular light guide. The image passes through the GRIN lens and focuses on the CCD image sensor (about 250,000 pixels). Images on the CCD are converted into color video signals and recorded on a videotape. The tissue is illuminated by light from a halogen lamp, which is transmitted through light guides surrounding the GRIN lens. A green filter was used to accentuate the contrast between the image of the vessel and the surrounding tissue. The spatial resolution of this system is approximately 5 µm for the x200 objective with GRIN lens using a United States Air Force 1951 test target. The maximum depth of field is about ± 250 µm. The time sequential images were stored on a Macintosh II$_{fx}$ (Apple Computer, Cupertino, CA) using a data translation 24-bit analog-to-digital color video board.

Fig. 9 shows typical images of a subendocardial arteriole and venule at end-diastole and end-systole. Diameter of subendocardial arterioles and venules decreased from end-diastole to end-systole by about 20%. However, we usually did not observe any collapse of the vessels during systole. From these results, it was suggested that the coronary arterioles and venules in the deep myocardial layer are pulsated during the cardiac cycle, and can be the origin for the systolic retrograde and diastolic anterograde coronary arterial blood flow, and also the systolic forward venous flow. If the systolic arterial retrograde flow increases by the augmentation of vascular pulsation, it is followed by the increase in the resistance of the subendocardial arterioles in the next diastole, especially early diastole. Thus the coronary slosh phenomenon is probably an important factor for explaining the systolic-diastolic interaction. Of course, the coronary systolic-diastolic interaction is influenced by the ability of vessels to recoil during diastole and the diastolic function of the left ventricle.

MYOCARDIAL MECHANICS AND CONTROL OF INTRAMYOCARDIAL BLOOD FLOW

Recent studies on the isolated coronary arteries and arterioles indicated that the vessels react to the instantaneous changes in both blood flow and pressure (26-28). Thus, it is unlikely that the intramyocardial resistance vessels only passively follow the mechanical changes of the surrounding myocardium without responding to these mechanical stimuli using inherent endothelial and myogenic mechanisms.

The local arterial dilatation that is evoked by flow is mediated by the endothelium-derived relaxing factors (EDRFs), and now the importance of mechano-physiological role of nitric oxide (NO) radical produced from L-arginine (29) is widely recognized. For example, EDRF release in response to the mechanical stimulus of hydrodynamic shear stress is greater when flow is pulsatile than steady (30, 31), and is modulated by the amplitude of the pulse pressure in isolated vessel segments (32). Accordingly, it is possible that the oscillatory flow related to the coronary slosh phenomenon influences endothelial vascular regulation by EDRF. Because the mechanical environment of coronary vessels is different between the superficial and deep myocardial layers, endothelial regulation of the vascular tone in the beating heart can be different between the two layers. In our preliminary study, the effect of L-NMMA on vascular response was different between subendocardium and subepicardium during reactive hyperemic response (33). On the other hand, the myogenic mechanism may also cause different vascular tone in the subepicardial and subendocardial layers: Kuo et al. (34) showed that the myogenic activity was higher in subepicardial than in subendocardial coronary arterioles, when isolated porcine coronary arterioles were studies. Goto et al. (35) reported that the different pulsation amplitudes of the arterioles, which exist between the subepicardium and subendocardium, caused differences in the mechanical characteristics of the vessels, i.e., the compliance of the coronary arteriole was larger with greater amplitude of the pulsatile transmural pressure than with smaller amplitude of the pulsatile transmural pressure. Furthermore, raising the amplitude of the pulsatile transmural pressure caused the mean vascular diameter in a steady state to increase (36). The latter finding indicates that the vasodilating effects of the pulsation may compensate for the extravascular compressing effects in the subendocardium. Further investigation is needed to understand the mechanisms and mechano-physiological and mechano-pathophysiological importance of the vascular regulations in the different layers of the beating myocardium, since the mechanical stresses may alter the structural and functional properties of cells in intramyocardial blood vessels at the cellular, molecular and genetic levels, leading to responses to a sustained mechanical environment of the heart.

ACKNOWLEDGMENTS

This study was supported by Grant-in-Aid 05454278 for General Scientific Research (B), Grant-in-Aid 05557043 for Developmental Scientific Research (B) from the Ministry of Education, Science, and Culture, Japan and by Research Project Grant (No 6-201) from Kawasaki Medical School.

REFERENCES

1. Harvey, W., 1628, Exercitatio Anatomica De Motu Cordis et Sanguinis in Animalibus, Fitzer, Frankfurt.
2. Butterfield, H., 1957, The origins of modern science 1300-1800, G. Bell and Sons Ltd, London.

3. Scaramucci, J., 1695, Theoremata familiaria viros eruditos consulentia de variis physico-medicis lucubrationibus juxta leges mecanicas. Apud Joannem Baptistam Bustum, pp 70-81.
4. Hoffman, J. I. E., 1987, A critical view of coronary reserve, Circulation 75(suppl I):I-6-I-11.
5. Marcus, M. L., 1983, Methods of measuring coronary blood flow, In: The coronary circulation in health and disease. New York, McGraw-Hill, Inc., p 25-61.
6. Marston, E. L, Barefoot, C. A, and Spencer, M. P., 1959, Non-cannulating measurements of coronary blood flow, Surg. Forum 10:636.
7. Kolin, A., Ross, G., Gaal, P., and Austin, S., 1964, Simultaneous electromagnetic measurement of blood flow in the major coronary arteries, Nature 203:148-150.
8. Hartley, C. J, and Cole, J. S., 1974, An ultrasonic pulsed Doppler system for measuring blood flow in small vessels, J. Appl. Physiol. 37:626-629.
9. Chilian, W. M, and Marcus, M. L., 1982, Phasic coronary flow velocity in intramural and epicardial coronary arteries, Circ. Res. 50:775-781.
10. Kajiya, F., Hoki, N., Tomonaga, G., and Nishihara, H., 1981, A laser-Doppler-velocimeter using an optical fiber and its application to local velocity measurement in the coronary artery, Experientia 37:1171-1173.
11. Kajiya, F., Ogasawara, Y., Tsujioka, K., Nakai, M., Goto, M., Wada, Y., Tadaoka, S., Matsuoka, S., Mito, K., and Fujiwara, T., 1986, Evaluation of human coronary blood flow with an 80 channel 20 MHz pulsed Doppler velocimeter and zero-cross and Fourier transform methods during cardiac surgery, Circulation 74(suppl III):III-53-III-60.
12. Wilson, R. F., Laughlin, D. E., Ackell, P. H., Chilian, W. M., Holida, M. D., Hartley, C. J., Armstrong, M. L., Marcus, M. L., and White, C. W., 1985, Transluminal, subselective measurement of coronary artery blood flow velocity and vasodilator reserve in man, Circulation 72:82-92.
13. Canty, J. M. Jr., and Brooks, A., 1990, Phasic volumetric coronary venous outflow patterns in conscious dogs, Am. J. Physiol. 258:H1457-H1463.
14. Kajiya, F., Tsujioka, K., Goto, M., Wada, Y., Tadaoka, S., Nakai, M., Hiramatsu, O., Ogasawara, Y., Mito, K., Hoki, N., and Tomonaga, G., 1985, Evaluation of phasic blood flow velocity in the great cardiac vein by a laser Doppler method. Heart Vessels 1:16-23.
15. Kajiya, F., Tomonaga, G., Tsujioka, K., Ogasawara, Y., and Nishihara, H., 1985, Evaluation of local blood flow velocity in proximal and distal coronary arteries by laser Doppler method, J. Biomech. Eng. 107:10-15.
16. Kajiya, F., Tsujioka, K., Ogasawara, Y., Hiramatsu, O., Wada, Y., Goto, M., and Yanaka, M., 1989, Analysis of the characteristics of the flow velocity waveforms in left atrial small arteries and veins in the dog, Circ. Res. 65:1172-1181.
17. Kajiya, F., Tsujioka, K., Ogasawara, Y., Wada, Y., Matsuoka, S., Kanazawa, S., Hiramatsu, O., Tadaoka, M., Goto, M., and Fujiwara, T., 1987, Analysis of flow characteristics in poststenotic regions of the human coronary artery during bypass graft surgery, Circulation 76:1092-1100.
18. Fujiwara, T., Kajiya, F., Kanazawa, S., Matsuoka, S., Wada, Y., Hiramatsu, O., Kagiyama, M., Ogasawara, Y., Tsujioka, K., and Katsumura, T., 1989, Comparison of blood-flow-velocity waveforms in different coronary artery bypass grafts, Circulation 78:1210-1217.
19. Hoffman, J. I. E., and Spaan, J. A. E., 1990, Pressure-flow relations in coronary circulation, Physiol. Rev. 70:331-390.
20. Kimura, A., Hiramatsu, O., Yamamoto, T., Ogasawara, Y., Yada, T., Goto, M., Tsujioka, K., and Kajiya, F., 1992, Effect of coronary stenosis on phasic pattern of septal artery in dogs, Am. J. Physiol. 262:H1690-H1698.
21. Goto, M., Flynn, A. E., Doucette, J. W., Kimura, A., Hiramatsu, O., Yamamoto, T., Ogasawara, Y., Tsujioka, K., Hoffman, J. I. E., and Kajiya, F., 1992, Effect of intracoronary nitroglycerin administration on phasic pattern and transmural distribution of flow during coronary artery stenosis, Circulation 85:2296-2304.
22. Ashikawa, K., Kanatsuka, H., Suzuki, T., and Takishima, T., 1986, Phasic blood flow velocity pattern in epimyocardial microvessels in the beating canine left ventricle, Circ. Res. 59:704-711.
23. Kanatsuka, H., Lamping, K. G., Eastham, C. L., Dellsperger, K. C., and Marcus, M. L., 1989, Comparison of the effects of increased myocardial oxygen consumption and adenosine on the coronary microvascular resistance, Circ. Res. 65:1296-1305.
24. Nellis, S. H., and Whitesell, L., 1989, Phasic pressures and diameters in small epicardial veins of the unrestrained heart, Am. J. Physiol. 257:H1056-H1061.
25. Yada, T., Hiramatsu, O., Kimura, A., Goto, M., Ogasawara, Y., Tsujioka, K., Yamamori, S., Ohno, K., Hosaka, H., and Kajiya, F., 1993, In vivo observation of subendocardial microvessels of the beating porcine heart using a needle-probe videomicroscope with a CCD camera, Circ. Res. 72:939-946.

26. Griffith, T. M., and Edwards, D. H., 1990, Myogenic autoregulation of flow may be inversely related to endothelium-derived relaxing factor activity, Am. J. Physiol. 258:H1171-H1180.
27. Kuo, L., Chilian, W. M., and Davis, M. J., 1990, Coronary arteriolar myogenic response is independent of endothelium, Circ. Res. 66:860-866.
28. Koller, A., and Kaley, G., 1991, Endothelial regulation of wall shear stress and blood flow in skeletal muscle microcirculation, Am. J. Physiol. 260:H862-H868.
29. Moncada, S., Palmer, R. M. J., and Higgs, E. A., 1991, Nitric oxide: physiology, pathophysiology and pharmacology, Pharmacol. Rev. 43:109-142.
30. Pohl, U., Busse, R., Kuon, E., and Bassenge, E., 1986, Pulsatile perfusion stimulates the release of endothelial autacoids, J. Appl. Cardiol. 1:215-235.
31. Rubanyi, G. M., Romero, J. C., and Vanhoutte, P. M., 1986, Flow-induced release of endothelium-derived relaxing factor, Am. J. Physiol. 250:H1145-H1149.
32. Hutcheson, I. R., and Griffith, T. M., 1991, Release of endothelium-derived relaxing factor is modulated both by frequency and amplitude of pulsatile flow, Am. J. Physiol. 261:H257-H262.
33. Yada, T., Hiramatsu, O., Tachibana, H., Matsumoto, T., Toyota, E., Goto, M., Ogasawara, Y., Tsujioka, K., and Kajiya, F., 1994, Subendocardial arteriole has larger vasodilatory capacity and systolic-to-diastolic pulsation amplitude than subepicardial arteriole, Circulation (abstract), 90:I-266.
34. Kuo, L., Chilian, W. M., and Davis, M. J., 1988, Myogenic activity in isolated subepicardial and subendocardial coronary arterioles, Am. J. Physiol. 255:H1558-H1562.
35. Goto, M., VanBavel, E., Giezeman, M. J. M. M., and Spaan, J. A. E., 1993, Mechanical properties of coronary arterioles under pulsation, in Maruyama Y. et al (eds): Recent advances in coronary circulation, Tokyo, Springer-Verlag, 182-188.
36. Goto, M., Giezeman, M. J. M. M., VanBavel, E., and Spaan, J. A. E., 1992, Increase in amplitude of pulsatile transmural pressure dilates coronary arterioles, Circulation (abstract) 86:I-508.

14

NONLINEAR MODELS OF CORONARY FLOW MECHANICS

Jos. A. E. Spaan

Deparment of Medical Physics
Academic Medical Center
University of Amsterdam
Meibergdreef 15, 1105 AZ Amsterdam, The Netherlands

INTRODUCTION

Physiology

The coronary circulation is the vascular system that supplies the heart with blood and thereby with oxygen and substrates so it can performs its task. The task of the heart is elementary to life and cessation of its performance for only a few seconds will result in unconsciousness. Cessation of blood supply to the brain exceeding 2 to 3 minutes results in irreversible brain damage. The heart itself is also vulnerable to insufficient blood supply. The body is rather inconsiderate in its demand for perfusion and the circulatory control systems drive the heart to accommodate this demand. As a result there may be circumstances that coronary blood flow is not sufficient to deliver oxygen to the heart so it can perform its task. Mostly this is the case as a result of a disease process that either reduces the upper limit of blood supply as in atheroscleroses or increases the mass of the heart to an extent that the demands for oxygen can not sufficiently be met as in hypertrophy.

The source for blood supply to the heart muscle is the aorta. Notwithstanding a higher aortic pressure in systole than in diastole, flow into the main coronary arteries is lowest in the contraction phase of the heart. One speaks of systolic impediment of coronary arterial flow. However, the pattern of the outflow signal of the coronary circulation is approximately 180° out of phase with the arterial flow signal, being high in systole and low in diastole. In the beginning of this century it was therefore speculated that, because of the increased systolic outflow, heart contraction would facilitate coronary perfusion by a massaging effect. However, measurements, now 40 years ago, demonstrated that time averaged flow increased in a period of cardiac arrest (50). Hence, although the systolic impediment of flow in itself does not necessarily imply that contraction reduces the time averaged perfusion, contraction of the heart indeed forms an extra resistance to flow.

In a healthy heart the upper limit for coronary flow is four to five times higher than the flow needed at rest. The coronary circulation has a control system to adjust the flow to the needs of the heart. Through local feedback the smooth muscle tone in small arteries with a diameter lower than 400 μm is adjusted such that flow will not exceed the needs by too large an amount. This is demonstrated by the fact that oxygen extraction from blood by the myocardium is about 75%. Any increase in this work load of the heart resulting in an increased demand for oxygen must therefore be met by an increase in coronary blood flow. When the possibility to adjust smooth muscle tone is already exhausted to compensate for pathological factors that inhibit flow, coronary flow can not further be adjusted. The range of flow control is exceeded and coronary blood flow will be the result of the balance between two mechanical forces: the driving pressure and heart contraction.

Engineering

The interaction between heart contraction and coronary blood flow is complicated because intramyocardial blood volume is highly variable. This interaction is determined by how well the vascular volume can respond to intraluminal pressure variations and compression by contracting muscle fibers. The relation between pressure and volume is referred to as compliance. Compliance is not constant because of the non-linear pressure-volume relationships of vessels and because of material properties of the surrounding tissue which change continuously in the cardiac cycle thereby altering local stress levels. Because of the varying vascular volume the resistance of intramural blood vessels is highly variable as well. A further complication for experimental studies is that the flow control system is rather fast so that interventions directed to alter flow by a mechanical factors are failing because responses are compensated for by changes in smooth muscle tone.

The complicated and not yet well understood nature of the interactions between cardiac contraction and coronary flow on the one hand and the poor observability of the myocardial microcirculation on the other hand is an invitation to study the system with models. The emphasis here should be on study. The basic mechanisms of interaction are not known but obviously one can make educated guesses with respect to their nature. A model can then be constructed to test whether such a guess, or more scientifically spoken a hypothesis, results in a good description of the relationship between measurable variables. The art is to make a model as simple as possible so as not to introduce too many parameters. The requirement should not be perfection of the predictions but discriminating power between mechanisms. The models should aim to predict at least one measurable set of data not used to derive the model structure. In other words, they need to have a predictive power. At this moment modeling all the different mechanisms involved in coronary flow mechanics and control in one comprehensive model is not useful because of the uncertainty with respect to the accuracy correctness of practically all model elements.

Outline of the Chapter

In this chapter some mechanical factors determining coronary blood flow will be analyzed by simple models. Two classical models describing the effect of heart contraction on coronary flow will be presented first. These models are the extravascular resistance model and the waterfall model. The reason is not only their historical importance but also their relevance to more recent concepts. Since these models consider systole and diastole as independent events they can be discussed before discussing the flow in the non-contracting myocardium. Pressure-flow relations in the arrested heart will then be discussed with emphasis on the volume dependence of resistance. This dependence forms a serious complication in deriving a hydrodynamic resistance value from a non-linear pressure-flow relation. After having discussed the perfusion of the arrested heart the focus of this chapter will be directed again to the effect of heart contraction on coronary flow. However, now systole and diastole will no longer be treated as independent events. It will be discussed that normally systole will expel blood volume from the intramural vessels which is replaced in diastole. Volume variations are related to resistance variations and thereby form a mechanism by which coronary flow is impeded.

EXTRAVASCULAR RESISTANCE AND WATERFALL MODEL

The extravascular resistance concept was introduced in the forties (26). The idea was that in diastole and especially at the end of diastole, there would be no effect of cardiac contraction on coronary arterial flow. Flow would be determined by the pressure difference between arteries and veins and the resistance of the vascular bed alone. Without specifying a mechanism it was assumed that contraction would add to the coronary resistance and this amount was called "extravascular resistance". This extravascular resistance model and its prediction for coronary flow pulsations is elucidated in figure 1. In the top the effect of contraction is represented by the switch. Diastole and systole have a resistance in common, the diastolic resistance, R_{dia}. In diastole it is the sole resistance but in systole a resistance is added, the extravascular resistance, R_{extvas}. In order to bring across the characteristics of the model behavior the left ventricular pressure, P_{LV}, wave is represented by a square wave and the arterial pressure, P_a, is assumed to be constant. For $P_a = 50$ mmHg the diastolic resistance was given an arbitrary value and $R_{ex.\ vas}$ was chosen such that a substantial reduction of arterial flow was achieved. Resistance values were then kept constant but P_a increased in order to demonstrate the effect of arterial pressure on the coronary arterial flow pulsations. Note that these pulsations increase with the level of perfusion pressure.

The prediction of pulsatile flow by the extravascular resistance model is obvious. However, its predictive value is low because there is no mechanistic basis for the choice of the extravascular resistance value. Both resistance values were chosen such that the model prediction would agree reasonably with coronary arterial flow pulsations. On the other hand, the model has predictive value for the increase in flow pulsations with the increase in perfusion pressure since the extravascular resistance was not altered to achieve this. Although simple, this is an important conclusion since it takes away the exclusivity of a different and recent model for the prediction of increase in flow pulsatility with perfusion pressure (44). This recent model is based on the elastance concept which will be discussed in more detail below. In the elastance model it is assumed that the increased flow pulsatility is the result of

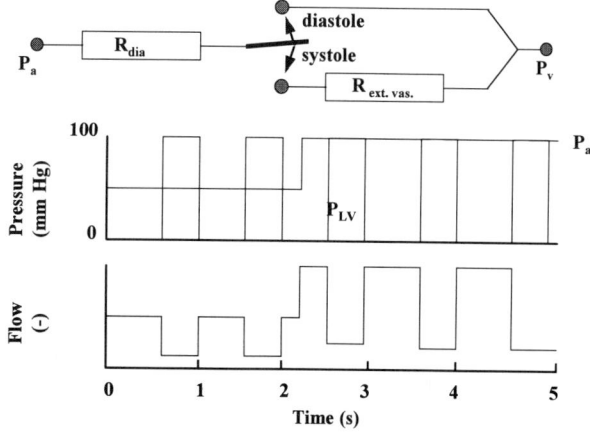

Figure 1. The extravascular resistance concept. It is assumed that in diastole, given a certain arterial-venous pressure difference, the flow is determined by the resistance of the vasculature alone undisturbed by cardiac contraction. In systole the flow is assumed to be inhibited by an extravascular resistance. Note that the model predicts the observed increase in flow pulsations with increasing inlet pressure.

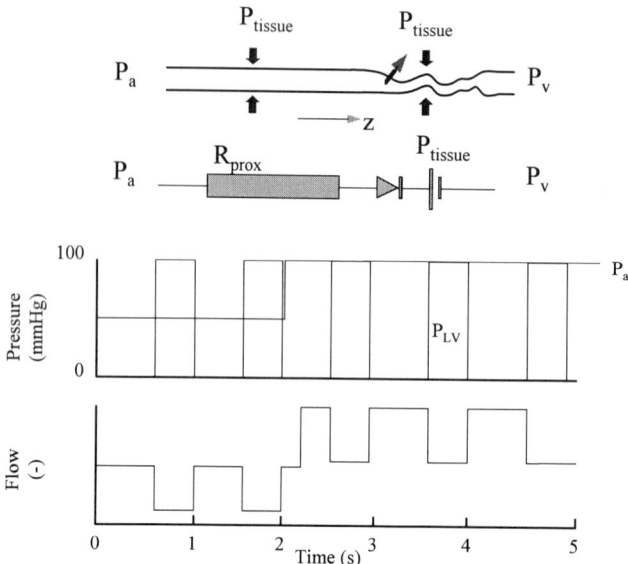

Figure 2. The waterfall model and its prediction for the pulsatility of the coronary arterial flow. The waterfall model assumes that vessels are submersed in tissue as were it in a fluid. The external pressure compresses the vessels when it exceeds venous pressure. The flow is then determined by the difference between inlet pressure and the external pressure and the resistance proximal to the collapse point. The systolic external pressure is assumed to decrease from left ventricular pressure at the sub-endocardium to atmospheric pressure in the sub-epicardium. The prediction on flow pulsations is demonstrated for the case where the myocardial wall is assumed to be devided in 10 layers. Note that the flow pulsations do not increase with increased inlet pressure.

an increased filling of the microcirculation at a higher perfusion pressure by which the contraction may generate more pressure in the intramyocardial vessels.

There is a simple reason for rejecting the extravascular resistance model to explain coronary venous flow since coronary venous flow is high in systole and low in diastole. Such a phase shift in arterial and venous flow pulsations can only be explained if the volume changes in the intramural microcirculation are taken into account. Considerations of these volume changes are also a prerequisite for the explanation of reversed systolic flow often observed in coronary arterial flow (17,59). However, notwithstanding these limitations the extravascular resistance model demonstrates some effects of reality because with a decreased intramural blood volume in systole, systolic resistance will be higher than in diastole.

The waterfall model was applied to the coronary circulation by Downey and Kirk (23) in analogy to its application in lung physiology (48). This model has some relation to the extravascular resistance model since again it was assumed that in diastole the coronary flow would be independent of systole. However, the effect of systole was assumed to work in a rather different way. It was assumed that the intramural vessels were submersed in fluid having a pressure dependent on the pressure in the left ventricular cavity. This pressure was called tissue pressure. The model is elucidated in figure 2. The assumption is that if tissue pressure is lower than venous pressure all vessels would be open and coronary flow is determined by $P_a - P_v$ divided by the diastolic resistance. However, if P_{tissue} is larger than P_v the intramural vessels at the low pressure side would collapse and locally the intraluminal pressure would be equal to and maintained at P_{tissue}. The flow would then be determined by $P_a - P_{tissue}$ divided by the diastolic resistance and be independent of venous pressure. That is why the name "waterfall" is apropriate because flow across the fall is also independent of its height and only dependent on upstream conditions. The electrical analogue used to describe the waterfall effect is shown in the middle panel. Because it was assumed that P_{tissue} would decrease from left ventricular pressure at the subendocardium to atmospheric at the subepicardium the effect of contraction on coronary flow would be more dominant at the subendocardium than at the subepicardium.

The waterfall model had considerable predictive power for flow distribution in the heart. Cardiac contraction has a predominant effect on flow at the subendocardium and almost no effect at the subepicardium which was well explained by the higher tissue pressure in the inner layer of the myocardium. Moreover, the model predicted a parallel shift of mean pressure-mean flow curves in the vasodilated coronary bed when perfusion pressure exceeded systolic left ventricular pressure. The model was further in line with experiments directed to measure tissue pressure by introducing a pressure transducer in the left ventricular wall. However, a shortcoming of the waterfall model in explaining the magnitude of the arterial flow pulsations remained unnoticed (60). The effect of cardiac contraction on these flow pulsations is demonstrated in figure 2. This simulation result comes from a layered model of the left ventricular wall with the variations of tissue pressure over the left ventricular wall as discussed above. Because of the small effect of contraction on the subepicardial layers the predicted pulsations are about half of those by the extravascular resistance model. Also, the waterfall model has no correct prediction of venous outflow pulsations and is not able to predict retrograde coronary arterial flow.

At present there is not much reason to support the waterfall model in predicting the effect of contraction on coronary flow in a normal beating heart (34). As we will discuss below, there is not sufficient time to displace the blood needed to bring vessels to collapse. However, collapse of vessels by extravascular forces is conceivable when the left ventricular wall is overdistended as in heart failure and also in epicardial veins which may be compressed between the outermost muscle layer and the epicardium and/or pericardium (67,68).

DIASTOLIC PRESSURE FLOW RELATIONS

Vasodilation

First coronary arterial pressure-flow relations in the vasodilated bed are discussed. Active smooth muscle tone is a complicating factor in the determination of pressure-flow relations because of the simple reason that it takes time to obtain them and that tone may change during the process of measurement. Pharmacological vasodilation allows the study of pressure-flow lines without the effect of tone. Furthermore, a normal diastole is too short to measure pressure-flow relations not influenced by contraction. Different techniques have been developed to study coronary haemodynamics during periods of over 10 s of cardiac arrest such as during stimulation of the vagal nerve (50) and destruction of the bundle of His. Pressure-flow relations measured during prolonged diastole are referred to as diastolic pressure-flow relations.

As demonstrated in figure 3 diastolic pressure-flow relations at vasodilatation are not linear but curved, especially at lower perfusion pressures (40). The curve also exhibits a positive intercept at the pressure axis. This intercept is small but significantly above right atrial pressure and is referred to as the pressure at zero flow, P_{ZF}. The possible physiological significance of P_{ZF} has been extensively discussed in the literature (39,40,54,55). It has been suggested that it reflects a functional anatomical site of collapse were the pressure is maintained at that value when blood flows (8). Coronary resistance should in that case be calculated by $(P_a-P_{ZF})/R_{pr}$, with P_a = the arterial pressure and R_{pr} the resistance proximal to

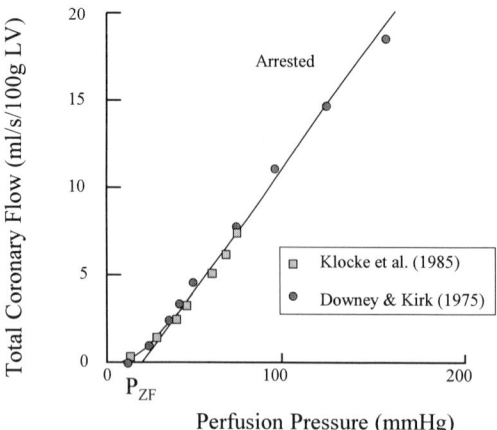

Figure 3. Diastolic pressure-flow relations measured in anesthetized dogs with artificially prolonged diastoles. Note that these curves are nonlinear and curved at low pressures. Pressure-flow data are sometimes extrapolated to obtain a so called pressure at zero flow, P_{ZF}. that results in an overestimation of P_{ZF}.

the point of collapse. This explanation is therefore according to the waterfall model discussed above where it is now assumed that collapse occurs also during diastole. Note that in this case P_{ZF} is the pressure measured at zero flow but reflects the back pressure for arterial flow at positive values.

There are two fundamental complications with the interpretation of pressure-flow relations. The first is the functional meaning of P_{ZF}. Undoubtedly, micro-vessels collapse if luminal pressure decreases below a certain level and therefore P_{ZF} may reflect the threshold pressure for collapse at zero flow. However, that is no proof that the collapse continues to exist when flow is restored which is a necessity for maintaing the assumed P_{ZF}. In a nice study (6) it was demonstrated that arterioles decreased in diameter by pressure reduction and that only those with a diameter in the order of 10 μm would collapse at a pressure above venous pressure. However, these vessels opened again when flow was restored. The second complication is the pressure dependence of resistance. It has been argued that when blood flows there will be no collapse of intramural vessels but that with decreasing pressure resistance will increase thereby slowing down the emptying of the microcirculation (18,34,54). Since pressure-flow relations are often measured with a continuously decreasing pressure it would seem that flow stops at a finite pressure but that the P_{ZF} found would be lower when the rate of decrease of pressure would have been zero (24).

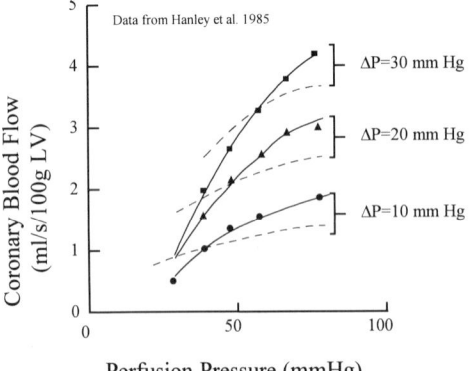

Figure 4. Pressure dependence of coronary resistance in the arrested heart. In a supported isolated heart the arterial-venous pressure difference was kept constant at increasing perfusion pressure. If resistance was constant, flow should remain constant at a given pressure difference. The increase in flow demonstrates that vascular resistance is decreasing with pressure. The broken lines represent much less pressure dependency of resistance and are used to calculate the pressure-flow line drawn in figure 3 (redrawn from (29)).

Pressure Dependency of Coronary Resistance

It has clearly been shown by Hanley et al (29) that the coronary circulation in cardiac arrest is pressure dependent which is demonstrated in figure 4. They perfused an isolated heart at a number of constant arterial-venous pressure differences by varying both these pressures independently. At a larger arterial-venous pressure difference, the flow was higher but not proportionally higher especially at the higher arterial pressure. Moreover, the higher the arterial pressure the higher the flow at a constant pressure difference. This figure also shows the curves (dotted one) used to calculate the fit to the experimental pressure-flow data of Downey and Kirk (23) and Klocke et al (40) in figure 3. Note that pressure dependency of resistance reflected by those dotted curves in figure 4 is less than the data shown in this figure. The data shown are from the experiment exhibiting the highest pressure dependency of resistance in that data set. In any case, one may not neglect the pressure dependency of resistance in the interpretation of diastolic pressure-flow relations. Often the slope of the pressure-flow, P-Q, relation is interpreted as the conductance of the vascular bed. Although the units of the slope are equal to conductance (inverse of resistance), it cannot be set equal to the haemodynamic conductance. When conductance G is a constant then, Ohm's law simply states that:

$$\frac{\Delta Q}{\Delta P} = G$$

where $\Delta Q/\Delta P$ is the slope of the P-Q relation which per definition then passes through the origin. However, when the conductance is pressure dependent it has to be expressed as:

$$\frac{\Delta Q}{\Delta P} = G + \frac{\Delta G}{\Delta P} P.$$

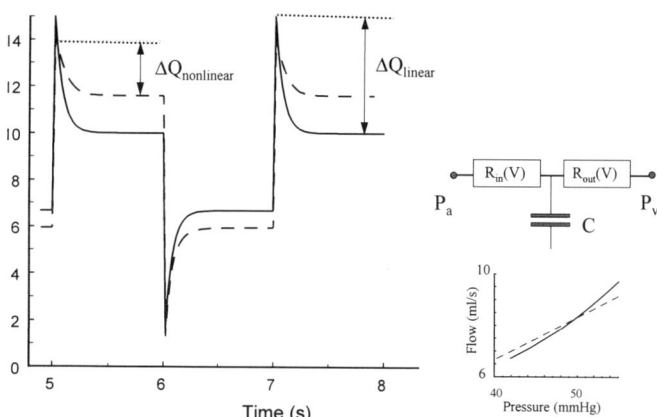

Figure 5. Demonstration of the effect of volume dependence of resistance on the transients of inflow resulting from step changes in inlet pressure. The insert at the bottom right shows the inlet pressure-flow relations applying to the steady state. For the linear model (broken lines), R_{in} and R_{out} are constant resulting in a curve through the origin. For the nonlinear model the relationship is curved (solid line). Note that the changes in pressure were around the same working point chosen a pressure of 50 mmHg and per definition resistances in both cases were the same at this pressure. Note that the overshoots and undershoots are diminished when the resistances are volume dependent.

Hence, the slope of the pressure flow curve is not simply equal to the haemodynamic conductance but a term has to be added that depends on the absolute value of pressure and the pressure dependency of conductance. The conductance that carries a relationship to the physical dimensions of the vascular bed and the viscosity of blood is G and is the ratio between actual flow and pressure.

Neglecting the pressure dependency of resistance can also result in misinterpretation of dynamic data. In figure 5 the flow responses to step changes in perfusion pressure are shown which were calculated from a simple R-C-R network in which R depends on the load of C. P_a was altered as a square wave changing between 40 and 60 mmHg. The network is depicted at the right of this figure together with the steady state pressure flow relations. The dotted line represents the linear system in which the resistances are constant and therefore this line passes through the origin since it is assumed that $P_V = 0$ mmHg. The solid line was obtained by assuming that $R_{in} = R_o + KR*V$ where $R_o =$ constant, V = volume and KR = the sensitivity of resistance to volume in the compliance. R_o was determined such that flow at $P_a = 50$ mmHg was equal for the two models. Moreover, at Pa=50 mmHg the ratio of proximal resistance to total resistance amounted to 1/3.

Obviously, the steady state values of the nonlinear model are further apart than those of the linear model as dictated by the curves in the lower right panel. However, the overshoot for each step increase is much less with the nonlinear model than with the linear model. With the linear model, R_{in} is higher at the beginning of the pressure increase than its reference value at 50 mmHg, reducing the peak flow response after the pressure increase. During the flow response resistance is decreasing such that the new steady state in flow is higher than predicted with the linear model. Hence, the reduction in peak of the flow response and increase of flow plateau with an increase in pressure makes the total range of flow decay considerably less in the nonlinear model than in the linear model. This holds notwithstanding the fact that the compliance and average resistance in both models are equal. A similar reasoning applies for the flow response to a pressure decrease.

We have reported earlier that through a different choice of parameters and resistance-volume relationships the overshoot may disappear almost completely (34). Hence, by manipulating the different relations many different responses of flow to a pressure step can be obtained.

This example demonstrates that it is dangerous to apply linear system analysis to the dynamic flow responses although publications doing so can be found (13,14,46). The problem is not so much that linear analysis is applied since in most cases a nonlinear system can be linearized within a range around a working point. The problem lies in the interpretation of the parameter values obtained from the linear analysis. In the example at hand the mean flow for the nonlinear case is not so much different from that of the linear model. This implies that the ratio between mean flow and mean pressure results in an estimate for the overall haemodynamic conductance for both models. However, interpreting the slope of the pressure-flow relation as a conductance is wrong since the slope is not dependent on parameters of the vascular bed alone but also on absolute pressure and the sensitivity of resistance to pressure and volume

Autoregulation and Diastolic Pressure Flow Lines

In coronary beds with autoregulation linear diastolic pressure-flow relations have also been demonstrated, with P_{ZF} values up to 50 mmHg (8,20). Again, these results have been explained in terms of the waterfall model, with the P_{ZF} values to reflect the pressure maintained at a collapse point. This pressure would be the result of the summation of tissue pressure and smooth muscle tone and therefore would represent the collapse of arterioles. Again, the same reservations should hold here as in the case of vasodilatation. The fact that

Nonlinear Models of Coronary Flow Mechanics

P_{ZF} exists does not imply that it is the back pressure for flow when flow is at a normal level. Vessels would then be open and flow is determined by the normal vascular parameters as diameter and wall elasticity.

The interpretation of pressure flow lines obtained with smooth muscle tone intact are complicated by the fact that vascular tone may change, either by constriction because of a reduction in oxygen consumption (metabolic flow adaptation) or dilation resulting from the reduction of pressure (autoregulation). Therefore, in experimental studies the pressures had to be reduced rather rapidly, in the order of 10 mmHg/s, which makes the measurement vulnerable to capacitance effects (24,54). In relation to these capacitance effects it is a pity that, in most studies, the pressure was allowed to decrease only to the level where flow would stop but was prevented from decreasing further. In one of these studies where pressure was allowed to decrease below "P_{ZF}" there was a considerable retrograde flow for a period of time (62). This makes P_{ZF} rather artificial since it represents simply the crossing of a pressure-flow line with the pressure axis and not an endpoint of this curve.

It is always easy to criticize and to explain why a certain method of analysis is wrong and it is more difficult to point out the right way for doing an analysis. The problem with the coronary circulation is that so many parameters are involved. The RCR model is much too simple, since at least two characteristic time constants are involved in the flow response to a change in pressure (18,35,60). So, at least two compartments should be considered in describing this response. Parameter estimation should then be performed to establish the distribution of pressure dependent compliances and resistances not only by their average value at different pressures but also by their sensitivities to pressure change at those different pressure levels. Some attempts are made in that direction by determining transfer functions between perfusion pressure on the one hand and arterial flow and microvascular volume on the other hand (16,19). However, more direct measurements of microvascular dimensions and pressure are required to arrive at better estimates of these distributed parameters.

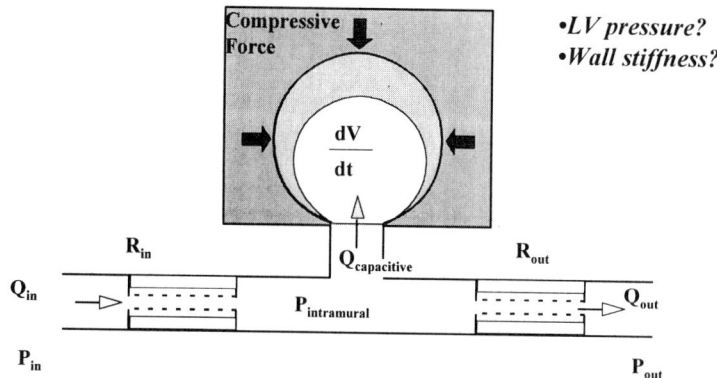

Figure 6. Demonstration of the effect of volume dependence of resistance on the transients of inflow resulting from step changes in inlet pressure. The insert at the bottom right shows the inlet pressure-flow relations applying to the steady state. For the linear model (broken lines), R_{in} and R_{out} are constant resulting in a curve through the origin. For the nonlinear model the relationship is curved (solid line). Note that the changes in pressure were around the same working point chosen a pressure of 50 mmHg and per definition resistances in both cases were the same at this pressure. Note that the overshoots and undershoots are diminished when the resistances are volume dependent.

HEART CONTRACTION AND CORONARY FLOW

In the section on extravascular resistance and the waterfall model it was already explained that models not taking into account the volume changes in the intramural vessels fall short in explaining some essential features of coronary arterial flow patterns. In the first place, these models have no predictive value for coronary venous flow. In principle, coronary venous flow would be in phase with arterial flow according to these models while in fact the coronary venous flow is about 180° out of phase with the coronary arterial flow waveform. Apart from the requirement that a good model of the coronary circulation should explain the essentials of coronary arterial and venous pressure and flow waveforms it also should explain the flow distribution through the different layers of the myocardial wall. It has been well documented that the subendocardium is more vulnerable to ischemia than the subepicardium. Obviously, a relation is sought with the relative underperfusion of the subendocardium especially at higher heart rates (32,33).

The Intramyocardial Pump Model

The intramyocardial pump model (60) demonstrated in figure 6 couples phasic flow changes to intramyocardial blood volume changes. The model applies to an intramural vascular compartment and is similar to the RCR model in figure 5. Contraction compresses the intramural vessels resulting in a reduction of vascular volume. This volume is reduced by an increase in outflow through the distal channel of the compartment and a reduction in inflow through the proximal channel. Note that when the compression results in an intramural blood pressure higher than the inflow pressure, flow at the entrance can become retrograde. The model expresses simply the law of mass conservation: the rate of change of volume is the difference between inflow and outflow. The rational consequence of a diminishing vascular volume is that vascular resistance increases. Hence, vascular volume and therefore resistance are not only dependent on arterial pressure but also on the state of contraction of the myocardium.

The amount of volume that can be expelled from the intramural vessels in each heart systole and can be restored each diastole depends on the inlet and outlet resistances of the compartment. With high resistances it will take more time to displace a certain volume than with low resistances. It has been shown that when the heart is arrested it takes about 1.5 to 1.8 s for the intramural blood volume increase to be completed for 67% (66). Hence, the time needed to displace volume in the intramural vascular space takes longer than the duration of diastole and systole. Consequently, since volume restored in diastole depends on the volume expelled in systole, systole and diastole are phases that can not be studied independently from each other. The effect of a reduced volume is an increase of resistance explaining the impediment of heart contraction on time averaged myocardial perfusion (3,12).

Although concomitant changes of volume and resistance of the intramural vascular bed are accepted in the analysis of coronary blood flow mechanics a clear picture does not exist yet of the mechanism of squeezing the intramural vessels. Before discussing some possible mechanisms, the effect of cardiac contraction on flow distribution will be discussed first.

Contraction and Flow Distribution

The differential effect of cardiac contraction on perfusion distribution in the myocardial wall is demonstrated in figure 7 (5). Flow into the different layers of the heart was

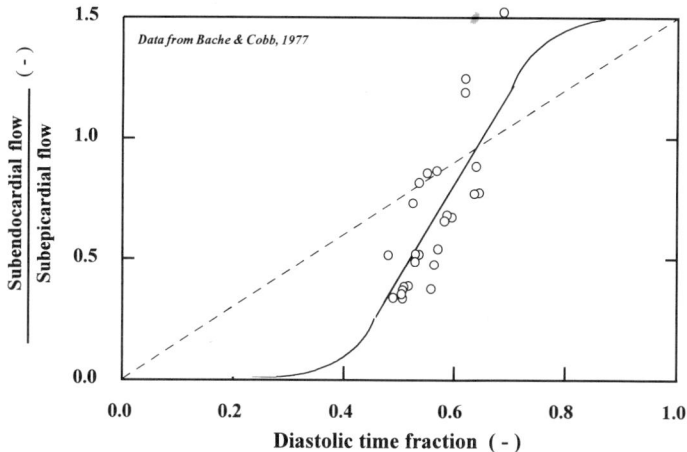

Figure 7. The effect of heart rate on the distribution of flow in the myocardium. With increasing heart rate the time fraction that the heart is in diastole diminishes. Note that flow in the subepicardium is hardly affected by heart contraction and that changes are therefore due to changes in subendocardial perfusion. The line of unity represents the case where subendocardial flow would be determined by diastole alone assuming that in an arrested heart, the time fraction is unity, subendocardial flow is 50 % higher than subepicardial flow. The data show that subendocardial flow diminished more than proportionally with diastolic time fraction and thereby the detrimental effect of a high heart rate on subendocardial perfusion. The sigmoid curve is a hypothetical relationship (redrawn from (57))

measured with the microspheres technique. Radioactive labeled microspheres were injected into the coronary artery and distributed with flow. After excision the heart was cut into pieces and the radioactivity of each piece is measured. All data points in figure 7 were obtained at a rather constant mean aortic pressure of 100 mmHg. On the vertical axis the ratio between subendocardial and subepicardial flow is given. One should take into consideration, however, that subepicardial flow, the flow in the outer layer of the heart wall, is hardly affected by heart contraction. Some recent studies even demonstrate a small increase of sub-epicardial perfusion with heart rate (22). Consequently, the change in the ratio given in figure 7 is completely due to flow variations in the inner layer of the heart wall. On the horizontal axis the diastolic time fraction has been plotted since one expects a positive correlation between flow and time that the heart is relaxed.

From figure 7 it is clear that when the heart is arrested, i.e., at a diastolic time fraction = 1, the inner layer of the myocardium has a much lower resistance to flow than the outer layer. This lower resistance is attributed to the higher microvasular volume that is found in the arrested heart (70). With increasing heart rate the fraction of the time that the heart is in diastole is diminishing and subendocardial flow is decreasing to less than 50 % of the subepiacardial value. In this particular study (5) the heart rate was increased up to 200 beats/min. Hence, subendocardial flow decreases by about a factor of three when the heart is brought from arrest to its maximum heart rate.

Apart from the strong effect of contraction on subendocardial flow figure 7 demonstrates two other important features. It is known from other studies that subendo/subepi flow ratio is about 1.5 in the arrested heart (21,70). This suggests therefore that subendocardial flow is much less influenced by a certain decrease in diastolic time fraction at a low heart rate than by the same decrease at a higher heart rate. Therefore the sigmoid curve has been drawn by eye through the data points in figure 7 as to schematically illustrate the relation between subendo-subepicardial flow ratio. The interpretation according to the intramyocardial pump model would be that when contractions are not succeeding each other too rapidly the beating of the heart has only a minor influence on myocardial perfusion. In contrast, for lower diastolic time fractions there is not suffucient time for the intramural vessels to refill and flow will be seriously impeded (34)

Direct Contraction Effect on Coronary Flow

As explained above, the road to a rational approach for understanding the effect of contraction on flow is via volume and volume coupled resistances in the intramural vascular bed. However, what is the force at the basis of the volume changes? It is most likely that there is a direct interaction between the contracting myocytes and blood vessels that forces the blood out of the microcirculation. This conclusion is based on observations that the wave form of coronary arterial flow is hardly affected by the volume displaced in the left ventricle and/or pressure developed by it (43,45). In addition, the distribution of flow in the left ventricular wall is not much different between normally beating or empty beating hearts (64).

Myocardial cells change mechanical and geometrical properties during contraction and one would expect the interaction between contraction and flow to be very dependent on pressure generated in and deformation of the left ventricle. A concept that may be useful in describing the direct interaction between contraction and coronary flow is the elastance concept (43,69) which is illustrated in figure 8. The concept is formulated in analogy to left ventricular mechanics. In the left ventricle pressure is generated by contraction until the aortic valve is opened. After that the ventricle expels volume throughout the biggest portion

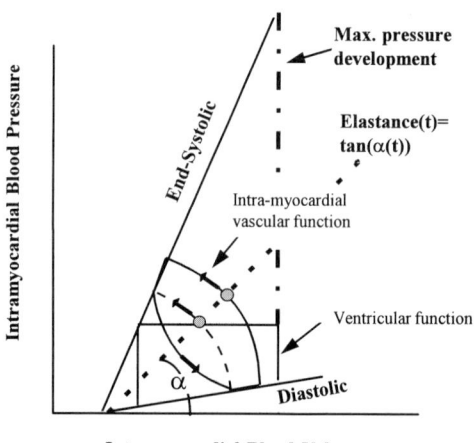

Figure 8. The elastance concept for left ventricular and coronary flow mechanics. From left ventricular mechanics it is known that the pressure that can be maximally generated by contraction depends on its filling in end-diastole. With an intact aortic valve the left ventricle will start to empty when ventricular pressure exceeds aortic pressure. The degree to which it can empty depends on the end-systolic pressure-volume curve. With relaxation left ventricular pressure drops at constant volume and the ventricle will be filled during diastole. This ventricular function can be described by a time-varying elastance (61). Intramural elastance is here defined in a simplified way as the slope of the instantaneous pressure-volume relation indicated by the dotted line, which is determined by connecting points from the same time in the cycle but on curves obtained at different end-diastolic volumes. Application of the elastance concept to the intramural vessels explains the possible increase of pressure in these vessels upon contraction. This pressure will reduce blood volume which is replaced in diastole. This schematic demonstrates that with a higher degree of filling a larger systolic pressure is generated in the intramural vessels (redrawn from (56)).

of systole. The aortic valve closes again and the ventricle relaxes in the isovolumic relaxation phase until left ventricular pressure becomes lower than atrial pressure and the mitral valve opens. The ventricle fills again until the end of diastole. With the onset of systole the cycle restarts. The ventricular pressure-volume loop describing these events in a heart cycle is depicted schematically in figure 8. Obviously, the exact course that this pressure-volume loop is taking depends on the loading conditions of the heart. If the aortic pressure is high, more pressure has to be generated before the valve opens and the ventricle starts to expel volume. The result of this higher pressure is that less volume can be expelled. As a first approximation, the pressure-volume coordinates at the end of systole are on a single line independent of the afterload. This curve is the so-called end-systolic pressure volume curve. Hence the function of the ventricle is determined by the diastolic and end systolic pressure-volume lines. Which loop the pressure-volume takes during the heart beat is determined by the amount of filling of the ventricle at end-diastole and the afterload of the ventricle.

The term elastance concept stems from the following. The pressure-volume coordinates during a heart cycle together with a fixed point at the pressure axis determine a line having a slope which can be taken as the elastance of the ventricular wall. In figure 7 this sloping line is dotted. Hence, the elastance of the ventricle is changing as a function of time, especially during the contraction phase. This time-elastance relation is quite independent of the loading conditions. The elastance reaches its maximal value at the end of systole when it equals the slope of the end-systolic pressure-volume relation. When the heart is contracting more forcefully or, in other words, its contractility has increased, the end-systolic elastance will be higher and a higher pressure can be generated at the same end diastolic filling of the ventricle.

The elastance concept applied to the intramural microvasculature assumes a similar interaction between intramural vascular space and elastance of the wall. In diastole the intramural vessels are filled which is consistent with a high arterial inflow and low venous outflow. In systole, pressure is generated in the intramural vessels dictated by the volume that is collected in them during diastole. Because of the absence of valves in the coronary circulation volume will be expelled continuously in the contraction phase at a rate which is determined by the resistances in the arterial and venous side of the coronary circulation. When the force of contraction fades away in the last part of systole intramural vascular pressure will decrease and depending on the balance of intramural and extramural blood pressures, arterial inflow will increase again. As is suggested in figure 8 pressure and volume of the intramural vessels will follow coordinates forming a loop similarly to that of the left ventricular cavity. Obviously, the schematic shown is purely didactic. The diastolic and systolic pressure volume lines of intramural vessels and the left ventricular cavity are assumed to be scaled in order to facilitate the comparison.

The simplicity of the concept is appealing especially since many experimental observations are in agreement with it. Before describing some of these observations that are in favor of such a concept it should, however, be underlined that there are many conceptual problems in the analogy between the intramural vessels and ventricle. The ventricle is surrounded by contiguous fibers while the microvessels are in between fibers. Furthermore, it is probably wrong to assume that the elastance of the ventricular wall as a whole would resemble that of the surroundings of the vessels. Moreover, there are observations difficult to reconcile with the elastance hypothesis. The most striking conflict is the absence of an effect of cardiac contraction on the time-averaged flow in the sub-epicardium. The microvessels in the sub-epicardium should be influenced by surrounding myocytes quite similarly to the sub-endocardial vessels since there are hardly differences in structure. However, many models have a core that will stand in time despite many shortcomings. The major point made by Westerhof and colleagues (42,43,45,69) that will prove remaining is that the pressure in the microvessels generated by contraction is dependent on the volume that has been accumulated in them in diastole.

When itemizing the observations that are in favor of the elastance concept one has to start with the effect of contractility on coronary arterial flow. When contractility is increased independent of left ventricular filling, the systolic-diastolic flow differences increase. Secondly, when coronary perfusion pressure increases at constant contractility the diastolic-systolic flow differences increase as well. According to the elastance concept this would be the result of a larger degree of filling of the intramural vessels in diastole. Consequently, the intramural blood pressure generated in systole would be larger, resulting in a larger opposing force and in a lower systolic inflow. The elastance concept also predicts higher pulsations in perfusion pressure at a higher flow when the coronaries are perfused at constant flow, i.e. systolic flow = diastolic flow. Again, at a higher flow the microvessels will be more filled and a higher intramural blood pressure will be generated showing up in higher pulsations in the pressure in the coronary arteries.

Elastance Concept and the Coronary Venous System

Obviously, contraction has an influence on the intramural veins as well. We have studied this by cannulating a small epicardial vein close to the location where it drains transmural veins. The bundle of His was destroyed with formaline and the ventricles were paced electrically. After cessation of pacing contraction could be delayed for some seconds. Typical results of these measurements are shown in figure 9.

Two interventions are shown in the figure, one when coronary arterial tone was present and one when the coronary arterioles were dilated by adenosine. Perfusion pressure was kept constant during the interventions but was lower in the case of adenosine. Because of cannulation the vein was obstructed making the pulse on the venous pressure visible. The obstruction of such a small vein can not be considered serious for the venous drainage of the

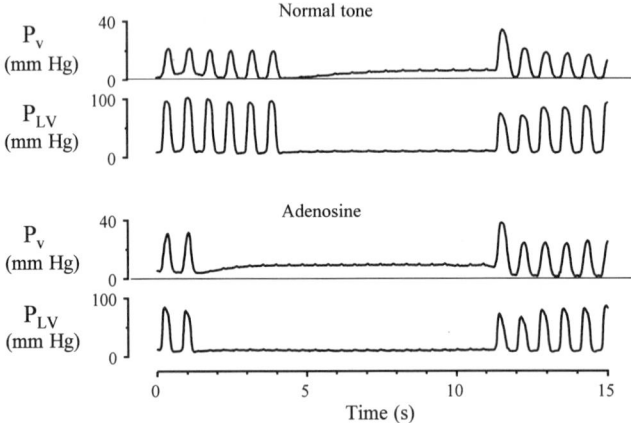

Figure 9. Effect of contraction on the pressure wave in a cannulated epicardial vein with intact tone (top panels) and after vasodilation (bottom panels). Measurements are from a cannulated left main stem preparation in an anesthetized goat. The bundle of His has been destroyed by local injection of formalin. Long diastoles are induced by cessation of pacing. Note that in the first moments of cardiac arrest venous pressure increases reflecting the filling of intramural microvessels. This filling occurs at a faster rate with vasodilation. The systolic pressure generated in the first beat after the long diastole is higher than before arrest even with a decreased systolic left ventricular pressure (not shown). The effect of degree of filling on the generation of pressure in intramural veins agrees with a basic assumption of the elastance concept (Redrawn from (65)).

heart because there are many anastomoses in the venous system that take over the function of the cannulated vein (53).

In the first seconds after arrest the coronary venous pressure increases. Since during this period intramural blood volume increases as well (66) the venous pressure increase can be interpreted as being the result of the filling of the intramural vessels. Note that the increase in venous pressure is faster with vasodilation than with intact tone reflecting the lower resistance opposing flow into the intramyocardial microvessels in that case. When the heart resumes beating, the first systolic venous pressure pulse generated is higher than before the start of cardiac arrest. This higher systolic pressure must be interpreted as the result of higher degree of filling of intramural blood vessels causing a larger effect of contraction. Indeed, a good correlation was found between the systolic venous pressure generation and the end-diastolic venous pressure. As explained above, according to the elastance hypothesis the degree of filling of the intramural vessels is the main determinant of pressure generation when contractility is constant. Moreover, the systolic venous pressure appeared to be independent of the level of pressure generated in the left ventricle (not shown in figure 9). Hence, these experiments clearly show that intramural venous pressure in systole is generated by a direct interaction between blood volume and contracting myocytes in the direct surroundings of intramural vessels.

From a mechanistic point of view the direct contraction effect is very useful in regulating coronary venous blood flow distribution. Since the epicardial venous pressure is very low, coronary venous flow may easily be obstructed locally by compression. This obstruction then leads to a higher intramural blood volume which will directly increase the pressure generation in the intramural veins by the contracting muscle. Hence, each obstruction of outflow is immediately counteracted by an increased pressure for outflow. The extent to which a local mechanism for outflow control will have a regulating effect on the local inflow is not clear but it is certainly not unlikely.

Compartmentalization of Tissue Pressure

The concept of tissue pressure has played a dominant role in the field of coronary research for decades (25,34). Models of left ventricular mechanics demonstrated equilibrium between radial wall stress and left ventricular pressure at the endocardium and thoracic pressure at the epicardium. This radial wall stress was equated with tissue pressure which then in systole would decrease from left ventricular pressure at the subendocardium to zero at the subepicardium (4). The interaction with the blood vessels was understood by assuming that these vessels were exposed to tissue pressure as if they were submersed in fluid having that pressure. In fact, this line of thinking was at the basis of the waterfall model discussed above. It has been attempted to measure tissue pressure by introducing sensors in the myocardial wall. In earlier studies it was tried to elucidate the effect of tissue pressure on vessels by pulling desected vein segments through the wall which were perfused artificially. It was measured to which degree flow was obstructed by the contraction effect on these vessels (47). In other experiments pressure transducers were inserted into the myocardial wall and again the implicit assumption was that tissue pressure could be measured as fluid pressure(1,2,51). The fact that systolic tissue pressure at the subendocardium was close to left ventricular pressure represented strong evidence for the tissue pressure model. However, in the early days it was already noted that tissue pressure so measured persisted when the heart was beating empty (7). This was interpreted by many as an artifact of the strongly deforming myocardium when the left ventricle is empty. However, both earlier and more recent studies recognized that tissue pressure is dependent on the method by which it is measured (11,30,49).

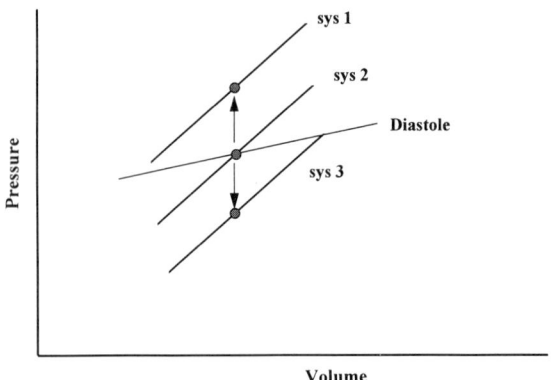

Figure 10. Simplified pressure-volume relationships that illustrate the possible results of contraction on a compartment within the myocardium. The circle on the diastolic relation reflects the values of pressure and volume of that compartment in diastole. Contraction may alter the pressure-volume relationship in many ways from which 3 possibilities are shown. Sys1 reflects the situation explaining the pulsatility of coronary arterial inflow. Contraction generates pressure expelling fluid from this compartment. Sys3 corresponds to observations on the epicardial lymphatic pressure. Pressure in that compartment is actually reduced by contraction. Sys2 corresponds to the situation that the compartment is not affected by contraction, obviously this true only if the volume remains constant and is not altered by flow from or to other compartments.

The problems with tissue pressure are not limited to the technique of measurement but are also conceptual. The heart tissue has many compartments that are filled with fluid, blood vessels, lymph vessels, interstitium etc. In all these compartments a different pressure may be generated by contraction depending on how these compartments are submersed in and connected to the surrounding tissue elements. The spectrum of possibilities is demonstrated in figure 10 on the basis of pressure-volume relations of one compartment. Assume that pressure and volume in diastole have a certain value corresponding to the circle on the diastolic pressure-volume curve. Contraction moves that compartment to a different pressure-volume curve by changing its material properties. Three different possibilities are arbitrarily chosen and are denoted sys1 to sys3. Curve sys1 dictates that pressure in the compartment will rise if volume remains constant. In case curve sys2 applies pressure in the compartment will stay unaltered at constant voume. In case of curve sys3 pressure in the compartment will decrease upon contraction.

Curve 1 is in agreement with the earlier discussion of the elastance model and the effects of heart contraction on coronary arterial flow and venous pressure. Curve sys3 predicts that pressure in the compartment would decrease with contraction. This seems at odds with the general idea that contraction should squeeze the compartments in the ventricular wall. However, a switch from a diastolic curve to a systolic curve predicting a lower fluid pressure in systole is consistent with our measurements on lymph pressure (27,28). Epicardial lymph pressure is often higher in diastole than in systole. Moreover, the systolic lymph pressure wave has two phases (58). In early systole the lymph pressure rises due to an increase in left ventricular pressure. However, in mid-systole the pressure follows a different course and the maximal lymph pressure in mid-systole is lower than in early systole. It was concluded that contraction even shielded the lymph system from compression by left ventricular pressure. Also, observations of endocardial microvessels demonstrate that small arteries and venules remain open during contraction and the diameter reduction in systole is smaller with a smaller size of the vessel (31,37).

The possible protection of blood capillaries in the myocardium during systole was predicted from anatomical studies on the structure of collagen struts connecting the capillary walls to their surrounding myocytes (9,15). It was also the conclusion from a study on the effect of aortic clamping on the arterial inflow pattern (41). The increased systolic left ventricular pressure had only an effect in early systole and not in mid-systole.

The observation that lymph pressure in systole may be lower than in diastole is not in contradiction with the elastance concept. The elastance hypothesis only expresses that the pump function of intramural vessels is described by appropriate pressure-volume loops of the intramural compartments. Apparently, the pressure-volume curve for intramural lymph vessels in contracted myocardium is below that of relaxed myocardium.

Epilogue

In the steady progression of research on the coronary circulation models have played an increasingly important role. Physiologists have always felt the need for quantification of the effect of contraction on coronary flow. One may picture the following developments in history.

Although the idea that systolic and diastolic events must be interrelated is very old (52) it has been abandoned in the first decades when coronary arterial inflow could be measured accurately by electro magnetic flow meters. The reason for that was that the responses of arterial flow to perturbations in pressures seemed rather fast so that it was reasonable to assume that a steady state was reached at the end of diastole. The discussion on time constants is still going on, since diastolic coronary flow is going much faster to a steady state than one would expect on the basis of a large intramyocardial compliance (10,38). The problem is that those conclusions are based on observations of arterial inflow alone. Coronary venous flow observations tell a different story since venous flow responses to arterial or contraction interventions are much slower (36,66). However, time constants for arterial flow may seem fast because of the interaction of capacitance effects and a concomitantly changing resistance.

On the assumption of fast time constants the extravascular resistance model and waterfall model were formulated. Since coronary resistance is changing during the cardiac cycle the extravascular resistance model is still an extreme with a useful message namely that diastolic-systolic flow differences may be the result of resistance variations. The waterfall model has serious shortcomings but formed for a long time the paradigm for explaining the flow distribution in the myocardium. The concept is still important since waterfall behavior has been demonstrated in epicardial veins (63).

The linear intramyocardial pump model recognized the larger time constants for changing intramural blood volume. It was primarily directed at the explanation of the pulsatility of coronary arterial and venous flow on the basis of intramural blood volume variations. Including the volume dependence of resistance, which makes the model nonlinear, gave it predictive power for the blood flow distribution. However, it predicts that subendocardial veins are more compressed than experimentally observed (31). The elastance model is basically an intramyocardial pump model but points to time varying elastance as the generator for the pump. This model makes a direct link between changing material properties of the heart muscle and the intramural blood vessels (69).

All these models are very simple in their representation and focus on a certain mechanism but their beauty is in their simplicity. Most likely we oversee yet another mechanism and a new simple model will emerge. Since in reality different mechanisms interact there is a need of models describing these interactions. However, a large disadvantage of models based on many interactions is that one has to be very careful how the different mechanisms are represented. In the end, it is experimental observations that decide which postulated mechanisms are really functional. Because of progress in experimental techniques these observations will be done at a level of greater detail as for example observations of microvessels at the subendocardium (31) or even intramurally. It is the interaction between experiment and model development that will help us to understand the interaction between cardiac contraction and coronary flow.

REFERENCES

1. Armour, J.A., and Randall, W.C. 1995, Canine left ventricular intramyocardial pressures. Am J Physiol 220: 1833-1839.
2. Armour, J.P., and Randall, W.C. 1971, Canine left ventricular intramyocardial pressure. Am J Physiol 220: 1833-1839.
3. Arts, T., and Reneman, R.S. 1985, Interaction between intramyocardial pressure (IMP) and myocarcial circulation. J Biomed Eng 107: 51-56.
4. Arts, T., Veenstra, P.C., and Reneman, R.S. 1982, Epicardial deformation and left ventricular wall mechanics during ejection in the dog. Am J Physiol 243: H379-H390.
5. Bache, R.J., and Cobb, F.R. 1977, Effect of maximal coronary vasodilation on transmural myocardial perfusion during tachycardia in the awake dog. Circ Res 41: 648-653.
6. Baez, S., Feldman, S.M., and Gootman, P.M. 1977, Central neural influence on precapillary microvessels and sphincter. Am J Physiol 233: H141-H147.
7. Baird, R.J., Goldbach, M.M., and de la Rocha, A. 1972, Intramyocardial pressure: the persistence of its transmural gradient in the empty heart and its relationship to myocardial oxygen consumption. J Thorac Cardiovasc Surg 64: 635-646.
8. Bellamy, R.F. 1978, Diastolic coronary pressure-flow relations in the dog. Circ Res 43: 92-101.
9. Borg, T.K., and Caulfield, J.B. 1981, The collagen matrix of the heart. Federation Proceedings 40: 2037-2041.
10. Bouma, P., Sipkema, P., and Westerhof, N. 1993, Coronary arterial inflow impediment during systole is little affected by capacitive effects. Am J Physiol Heart Circ Physiol 264: H715-H721.
11. Brace, R.A., and Guyton, A.C. 1979, Interstitial fluid pressure: capsule, free fluid, gel fluid, and gel absorption pressure in subcutaneous tissue. Microvasc Res 18: 217-228.
12. Bruinsma, P., Arts, T., Dankelman, J., and Spaan, J.A.E. 1988, Model of the coronary circulation based on pressure dependence of coronary resistance and compliance. Basic Res Cardiol 83: 510-524.
13. Canty, J.M.J., Klocke, F.J., and Mates, R.E. 1985, Characterization of capacitance-free pressure—flow relations during single diastoles in dogs using an RC model with pressure-dependent parameters Pressure and tone dependence of coronary diastolic input impedance and capacitance. Circ Res 248: H700-H711.
14. Canty, J.M.J., Klocke, F.J., and Mates, R.E. 1985, Pressure and tone dependence of coronary diastolic input impedance and capacitance. Am J Physiol 248: H700-H711.
15. Caulfield, J.B., and Borg, T.K. 1979, The Collagen Network of the Heart. Lab Invest 40,No.3: 364-372.
16. Chan, C., Wentzel, J., Spaan, J.A.E. 1995, Tissue thickness is a measure of coronary vascular volume in canine interventricular septa. Federation Proceedings A4910[Abstract].
17. Chilian, W.M., and Marcus, M.L. 1982, Phasic coronary flow velocity in intramural and epicardial coronary arteries. Circ Res 50: 775-781.
18. Chilian, W.M., and Marcus, M.L. 1984, Coronary venous outflow persists after cessation of coronary arterial inflow. Am J Physiol 247: H984-H990.
19. Cornelissen, J.M., Chan, C., Wentzel, J., et al. 1995, Pressure dependence of transfer functions between coronary vascular pressure, flow and volume. Federation Proceedings A4914[Abstract].
20. Dole, W.P., and Bishop, V.S. 1982, Influence of autoregulation and capacitance on diastolic coronary artery pressure-flow relationships in the dog. Circ Res 51: 261-270.
21. Domenech, R.J. 1978, Regional diastolic coronary blood flow during diastolic ventricular hypertension. Cardiovasc Res 12: 639-645.
22. Doucette, J.W., Goto, M., Flynn, A.E., Austin, R.E., Jr., Husseini, W.K., and Hoffman, J.I.E. 1993, Effects of cardiac contraction and cavity pressure on myocardial blood flow. Am J Physiol Heart Circ Physiol 265: H1342-H1352.
23. Downey, J.M., and Kirk, E.S. 1975, Inhibition of coronary blood flow by a vascular waterfall mechanism. Circ Res 36: 753-760.
24. Eng, C., Jentzer, J.H., and Kirk, E.S. 1981, Coronary capacitive effects on estimates of diastolic critical closing pressures. Basic Res Cardiol 76: 559-563.

25. Feigl, E.O. 1983, Coronary physiology. Circ 63: 1-205.
26. Gregg, D.E., and Green, H.D. 1940, Registration and interpretation of normal phasic inflow into a left coronary artery by an improved differential manometric method. Am J Physiol 130: 114-125.
27. Han, Y., Vergroesen, I., Goto, M., Dankelman, J., Van der Ploeg, C.P.B., and Spaan, J.A.E. 1993, Left ventricular pressure transmission to myocardial lymph vessels is different during systole and diastole. Pflügers Arch 423: 448-454.
28. Han, Y., Vergroesen, I., and Spaan, J.A.E. 1993, Stopped flow epicardial lymph pressure is affected by left ventricular pressure in anesthetized goats. Am J Physiol (Heart Circ Physiol) 264: H1624-H1628.
29. Hanley, F.L., Messina, L.M., Grattan, M.T., and Hoffman, J.I.E. 1984, The effect of coronary inflow pressure on coronary vascular resistance in the isolated dog heart. Circ Res 760-772.
30. Heineman, F.W., and Grayson, J. 1985, Transmural distribution of intramyocardial pressure measured by micropipette technique. Am J Physiol 249: H1216-H1223.
31. Hiramatsu, O., Goto, M., Yada, T., Kimura, A., Tachibana, H., Ogasawara, Y., Tsujioka, K., and Kajiya, F. 1994, Diameters of subendocardial arterioles and venules during prolonged diastole in canine left ventricles. Circ Res 75: 393-399.
32. Hoffman, J.I.E. 1984, Maximal coronary flow and the concept of coronary vascular reserve. Circ 70: 153-159.
33. Hoffman, J.I.E. 1987, A critical view of coronary reserve. Circ 75 Suppl I: 6-11.
34. Hoffman, J.I.E., and Spaan, J.A.E. 1990, Pressure-flow relations in coronary circulation. Physiol Rev 70: 331-389.
35. Judd, R.M., Resar, J.R., and Yin, F.C.P. 1993, Rapid measurements of diastolic intramyocardial vascular volume. Am J Physiol 265: H1038-H1047.
36. Kajiya, F., Tsujioka, K., Goto, M., Wada, Y., Chen, X.L., Nakai, M., Tadaoka, S., Hiramatsu, O., Ogasawara, Y., Mito, K., et al. 1986, Functional characteristics of intramyocardial capacitance vessels during diastole in the dog. Circ Res 58: 476-485.
37. Kajiya, F., Yada, T., Kimura, A., Hiramatsu, O., Goto, M., Ogasawara, Y., and Tsujioka, K. 1993, Endocardial coronary microcirculation of the beating heart. Adv Exp Med Biol 346: 173-180.
38. Katz, S.A., and Feigl, E.O. 1988, Systole has little effect on diastolic coronary blood flow. Circ Res 62: 443-451.
39. Klocke, F.J., Mates, R.E., Canty, J.M.J., and Eclit, A.K. 1985, Response to the article by Spaan on 'Coronary diastolic pressure- flow relation and zero flow pressure explained on the basis of intramyocardial compliance' (Circ. Res. 56: 293—309,1985). Circ Res 56: 791-792.
40. Klocke, F.J., Mates, R.E., Canty, J.M.J., and Eclit, A.K. 1985, Coronary pressure-flow relationships. Controversial issues and probable implications. Circ Res 56: 310-323.
41. Kouwenhoven, E., Vergroesen, I., Han, Y., and Spaan, J.A.E. 1992, Retrograde coronary flow is limited by time-varying elastance. Am J Physiol 263: H484-H490.
42. Krams, R., Sipkema, P., and Westerhof, N. 1989, Can coronary systolic-diastolic flow difference be predicted by left ventricular pressure of time varying intramyocardial elastance? Basic Res Cardiol 84: 149-159.
43. Krams, R., Sipkema, P., and Westerhof, N. 1989, The varying elastance concept may explain coronary systolic flow impediment. Am J Physiol 257: H1471-H1479.
44. Krams, R., Sipkema, P., and Westerhof, N. 1990, Coronary oscillatory flow amplitude is more affected by perfusion pressure than ventricular pressure. Am J Physiol 258: H1889-H1898.
45. Krams, R., Sipkema, P., Zegers, J., and Westerhof, N. 1989, Contractility is the main determinant of coronary systolic flow impediment. Am J Physiol 257: H1936-H1944.
46. Lee, J., Chambers, D.E., Akizuki, S., and Downey, J.M. 1984, The role of vascular capacitance in the coronary arteries. Circ Res 55: 751-762.
47. Meer, J.J.v., Reneman, R.S., Schneider, H., and Wieberdink, J. 1970, A technique for estimation of intramyocardial pressure in acute and chronic experiment. Cardiovasc Res IV: 132-140.
48. Permutt, S., and Riley, R.L. 1963, Hemodynamics of collapsible vessels with tone: the vascular waterfall. J Appl Physiol 18: 924-932.
49. Rabbany, S.Y., Kresh, J.Y., and Noordergraaf, A. 1989, Intramyocardial pressure:interaction of myocardial fluid pressure and fiber stress. Am J Physiol 257: H357-H364.
50. Sabiston, D.C.j., and Gregg, D.E. 1957, Effect of cardiac contraction on coronary blood flow. Circ 15: 14-20.
51. Salisbury, P.F., Cross, C.E., and Rieben, P.A. 1962, Intramyocardial pressure and strength of left ventricular contraction. Circ Res 10: 608-623.
52. Scaramucci, J. 1695, De motu cordis, theorema sextum. in: Theoremata familiaria de physico-medicus lucubrationibus Iucta leges mecanicas, Anonymous.

53. Scharf, S.M., Bromberger-Barnea, B., and Permutt, S. 1971, Distribution of coronary venous flow. J Appl Physiol 30: 657-662.
54. Spaan, J.A.E. 1985, Coronary diastolic pressure-flow relation and zero flow pressure explained on the basis of intramyocardial compliance. Circ Res 56: 293-309.
55. Spaan, J.A.E. 1985, Response to the article by Klocke et al. on "Coronary pressure-flow relationships: controversial issues and probable implications" (Circ.Res. 56: 310-323, 1985). Circ Res 56: 789-792.
56. Spaan, J.A.E. 1991, Interaction between contraction and coronary flow: Theory. 6, in: Coronary blood flow; Mechanics, Distribution, and Control. ; Spaan JAE, editors.Dordrecht/Boston/London: Kluwer Academic Publishers, p. 131-62.
57. Spaan, J.A.E. 1991, Interaction between contraction and coronary flow: Experiment. 7, in: Coronary blood flow; Mechanics, Distribution, and Control. ; Spaan JAE, editors.Dordrecht/Boston/London: Kluwer Academic Publishers, p. 163-92.
58. Spaan, J.A.E. 1995, Mechanical determinants of myocardial perfusion. Basic Res Cardiol 90: 89-102.
59. Spaan, J.A.E., Breuls, N.P.W., and Laird, J.D. 1981, Forward coronary flow normally seen in systole is the result of both forward and concealed back flow. Basic Res Cardiol 76: 582-586.
60. Spaan, J.A.E., Breuls, N.P.W., and Laird, J.D. 1981, Diastolic-systolic coronary flow differences are caused by intramyocardial pump action in the anesthetized dog. Circ Res 49: 584-593.
61. Suga, H., Sagawa, K., and Shoukas, A.A. 1973, Load independence of the instanteneous pressure-volume ratio of the canine left ventricle and effects of epinephrine and heart rate on the ratio. Circ Res 32: 314-322.
62. Tomonaga, G., Tsujioka, K., Ogasawara, Y., et al. 1984, Dynamic Characteristics of diastolic pressure-flow relation in the canine coronary artery. in: The Coronary Sinus. ; Mohl W, Wolner E, andGlogar D, editors.Darmstadt: Steinkopff Verlag, p. 79-85.
63. Uhlig, P.N., Baer, R.W., Vlahakes, G.J., Hanley, F.L., Messina, L.M., and Hoffman, J.I.E. 1984, Arterial and venous coronary pressure-flow relations in anesthetized dogs. Evidence for a vascular waterfall in epicardial coronary veins. Circ Res 55: 238-248.
64. VanWinkle, D.M., Swafford, A.N., and Downey, J.M. 1991, Subendocardial coronary compression in beating dog hearts is independent of pressure in the ventricular lumen. Am J Physiol 261: H500-H505.
65. Vergroesen, I., Han, Y., Goto, M., and Spaan, J.A.E. 1994, Cardiac contraction and intramyocardial venous pressure generation in the anaesthetized dog. J Physiol London 480: 343-353.
66. Vergroesen, I., Noble, M.I.M., and Spaan, J.A.E. 1987, Intramyocardial blood volume change in first moments of cardiac arrest in anesthetized goats. Am J Physiol 253: H307-H316.
67. Watanabe, J., Levine, M.J., Bellotto, F., Johnson, R.G., and Grossman, W. 1990, Effects of coronary venous pressure on left ventricular diastolic distensibility. Circ Res 67: 923-932.
68. Watanabe, J., Maruyama, Y., Satoh, S., Keitoku, M., and Takashima, T. 1987, Effects of the pericardium on the diastolic left coronary pressure-flow relationship in the isolated dog heart. Circ 75: 670-675.
69. Westerhof, N. 1990, Physiological hypotheses-Intramyocardial pressure. A new concept, suggestions for measurement. Basic Res Cardiol 85: 105-119.
70. Wüsten, B., Buss, D.D., Deist, H., and Schaper, W. 1977, Dilatory capacity of the coronary circulation and its correlation to the arterial vasculature in the canine left ventricle. Basic Res Cardiol 72: 636-650.

15

SIMULATION OF FORCED BREATHING MANEUVERS

James J. Shin,[1] David Elad,[2] and Roger D. Kamm[1]

[1] Fluid Mechanics Laboratory
Department of Mechanical Engineering
Massachusetts Institute of Technology
Cambridge, Massachusetts 02139
[2] Department of Biomedical Engineering
Faculty of Engineering
Tel Aviv University
Tel Aviv 69978, Israel

BACKGROUND

Numerous theoretical and computational models have been developed over the years for the purpose of simulating pulmonary air flow in circumstances ranging from normal breathing [21, 29, 49, 50, 51] to forced expiration [8, 10, 11, 12, 14, 36, 40, 59] to high frequency oscillation [15, 27] (see 20, 52, 53, and 65 for recent reviews). These models were typically based on airway geometry as determined by one of several available morphometric descriptions [23, 24, 62], and model the problem of a complex, three-dimensional flow through a compliant network using various simplifications. As the years have progressed, the models have become less restrictive, but still rely upon considerable idealizations.

Models of normal breathing often treat the airway system as a rigid network of branched tubes [21, 29, 49, 50]. The flow analysis in these models takes into account the fluid dynamic non-linearities associated with convective acceleration and the influence of boundary layer development and secondary flows due to curvature on flow resistance, but does not include unsteady acceleration nor allow for continuous changes in lung volume with inspired/expired flow. Consequently, each calculation with these models represents a single flow rate at a particular lung volume and the process of breathing is represented by a sequence of quasi-steady flows.

Simulations of small volume, high frequency flow oscillation have been developed to better understand and interpret the measurement of airway impedance by the forced oscillation technique [53] and for the determination of airway cross-sectional area by the acoustic impedance method [15, 27]. Such models include unsteady acceleration but are restricted to small amplitude, linear flow conditions, assuming that the airway tree can be treated essentially as a collection of straight tube segments, neglecting the influence of convective acceleration, boundary layer development, secondary flows and the like. Consequently, no models of this type are capable of treating the non-linear effects that accompany the large amplitude flows of high frequency ventilation (HFV).

Other models have been developed for the purpose of simulating a forced expiration [8, 10, 11, 12, 14, 36, 40], taking into account airway branching, airway compliance, and flow non-linearities, but *not* the effects of gas acceleration. Of these, the model of Lambert and co-workers [36, 37, 40] has been particularly successful in reproducing many of the phenomena associated with a forced vital capacity (FVC) maneuver (a forced expiration from maximal or total lung capacity (TLC) to minimal or residual volume (RV)). As in the normal breathing models described above, however, these too are restricted to the simulation of steady flow at a fixed lung volume and therefore lack the capability of simulating, in a single computation, a complete FVC maneuver. Perhaps more importantly, these models are based on sparse data on the properties of small airways; due to the exquisite sensitivity of the predictions on small airway characteristics, this becomes a serious issue.

Despite such reservations, these previous models have taught us much of what is currently known about the physics of a forced expiration and the factors that limit flow. The underlying theory for flow limitation is based on the works of Shapiro [55], Dawson and Elliott [7], Oates [46], Bonis and Ribreau [4], and others. As in compressible flow through a variable area channel, flow through a collapsible tube can become choked in the sense that the flow rate becomes insensitive to downstream pressure once that pressure is reduced below some critical value. This condition is termed "flow limitation". When flow limitation is achieved, there exists at some location in the tube or airway tree, the condition that mean flow velocity, U, equals the speed of wave propagation, c — this is termed the "choke point" or "flow limitation site" (FLS). When this occurs, further reductions in downstream pressure are unable to propagate upstream past the FLS and the flow rate is therefore unaffected. This theory is the foundation on which all previous models of forced expiration are built, the differences being largely in terms of the method of implementation and the manner in which the pulmonary network is represented.

In a normal forced expiration, forcing is produced by a vigorous contraction of the respiratory muscles rather than a reduction in downstream pressure. The net effect remains the same, however, since respiratory muscle activation leads initially to a more or less uniform increase in gas pressure within the lung and flow results as the pressure gradient, initially residing in the vicinity of the trachea, propagates toward the lung periphery, accelerating the flow as it does. This produces a rapid initial increase in flow rate as measured at the mouth which soon leads to the formation of the FLS. Flow remains choked during most of the ensuing expiration, becoming unchoked again near end expiration as lung volume approaches RV.

In this chapter, previous models for forced expiration will be reviewed with an aim toward identifying their most critical limitations. In addition, a new computational model is presented that eliminates some of the limitations of previous models and is capable of providing new physical insight into certain respiratory flow phenomena. The model is flexible in that it allows simulation of all the different types of flow discussed above, including effects of both convective and temporal acceleration. It couples flow at the mouth to changes in lung volume and thereby allows simulation of complete respiratory maneuvers in a single calculation. The effects of lung tissue and chest wall are included for the purpose of simulating measurements of respiratory impedance although this topic is not explicitly discussed. The model also incorporates recent measurements of changes in airway wall morphometry [63, 64] that accompany asthma or chronic obstructive pulmonary disease (COPD) and the influence of smooth muscle constriction so that the effects of disease can be simulated in a more realistic manner. This approach provides a tool for exploring normal physiology or for obtaining a better appreciation of the link between underlying pathology and the macroscopically observed manifestations of the disease. The need still exists, however, for further refinements in the analysis and for new mechanical and morphometric data on which to base the model.

The purpose of this chapter is two-fold. One is to present new and previous computational models of forced breathing and illustrate their uses and predictions. The other is to identify the limitations that remain and to point out the areas of research that need further exploration if existing deficiencies are to be eliminated.

DESCRIPTION OF THE COMPUTATIONAL MODEL

The present model is based on previous work by our group on simulation of a FVC maneuver [10, 12, 14], and on methods for solving the unsteady flow equations in a compliant tube [35]. The model is comprised of several components which include: (i) the equations governing one-dimensional flow through a network of compliant tubes, (ii) upper airway resistance, (iii) constitutive relations describing airway compliance, (iv) distributions of airway cross-sectional area, airway stiffness, and number of airways, and (v) simulation of respiratory muscle effort and the mechanical properties of lung tissue and chest wall. In addition, modifications are made to mimic pathological states: the airway wall thickening, airway liquid accumulation, and smooth muscle constriction associated with asthma and COPD. These are merged into a single computational algorithm that can be used to simulate a variety of respiratory flows. Unless otherwise specified, the simulations are assumed to occur at body temperature (37 °C) with the density and kinematic viscosity of air given by $\rho = 1.2$ kg/m^3 and $\nu = 1.5 \times 10^{-5}$ m^2/sec, respectively.

The geometry of the airways is continuously changing with time during breathing, thus the flow of gas through the airways is modeled using the one-dimensional governing equations for flow through a single compliant tube whose cross-sectional area is the sum of the areas of all parallel branches located a given distance from the mouth [10] and whose dimensions vary with time. The equations for conservation of mass and momentum are derived for an unsteady, one-dimensional (1-D) fluid flow through a collapsible tube whose total axial length varies with time. Gas compressibility is assumed negligible [11]. To simplify the analysis, we introduce a Lagrangian (material) coordinate, ξ, which is fixed to the tube wall and defined in such a way that the ends of the network always correspond to $\xi = 0$ (alveolar end) and $\xi = 1$ (upper end of trachea) (see [10], Fig. 1). The Eulerian (laboratory) coordinate, x, is fixed in space and is related to the Lagrangian coordinate by:

$$x = x(t, \xi) = L(t)\xi \qquad (1)$$

and therefore,

$$\left(\frac{\partial x}{\partial t}\right)_\xi = \xi \dot{L} \qquad \left(\frac{\partial x}{\partial \xi}\right)_t = L(t) \qquad (2)$$

where L(t) is the time-varying length of the airway network and $\xi\dot{L}$ is the speed of a material point at x in the laboratory frame. One therefore obtains:

$$\left(\frac{\partial Z}{\partial \xi}\right)_t = L\left(\frac{\partial Z}{\partial x}\right)_t \qquad (3)$$

$$\left(\frac{\partial Z}{\partial t}\right)_\xi = \xi \dot{L} \left(\frac{\partial Z}{\partial x}\right)_t + \left(\frac{\partial Z}{\partial t}\right)_x \tag{4}$$

where $Z(x, t)$ is any physical quantity.

Conservation of Mass

Conservation of mass for an incompressible fluid is imposed by relating the rate of change of airway cross-sectional area to the net accumulation due to a gradient in volume flow rate. The resulting equation in the Eulerian frame is:

$$\left(\frac{\partial A}{\partial t}\right)_x + \left(\frac{\partial (UA)}{\partial x}\right)_t = 0 \tag{5}$$

where U is the local mean axial velocity and A is the summed cross-sectional area of all airways a distance x from the periphery. Introducing Eqs. 3 and 4 into Eq. 5 gives the continuity equation in the Lagrangian frame:

$$\left(\frac{\partial A}{\partial t}\right)_\xi + (U - \xi \dot{L})\frac{1}{L}\left(\frac{\partial A}{\partial \xi}\right)_t + \frac{A}{L}\left(\frac{\partial U}{\partial \xi}\right)_t = 0 \tag{6}$$

For any large-amplitude breathing maneuvers, it can be shown that gas velocity is much larger than the axial movement of the airway wall; thus, $U > \xi \dot{L}$ and Eq. 6 becomes:

$$L\left(\frac{\partial A}{\partial t}\right)_\xi + \left(\frac{\partial (UA)}{\partial \xi}\right)_t = 0 \tag{7}$$

Conservation of Momentum

The momentum conservation equation incorporates the contributions to the total pressure drop due to unsteadiness or temporal acceleration, convective acceleration, and frictional resistance. Pressure changes due to convective acceleration are calculated on the assumption that the velocity is uniform over the cross-section at all locations, resulting in the following form of the momentum equation in the Eulerian frame:

$$\left(\frac{\partial U}{\partial t}\right)_x + \frac{\partial}{\partial x}\left(\frac{U^2}{2} + \frac{P_{aw}}{\rho}\right)_t + F\left(U - \left(\frac{\partial x}{\partial t}\right)_\xi\right)^2 = 0 \tag{8}$$

Transforming again into the Lagrangian frame, we obtain:

$$L\left(\frac{\partial U}{\partial t}\right)_\xi + \frac{\partial}{\partial \xi}\left(\frac{U^2}{2} + \frac{P_{aw}}{\rho}\right)_t + LFU^2 = 0 \tag{9}$$

Here, P_{aw} is internal airway pressure, and $F = sf_T/2A$ characterizes the contribution due to shear stress at the wall where s is the airway perimeter. The first term represents local gas acceleration, the second, convective acceleration, and the last, wall friction due to viscous effects. The friction coefficient used here is based on experiments conducted in a rigid, multi-generation model of the central airways [6] and has the following form:

$$f_T = \begin{cases} \frac{16}{Re}\left(0.556 + 0.067\sqrt{Re}\right) & Re \geq 55 \\ \frac{16}{Re} & Re < 55 \end{cases} \quad (10)$$

It represents fully-developed viscous flow at low values of the local Reynolds number (Re = Ud/ν where d is the local airway diameter) and exhibits the influence of curvature, boundary layer development and turbulence at progressively higher Re. It is worth noting that this expression is subject to error for oscillations which are simultaneously of high frequency and small tidal volume [31] for which flow resistance may be better characterized by the expression for oscillatory pipe flow [31, 66].

Airway Segment Lengths

The length of each airway is assumed to vary as the cube root of lung volume, V_L. Accordingly,

$$L = L_0 \lambda^{1/3} \quad (11)$$

where λ is dimensionless lung volume ($\lambda = V_L/V_{L0}$, V_{L0} equals to the lung volume at 35% TLC) and L_0 is the total length of the airway network at V_{L0}. From Weibel's [63] morphometric data for 17 generations measured at 75% TLC, we obtain L = 20.6 cm.

Upper Airway Resistance

The pressure drop across the glottis due to the formation and subsequent dissipation of a turbulent jet downstream of the constriction is described in a manner suggested by Jaeger and Matthys [28], taking into account the dependence of glottal aperture on lung volume [57]. The glottal pressure drop ΔP_g may then be expressed as:

$$\Delta P_g = \frac{1}{(C_d)^2} \frac{\rho U^2}{2} \quad (12)$$

where Re and U are computed based on the velocity and opening width of the glottis, and C_d is a dimensionless function of Reynolds number as given in [10].

Airway Compliance

The airway compliance is described by a relationship between the local cross-sectional area and the transmural pressure difference acting across the airway wall; internal airway pressure, P_{aw}, minus the pressure acting external to the airway, P_{ext}. For intrathoracic airways, $P_{ext} = P_A$, where P_A is alveolar pressure, and for extrathoracic airways, $P_{ext} = P_{atm}$, where P_{atm} is atmospheric pressure. Based on Elad et al. [9] the airway pressure-area law is characterized by

$$\Pi = \frac{P_{aw} - P_{ext} - P_0}{K_p} = \left(\frac{A}{A_0}\right)^{0.5} - \left(\frac{A}{A_0}\right)^{-0.2} \quad (13)$$

where K_p is the effective wall stiffness, P_0 and A_0 are reference pressure and cross-sectional area, respectively. K_p and P_0 are specified functions of λ and ξ. The form of Eq. 13 is based on measurements made in canine airways during quasi-static changes in transmural pressure [58]; consequently, the effects of airway wall acceleration and the viscoelastic properties of airway wall and the surrounding parenchyma are not included. As will be pointed out below, P_0 is taken to be equal to the negative of the elastic recoil pressure of the lung, $-P_{el}$. Accordingly, the left-hand-side of Eq. 13 represents the difference between airway pressure, P_{aw}, and pleural pressure (P_A-P_{el}) as in [36]. The difference, however, is that K_p in Eq. 13 depends on lung volume reflecting the fact that as lung volume falls, the structure surrounding the airway becomes more compliant.

Distributed Properties

The parameters that specify airway stiffness ($K_p(\xi, \lambda)$), reference cross-sectional area ($A_0(\xi)$), reference pressure ($P_0(\lambda)$), and number of parallel airways ($N(\xi)$) are represented as smooth functions of distance along the airway tree. For the central airway network, these are given by the following expressions, as discussed in [8, 9, 10].

$$A_0(\xi) = A_w(\xi)\left(0.78 + 0.211\tanh(8(\xi - 0.25))\right) \quad (14)$$

$$N(\xi) = 1.0387(\xi + 0.01)^{-2.4} - 50.0e^{(-5.9766\xi)} \quad (15)$$

$$K_p'(\xi, \lambda) = K_{p00}\left[1 + 0.078(\lambda^{2.7} - 1)\right]\left[1 + 2.25(1 + \tanh(8(\xi - 0.4)))\right] \quad (16a)$$

$$K_p(\xi, \lambda) = K_p'(\xi, \lambda_{max}) + [1 - G(\xi)]\left[K_p'(\xi, \lambda) - K_p'(\xi, \lambda_{max})\right] \quad (16b)$$

$$P_0(\lambda) = -200(\lambda^{2.5} - 1) = -P_{el} \quad [N/m^2] \quad (17)$$

where $A_w(\xi) = 2.3 \times 10^{-4}(\xi+0.01)^{-0.9} - 1.4 \times 10^{-3}e^{(-3.9844\xi)}$ is a curve-fit of Weibel's data [63] at $V_L = 75\%$ TLC. K_{p00} is the effective wall stiffness of generation 17 in the Weibel model at V_{L0} and is estimated to be 1692 N/m² from the data of Martin and Proctor [43].

The expressions for A_0, P_0, and K_p have been modified from [10]. These new expressions for P_0 and A_0 provide a satisfactory fit to all the data of Takishima et al. [58], and also give physical meaning to the reference condition: A_0 is the airway cross-sectional area that exists when $P_{aw} = P_A$ and lung elastic recoil pressure is zero ($\lambda = 1$), and is modified to take into account the fact that the extra-parenchymal airways require no correction for lung volume. The distribution of K_p has been modified by Eq. 16b to ensure that tracheal wall stiffness is essentially independent of ξ and λ recognizing that tracheal stiffness remains constant as lung volume falls. $\lambda_{max} = 2.857$ corresponds to TLC and $G(\xi) = 0.5(1 + \tanh(8(\xi - 0.4)))$.

Because of the different treatment in the periphery (see below), it is possible and more realistic to employ an asymmetric morphometric model [23, 24] for these airways. The matching between central (compliant) and peripheral (non-compliant) segments is assumed at the level of the 1.8 mm diameter airways (generation 8 in the Weibel model and order 16 in the model of Horsfield and Cumming). A complete description of airway geometry and dimensions is given in Table 1. In the peripheral segment, airways are treated as discrete "resistors" and the resulting network is analyzed as a branched electrical system; the length and diameter of each airway, and therefore its resistance as well, is based on the Horsfield and Cumming model, but modified as described below to account for changes in lung volume, airway wall thickening due to disease, and smooth muscle constriction.

Respiratory Muscle Forcing and Terminal Impedance

In order to simulate a FVC maneuver, it is necessary to specify pleural pressure (P_{pl}) as a function of both time and lung volume to mimic the forces generated during expiration by respiratory muscle (chest wall and diaphragm) activation. The present approach is based on experiments by Hyatt and Flath [25] who measured esophageal pressure (assumed to be a measure of P_{pl}) and lung volume during dynamic maximal effort at different levels of lung inflation. These data provide the dependence of P_{pl} on lung volume. To account for the finite

Table 1. The following table provides the dimensions of conducting airways for both symmetric and asymmetric airway models developed by Weibel [60] and Horsfield and Cumming [23, 24]. Generation 0 is the trachea, and order 0 represents the alveolar zone. δ specifies the difference in order number between the two daughter branches arising from a parent of order n. Number of endings represents the number of terminal airways that originate from a single airway of given order.

Generation	Diameter (mm)	Length (mm)	Area (cm^2)	Order	δ	Diameter (mm)	Length (mm)	# of Endings
0	18.0	120	2.5					
1	12.2	47.6	2.3					
2	8.3	19.0	2.1					
3	4.8	6.5	1.5					
4	3.9	10.9	1.8					
5	3.0	9.2	2.3					
6	2.4	7.7	2.9					
7	2.0	6.5	3.7	17	3	2.00	6.3	15300
8	1.6	5.5	5.1	16	3	1.80	5.17	11200
9	1.3	4.6	7.0	15	3	1.60	4.80	8190
				14	2	1.40	4.20	6140
				13	2	1.10	3.60	4090
				12	1	0.95	3.10	3070
				11	0	0.76	2.50	2040
				10	0	0.63	1.10	1020
				9	0	0.53	1.31	512
				8	0	0.48	1.05	256
				7	0	0.43	0.75	128
				6	0	0.80	0.59	64
				5	0	0.80	0.48	32
				4	0	0.80	0.48	16
				3	0	0.80	0.48	8
				2	0	0.80	0.48	4
				1	0	0.80	0.48	2
				0	0	0.80	0.48	1

rate of contraction of the respiratory muscles (about 0.2 s to produce maximal tension), a time-dependent rise is introduced at the onset of expiration. Based on available data [54], P_{pl} is estimated by the following expression:

$$P_{pl} = 12000\left(1 - e^{-5t}\right)\cos\left(\frac{\pi(3-\lambda)}{7.2}\right) \quad [N/m^2] \quad (18)$$

Similar considerations apply to coughing. In this case, a time-varying impulse-type decrease is imposed in transpulmonary pressure at $\xi = 1$, produced, for example, by a sudden opening of the glottis with elevated P_{pl} [41]. The time dependence assumed for this purpose is given by:

$$P_{pl} = \begin{cases} P_c e^{-(150(t-0.02))^2} & 0.00 \leq t < 0.02 \\ P_c & 0.02 \leq t < 0.05 \\ P_c e^{-(60(t-0.05))^2} & 0.05 \leq t < 0.10 \end{cases} \quad (19)$$

where P_c is the maximum possible pleural pressure that can be exerted at a given lung volume, corresponding to the value obtained from Eq. 18 when $t \to \infty$. The time scale of 0.02 s for the initiation of a cough was estimated from the experimental results of Leith et al. [41].

Upstream pressure in the most peripheral airways (at $\xi = 0$) is equal to P_A and is computed from $P_A = P_{pl} + P_{el}$. For the present computation, we approximated P_{el} from experimental data [1] by:

$$P_{el} = 200\,(\lambda^{2.5} - 1) \quad [N/m^2] \quad (20)$$

To simplify the numerical procedure, it is useful to have a single expression for P_{ext} in Eq. 13 for both intra- and extrathoracic airways. Accordingly P_{pl} is modified by introducing an axial distribution yielding the following expression for P_{ext}:

$$P_{ext}(t,\lambda,\xi) = [P_{pl}(t,\lambda) + P_{el}]\,[(0.5\tanh(20\,(0.85-\xi))+0.5)] \quad (21)$$

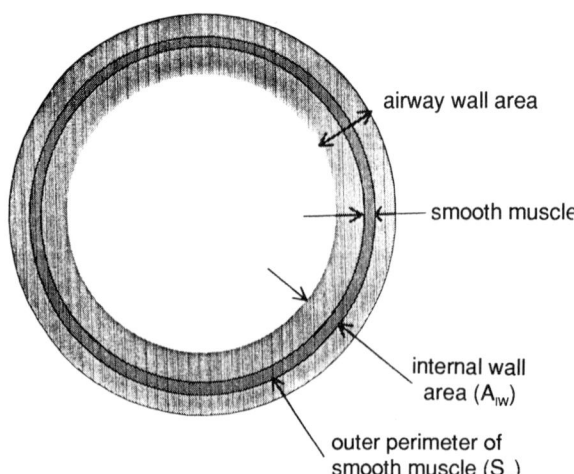

Figure 1. The cross-sectional view of a peripheral airway showing the relevant parameters in calculation of airway obstruction.

Simulation of Forced Breathing Maneuvers

This ensures that $P_{ext} \cong P_A$ in intrathoracic airways and $P_{ext} \cong 0$ for extrathoracic airways where all pressures are measured with respect to P_{atm}.

While not presented in this chapter, the present model can also be used to simulate oscillatory flow of either large or small amplitude. When oscillatory flows are generated by an external source such as a piston or loudspeaker placed at the mouth as in the measurement of respiratory impedance or the use of high frequency mechanical ventilation, the dynamics of the chest wall also need to be taken into account. This is not necessary when flows are generated by the action of respiratory muscles since these flows are excited by empirically-based variations in P_{pl} which already reflect the dynamic influence of the chest. Because high frequency flows are generally of small amplitude, the model employs a relatively simple description of chest wall effects, employing constant values for resistance, inertance, and compliance. Consistent with this approximation, the alveolar-to-body surface pressure difference is related to lung volume and its time-derivatives according to:

$$P_{bs} - P_A = R_{cw}\dot{V}(t) + I_{cw}\frac{\partial \dot{V}(t)}{\partial t} + \frac{1}{C_{cw}}\int_0^t \dot{V}(t)dt \tag{22}$$

reflecting the effects of lung and chest wall resistance (R_{cw}), inertance (I_{cw}), and compliance (C_{cw}), respectively. Following the study by Peslin et al. [53], $R_{cw} = 1.10 \times 10^{-3}$ cmH$_2$O/ml/s, $I_{cw} = 2.1 \times 10^{-6}$ cmH$_2$O/ml/s^2, and $C_{cw} = 20.8$ ml/cmH$_2$O. $\dot{V}$ is the expiratory volume flow rate; in normal conditions, pressure at the body surface, P_{bs}, equals P_{atm}. For cases of large amplitude flow, non-linear forms could be introduced with little difficulty.

Partitioning of the Airway Network

For reasons of computational economy, the airway network is partitioned into two segments, one representing the small, peripheral airways and another representing the remaining, more central airways. It is assumed that the airways in the peripheral segment change dimension as a function of lung volume alone, as described below, independent of local transmural pressure, whereas the central segment exhibits normal compliance as described by Eq. 13. The rationale for this approach, validated by comparison to simulations in which the entire network was assumed to be compliant, is that despite their greater compliance, the change in cross-sectional area of the peripheral airways resulting from changes in transmural pressure is typically small compared to the change they experience due to changes in lung volume. The transition between the central and peripheral segments is assumed at the level of the 1.6 to 1.8 mm diameter airways, between generation 8 of the Weibel model and order 16 of the Horsfield and Cumming model (see Table 1) or $\xi = 0.134$ in our notation.

The pressure at the entrance to the central airways (at generation 8) is related to the instantaneous volume flow rate, the geometry of the peripheral airway tree, and alveolar pressure. The condition used to calculate the viscous contribution to the difference between alveolar and central airway pressure is based on the same expiratory flow pressure drop measurements used in obtaining Eq. 10 [6]. When expressed in terms of a flow resistance ($\Delta P/\dot{V}$) this results in the following expression for the resistance of a single airway of generation n:

$$R_n = \begin{cases} R_p(0.556 + 0.067\sqrt{Re}) & Re \geq 55 \\ R_p & Re < 55 \end{cases} \tag{23}$$

where $R_p = 8\pi\mu L/A^2$ is the resistance of a uniform tube conveying Poiseuille flow. Defining S_n as the equivalent resistance of an airway of order n and all the airways that branch from it up to the alveoli, and T_n as the equivalent resistance of two S_n in parallel, the following expressions are obtained:

$$T_n = \frac{S_n S_{n-\delta}}{S_n + S_{n-\delta}} \tag{24}$$

$$S_n = T_{n-1} + R_n \tag{25}$$

where δ specifies the difference in order between two daughter branches; for a symmetric bifurcation, $\delta = 0$. The resistance R_n of each airway can be determined either from Weibel's symmetric model or from the Horsfield and Cumming asymmetric model.

Neglecting the contribution of temporal acceleration, the pressure difference across the peripheral airways is calculated as the sum of the contributions due to friction, as determined above, and convective acceleration. Accordingly, the pressure drop from the alveoli to an airway of order n is given by:

$$P_A - P_n = S_n \dot{V}_n + \tfrac{1}{2}\rho U_n^2 \tag{26}$$

where U_n is the velocity and $\dot{V}_n$ is the volume flow rate through a single airway of order n ($\dot{V}_n = N_{en}\dot{V}/N_T$, N_T is the total number of alveolar endings and N_{en} is the number of alveolar endings originating from a single airway of order n).

Airway Obstruction

In a recent morphometric study, Wiggs et al. [63, 64] demonstrated that airway obstruction occurs as a result of a combination of two factors (thickening of the airway wall and shortening of the smooth muscle) and reported data for the axial distribution of "internal wall area" (A_{IW}, Fig. 1) in normals and in subjects with COPD and asthma. Morphometric studies have shown that A_{IW} remains constant with respect to changes in lung volume [30, 64]. To account for airway narrowing, it is assumed that the axial distribution of airway cross-sectional area $A_0(\xi)$ for normal, non-constricted airways (Eq. 14 for the central airways and Horsfield's data for the peripheral airways) is the lumenal area in Fig. 1 and is surrounded by normal wall area, A_{IW}. The data for normal A_{IW} of Wiggs et al. are then used to calculate

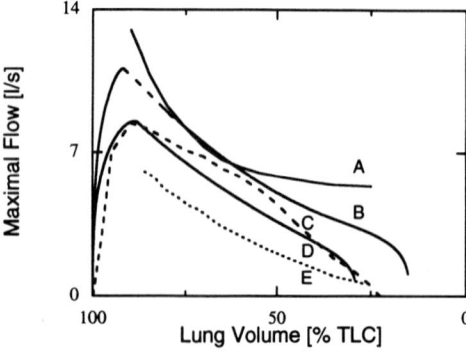

Figure 2. Maximal flow-volume curves from a forced vital capacity maneuver. Curve A: Elad et al. [12]. Curve B: present model with 0% smooth muscle shortening. Curve C: typical curve from a healthy human subject [25]. Curve D: present model with 20% smooth muscle shortening. Curve E: Lambert et al. [39].

an effective outer perimeter (S_m) of smooth muscle for each airway at TLC on the assumption that the airways are circular in cross-section.

To determine the reduction in lumenal cross-sectional area associated with changes in lung volume, different approaches are used for the central and peripheral airways. In the periphery, it is assumed that S_m varies as the cube root of lung volume, and that A_{IW} remains constant. Thus, as lung volume falls, so does S_m, and A_{IW} is forced into a smaller and smaller lumenal region. In this respect, the internal wall area can be thought of as including the airway liquid lining, except that airway closure due to meniscus formation [22, 32] is not allowed in the present approach. In the central airways for a normal lung, the original approach in [10] is employed for determining lumenal area as characterized by $A_0(\xi)$ (Eq. 14). This is based on the assumption that the changes in cross-sectional area due to both airway compliance and changes in lung volume, based as before on the measurements of Takishima et al. [58] in dogs, already account for the effects of normal wall area.

To model the diseased state, the description for the peripheral airways is modified in two ways. First, to account for increased A_{IW}, the difference between normal and diseased values is computed using the data correlations from Wiggs et al., assumed to be independent of lung volume, and subtracted from the internal cross-sectional area for each airway, thus producing additional airway obstruction at all lung volumes. Second, to simulate smooth muscle effects, the presumed degree of smooth muscle shortening (0 - 40%) is multiplied by the percentage of the perimeter containing smooth muscle [63, 64], and the smooth muscle perimeter S_m for each airway is reduced by that amount. For example, for 20% muscle constriction in a peripheral airway that contains smooth muscle over 60% of its circumference, the "constricted" perimeter is $S_m^c = [(1 - 0.2)(0.6) + (1)(1 - 0.6)]S_m$. These two effects, wall thickening and smooth muscle constriction, thereby act in concert to reduce lumenal area.

In the central airways, the difference between the normal and thickened A_{IW} is determined as a function of generation and subtracted from the lumenal cross-sectional area computed at every step in the simulation. The percent smooth muscle shortening is once again multiplied by the percentage of the circumference containing smooth muscle, and the resulting factor is used to compute a new value for A_0. For example, 20% shortening of smooth muscle in the trachea (where smooth muscle is found around 33% of its circumference [63]) will result in $A_0^c = [(1 - 0.2)(0.33) + (1)(1 - 0.33)]^2 A_0$. As in the peripheral airways, wall thickening and smooth muscle shortening have the effect of forcing more wall tissue into a smaller central lumen, and through their interaction, accentuating the extent of airway obstruction. For lack of experimental data, no account is taken of the change in airway stiffness that this might produce.

Numerical Procedures

In the numerical simulation, the calculation is divided into several distinct elements involving (i) chest wall mechanics and lung elastic recoil, (ii) the peripheral airways, (iii) the distributed central airways, and (iv) the upper airways including the glottis. Items (i) and (ii) are used to determine the peripheral boundary condition for the central airway calculation. The peripheral airways resistance is computed either for symmetric or asymmetric networks as described above. The upper airway resistance is accounted for using Eq. 12 to relate airway pressure at the trachea to mouth pressure. Flow in the central airway network (item iii) is represented by a set of partial differential equations that are solved by a standard finite difference method using the two-step MacCormak scheme as described in [35].

The boundary conditions are handled differently for forced expiration and forced oscillation maneuvers. In a FVC maneuver or coughing, mouth pressure is assumed to equal atmospheric while alveolar pressure is determined from the lung-volume dependent function for pleural pressure (Eq. 18) and elastic recoil pressure (Eq. 20). Flow rate is computed from the velocity and cross-sectional area at the downstream end of the central airways (tracheal outlet). Using the expressions for upper and peripheral airway resistances, one can determine the upstream and downstream boundary conditions of the compliant central airways. If one wished to simulate forced oscillation or high frequency ventilation, flow rate at the mouth could be assumed to vary sinusoidally and body-surface pressure can be taken to be atmospheric. Alveolar pressure is then determined from the expression for chest wall mechanics (Eq. 22). The computation proceeds in all other respects as in forced expiration maneuvers.

MODEL PREDICTIONS

Predicted Flow-Volume Curves in Normal Subjects

In this section, predictions of the computational models for flow limitation are presented with an aim toward elucidating the phenomena that determine the form of the flow-volume curve. A typical curve for a normal subject is given by curve C in Fig. 2. Actual

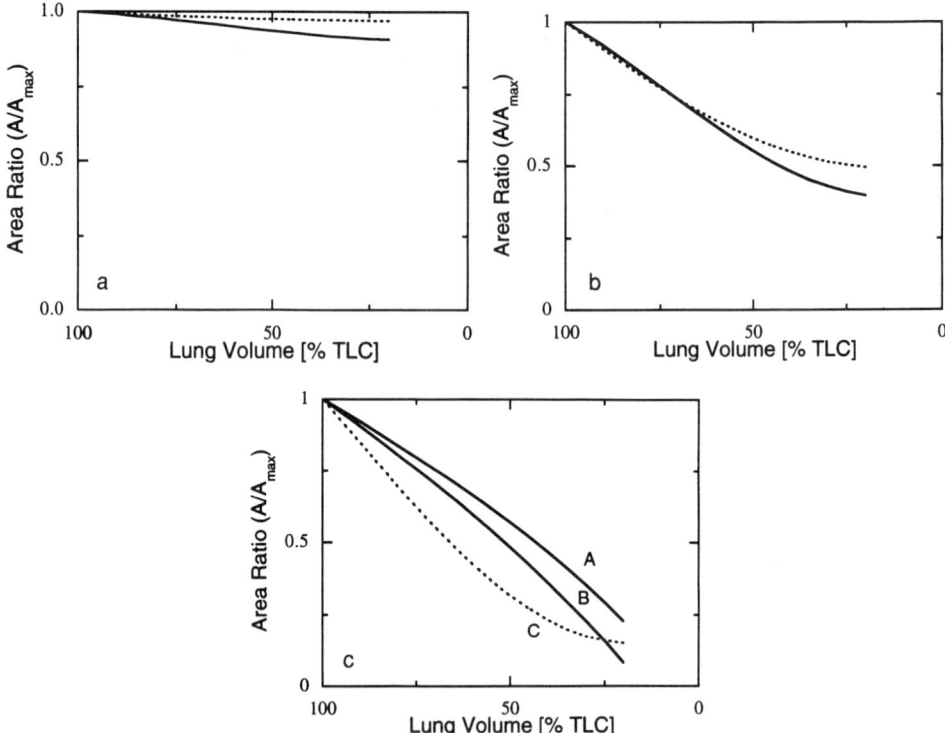

Figure 3. Airway cross-sectional area, normalized by the value at TLC, plotted as a function of lung volume during a quasi-static exhalation (e.g., $P_{aw} = P_A$). Dashed line: Lambert et al. [36]. Solid line: present model with 20% smooth muscle shortening ("normal"). (a) Trachea ($A_{max} = 2.15$ cm^2 for Lambert et al. and $A_{max} = 2.12$ cm^2 for present model). The normal case corresponds to the midpoint of the trachea. (b) Sixth generation airway ($A_{max} = 3.81$ cm^2 Lambert et al. and $A_{max} = 3.02$ cm^2 for present model). (c) Twelfth generation airway. Curve A: present model with no smooth muscle constriction ($A_{max} = 25.57$ cm^2). Curve B: normal ($A_{max} = 14.32$ cm^2). Curve C: Lambert et al. [36] ($A_{max} = 14.53$ cm^2).

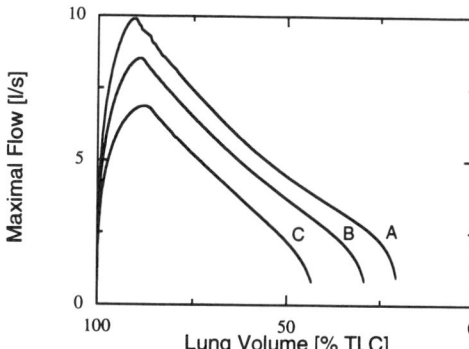

Figure 4. Flow-volume curves from a FVC maneuver showing the effect of smooth muscle constriction in the present model. Curve A: 10% shortening. Curve B: 20% shortening ("normal"). Curve C: 30% shortening.

flows can vary over quite a wide range depending on factors such as patient size, age and sex. The shape of the curve is also variable, but typically falls in a nearly linear fashion with decreasing lung volume following an early initial peak. The flow-volume curve for a given subject might exhibit one or more somewhat abrupt changes in slope or even in flow rate that are often quite repeatable [45].

Predicted flow-volume curves are shown in Fig. 2 for three models: Lambert et al. [36, 39] (curve E), Elad et al. [10, 12] (curve A), and the present model (curve B). Although the models differ in many respects, perhaps the single most important factor contributing to the shape of the curve has to do with the way in which airway cross-sectional area depends upon transmural pressure and lung volume; the "interdependence effect". In what follows, these differences will be analyzed in some detail.

In the previous model of Lambert et al. [36, 39], measurements of the pressure-area relationships from excised human central airways for positive transmural pressures [26] were used. These data were then extrapolated, both to negative transmural pressure and out into the small airways and adjusted so that an optimal match was obtained between the model predictions and actual measurements. This model produced the flow-volume curve labeled E in Fig. 2 and was subsequently used to explore a wide variety of related phenomena. This compares favorably to the typical human curve (curve C in Fig. 2.). While this model was capable of reproducing a wide variety of experimental observations and led to considerable insight into the mechanisms of flow limitation, its predictive capabilities were limited by the lack of complete mechanical data on small airways in normal subjects, and the complete absence of such data for diseased airways.

In the earlier models by Elad et al. [10, 12], an attempt was made to rely on independent measurements for all of the critical parameters. To this end, the data of Takishima et al. [58] were used to obtain the complete pressure-area law including the range of negative transmural pressure. The data of Martin and Proctor [43] provided an estimate for the difference in compliance between the large and small airways. These data, however, provided no guidance as to the *shape* of the stiffness distribution along the airway tree and still gave only a rough measure of the mechanical properties of the smallest airways. As seen in Fig. 2, curve A, this approach led to predicted flow rates at low lung volume which were markedly higher than observed and which never approached zero, presumably due in part to an underestimation of the increase in small airways resistance with reducing lung volume.

Since both of these models lacked unsteady acceleration and a coupling between lung volume and expired gas volume, an entire flow-volume curve could only be constructed by a sequence of quasi-steady simulations, each at a different fixed lung volume. Furthermore, in neither case could one be fully confident that the characteristics of small airways were accurately portrayed and, as a result, predictions from such models on the influence of small airways are highly speculative. This is particularly unfortunate since the FVC maneuver is most useful in the diagnosis of various obstructive diseases, many of which primarily affect the small airways. Consequently, the link between the primary pathology and the macroscopic observations had not effectively been made.

The primary impetus for the present study was the development of a model for the simulation of an entire forced expiration which also incorporates recent morphometric data to produce a more realistic representation of small airway behavior than was possible in earlier models. In addition, the present model, as described in detail above, introduces temporal acceleration and couples expiratory flow to changes in lung volume, thereby producing a complete flow-volume curve (shown in Fig. 2, curve B) in a single calculation. The new approach for treating small airways resistance reduces the flow rate at low V_L (compared to the earlier models by Elad et al.) although the relatively abrupt change in slope produced by the model near RV is not typically seen in human flow-volume curves. (Note that the dashed line just following the peak is a period of some numerical oscillation associated with the establishment of the FLS.) The modified approach for the small airways is essentially based on two assumptions: (i) in the periphery, airway cross-sectional area is predominantly a function of lung volume rather than transmural pressure due to the relatively small pressure drops that occur across these airways and (ii) as lung volume changes, there exists some dimension in the airway wall which varies as $V_L^{1/3}$; that dimension is assumed to be the diameter to the outside of the smooth muscle layer. These assumptions, when taken together, circumvent the need for *mechanical* measurements on small airways in that morphometric data provide all the needed information. They lead, however, to complete neglect of changes in small airway caliber associated with the upstream pressure drop which inevitably produces errors which grow with a rise in peripheral resistance. While not an ideal solution, this was felt to be both an expeditious and reasonable approach in view of the limited data currently available.

Effects of Airway Cross-Sectional Area and Smooth Muscle Constriction

These modifications improve the shape of the flow-volume curve, but discrepancies are still apparent at low V_L consistent with an underestimation of small airways resistance. One potential explanation for these discrepancies is revealed when cross-sectional area is plotted as a function of lung volume for several representative airways for the Lambert model which generates more realistic flow at low V_L, and the present model (Fig. 3). In the central airways (Figs. 3a, 3b), the present model exhibits a somewhat more rapid reduction in area as V_L falls but the differences are small. In the periphery, however, as shown here for the twelfth generation (Fig. 3c), the relationships used in Lambert's model predict consistently smaller cross-sectional areas at any given lung volume (e.g., ~ 50% at V_L = 50% TLC).

That these differences are important can be seen by a simple "experiment". Results more consistent with Lambert's can be obtained if the smooth muscle in the present model is allowed to constrict. In Fig. 4, the effect of 10%, 20%, and 30% shortening can been seen, compared to the original curve from Fig. 2 (0 % shortening, curve B). A prediction more consistent with the "typical" flow-volume curve seen in normal subjects results from about a 20% constriction producing peak flows of about 8.5 l/s and residual volume (RV) of about 25% TLC (see also Fig. 2, curve D). At the same time, the pressure-area curve in generation twelve which is also shown in Fig. 3c (curve B) is seen to be closer to that of Lambert.

Why is it that typical flow volume curves from normal subjects are obtained when airway dimensions are reduced by an amount consistent with approximately a 20% constriction in airway smooth muscle? One obvious possibility is that airway smooth muscle *does* exhibit a degree of basal tone sufficient to produce significant airway narrowing. Some recent experiments (Mitzner, personal communication) in which airways of various species have

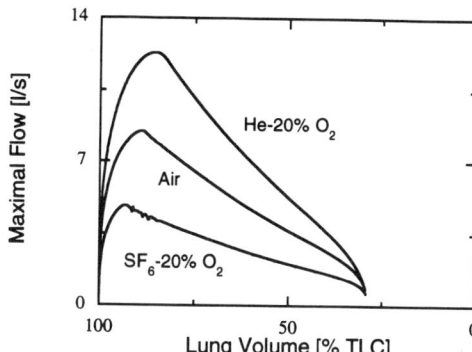

Figure 5. Flow-volume curves from a FVC maneuver showing the effect of changing gases. Gas properties given in text.

been imaged using high resolution CT show that airway cross-sectional area does in fact increase when a smooth muscle relaxant such as atropine is introduced. For example, increases in airway diameter on the order of 30% or more have been observed in dogs. If basal tone of comparable magnitude exists in humans, this could explain the need for smooth muscle constriction to produce more reasonable flow rates at low lung volume in the model. On the other hand, studies have shown that atropine has relatively little influence on human flow-volume curves [60].

It is therefore possible that while this model fits observation, it may do so for the wrong reasons and that basal tone does not exist to such an extent in humans. One possible alternative explanation is associated with the fact that the data for small airway dimensions are obtained from fixed tissue which produces a variable degree of shrinkage due to dehydration. This fixation artifact may have little effect on the length of the basement membrane which is used to determine the relaxed perimeter of the airway, but could lead to an underestimation of wall thickness. If, consistent with this, the thickness of the region inside the smooth muscle were increased, the effect would be similar to that of smooth muscle constriction. Second, the assumption made in the model that the resting diameter outside of the smooth muscle varies as $V_L^{1/3}$ is admittedly somewhat arbitrary and, as noted above, neglects the changes in cross-sectional area associated with the drop in airway pressure due to flow. This remains one of the greatest uncertainties in all of the models mentioned and will remain so until a rationale description of small airway mechanics is obtained, based on a combination of morphometric measurement and structural modeling and/or direct mechanical measurement.

Another alternative explanation relates to the heterogeneity that exists in all lungs and which increases with disease. Although some degree of asymmetric branching has been

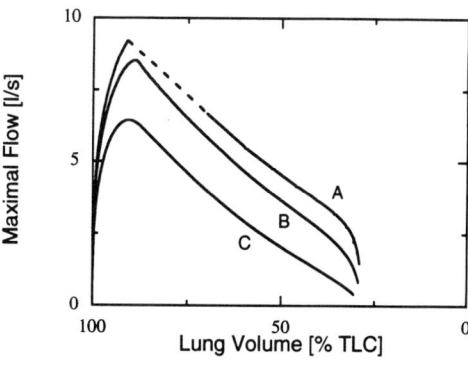

Figure 6. Flow-volume curves from a FVC maneuver showing the effect of gas viscosity alone. Curve A: normal model with gas five times less viscous than air. Curve B: normal model with air. Curve C: normal model with gas five times more viscous than air.

introduced into the present model, heterogeneity results from many factors which the present model does *not* include. As lung volume approaches RV during a forced expiration airway closure will ultimately influence expiratory flows, most likely in a highly nonhomogeneous manner. In a model of this effect, Elad and Einav [13] demonstrated significant flow reduction at low V_L. In that study, the effects of a progressive increase in the number of closed parallel peripheral airway segments was mimicked by a simultaneous reduction in the number of airways available for flow and a corresponding reduction in the total cross-sectional area. Using several hypothesized distributions of closure with V_L, it was possible to produce a reduction in flow rate as V_L approached RV which had a more realistic appearance than the prediction obtained without closure. Their model, however, mimicked this *heterogeneous* phenomenon by altering the parameters of the *homogeneous* model and was forced to rely on somewhat arbitrary estimates of the range of lung volumes over which airway closure occurred, there being no reliable data on which to base the model.

To summarize, it is fair to say that the *definitive* measurements on small airways that would allow one to distinguish between these various possibilities have not yet been made. When they are, it may well be necessary to revise the current approach. In the meantime, the present model with 20% smooth muscle constriction, while still being somewhat arbitrary in its detail, seems most consistent with existing data and least dependent on arbitrary assumptions. This will be taken as the "normal" model in all subsequent calculations and discussion.

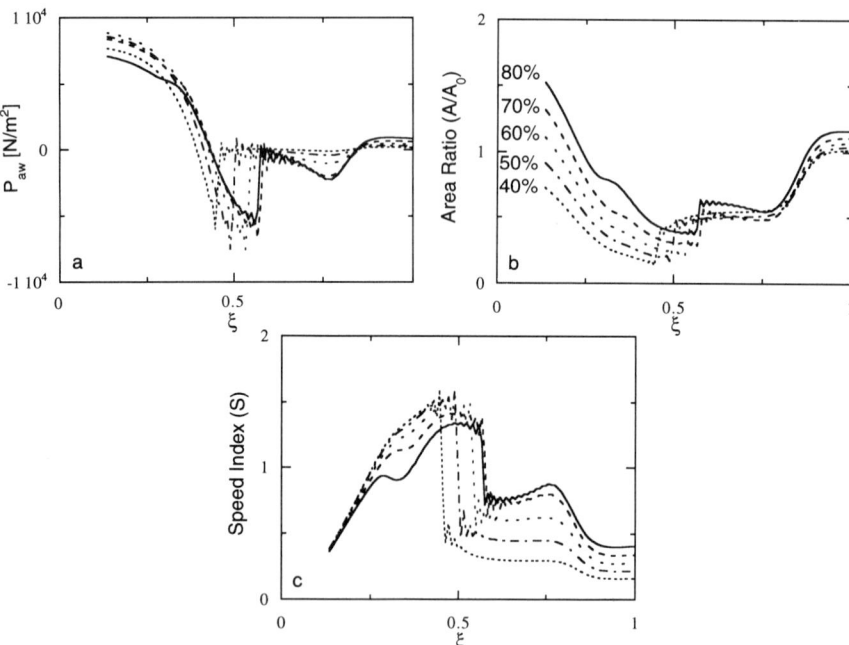

Figure 7. Axial distributions of (a) internal airway pressure, P_{aw}, (b) total airway cross-sectional area, normalized by A_0, (c) speed index, $S = U/c$. All curves begin at the upstream end of the central airway segment, defined as the eighth generation in the Weibel model (see Table 1). Each curve corresponds to a different lung volume as identified in (b). The abrupt increase in P_{aw} and A/A_0, and decrease in speed index S marks the location of the elastic jump. Note that the trachea extends from $\xi = 0.55$ to $\xi = 1$.

Effects of Changes in Gas Physical Properties

Gases other than air have been used for the purpose of investigating the phenomena responsible for flow limitation, and to distinguish in the clinical setting between peripheral and central airway obstruction. The rationale for this clinical application is the belief that less dense gases will produce marked changes in the flow-volume curve of an individual with central airway obstruction but will exert much less of an effect in the presence of peripheral obstruction. Even normal subjects, however, exhibit considerable variability in their response to breathing He-20% O_2 suggesting that the actual situation is quite complex.

Models of forced expiration have been used to examine a variety of observations made during maximal flow experiments in human or animal studies. For example, it had been observed that gas physical properties influence flow-volume curves in a number of ways. The most obvious influence, and the best documented, is the effect of gas density on maximal flow [56]. The primary cause of this effect resides in the role of density in establishing the speed at which long waves propagate along a compliant tube. Wave speed can be expressed as [55]:

$$c = \left(\frac{A}{\rho} \frac{d(P_{aw} - P_{ext})}{dA} \right)^{\frac{1}{2}} \qquad (27)$$

where the dispersive effects of wall mass and viscoelasticity have been neglected. Due to the inverse square root dependence of wave speed on gas density, maximal flows are higher for light gases than dense ones [56], an effect which is readily captured by the models of both Lambert [36, 39] and Elad et al. [10]. In addition, there is a tendency for the effect of density to diminish as lung volume falls which only Lambert reproduces. This has been explained by a combination of an increasing influence of small airways resistance which is dominated by the effects of gas viscosity, and the tendency of the flow to become unchoked at low lung volumes [42].

Another influence of gas properties on maximal flow enters via the pressure drop upstream of the FLS. This pressure drop primarily acts to influence the cross-sectional area at the site of flow limitation. A greater pressure drop causes the cross-sectional area to be smaller, thereby reducing the flow rate at the FLS (which equals the product of wave speed and local cross-sectional area). The influence of density is manifested by the generation and subsequent dissipation of turbulence and secondary flows, and the blunting of velocity profiles due to boundary layer phenomena and convergent flow effects. Lambert [38] shows that constriction of small airways reduces density dependence whereas constriction of larger airways increases it. Gas viscosity enters directly via an increase in wall shear stress at a given flow rate. Increasing viscosity has been shown to reduce maximal flows, having the greatest (fractional) influence at low lung volumes when the frictional resistance of the small peripheral airways becomes relatively more important.

The "normal" model just described can also be used to examine the effects of gas density and viscosity. In the first example (Fig. 5), two gas mixtures which have been used in humans, He-20% O_2 ($\rho = 0.4$ kg/m^3, $\nu = 0.53 \times 10^{-4}$ m^2/s) and SF_6-20% O_2 ($\rho = 5.0$ kg/m^3, $\nu = 0.03 \times 10^{-4}$ m^2/s) are compared to results with air. The trends exhibited — a large difference at maximal flow which diminishes as V_L falls — are qualitatively and quantitatively consistent with previous observations [56, 68]. For example, the predicted increase in flow rate at 50% vital capacity is 48% with He-20% O_2 which compares favorably with measurements of 57% [56] and 41% [5] from the literature.

In the results of Fig. 5, it is somewhat difficult to separate the effects due to density change from those associated with a change in viscosity since both properties vary between gas mixtures. To isolate the effect of gas viscosity, simulations have been performed with viscosity increased and decreased by a factor of five (Fig. 6). Compared to the previous example, the change in peak flow is reduced while the differences at low V_L are accentuated, reflecting the increased contribution of small airways resistance (largely a viscous phenomenon) as lung volume is reduced. In particular it is interesting to note the relatively small effect produced by a five-fold reduction in viscosity, an indication of the dominant influence of density-dependent flow phenomena in the normal lung, at least at the higher lung volumes.

Gas compressibility might also be expected to influence expiratory flow rates, especially in view of the fact that the highest air flow speeds predicted by this model are approaching a Mach number of 0.3 [11]. As one might expect, compressibility begins to play a role at these speeds, but the effect on volume flow rate is relatively minor as demonstrated by Elad et al., [11] who examined the combined influence of airway compliance and gas compressibility. Larger effects are observed in terms of the location of the FLS, but these are apparently not sufficient to alter the overall appearance of the flow-volume curve.

Spatial Distributions

Further insight can be gained by examining the details of the solution more closely, using again the parameters of the "normal" model. Fig. 7 shows the distributions of area ratio (A/A_0), internal airway pressure (P_{aw}) and speed index ($S = U/c$) at several lung volumes. Note that these distributions are restricted to the central airways (generations 0 - 8), and exclude the periphery where the airways are treated as "rigid" in the sense that their cross-sectional areas are determined solely by lung volume and the properties of the airway wall. The region contained in these central airways does however represent 87% of the total length of the airway tree.

Airway pressure, P_{aw} (Fig. 7a), is seen to fall from a maximum positive value at the peripheral end, to a negative value of comparable magnitude before rather abruptly attaining a value close to ambient pressure where it remains throughout the trachea. In general, the magnitudes, both positive and negative, fall as the lung empties, consistent with the fall in driving pressure (P_A). The area ratio (Fig. 7b) reflects these changes in P_{aw}, except for the rise at ξ greater than about 0.8 which corresponds to a reduction in external pressure from the intrathoracic to extrathoracic trachea. The *shape* of the curves for P_{aw} and A/A_0 can be best understood in context of the distribution of speed index, S. Flow in the periphery begins subcritical, then accelerates to supercritical speed at a location $\xi \cong 0.4$ (in the first generation near the main carina) at high lung volumes, then moves toward the periphery, settling at a position around $\xi = 0.25$ (the fourth generation) where it remains for most of the remainder of expiration. The region of supercritical flow which is also a region of high collapse, terminates in an elastic jump (an abrupt reduction in S from supercritical to subcritical speeds analogous to a shock wave in gas dynamics) which begins in the trachea, then moves upstream (toward the periphery) as lung volume falls. Associated with this elastic jump is an abrupt change in all variables, notably an increase in A/A_0 to a somewhat less collapsed configuration.

These results are qualitatively consistent with experimental observations, primarily in dogs, but differ in some respects. It has been observed that the FLS first forms in the vicinity of the trachea, but then migrates upstream as expiration proceeds [48]. The simulation is roughly consistent with this pattern, although the FLS appears to remain at about the fourth generation over much of expiration with no obvious tendency to continue its movement toward periphery as lung volume approaches RV. In this connection, it is important to note that the FLS will be located where the product of cross-sectional area and

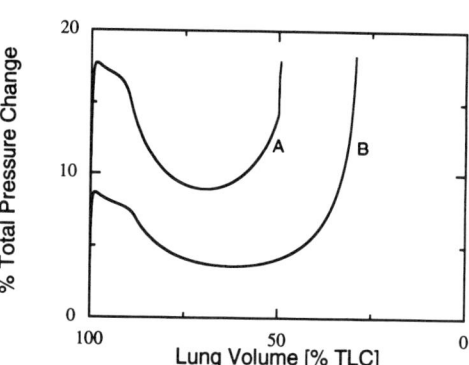

Figure 8. Percentage of the total (alveolar-to-mouth) pressure drop residing in the peripheral airways (upstream of generation eight). Curve A: asthma, 20% smooth muscle shortening. Curve B: normal. The high resistance near TLC occurs during formation of the FLS. The increase at low V_L corresponds to the narrowing, and eventual closure, of the small airways.

wavespeed (Ac) attains a minimum. To a first approximation, Ac varies as $A_0(K_p)^{1/2}$. In the model, the minimum in $A_0(K_p)^{1/2}$ remains fixed at a location between $\xi = 0.2$ and 0.25. In the real lung, the stiffness of extra-parenchymal airways would be less sensitive to changes in lung volume than that of airways subjected to parenchymal tethering; one would therefore anticipate that the decrement in K_P with falling V_L would be greatest in the periphery and least in the trachea. This would cause the choke point to move peripherally during expiration and for the flow to subsequently fall more rapidly. The ξ-dependent terms in Eqs. 14 and 16 are an attempt to capture this effect. What is lacking, however, are the anatomical or mechanical data needed to establish the correct form of the spatial variation.

It has also been argued that airway geometry, dimensions, and mechanical properties, rather than being smooth and monotonic functions of distance, are strongly influenced by factors such as the local mechanics in the vicinity of a bifurcation, the cartilage plates present in the central airways, and variations in airway wall thickness. One consequence of this is that parameters such as airway cross-sectional area and airway stiffness are likely to be spatially non-uniform. For example, one might expect each bifurcation to contain a local minimum in stiffness. The implications of this may be quite important in that such a minimum can serve as a site where the FLS can become "frozen" for some range of V_L, preventing it from migrating smoothly toward the lung periphery. It has also been shown [12] that if several weak points exist in series along a flow path, flow might pass through S = 1 several times. If several such sites exist, the possibility arises that as the lung empties, the different sites might compete to determine the rate of emptying from that portion of the lung located upstream from it. If the peripheral airways become relatively more compliant as lung volume falls, as mentioned above, the choke point might therefore travel toward the periphery in a discrete fashion jumping from one location to another as suggested in [44]. Furthermore, as

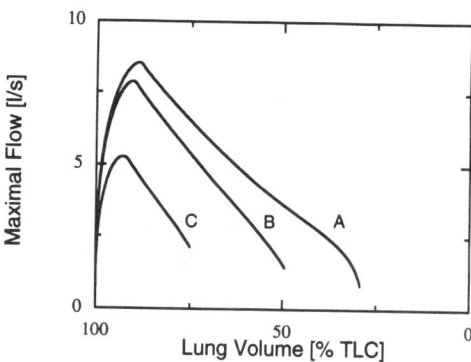

Figure 9. Flow-volume curves from a FVC maneuver simulating asthma by a thickening of the airway wall as measured in [64]. Curve A: normal. Curve B: 20% shortening. Curve C: 30% shortening.

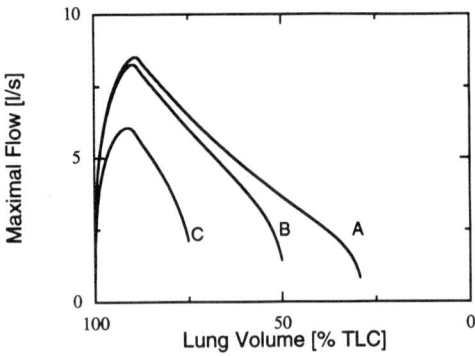

Figure 10. Flow-volume curves from a FVC maneuver showing simulating COPD by a thickening of the airway wall as measured in [64]. Curve A: normal. Curve B: 20% shortening. Curve C: 30% shortening.

the site of flow limitation moves upstream, the number of parallel pathways increases and more choke points are consequently necessary to produce choked flow.

Recognizing the potential importance of parallel inhomogeneities on flow phenomena, Lambert [40] constructed a two-compartment model in which he could examine the influence of non-uniform peripheral compliances or resistances. Even this simple model was capable of predicting differences in alveolar pressures in approximate agreement with experimental measurement [17]. This model was used to demonstrate that such differences would increase when gas viscosity was increased or gas density decreased.

A second feature of these results is the relatively uncollapsed configuration of the trachea, most of which lies downstream of the elastic jump. This feature too is strongly influenced by the axial distribution of stiffness used in the model. Experience with a variety of different distributions (see e.g., Elad et al. [12]) led to the conclusion that while the location of the elastic jump can vary considerably, the effect this has on the flow-volume curve is minimal. This would be expected since, once flow limitation is established, the solution upstream of the FLS becomes decoupled from downstream events and simply adjusts to satisfy the externally imposed boundary conditions.

Unsteadiness

While the discussion of the previous section is consistent with steady collapsible tube theory as well as with in vitro experiments in water [34], it fails to recognize an important but often overlooked feature of a forced expiration or cough. The flow-volume curves as normally represented and used clinically are low-pass filtered with a cut-off frequency as low as about 20 Hz. Unfiltered, these traces exhibit oscillations of large amplitude [62]. In recent experiments flowing air through a tube which, like the lung, becomes less compliant in the direction of flow [33], the tube exhibited flutter oscillations under conditions in which flow limitation would have been expected. These oscillations were similar in character to those observed in other experiments in compliant tubes [3, 19]. The observation of oscillation coincident with flow limitation has also been observed in humans [18]. The relatively good agreement between predictions of existing models which lack the capability of capturing these flow-induced instabilities, and the vast majority of phenomena seen in human or animal lungs suggests that oscillations per se might not be a critical factor in determining the time-mean maximal flow. This would be true, for example, if the oscillations were confined to the region downstream of the FLS and only occurred when the flow was limited. In that case, the oscillations would only need to be considered when examining phenomena in the central airways, at locations downstream of the FLS. At this point, it is unclear to what extent flow-induced oscillations influence the shape of the flow-volume curve or the observed

Simulation of Forced Breathing Maneuvers

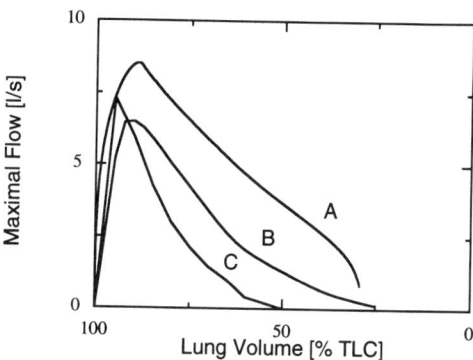

Figure 11. Flow-volume curves from a FVC maneuver showing the effect of parallel heterogeneity in the degree of smooth muscle constriction. Curve A: normal. Curve B: with 50% of the lung experiencing 30% shortening, 30% of the lung with 20% shortening, 10% of the lung with 10% shortening and 10% of the lung with no constriction. The resulting curve was smoothed to eliminate the discontinuities in slope associated with the discrete nature of the simulation. Curve C: a typical asthmatic curve.

distributions of cross-sectional area and velocity. Their inclusion in a model of forced expiration would require not only a simulation of the effects of wall acceleration and viscoelasticity, but perhaps a full three-dimensional solution of the Navier-Stokes equations as well. It is clear, however, that further studies are needed to clarify our understanding of these effects that raise serious questions about the applicability of the present formulation of wave speed flow limitation theory in the presence of potentially large amplitude oscillations.

Partitioning of Flow Resistance

It is also informative to consider the separate contributions of the peripheral and central airways to total flow resistance to determine which region of the lung exerts the dominant influence. For the purpose of presentation, the division is taken to be at generation eight in the Weibel model. As shown in Fig. 8, at high to intermediate V_L, the central airways account for most of the alveolar-to-mouth pressure drop. As lung volume falls, the contribution of the peripheral airways becomes increasingly significant, comprising a fraction approaching 20% of the total at a lung volume of 30% TLC. The calculation is terminated at this point since the assumption of this analysis that peripheral airway dimensions are unaffected by changes in transmural pressure becomes less valid. Although a small fraction of the total, the resistance of these airways exerts considerable influence over the shape and

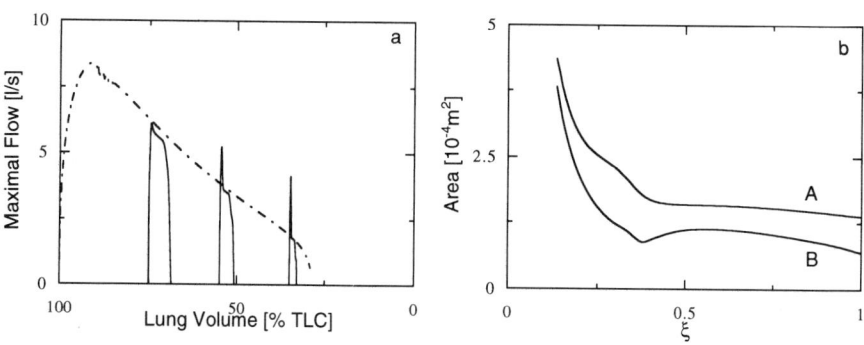

Figure 12. Simulations of coughing. (a) Flow traces from cough, initiated at three different lung volumes, superimposed on a normal flow-volume curve. Flows above the flow-volume envelope are termed "supramaximal" flows. (b) Distributions of cross-sectional area before (curve A) and during (curve B) a cough at 75% TLC.

magnitude of the flow-volume curve. This will become evident in the context of disease simulations discussed next.

Lungs with Obstructive Disease

Most of the discussion thus far has focused on gaining a better understanding of normal, healthy lungs. While some previous studies have addressed the effect on flow of various perturbations thought to mimic diseased conditions (e.g. [12, 13, 36]), these were largely hypothetical and based on reasoned speculation rather than on direct measurement or fundamental principles. Recently, however, quantitative anatomical data from obstructed lungs have become available [64] that allow for realistic simulation and study of the fundamental relationship between airway pathology and macroscopically observed flows. These data from subjects with asthma and COPD have been employed in a model of normal breathing at fixed lung volume [61] for the purpose of studying the relationship between airway wall thickening, smooth muscle constriction, and elevated airways resistance.

One new feature of the present model is its capability to simulate the pathologic effects of airway wall thickening and smooth muscle constriction, the results of which are illustrated in Fig. 9 for asthma and Fig. 10 for COPD. These results incorporate airway wall thickening observed morphologically; the curves shown correspond to various degrees of smooth muscle shortening, up to a maximum of 30%. Using the dimensions of the thickened asthmatic airway wall and 20% smooth muscle shortening produces curve B in Fig. 9 with a peak flow approaching 8 l/s and residual volume of approximately 40% TLC. Allowing for 30% shortening of the smooth muscle, the peak flow drops to about 5 l/s and RV rises to about 70% TLC. Maximal shortening, which is thought to be in the range of 40% [64] leads to such a degree of narrowing that the calculations become unstable immediately due to the high degree of peripheral obstruction. Asthma also produces a shift in the distribution of flow resistance. As seen in Fig. 8, the fractional pressure drop in the peripheral airways increases by about a factor of two even for a relatively minor degree of smooth muscle constriction in the asthma model.

Similar results are obtained with simulations of COPD as shown in Fig. 10 although the amount of flow reduction for a given degree of smooth muscle constriction is somewhat smaller than with asthma because the increase in wall thickness is not as great. In addition, the shape of the flow-volume curve takes on a different appearance, becoming slightly concave downward.

The reductions in maximal flow and the tendency for RV to rise are generally consistent with clinical observation, as is the amount of smooth muscle constriction needed to produce such changes. However, in addition to a reduction in peak flow and an increase in RV, features that *are* reproduced in the model, the flow volume curves obtained in patients with either disease also exhibit a shape that is distinctly convex upward as shown in Fig. 11, curve C. In none of the simulations of asthma or COPD was there any indication of such a pattern. In fact, the tendency in COPD was, if anything, the reverse. There are at least two possible explanations for this effect. One pertains to the pressure-area law itself. The airway wall thickening in asthma or COPD might also be expected to increase airway stiffness, especially in the periphery. A stiffening of the airways at low lung volume would tend to reduce the rate at which the maximal flow rate falls, producing the observed concave-upwards appearance.

It is also known that the degree of heterogeneity increases with diseases such as asthma and COPD. One of the most common complications in obstructive disease is a reduction in the partial pressure of oxygen in the blood resulting from a mismatch between the distributions of ventilation and perfusion. Airway constriction itself is highly nonuniform as has been observed in animal experiments [16], and such heterogeneities have long been recognized as potentially contributing to the changes seen in the flow-volume curves in obstructive disease [44]. None of these effects can be reproduced with the present model, however, which allows only for asymmetric branching in the periphery of an otherwise homogeneous and symmetric lung. To simulate these effects would require a truly nonhomogeneous model capable of capturing the real influence of inhomogeneity on ventilation and, in particular, on the shape of the flow-volume curve. This remains one of the most critical features lacking in the current disease model.

The effect of heterogeneity can be crudely represented with the present model, however, by superimposing the flow-volume curves with different degrees of smooth muscle constriction and weighting the curves according to the fraction of lung assumed to experience a given amount of constriction. The example shown in Fig. 11 was obtained by this procedure, merely by dividing the lung into four regions with different degrees of constriction. This results in a curve with a steep negative slope at high V_L during the time that all regions of the lung are emptying, which decreases at lower V_L when the contributions from the constricted regions diminish. (Such a superposition is possible only when the flow limiting sites are upstream of the trachea, as they are in this case.)

Coughing

A cough was simulated at various lung volumes by means of a transient increase in pleural pressure as described by Eq. 19. As seen in Fig. 12a in which three coughs are superimposed on the normal flow volume curve, the peak in flow rate exceeds the maximal flow envelope (supra-maximal flow) only at the two lower lung volumes, by an amount as great as 2.5 l/s. On examination of the cross-sectional area traces before and during cough (Fig. 12b for cough at V_L=75% TLC), coughing is seen to produce significant transient collapse of the central airways in response to the sudden fall in transmural pressure. The simulation results also show that at no time during the cough does flow rate at the choke point exceed the level attained during a FVC maneuver at the same lung volume. Therefore, the supra-maximal flow produced in the simulation is caused entirely by the volume expelled from the collapse of airways mouthward of the FLS, contrary to the conclusion reached by Pedersen et al. [47]. This initial transient gives rise to a short period of highly damped oscillations before settling down to the flow-volume envelope. As pressure is released near the end of the cough, the flow remains choked while the central airways reinflate, causing the initial drop in flow rate measured at the mouth; the second, more dramatic fall in flow occurs when the pressure falls by an amount sufficient to cause the flow to unchoke and, in the absence of continued compression, return to zero.

Simulations of coughing produce the anticipated airway narrowing, even at relatively peripheral locations. However, the degree of tracheal collapse is less than might be expected (~25%) compared to what is normally observed. Several reasons why tracheal collapse may be underestimated in these simulations were given earlier in connection with Fig. 7. Recall that the degree of collapse in the trachea is largely influenced by the location at which the flow becomes sub-critical once again, which in the simulation occurs at the site of the elastic jump. Unlike nearly all other aspects of the calculation, this location is quite sensitive to the presumed *shape* of the K_P distribution about which we know very little. Furthermore, cough is an explosive maneuver in which airway collapse occurs in a small fraction of a second. Consequently, it is quite likely that effects associated with airway wall mass and perhaps wall viscoelasticity come into play. These could generate even greater and more prolonged supra-maximal flows as suggested by a comparison of the present prediction to actual measurement (see Fig. 5, P324 [41]).

SUMMARY AND FUTURE DIRECTIONS

Many advances have been made over the years in the simulation of forced breathing such as the FVC maneuver and coughing. These have led to a more fundamental appreciation of the physical mechanisms responsible for determining maximal flow and, in particular, the critical role of small airways. In this chapter, previous flow models were reviewed and a new model was introduced motivated by the need to elucidate the connection between alterations in the small airways and changes apparent in the measured flow-volume curves with obstructive disease. Using data on airway compliance obtained from dogs by Takishima and co-workers [58], morphometric models of Weibel [62] (for the central airways) and Horsfield and Cumming [23, 24] (small airways), and studies of airway wall area in healthy and diseased lungs of Wiggs et al. [63, 64], a more realistic representation of maximal flow at low lung volumes has been obtained. This improvement is due primarily to the introduction of an incompressible airway wall region that impinges upon the cross-sectional area available for flow, thereby reducing flow rates, especially at low lung volume. This effect is further accentuated by imposing a degree of basal tone in airway smooth muscle.

The simulations of asthma and COPD cause the anticipated reduction in flow and increase in RV, but fail to generate the concavity characteristic of disease. Several factors are discussed that might contribute to this apparent discrepancy including: heterogeneity of the disease process, an increase in small airway stiffness due to wall thickening, and a greater influence of distributed airway closure. These effects are almost certain to depress the low volume portion of the flow-volume curve, but at present, they are not sufficiently well understood to incorporate into the model with reasonable confidence.

The model, while improved over previous ones, still has a number of limitations that affect its realism and its predictive capabilities. Some of these are limitations in the anatomical and mechanical data needed to characterize the system; others are due to the simplifying assumptions used to generate a tractable model. Current models continue to be limited by the lack of detailed, quantitative information concerning the spatially-varying compliance of the airway network as a function of lung volume. In particular, the assumed distribution of K_p is arbitrary. Fortunately it exerts relatively little effect on the main output variable, but does influence other important factors such as the location of the choke point and its migration toward the periphery as lung volume falls, and the shape of the flow-volume curve as discussed extensively above.

Although we are now more confident of the distribution of cross-sectional area due to the new data from Wiggs and co-workers, we still lack a precise morphometric description of the small airways as a function of lung volume, especially in disease. Techniques are now becoming available, however, such as high resolution CT, that provide a means for measuring the size of human airways and provide critical data on the roles of airway wall remodeling and smooth muscle constriction in obstructive disease. In addition, to accurately predict changes in airway cross-sectional area requires a fundamentally-based description of airway wall mechanics which is not currently available. Airway closure which is now under intense study by several groups (see [20] for a recent review) needs to be better understood so that it too can be modeled more accurately. At present, we lack the data or understanding to simulate these effects in a realistic manner.

A second category of model limitations pertains to the simplifying assumptions used. In our view, one of the most significant of these is the assumption of branching and flow symmetry in the central airways. While some effects were examined in a two-compartment model by Lambert [40], a simulation of real lung behavior must allow for an even greater degree of heterogeneity. The model described in this chapter incorporates geometrical asymmetries in the small airways (peripheral to the 1.8 mm diameter airways), but this needs to be expanded to include the entire trachea-bronchial tree, especially to examine phenomena such as heterogeneous emptying.

Now that the model is capable of dealing with intrinsically unsteady flows, the associated effects of wall mass and viscoelasticity should also be included. While this is not technically difficult, there remains the issue once again of obtaining realistic values for the relevant parameters from experimental measurements. Recently, Aljuri et al. [2] have shown that the trachea exhibits significant viscoelastic behavior. They proposed that the effective increase in stiffness associated with rapid changes in cross-sectional area could cause the peak flow during a forced expiration or cough to be transiently higher than that predicted by models that fail to take wall viscoelasticity into account. This could further contribute to the appearance of supra-maximal flows during a cough. Lacking these effects, supra-maximal flows can only result from transient airway collapse downstream of the flow limiting site as noted above.

On a related point, these calculations predict the existence of a stable supercritical region terminated by an elastic jump. This description fails to take into account the large amplitude oscillations observed in expiratory flow coincident with airway flutter. This unsteadiness raises serious questions concerning even the fundamental validity of the present one-dimensional flow analysis in such cases and must certainly be addressed in future studies.

ACKNOWLEDGMENTS

The support of the National Heart, Lung, and Blood Institute (HL-33009) and the Freeman Foundation is gratefully acknowledged. In addition, we would like to thank Dr. Wayne Mitzner for providing us with his most recent unpublished measurements on airway dimension using high resolution CT and Prof. Ted Wilson for his comments on the manuscript.

REFERENCES

1. Agostoni, E., and Hyatt, R. E., 1986, Static behavior of the respiratory system. In: *Handbook of Physiology. The respiratory system. Mechanics of Breathing.* Bethesda, MD: Am. Physiol. Soc., sect. 3, vol. III, pt. 1, chapt. 7, p. 113-130.
2. Aljuri, N., Freitag, L., and Venegas, J., 1994, Non-linear dynamic properties of excised human tracheas, *Am. J. Respirat. Crit. Care Med.* 149:A74.
3. Bertram, C. D., Raymond, C. J., and Pedley, T. J., 1991, Application of nonlinear dynamics concepts to the analysis of self-excited oscillations of a collapsible tube conveying a fluid, *J. Fluids and Structures* 5:391-426.
4. Bonis, M., and Ribreau, C., 1977, Pressure-flow relationships in collapsible tubes, *Cardiovas. Pulmon. Dyn.* 92:459-466.
5. Castile, R. G., Hyatt, R. E., and Rodarte, J. R., 1980, Determinants of maximal expiratory flow and density dependence in normal humans, *J. Appl. Physiol.* 49:897-904.
6. Collins, J. M., Shapiro, A. H., Kimmel, E., and Kamm, R. D., 1993, The steady expiratory pressure-flow relation in a model pulmonary bifurcation, *J. Biomech. Eng.* 115:299-305.
7. Dawson, S. V., and Elliott, E. A., 1977, Wave-speed limitation on expiratory flow— a unifying concept, *J. Appl. Physiol.* 43:498-515.
8. Elad, D., Kamm, R. D., and Shapiro, A. H., 1987, Choking phenomena in a lung-like model, *Trans. ASME J. Biomed. Eng.* 109:1-9.
9. Elad, D., Kamm, R. D., and Shapiro, A. H., 1988, Tube law for the intrapulmonary airway, *J. Appl. Physiol.* 65:7-13.
10. Elad, D., Kamm, R. D., and Shapiro, A. H., 1988, Mathematical simulation of forced expiration, *J. Appl. Physiol.* 65:14-25.

11. Elad, D., Kamm, R. D., and Shapiro, A. H., 1989, Steady compressible flow in collapsible tubes: application to forced expiration, *J. Fluid Mech.* 203:401-418.
12. Elad, D., and Kamm, R. D., 1989, Parametric evaluation of forced expiration using a numerical model, *J. Appl. Physiol.* 11:192-199.
13. Elad, D., and Einav, S., 1989, Simulation of airway closure during forced vital capacity, *Annals of Biomed. Eng.* 17:617-631.
14. Elad, D., and Kamm, R. D., 1991, Modeling a forced expiration, *Comments Theor. Biol.* 2: 239-260.
15. Fredberg, J. J., and Hoenig, A., 1978, Mechanical response of the lungs at high frequencies, *Trans. ASME J. Biomech. Eng.* 100:57-66.
16. Fredberg, J. J., Ingram, R. H., Castile, R. G., Glass, G. M., and Drazen, J. M., 1985, Nonhomogeneity of lung response to inhaled histamine assessed with alveolar capsules, *J. Appl. Physiol.* 58:1914-1922.
17. Fredberg, J. J., Topulos, G. P., Nielan, G. J., and Glass, G. M., 1988, Inter-regional alveolar pressure differences are limited during forced deflation, *FASEB Journal*, 2:A1699.
18. Gavriely, N., Kelly, K. B., Grotberg, J. B., and Loring, S. H., 1987, Forced expiratory wheezes are a manifestation of airway flow limitation, *J. Appl. Physiol.* 62(6):2398-2403.
19. Gavriely, N., Shee, T. R., Cugell, D. W., and Grotberg, J. B., 1989, Flutter in flow-limited collapsible tubes: a mechanism for generation of wheezes, *J. Appl. Physiol.* 66(5):2251-2261.
20. Grotberg, J. B., 1994, Pulmonary flow and transport phenomena, *Ann. Rev. Fluid Mech.* 26:529-571.
21. Hardin, J. C., Yu, J. C., Patterson, J. L., and Trible, W., Jr., 1980, The pressure/flow relation in bronchial airways on expiration, In *Biofluid Mechanics*, edited by Schneck, D. J., Plenum Press, New York.
22. Halpern, D., and Grotberg, J. B., 1992, Fluid-elastic instabilities of liquid-lined flexible tubes, *J. Fluid Mech.* 244:615-632.
23. Horsfield, K., and Cumming, G., 1976, Morphology of the bronchial tree in the dog, *Respir. Physiol.* 26:173-182.
24. Horsfield, K., 1986, Morphometry of Airways. In: *Handbook of Physiology. The respiratory system. Mechanics of Breathing*. Bethesda, MD: Am. Physiol. Soc., sect. 3, vol. III, pt. 1, chapt. 7, p. 75-88.
25. Hyatt, R. E., and Flath, R. E., 1966, Relationship of air flow to pressure during maximal respiratory effort in man, *J. Appl. Physiol.* 21:477-482.
26. Hyatt, R. E., Wilson, T. A., and Bar-Yishay, E., 1980, Prediction of maximal expiratory flow in excised human lungs, *J. Appl. Physiol. : Respirat. Environ. Exercise Physiol.* 48:991-998.
27. Jackson, A. C., Butler, J. P., Millet, E. J., Hoppin, F. G., Jr., and Dawson, S. V., 1977, Airway geometry by analysis of acoustic pulse response measurements, *J. Appl. Physiol. : Respirat. Environ. Exercise Physiol.* 43:523-536.
28. Jaeger, M. J., and Matthys, H., 1989, The pattern of flow in the upper human airways, *Respir. Physiol.* 67:147-159.
29. Jaffrin, M. Y., and Kesic, P., 1974, Airway resistance: a fluid mechanical approach, *J. Appl. Physiol.* 36(3):354-361.
30. James, A. L., Pare, P. D., and Hogg, J. C., 1988, Effects of lung volume, bronchoconstriction, and cigarette smoke on morphometric airway dimensions, *J. Appl. Physiol.* 64(3):913-919.
31. Jan, D. L., Shapiro, A. H., and Kamm, R. D., 1989, Some features of oscillatory flow in a model bifurcation, *J. Appl. Physiol.* 67:147-159.
32. Johnson, M., Kamm, R. D., Ho, L. W., Shapiro, A. H., and Pedley, T. J., 1991, The nonlinear growth of surface-tension-driven instabilities of a thin annular film, *J. Fluid Mech.* 233:141-156.
33. Kamm, R. D., Patel, N., and Elad, D., 1993, On the effect of flow-induced flutter on flow rate during a forced vital capacity maneuver, *FASEB Journal 7* 3:53.
34. Kececioglu, I., McClurken, M. D., Kamm, R. D., and Shapiro, A. H., 1981, Steady, supercritical flow in collapsible tubes. Part 1. Experimental observations, *J. Fluid Mech.* 109:367-389.
35. Kimmel, E., Kamm, R. D., and Shapiro, A. H., 1988, Numerical solutions for steady and unsteady flow in a model of the pulmonary airways, *Trans ASME J. Biomech. Eng.* 110:292-299.
36. Lambert, R. K., Wilson, T. A., Hyatt, R. E., and Rodarte, J. R., 1982, A computational model for expiratory flow, *J. Appl. Physiol.* 52:44-56.
37. Lambert, R. K., 1984, Sensitivity and specificity of the computational model for maximal expiratory flow, *J. Appl. Physiol.* 57:958-970.
38. Lambert, R. K., 1986, Analysis of bronchial mechanics and density dependence of maximal expiratory flow, *J. Appl. Physiol.* 61(1):38-49.
39. Lambert, R. K., 1987, Bronchial mechanical properties and maximal expiratory flows, *J. Appl. Physiol.* 62(6):2426-2435.
40. Lambert, R. K., 1989, A new computational model for expiratory flow from nonhomogeneous human lungs, *Trans ASME J. Biomech. Eng.* 111:200-211.

41. Leith, D. E., Butler, J. P., Sneddon, S. L., and Brian, J. D., 1986, Cough. In: *Handbook of Physiology. The respiratory system. Mechanics of Breathing.* Bethesda, MD: Am. Physiol. Soc., sect. 3, vol. III, pt. 1, chapt. 7, p. 315-336.
42. Leith, D. E., and Mead, J., 1967, Mechanism determining residual volume of the lungs in normal subjects, *J. Appl. Physiol.* 23:221-226.
43. Martin, M. B., and Proctor, D. F., 1958, Pressure-volume measurements on dog bronchi, *J. Appl. Physiol.* 13:337-343.
44. Mead, J., 1978, Analysis of the configuration of maximum expiratory flow-volume curves, *J. Appl. Physiol. : Respirat. Environ. Exercise Physiol.* 44(2):156-165.
45. Mead, J., 1980, Expiratory flow limitation: a physiologist's point of view, *Federation Proc.* 39:2771-2775.
46. Oates, G. C., 1975, Fluid flow in soft walled tubes. Part 1. Steady flow, *Med. Biol. Eng. Comp.* 13:773-778.
47. Pederson, O. F., Lyager, S., and Ingram, R. H., Jr., 1985, Airway dynamics in transition between peak and maximal expiratory flow, *J. Appl. Physiol.* 59(6):1733-1746.
48. Pedersen, O. F., and Nielsen, T. M., 1977, The compliance curve for the flow limiting segments of the airway, *Acta Physiol. Scand.* 100:139-153.
49. Pedley, T. J., Schroter, R. C., and Sudlow, M. F., 1970a, Energy losses and pressure drop in models of human airways, *Respir. Physiol.* 9:371-386.
50. Pedley, T. J., Schroter, R. C., and Sudlow, M. F., 1970b, The prediction of pressure drop and variation of resistance within the human bronchial airways, *Respir. Physiol.* 9:387-405.
51. Pedley, T. J., Schroter, R. C., and Sudlow, M. F., 1977, Gas flow and mixing in the airways. In: *Bioengineering Aspects of the Lung.* J. B. West, ed. Marcel Dekker: New York, p. 163-265.
52. Pedley, T. J., and Kamm, R. D., 1991, Dynamics of gas flow and pressure-flow relationships. In: *The Lung: Scientific Foundations.* R. G. Crystal and J. B. West, eds. Raven: New York, p. 995-1010.
53. Peslin, R., and Fredberg, J. J., 1986, Oscillation mechanics of the respiratory system. In: *Handbook of Physiology. The respiratory system. Mechanics of Breathing.* Bethesda, MD: Am. Physiol. Soc., sect. 3, vol. III, pt. 1, chapt. 11, p. 145-177.
54. Rodarte, J. R., and Rehder, K., 1986, Dynamics of respiration. In: *Handbook of Physiology. The respiratory system. Mechanics of Breathing.* Bethesda, MD: Am. Physiol. Soc., sect. 3, vol. III, pt. 1, chapt. 10, p. 131-144.
55. Shapiro, A. H., 1977, Steady flow in collapsible tubes, *J. Biomech. Eng.* 99:126-147.
56. Staats, B. A., Wilson, T. A., Lai-Fook, S. J., Rodarte, J. R., and Hyatt, R. E., 1980, Viscosity and density dependence during maximal flow in man, *J. Appl. Physiol. : Respirat. Environ. Exercise Physiol.* 48(2): 313-319.
57. Stănescu, D. C., Pattijn, J., Clement, J., and Van De Woestijne, K. P., 1972, Glottis opening and airway resistance, *J. Appl. Physiol.* 32:460-466.
58. Takishima, T., Sasaki, H., and Sasaki, T., 1975, Influence of lung parenchyma on collapsibility of dog bronchi, *J. Appl. Physiol.* 38:875-881.
59. Thiriet, M., Bonis, M., Adedjouma, A. S., and Yvon, J. P., 1989, A numerical model of expired flow in a monoalveolar lung model subjected to pressure ramps, *Trans ASME J. Biomech. Eng.* 111:9-16.
60. Vincent, N. J., Knudson, R., Leith, D. E., Macklem, P. T., and Mead, J., 1970, Factors influencing pulmonary resistance, *J. Appl. Physiol.* 29:236-243.
61. Vincken, W. G., and Cosio, M. G., 1989, Flow oscillations on the flow-volume loop: clinical and physiological implications, *Eur. Respir. J.* 2:543-549.
62. Weibel, J. B., 1963, *Morphology of the Lung,* Academic Press, New York.
63. Wiggs, B. R., Moreno, R., Hogg, J. C., William, C., and Pare, P. D., 1990, A model of the mechanics of airway narrowing, *J. Appl. Physiol.* 69:849-860.
64. Wiggs, B. R., Bosken, C., Pare, P. D., James, A., and Hogg, J. C., 1992, A model of airway narrowing in asthma and in chronic obstructive pulmonary disease, *Am. Rev. Respir. Dis.* 145:1251-1258.
65. Wilson, T. A., and Hyatt R. E., 1991, Forced expiration. In: *The Lung: Scientific Foundations.* R. G. Crystal and J. B. West, eds. Raven: New York, p. 1021-1030.
66. Womersley, J. R., 1955, Method for the calculation of velocity, rate of flow and viscous drag in arteries when the pressure gradient is known, *J. Physiol. Lond.* 127: 553-563.
67. Wood, L. D. H., Engel, L. A., Griffin, P., Despas, P., and Macklem, P. T., 1976, Effect of gas physical properties and flow on lower pulmonary resistance, *J. Appl. Physiol.* 41:234-244.

16

RESPIRATORY MECHANICS AND NEW CONCEPTS IN MECHANICAL VENTILATION

Daniel Isabey, Laurent Brochard, and Alain Harf

Institut National de la Santé et de la Recherche Médicale
Inserm U296 - Physiologie Respiratoire
Services de Réanimation Médicale et d'Exploration
Fonctionnelle Respiratoire
Département de Physiologie, Hôpital Henri Mondor
94010 Creteil Cédex, France

INTRODUCTION

By definition, Mechanical Ventilation is an attempt to maintain or restore a normal rate of gas exchange in patients undergoing short or long term respiratory deficiency. This deficiency may be caused either by a chronic or an acute pulmonary disease, or be a consequence of anesthesia. Since its early age, let us say the beginning of this century, mechanical ventilation has been assigned to reproduce the modalities of normal spontaneous breathing, i.e., a tidal volume in the range V_T: 400-800 cm^3 (10-12 ml/kg) at an imposed frequency in the range f: 12-24 cpm (0.2-0.4Hz) (72, 73, 84). Delivering a controlled volume at a predetermined frequency is most often achieved by generating a constant flow during a fixed inspiratory time followed by a time limited passive exhalation (42, 98). This mode is called Controlled Mechanical Ventilation (CMV) which implicitly means that flow is controlled.

In principle, the CMV mode offers to the clinician the guarantee that the patient is insufflated with the desired ventilatory volume whatever the mechanical characteristics of the respiratory system. To satisfy such a criterion, two conditions appear essential: (i) to design ventilators whose inner impedance is considerably higher than the patient's mechanical impedance, (ii) to maintain a tight connection between patient and ventilator. In an attempt to satisfy the first (i) above condition, intense technological efforts have been performed by constructors during the two last decades. These efforts have undoubtedly resulted in a considerable improvement of ventilator performances. Most of modern CMV ventilators are now able to insufflate a predetermined gas volume or a constant flow independently of mechanical characteristics of patients. This requires to build flow generators insensitive to downstream pressure variations which, in patients treated in intensive care units, currently covers one order of magnitude (10-100 hPa). The more altered the mechanical properties, the higher the pressure needed to ventilate the respiratory system. To satisfy the second condition (ii) above, the tightness of connectors, the conventional method consists in intubating the trachea of the patient using an endotracheal tube (ETT) which includes, in adults only, an inflatable cuff located near the tracheal extremity of the tube.

The generalization of the CMV mode, at least in adults, results from the reliability of modern CMV machines as well as from the intubation with a cuffed ETT which guarantees tightness of the circuits (57, 107).

CMV, as optimized in modern ventilators, could have constituted a final stage of mechanical ventilation if the cost paid to force a predetermined volume (or flow) in the patient's respiratory system was not elevated in terms of pressure related effects (13, 108). Indeed, CMV has two major disadvantages : (i) the necessary adaptation of the patient to the machine and (ii) the risk of alveolar disruption resulting from excessive intrapulmonary pressure variations, or alveolar barotraumatisms.

It is precisely with the aim of reducing the risk of barotraumatisms inherent to CMV that a different method of mechanical ventilation called ventilation by High Frequency Oscillation (HFO) was proposed 15 years ago (18, 55, 113). A number of studies have been performed in the 1980s in an attempt to understand unsteady flow and gas transport phenomena in airways during oscillations at frequencies far above normal rates (5-50 Hz) (56, 81, 100, 112). Interestingly enough, HFO constitutes the first attempt to propose a mode of mechanical ventilation based on a concept which fundamentally differed from the classical concept of alveolar gas exchange by direct bulk convection through the anatomical dead space. However, the improvements expected in terms of both gas transport and gas exchange, were not sufficient to convince the clinical field of the interest of substituting HFO for CMV (86, 109, 114). Moreover, oscillatory volume as well as mean and variations in alveolar pressure appeared difficult to control from airway pressure variations, unless appropriate transfer functions are used (17, 32, 48, 106, 117).

Recently, the clinical field has undergone a continuous but irreversible evolution in terms of mechanical ventilation. This has been permitted by the development of new ventilatory modes compatible with a certain degree of patient's inspiratory activity (16, 42, 53, 67). These effort-compatible types of mechanical ventilation realize a partial ventilatory assistance. In most cases, a positive pressure created at airway opening, synchronized with the inspiratory effort of the patient, aims at reducing his inspiratory effort (11, 53, 69). This modality of partial mechanical ventilation is called Pressure Support (PS). PS differs from CMV since both pressure and inspiratory time are directly determined by the patient and not by the machine. The widespread interest of the PS mode is undoubtedly linked to the recent improvement of pressure ventilators which still require to be further improved. Thus, to maintain a flow-independent pressure, the internal impedance of perfect pressure generators must be zero (45) instead of infinite in case of perfect flow generator (82). For similar reasons, while aerodynamic resistance of the inspiratory circuit is unimportant during CMV, minimizing the circuit resistance appears crucial during PS. Then, knowledge of the mechanical characteristics of the respiratory system and of the circuit used including the ETT contribution appears essential with the Pressure Support mode. Incidentally, evaluation of passive mechanical characteristics of the ventilated respiratory system, which is already difficult during CMV, is not facilitated during PS because the patient contributes to the inspiratory gas motion. Thus, evaluation methods developed for CMV are not directly applicable to PS which explains why very few of them are today available (61).

The aim of the present chapter is first to present physiological principles which are behind spontaneous breathing and mechanical ventilation in humans. Then, we describe the different ventilatory modes proposed, their issue, and the methods developed to evaluate the mechanical characteristics of the respiratory system during controlled and assisted ventilation. Due to the difficulty of obtaining invasive information in clinical routines, development and integration of models of the mechanical behavior is highly desirable. To illustrate the

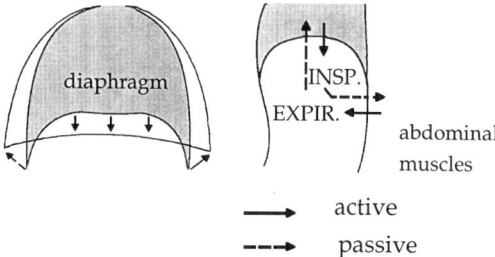

Figure 1. During inspiration, thoracic volume increases secondary to diaphragmatic contraction. By contrast, expiration is totally passive in quiet breathing conditions but abdominal muscles may be active in the expiratory phase during exercise, hyperventilation, and pulmonary lung disease (from West, J.B. (115)).

interactions between ventilator and patient, we also discuss the fluid mechanical basis of typical flow and pressure generators. Finally, due to the wide number of factors involved, classical servo-controlled systems fail to provide an appropriate adjustment of ventilatory parameters, especially during PS. Thus, mechanical evaluation of the respiratory system requires modelization procedures which could advantageously be integrated in a more general process of control, based on artificial intelligence. A recent attempt to automatically control the loop constituted by the couple patient-ventilator has been successfully tested during weaning from CMV (25). This suggests that real-time clinical and mechanical knowledge about the ventilated patient could advantageously be handled by a computer assisted machine.

PHYSIOLOGICAL BASIS OF SPONTANEOUS BREATHING AND MECHANICAL VENTILATION

The Cyclic Activation

In normal conditions, spontaneous breathing results from the cyclic activation of several muscles and particularly the diaphragm which is a thin dome-shaped sheet of muscle inserted into the lower ribs. As the diaphragm contracts, the abdominal content is pushed downward and forward and the rib cage is lifted. This effort causes an increase in inner thoracic volume and a negative intra-thoracic pressure deflection is created.

Inspiratory flow necessarily results from a positive pressure difference between the mouth and alveoli. Since buccal pressure remains always constant during spontaneous breathing, i.e., atmospheric, normal inspiration requires a decrease of intra-thoracic pressure. During spontaneous breathing, this intra-thoracic pressure deflection constitutes the driving pressure. By contrast, the totality (during CMV) or a part (during PS) of the driving pressure is developed by a machine.

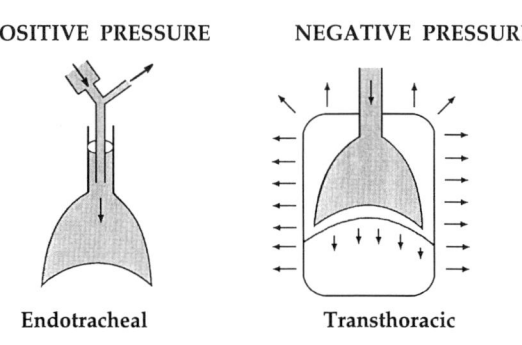

Figure 2. Two types of mechanical ventilation have been historically proposed: the widely used modern type in which positive pressure is applied to airways through an endotracheal tube (left panel); an early type, mostly abandoned today, in which negative pressure is applied around the thorax with a body box system (right panel).

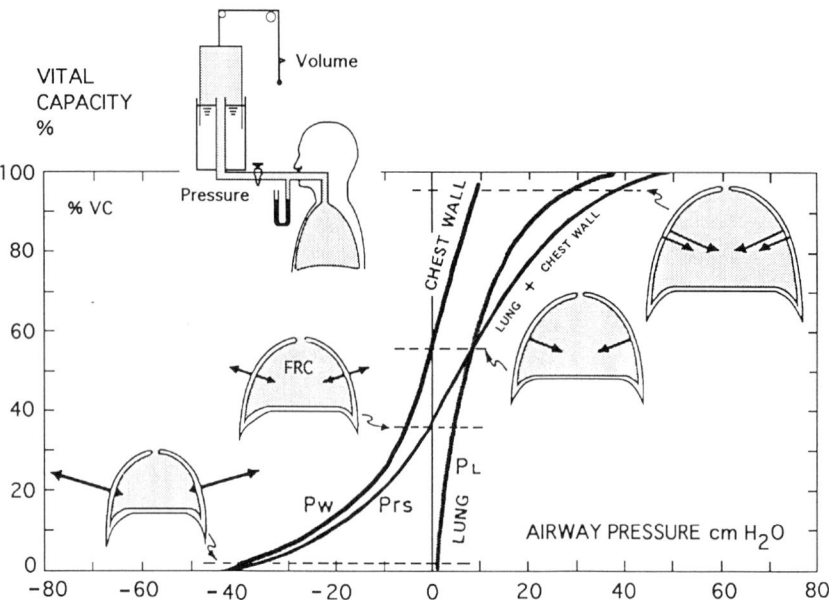

Figure 3. Typical static volume-pressure curves for the lung, the chest wall and the overall respiratory system (lung + chest wall) obtained in a normal sitting subject, during complete relaxation of muscles. The subject inspires (or expires) to a certain volume from the spirometer, the tap is closed and he relaxes his chest, so that airway pressure equilibrates elastic recoil. Prs: pressure exerted by the respiratory system is the sum of PL: pressure exerted by the lung and PW: pressure exerted by the wall, taken separately. Straight arrows indicate the direction of motion associated with spontaneous relaxation of each structure (from West, J.B., 115). Functional Residual Capacity (FRC) corresponds to the relaxation position of the overall structure (lung + chest wall). Lung volume is expressed in per cent of the vital capacity which is the maximal volume that can be inspired from FRC (or expired down to FRC).

Early modes of mechanical ventilation used a negative pressure applied around the thorax, e.g., iron lung, but have been rapidly abandoned secondary to impossible nursing and psychological problems associated with body box systems (107).

Modern modes of mechanical ventilation, including CMV or PS, use in common a positive intra-airway pressure. The connection between respirator circuit and airway opening is made with an endotracheal tube (ETT) or a face mask. With an endotracheal tube, tightness with patient's trachea is obtained by means of an inflatable balloon whereas a supple or inflatable pre-formed membrane is used in the case of a face mask.

Considering a given respiratory system ventilated with a fixed pattern of flow or pressure, spontaneous breathing (SB), mechanical ventilation (CMV), and pressure support (PS) would all require similar driving pressure, but the longitudinal pressure distribution will not be the same in each case (13). During CMV, the inspiratory airway pressure is positive and reaches its highest level compared to other modes. During PS, the airway pressure level is intermediate between the CMV pressure level and the SB pressure level.

By contrast to inspiration, expiration is entirely passive both during spontaneous breathing and mechanical ventilation, except in certain conditions such as exercise, hyperventilation, or pulmonary lung disease, where abdominal muscles are activated.

Whether inspiration is induced by the patient, the machine or both, the pressure gradient aims at overcoming retarding forces which are basically related to volume change and the time derivative of the volume change, as described in the two next paragraphs. Inertial

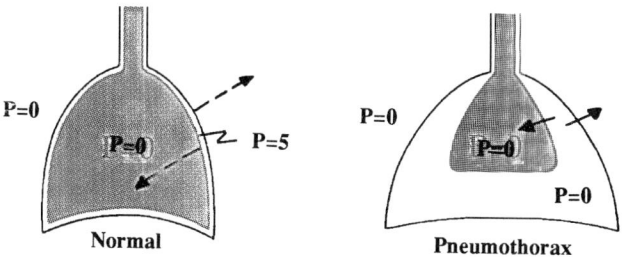

Figure 4. In normal conditions, the equilibrium state between the lung and the chest wall results in a −5hPa pressure in the pleural space (left panel). Pneumothorax corresponds to pathological conditions in which pleural pressure becomes suddenly atmospheric (right panel).

forces related to acceleration have to be considered at rates of ventilation clearly above normal rates, e.g., during ventilation by High Frequency Oscillations, but do not play an important role during CMV or PS.

Normal and Altered Elastic Properties of the Respiratory System

We present in this paragraph the quasi-static lung behavior, i.e., resistive forces related to time derivative of the volume change are negligible. Elastic properties of the respiratory system are classically characterized in resting conditions, i.e., when inspiratory muscles are relaxed, by a static strain-stress curve or relationship between volume variation and airway pressure (fig. 3).

The volume-pressure curve describing the overall respiratory system results from the addition of individual curves characterizing two intricate systems: the lung on one part and the chest wall on the other. These two systems are normally maintained in contact by the pleural surface. The relative motion of the lung into the thorax is facilitated by the presence of a few milliliters of liquid in the intrapleural space. Compliance, defined as the ratio between lung volume variation and pressure variation, represents the slope of volume-pressure curves. The lung and the chest wall have quite different compliant behaviors as shown by the different shapes of curves in fig. 3. The highest compliance is observed in a limited range of lung volume, e.g., 500ml above FRC in normal conditions, in agreement with the sigmoidal shape of volume-pressure curves. At high lung volume, compliance of the respiratory system tends to decrease because chest-wall expansion is limited. At low lung volume, compliance is also decreased because some residual volume of air trapped in alveoli cannot be exhaled secondary to airway closure.

Thus, due to the interaction between lung and thorax, the equilibrium position of the overall system does not correspond to the equilibrium states of each system taken separately. The relaxation volume of the chest wall is much higher than the relaxation volume of the lung. The lung volume corresponding to the relaxation position of the overall structure is called Functional Residual Capacity (FRC). At FRC with airways opened to atmosphere and muscles relaxed, the chest wall tends to expand to higher volume while the lungs tends to retract. These opposite recoil forces acting on the two systems attached to the pleura result in a slightly negative value of pleural pressure (−5 hPa). Thus, inspiration of 1 Liter of air in a respiratory system whose overall compliance is 100 ml/hPa requires an effort to decrease pleural pressure from −5 to −15 hPa. Incidentally, pneumothorax corresponds to a pathophysiological condition in which pleural pressure reaches atmospheric (115). Then, separation occurs between the lung and the chest wall and the lung retracts to its own resting volume

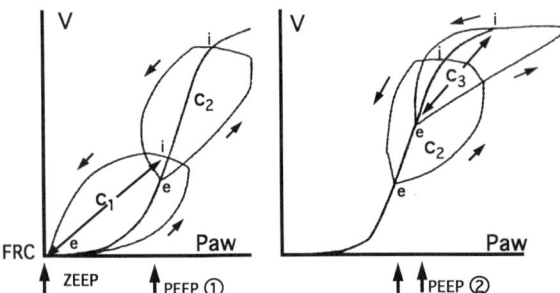

Figure 5. Static and dynamic compliance curves obtained in patients with acute respiratory disease. V, lung volume, is plotted on vertical axis. Paw, airway pressure, is plotted on horizontal axis. The non-linear character (sigmoidal shape) of static curves is increased compared to normal. Dynamic curves correspond to the loops (i: inspiration, e: expiration) and are obtained for different levels of Positive End-Expiratory Pressure (PEEP). ZEEP corresponds to Zero End-Expiratory Pressure. By definition, Functional Residual Capacity (FRC) corresponds to ZEEP conditions. Note that in patients with such altered compliance curves, it might be beneficial to ventilate with a moderate level of PEEP⟨1⟩, i.e., from a lung volume clearly above FRC but not too high like it is with PEEP⟨2⟩. With PEEP⟨1⟩, the patient is ventilated in a portion of the V-P curve where the slope (compliance) is the highest, e.g., C_2 (> C_1 and C_3) in the present case.

which is much smaller than FRC (see fig. 3). Besides this pathological state, spontaneous inspiration requires an effort to decrease pleural pressure below −5 hPa.

The lung is a highly distensible structure due to the layer of surfactant coating alveoli and to highly extensible chains of proteins called elastin, which constitute the lung parenchyma or deep lung. In normal lung, 90% of the lung volume resides in the parenchyma which is the zone of respiratory exchange including about 3.10^8 alveoli. A number of studies on mechanical behavior of lung tissue have been performed in excised normal lungs (see below) but application of these results to realistic in vivo conditions is hazardous because vascular perfusion, surfactant, and chest wall may play a role (38). Moreover, excised lungs tested were also normal lungs and the results are not necessarily applicable to abnormal lungs (58). Indeed, due to the difficulty of obtaining relevant models, there are very few studies in abnormal lungs, whereas the alterations of parenchyma properties and/or surfactant production result in drastic modifications of lung mechanical properties and are the cause of major lung pathologies (emphysema, fibrosis, respiratory distress syndrome...).

Compared to curves obtained in the normal subject (fig. 3), the non linear character of volume-pressure curves is increased in patients with acute respiratory disease, meaning that the compliance of the altered respiratory system shown in fig.5 is considerably reduced at high and low lung volumes. Thus, if these patients were mechanically ventilated from FRC, ventilation would not be efficient, due to the diminution of compliance at low lung volume. For similar reasons, mechanical ventilation at high lung volume would not be efficient either, however, the risk of barotraumatisms would be considerably increased at high lung volume. In patients with highly altered compliance curves, the solution is to ventilate from an end-expiratory volume clearly above FRC. This requires the generation in the ventilator circuit of a Positive End-Expiratory Pressure (PEEP), usually below 10 cmH$_2$O, such that the patient re-inspires at a pressure above atmospheric pressure which minimizes his inspiratory effort.

Abnormally elevated values of compliance are usually associated with emphysema while low values of compliance are related with lung fibrosis. Because the lung primarily acts as a network of parallel compartments in which the overall compliance is the sum of separate compliances (77), a reduction in compliance measured in a given patient may reflect

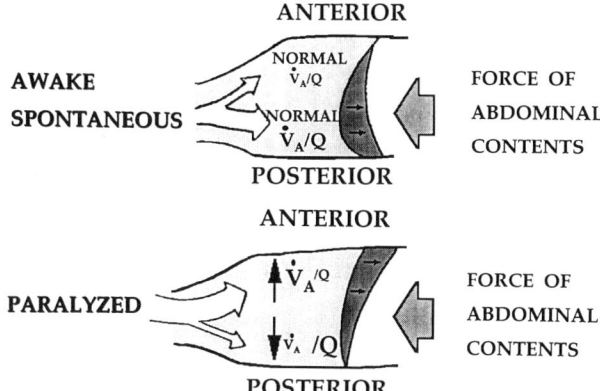

Figure 6. In supine position, a consistent relationship is maintained between ventilation and perfusion in awake patients (top) while this relationship is altered in paralyzed patients (bottom) (see text for explanation and ref. 34 and 89).

a reduction in the number of ventilated compartments, i.e., the lung volume available for ventilation is reduced (atelectasia). This phenomenon plays a crucial role in the reduction of compliance observed in patients with Acute Respiratory Distress Syndrome (ARDS). Thus, in clinical practice, reduction in compliance is not only associated with parenchyma alteration but also with the ventilation of a « small » lung, sometimes called in clinical practice « baby » lung.

Another important difference between CMV and spontaneous breathing concerns the alteration of the ventilation-perfusion relationship in supine paralyzed patients (34, 89). In awake spontaneous breathing, ventilation and perfusion remain closely matched in the supine position because a greater alveolar ventilation occurs in the posterior lung region which is also the most perfused region. This is because hydrostatic forces cause enhancement of posterior diaphragmatic movement as they apply to abdominal content whereas, the same forces also cause an increase in perfusion as they apply to the blood circuit. In paralyzed patients such as during CMV, ventilation is not matched anymore with perfusion. In such a case, the diaphragm does not contract, most of alveolar ventilation is anterior while perfusion remains posterior. This problem may be overcome at the expense of an increase in tidal volume, thus aggravating the risk of barotraumatisms with CMV.

Resistive Properties of the Respiratory System

The curves describing the relationship between lung volume and airway pressure in dynamic conditions exhibit an hysteresis which reflects a variety of phenomena: (i) the gas flow resistance in the tracheobronchial tree and connectors between patient and ventilator, (ii) the lung tissue resistance, and (iii) the chest wall resistance. Incidentally, the volume-pressure loops obtained during ventilation in severely ill patients (fig. 5) integrate these various resistive effects although the specific contribution of each factor is generally unknown. Most of the studies performed to evaluate the contribution of each factor have been performed in normal subjects in which parallel inhomogeneities are generally less predominant than during lung disease. The gas flow resistance in airways involves tridimensional aerodynamic phenomena that have often been studied in idealized geometry (79). The lung tissue resistance is associated with the irreversibility between alveolar recruitment and closure, and/or pseudoelastic or plastoelastic behavior of lung tissue (37). Its importance compared to other resistive effects is still under debate in normal humans (4, 5, 33, 65). As soon as there are significant differences in time constants between parallel compartments - a situation which characterizes most patients mechanically ventilated in the Intensive Care

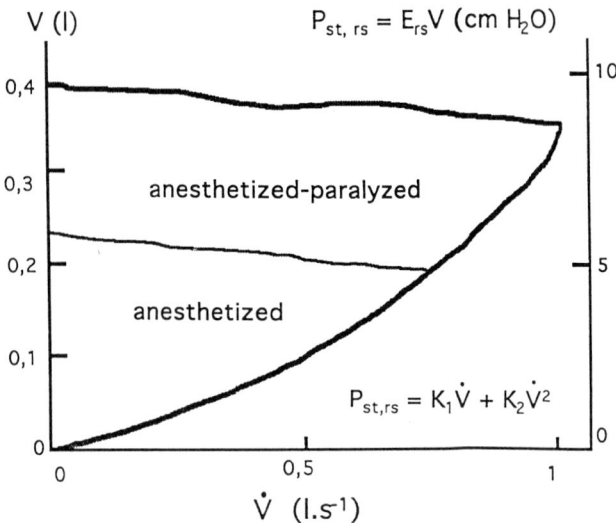

Figure 7. Relationships between volume in litre (left vertical axis) and flow rate in litre/second (horizontal axis), obtained during passive exhalation in a given subject under anesthetized and anesthetized-paralyzed conditions (from Behrakis et al. (7)). The elastic recoil pressure, (right vertical axis in cmH2O), is supposed to remain proportional to the lung volume V through the entire expiration (constant compliance E_{rs}). equilibrates the non-linear frictional pressure drop through the respiratory system including the endotracheal tube. This method, which requires interruption of CMV and complete muscle relaxation, is not suitable for monitoring respiratory mechanics, especially in patients with Acute Respiratory Distress Syndrome (ARDS) whose compliance is not constant over the entire tidal volume. Rohrer equation provides a satisfactory curve fitting of the non-linear feature but with poor fluid mechanical relevance.

Unit - it is very difficult to separate the specific contribution of series compartments: airways + connectors, tissue, chest-wall (see below and ref. 61).

Resistances of Airways and Endotracheal Tube: Aerodynamic resistance across airways has been given to represent up to 70% of the total respiratory system resistance in normal (non intubated) humans (30). Invasive and non invasive methods have been proposed to partition airway resistance within central and peripheral airways (41). Invasive methods consist in inserting a 2mm-catheter retrogradely into intra-thoracic airways but such a method has no place in the study of intact human subjects (66). Non-invasive methods consist in using gases of different physical properties (27) or differences in the frequency and flow dependence between central and peripheral airways (85). However, in addition to the questions raised by the analogue model proposed to describe the respiratory system, especially in altered lungs, the principal problem raised by the application of these non invasive techniques to intact human subjects remains the anatomic location of central and peripheral resistors (30).

Upper airway resistance which includes mouth, nose, pharynx and larynx represents a large part, i.e., up to 40%, of the total respiratory resistance, with a large variability between subjects and within each subject (90). When upper airways are by-passed by means of an endotracheal tube (ETT), the ETT contribution to total resistance remains significant (116). Moreover, a number of factors such as secretions, head or neck position, tube deformation are susceptible to increase the tube resistance when measured in the *in vivo* conditions (116). The concept of linear resistance is clearly insufficient to fully describe the marked flow-dependence of resistance observed both in airways and in endotracheal tubes. Rohrer equation given below constitutes an attempt to describe the non-linear behavior of the relationship between frictional pressure losses (ΔP) and flow rate ($\dot{V}$) through airways and endotracheal tube (91).

$$\Delta P = K_1 \dot{V} + K_2 \dot{V}^2$$

However, due to the poor fluid mechanical relevance of this equation, the coefficients K_1 and K_2 must be determined empirically and Rohrer equation fails to provide a sufficiently general understanding of flow resistive phenomena.

Still using the straight tube analogy, fluid mechanical considerations allow modification of the Rohrer equation to propose a physically meaningful pressure-flow relationship for a bronchus or the global airway system (d: diameter, L: length, of isolated bronchi or trachea (global system), μ: gas dynamic viscosity, ρ: gas density), as done by many authors (45, 50, 79, 104).

$$\Delta P = k \cdot \left(\frac{4}{\pi}\right)^n \cdot \left(\frac{L}{d^{3+n}}\right) \cdot \mu^{2-n} \cdot \rho^{n-1} \cdot \dot{V}^n$$

with:

n=1, k = 32 for Poiseuille flow;
n=1.5, k = 3.(L/d)$^{-0.5}$ for smooth entry in a straight tube;
n=1.75, k = 0.16 for hydraulically smooth turbulent flow;
n=2, k = 0.01-0.03, for turbulent flow in a rough tube (height of the roughness between 0.1%d and 3.3%d).

During the two last decades, a number of experimental and theoretical studies performed in branching models or airway casts have attempted to provide a consistent understanding of the aerodynamics of airway flow. Velocity profile measurements as well as pressure drop measurements in central airways have shown that, in normal breathing conditions (mouth flow rate $\leq$ 1 L/s), flow is essentially undeveloped, affected by centrifugal forces due to curvature, varying from turbulent in trachea (near peak flow rate) to laminar in most of bronchi (during the whole respiratory cycle (41, 45, 50, 79, 102, 104).

As for straight tubes, the slope of the relationship between the normalized frictional pressure drop and the Reynolds number :

$$\left(Re = \frac{4 \cdot \dot{V} \cdot \rho}{\pi \cdot d \cdot \mu}\right),$$

on Moody diagram (« n-2 », in the equation below) indicates the flow regime.

$$\frac{\Delta P}{0.5 \cdot \rho \cdot (4 \dot{V} / \pi d^2)^2 \cdot (L/d)} = 2 \cdot k \cdot Re^{n-2}$$

Values of « n » and « k » were found to be 1.5 and 10.5 respectively in a symmetrical branching network, at moderate values of Reynolds number (80, 102). A similar value of the slope « n » was found in experiments in an airway cast (45, 104) such as those reported in fig. 8 b. Surprisingly, this result means that tracheal Reynolds number correlates the normalized pressure drop through airways. A reasonable explanation is that flow distribution in networks with small values of the length over diameter ratio, e.g., L/d never exceeds 3.5 in the first 10 airway generations except in the trachea, has been found to be independent on tracheal flow and gas physical properties (43, 105).

During oscillatory flow in airways such as during HFO, the quasi-steady equation used above is not valid anymore, except near peak flow. The reason is that the degree of unsteadiness and therefore, the flow regime changes throughout the flow cycle from unsteady state near flow reversal to quasi-steady state near peak flow (47). A number of experimental

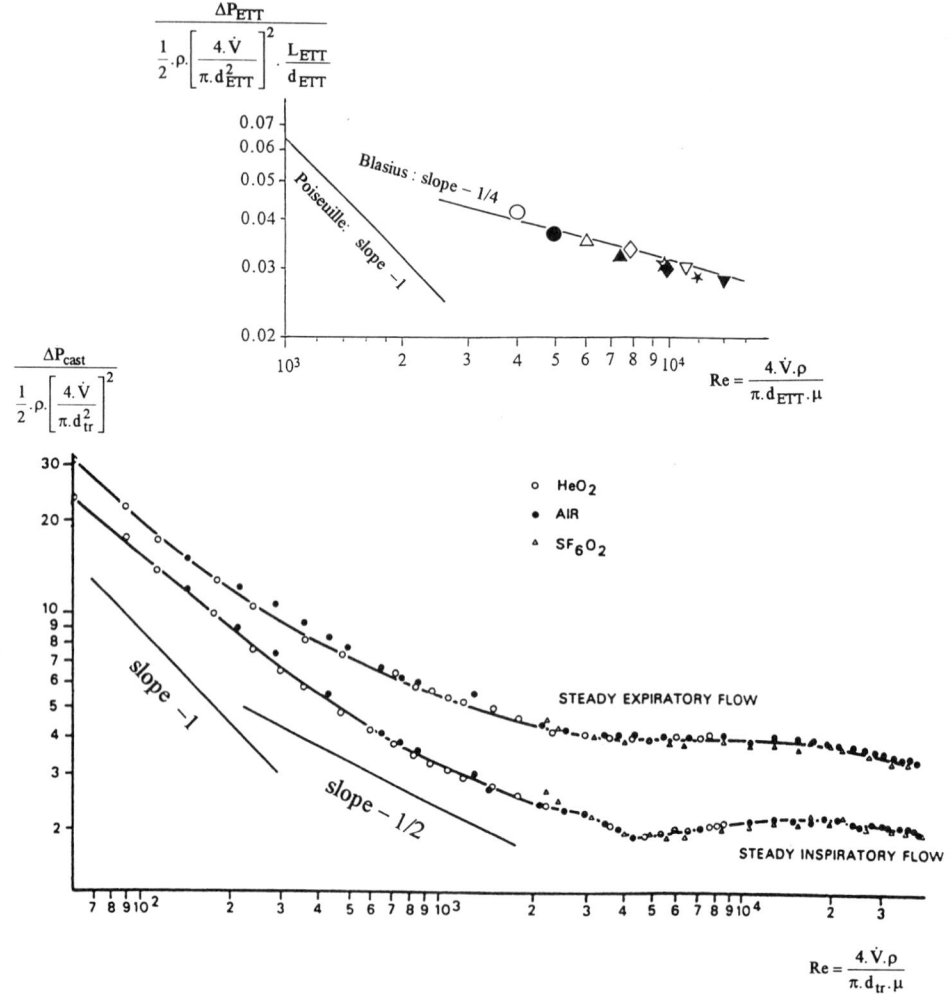

Figure 8. Log-log plots of the relationship between normalized pressure drop and Reynolds number, (a) in an endotracheal tube (Portex, inner diameter: d_{ETT}= 8.5mm) (from 60) and (b) in a cast of human central airways (from 45). Data in (a) were obtained in air with (open symbols) and without (closed symbols) a pressure catheter and for different flows. When a catheter was used, the hydraulic diameter used was: $d^* = d_{ETT} - d_{cath.}$. Data in (b) were obtained with gases of different physical properties, μ: dynamic viscosity, ρ: density [air; « 80%He-20%O_2 » mixture: ($\mu_{HeO_2}\backslash\mu_{air}$ = 1.19 and $\rho_{HeO_2}\backslash\rho_{air}$ = 0.33); « 80%SF_6-20%O_2 » mixture: $\mu_{SF_6O_2}\backslash\mu_{air}$ = 0.85 and $\rho_{SF_6O_2}\backslash\rho_{air}$ = 4.24]. The hydraulic diameter used in (b) was the tracheal diameter (d_{tr}). As in other studies of this type (104), a certain part of the pressure drop is due to kinetic energy changes associated with changes in cross-section area between tracheal level and bronchi level (5th airway generation), e.g., 10-20% in the present cast.

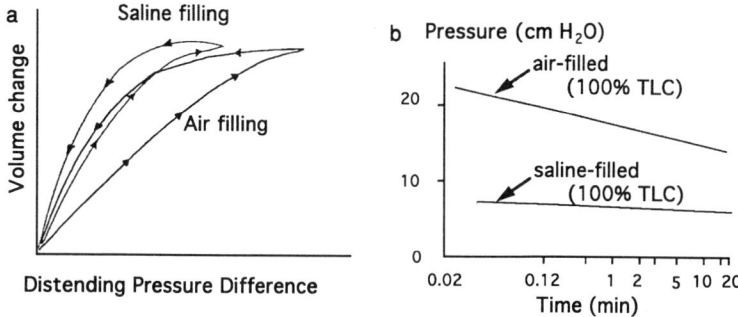

Figure 9. (a) Typical volume-pressure curves with air and saline filling in excised rabbit lungs. Arrows indicate inflation and deflation. Note that hysteresis does not disappear completely after saline filling. (b) Stress adaptation in normal cat lungs showing time course of transpulmonary pressure in air-filled and saline-filled cat lungs at total lung capacity. Saline-filled lungs show a smaller, but non negligible, stress adaptation compared to air-filled lungs (modified from Hoppin et al. (38)).

and theoretical studies (45, 47, 50, 51, 60, 79, 102,104) have shown that, in addition to Reynolds number reflecting the convective to viscous forces ratio, flow pattern and pressure drop through airways could be described by non-dimensional numbers proportional to centrifugal to convective inertia forces ratio :

$$\left(\text{Dean number: } \kappa = \left[\frac{d}{2.\chi}\right]^{0.5} \text{Re}, \chi = \text{radius of curvature}\right),$$

unsteady inertia to viscous forces ratio:

$$\left(\text{Womersley number} : \alpha = \frac{d}{2} \cdot \left[\frac{2.\pi.f.\rho}{\mu}\right]^{0.5}\right),$$

unsteady to convective inertia forces ratio:

$$\left(\text{Strouhal numbers: St} = \frac{\pi^2 . f . L . d^2}{2 . V_{peak}} = \frac{\pi . L . d^2}{V_T}\right)$$

at peak flow, and/or:

$$\varepsilon = \frac{\pi . L . d^2 . d\dot{V}(t)/dt}{4 . \dot{V}^2(t)} \text{ through the flow cycle}).$$

through the flow cycle). These numbers, usually estimated from quantities calculated in the trachea, averaged over the cross-section area, provide a satisfactory estimate of balance between phenomena occurring in a variety of conditions. To improve these predictions which are essentially issued from physical airway models, studies remain to be performed in models with fully controlled boundary conditions. This can be achieved using numerical models of the airway flow as done, for instance, in some recent studies (21, 110). However, Reynolds number is exceptionally considered in clinical practice, except to describe the ETT pressure drop as explained below.

During CMV, constant flow in adult size ETT (Re ≥3000) has been found to behave essentially as a fully developed and hydraulically smooth turbulent flow (above 0.5L/s) whose pressure-flow relationship can be described by the Blasius formula (case « n » = 1.75 in above equation) (60) instead of Rohrer equation (8). Fully turbulent flow has been assumed earlier to describe the oscillatory pressure-flow relationship in adult ETT during ventilation by HFO (35). Blasius formula can still be used when mucus is deposited at the ETT wall and/or if a pressure catheter is inserted with the aim of measuring distal pressure in the ETT (60). In the latter case, flow becomes annular which requires using a different diameter, in the formula above. The relationship between ETT pressure drop and flow in the presence of a pressure catheter (diameter:) may be written:

$$\Delta P = \frac{0.316}{2} \cdot \left(\frac{4}{\pi}\right)^{\frac{7}{4}} \cdot \mu^{\frac{1}{4}} \cdot \rho^{\frac{3}{4}} \cdot L \cdot (d_{ETT} - d_{cath.})^{-\frac{5}{4}} \cdot (d_{ETT}^2 - d_{cath.}^2)^{-\frac{7}{4}} \cdot \dot{V}^{\frac{7}{4}}$$

This formula has been used to determine after measurement of ΔP and $\dot{V}$, a mean ETT diameter (d_{ETT}) during mechanical ventilation (60). However, this hydraulic method, which requires the introduction of a pressure catheter invasively in the ETT, increases the risks inherent in intubation.

The acoustic reflection method which is non invasive, has recently been used to measure ETT patency during mechanical ventilation (111). This method, which classically uses a single-microphone in a 2m-long wave tube - to separate incident from reflective waves - has been modified by using two microphones instead of one, and a wave tube ten times shorter (63). In addition, air was used in the modified method instead of an HeO_2 mixture in the initial one. These modifications have facilitated non-invasive measurements of ETT patency in mechanically ventilated patients. Although the above hydraulic method and the acoustic reflection method were found equivalent for the detection of ETT area reduction caused by mucus deposition at inner wall, the superiority of the acoustic reflection method over the hydraulic diameter method is evident since acoustic reflection methods provide information concerning the location of the obstruction in the ETT (111).

Lung Tissue. According to Bachofen and Hildebrandt (2) and several other authors, the contribution of lung tissue to lung resistance is far from negligible, e.g., reaching about 40% in humans at normal breathing frequency. This type of result is in agreement with previous results obtained by Macklem and Mead at higher frequencies (66). Later studies in living open-chest dogs (65) have shown that, at breathing frequencies in the range 0.2-0.4Hz, tissue resistance had an even larger contribution (68-85%). Using a classical analogue model which comprises a resistance (airways) in series with a circuit (tissues) including in parallel a pure compliance (static compliance) and a viscoelastic element, Jonson et al. (52) have shown that, in healthy anesthetized humans, the magnitude of tissue resistance represented two times the inspiratory airway resistance, whereas the work dissipated through the viscoelasticity element represented 32% of the total energy loss in the respiratory system. The contribution of the chest wall has often been thought to be negligible in terms of resistance. Its estimation requires knowledge of pleural pressure. Some estimates performed in normal patients have shown that chest wall resistance represented about one-third of the pulmonary resistance during mouth breathing (90) and 27% of the total respiratory resistance in anesthetized paralyzed humans (24).

The classical equation of motion used by Mead (70) to describe the relationship between the pressure difference across tissue and change in lung volume during quiet breathing (inertial terms neglected and low flow rates) has the following form:

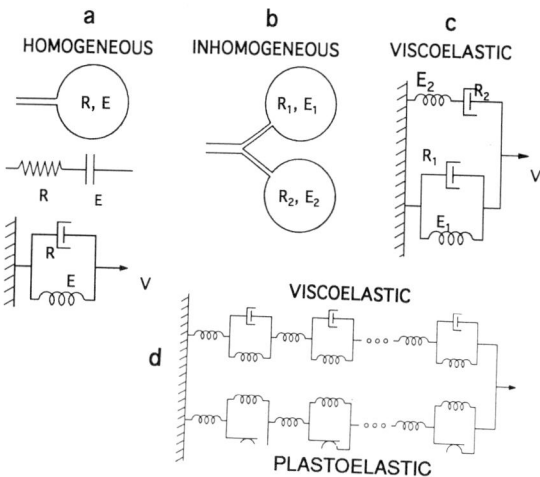

Figure 10. Various representations of the totality (or a part) of the respiratory system (airways, tissues including or not chest wall). In (a), the three representations: anatomic, electric (R-E circuit), rheologic (Maxwell element), of an homogeneous model correspond to the same first order differential equation. Electrical elements associated in series are subjected to the same strain, as are rheological elements associated in parallel. In (b), anatomic representation of the inhomogeneous model which corresponds to two electrical different elements (R1/E1≠R2/E2) in parallel or two rheological elements in series (not represented). In (c), rheologic representation of the respiratory system including a viscoelastic component (Maxwell element, E1-R1) in parallel with a Voigt element (R2-E2). In (d), model comprising a distribution of Maxwell elements in series with elastic elements in order to describe pulmonary stress adaptation and a substantial portion of hysteresis (70).

$$P_{tiss.} = E_{dyn} V + R_{ti} \frac{dV}{dt}$$

($P_{tiss.}$ = pressure difference across lung tissue = difference between alveolar gas pressure and pleural pressure, E_{dyn} is the dynamic compliance and R_{ti} is the tissue resistance).

Hysteresis behavior of the stress-strain relationship as well as the stress adaptation have been reported in early experiments on excised lung (fig. 9). Although most of hysteresis and time-dependent behavior were eliminated by saline filling, some remained, indicating that the air-tissue interface was not the only cause of imperfect elasticity. Energy dissipation associated with cycling lung expansion does not depend on the rate of expansion but on the amount of expansion (6) indicating that lung tissue behavior cannot be described with the classical notion of viscous stress. Factors such as the geometric irreversibility in the number of open lung units and non-elastic behavior of tissues, have also been suspected of playing a role (38,58).

Depending on whether airway or tissue resistance is emphasized, one of the representations given in fig.10 is used. Model (d) predicts dynamic elastance and frequency dependence but cannot account for 1/3 of the energy loss (hysteresis area). Replacement of viscous dashpots by dry-friction (Coulomb) elements generates static hysteresis and dependence of amplitude in response to oscillations. The role of visco-elastic process versus plastic process remains ambiguous, meaning that none of these two models satisfactorily describes the lung tissue behavior. In a recent study, Fredberg and Stamenovic (33) pointed out that the contribution of tissue resistance has been controversial because it is sensitive to a number of host factors such as frequency, tidal volume, and volume history. They proposed to use the concept of structural damping in which the dissipative and elastic processes are coupled at the level of each element of the structure (stress-bearing element). Thus, the basic equation above has to be replaced by the following equation:

$$P_{tiss.} = (E_{dyn} + j.\eta .E_{dyn})V$$

The imaginary term characterizes the dissipative process. η was called the lung tissue hysteresistivity and characterizes the nature of the structure but not the amount participating. Contrary to previous models, this model constitutes a valuable attempt to relate microscale to macroscale phenomena. However, the present concept, still under validation, has not been applied yet to differentiate abnormal from normal lungs.

Modelized Responses of the Ventilated Respiratory Sytem

It must be emphasized that assessment of pulmonary mechanics requires assessment of transpulmonary pressure, i.e., the difference between airway opening pressure and pleural pressure. Pleural pressure is practically obtained with an esophageal balloon which cannot be routinely used. By contrast, airway pressure () can be easily measured in common practice allowing assessment of respiratory mechanics rather than pulmonary mechanics. Practically, it is assumed that changes in respiratory mechanics essentially reflect changes in pulmonary mechanics. Respiratory mechanics is usually described on the basis of the following constant parameter linear equation which takes into account an overall elastance (E) and an overall resistance (R):

$$P_{aw} = EV + R\dot{V}$$

Depending on whether the author emphasizes flow resistive or tissue resistive phenomena, representations of the above equation may be different. These representations are: (i) the anatomical model, in which a resistance and an elastance are in series, and thus

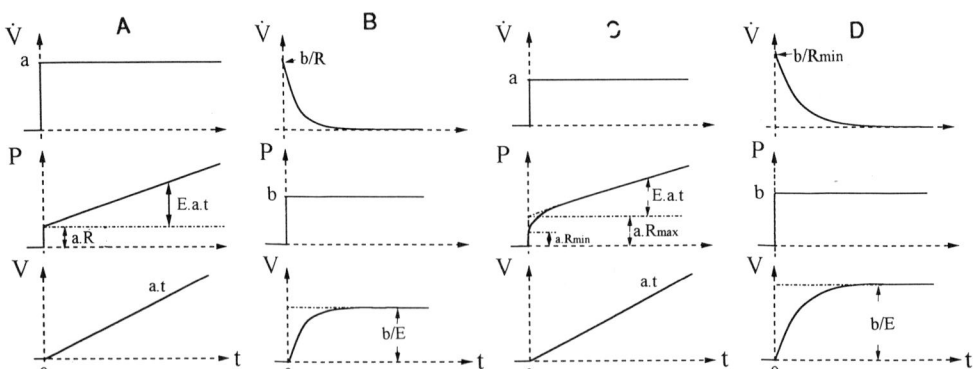

Figure 11. Typical responses of the respiratory system given by the homogeneous model (A and B) and the inhomogeneous model (C and D). In A and C, the input signal is a step increase in flow at t = 0, i.e., = a, or V = a.t, from t = 0); The response is a step increase in pressure followed by a linear increase. In B and D, the input signal is a step increase in pressure (P = b, from t = 0), the response is an exponential decay in flow and an exponential increase in volume up to a pressure plateau dependent of compliance. A and C simulate flow-Controlled Mechanical Ventilation (CMV). B and D simulate pressure controlled Mechanical Ventilation or Pressure Support (From Lorino and Harf (61)).

subjected to the same strain, (ii) the rheologic Voigt model in which a dashpot producing a velocity proportional to the load, and a linear spring are in parallel, and thus subjected to the same stress (61).

According to this equation, the time response to linear volume variations (V(t) = a.t from t = 0) such as during CMV, is a step increase in pressure followed by a linear increase:

$$P_{aw}(t) = R.a + E.a.t$$

This equation describes the shape of the airway pressure signal characterizing at first order the response of the overall respiratory system ventilated by CMV.

By contrast, the response to a transient increase in pressure ($P_{aw}(t) = b$ constant from t = 0) is a mono-exponential increase in volume with an asymptote value of b/E, and a mono-exponential decrease in flow, with a time constant (R/E).

$$V(t) = \frac{b}{E}\left(1 - e^{-Et/R}\right) \quad \text{and} \quad \dot{V}(t) = \frac{b}{R} e^{-Et/R}$$

These equations describe at first order the shape of volume and flow signals characterizing a type of pressure support in which the patient is assisted with a constant pressure at airway opening.

However, pathological lungs such as lungs with ARDS are mostly heterogeneous. The constant parameter linear equation above and its representations do not apply anymore.

Taking the Otis model (78) with two inhomogeneous compartments, i.e., $\tau_1 = R_1/E_1 \neq \tau_2 = R_2/E_2$, the time response to linear increase in applied volume (V = a.t, from t = 0), or a constant flow ($\dot{V}$ = a, from t = 0), produces an exponential increase in pressure instead of the abrupt change observed in the homogeneous model:

$$P_{aw}(t) = E.a.t + R_{max} a\left(1 - e^{-t/\tau}\right) + R_{min} a \left(e^{-t/\tau}\right) \quad \text{with}$$

$$E = \frac{E_1 E_2}{E_1 + E_2}; \quad \tau = \frac{R_{min}}{E}; \quad R_{min} = \frac{R_1 R_2}{R_1 + R_2}; \quad R_{max} = \frac{\frac{\tau_1}{E_1} + \frac{\tau_2}{E_2}}{E^2}$$

The larger the differences in time constants between parallel compartments

$$(\tau_1 = \frac{R_1}{E_1}, \tau_2 = \frac{R_2}{E_2}),$$

the larger R_{max}. Lorino and Harf (61) recently emphasized that the four mechanical parameters R_1, E_1, R_2, E_2, cannot be individually determined from the response of such an elementary inhomogeneous model to a linear volume variation, by contrast to the homogeneous model. Only lumped parameters, E, R_{min}, R_{max}, can be assessed in inhomogeneous systems with two or more elements.

The response to a step function of applied pressure ($P_{aw}(t) = b$ = constant from t = 0), results in a double-exponential increase in volume associated with a decrease in flow whose forms are:

$$V(t) = \frac{b}{E_1}\left(1 - e^{-t/\tau_1}\right) + \frac{b}{E_2}\left(1 - e^{-t/\tau_2}\right), \quad \text{and} \quad \dot{V}(t) = \frac{b}{R_1} e^{-t/\tau_1} + \frac{b}{R_2} e^{-t/\tau_2}$$

At t = 0, flow is maximum:

$$(\dot{V} = \frac{b}{R_{min}}).$$

Beyond a sufficient time, i.e., t > τ_1 and τ_2, volume tends towards a constant value:

$$(V = \frac{b}{e})$$

as is the case for the homogeneous model.

It is practically very difficult to assess with confidence the physiological meaning of the second parallel compartment. The reason is that viscoelasticity associated with pulmonary surfactant modifies the response of the homogenous lung (70) as do inhomogeneities between lung regions (61). For instance, the classical viscoelastic model of the respiratory system described by Bates et al. (4) and used by many authors such as Jonson et al. (52), actually consists of a Voigt element (E_1, R_1), i.e., the homogeneous element, associated with a Maxwell element (E_2, R_2) representing the pure viscoelastic component. The response to a step of applied flow ($\dot{V}(t)$ = a = constant from t = 0) may be expressed (61):

$$P_{aw}(t) = E_1.a.t + R_2.a.\left(1 - e^{-t/\tau_2}\right) + R_1.a$$

This equation resembles the equation proposed to describe the response of the inhomogeneous model. Thus, taking $R_{min} = R_1$, and $R_{max} = R_1 + R_2$, i.e., the difference $R_{max} - R_{min}$ represents the viscoelastic resistance. However, based on the two more elaborated models presented above, it appears that mechanical inhomogeneities between parallel elements and viscoelastic properties cannot be easily separated because they produce similar effects when studying the response of the respiratory system. This problem is accentuated in lungs with highly altered mechanical properties.

At the end of the inflation time (t = T_I), the pressure given by equation above is maximal.

$$P_{aw}(T_I) = E_1.a.T_I + R_2.a.\left(1 - e^{-T_I/\tau_2}\right) + R_1.a$$

According to this model, the response to a sudden airway occlusion performed at the end of inspiration (t = T_I), results in a pressure decrease with time. The difference between $P(T_I)$ and the new pressure $P_{aw}(t)$ is given by (4, 62):

$$P_{aw}(T_I) - P_{aw}(t) = R_1.a + R_2.a.\left(1 - e^{-T_I/\tau_2}\right)\left(1 - e^{-(t-T_I)/\tau_2}\right)$$

This equation shows that the different parameters of the model can be determined from the immediate decrease in pressure drop and from the plateau pressure value measured at a time representing at least $3.\tau_2$.

Measurement Methods during Mechanical Ventilation

Methods proposed to evaluate mechanical properties of the respiratory system during mechanical ventilation essentially concern the classical constant flow mode of ventilation (CMV) and very little pressure support modes (PS) whose development is much more recent. Indeed, development of methods of analysis dedicated to CMV only appeared in the last decade which is relatively recent compared to the CMV mode itself. The two methods proposed are the constant inspiratory flow method proposed by Zin et al. in 1983 (119) and the occlusion method whose application during CMV has been generalized after improvements proposed by Bates et al. in 1985 (5). These two methods are theoretically very close since they are based on the response of the respiratory system to a step applied flow similar in magnitude (flow suddenly rises from $\dot{V} = 0$ to $\dot{V} = a$ at $t = 0$ in the constant flow method, while flow suddenly drops from $\dot{V} = a$ to $\dot{V} = 0$ at $t = 0$, in the occlusion method as shown in fig. 12). Practically, the accuracy of both methods strongly depends on the technical performances of ventilators. Thus, predictive equations and the above solutions apply to both methods.

The constant inspiratory flow method assumes a linear relationship between the pressure applied by the ventilator and the changes in lung volume. Static elastance (E_{RS}) and total resistance (R_{RS}) are respectively obtained from the slope of the linear relationship and its intercept on the pressure axis (fig. 12). This method can be applied in the presence of mechanical inhomogeneities provided the analysis is performed on the linear part of the pressure increase (fig. 12). The main advantage of this method is that E_{RS} and R_{RS} can be measured in the course of mechanical ventilation.

The occlusion method requires a sudden occlusion created at airway opening during a constant inspiratory flow. As flow drops to zero, the pressure response follows a mono-ex-

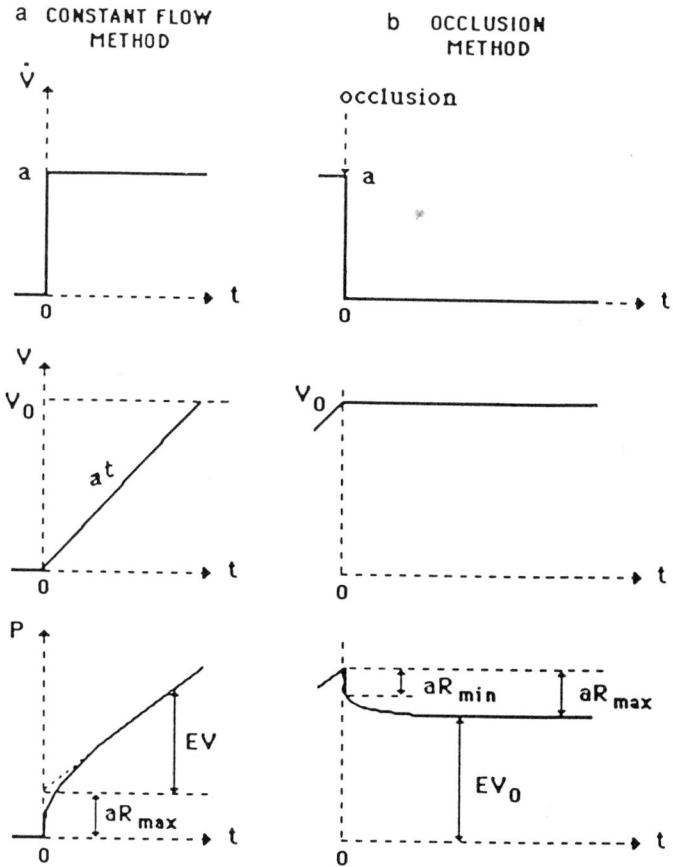

Figure 12. In (a), constant inspiratory flow method derived by Zin et al. (119). In (b) occlusion method improved by Bates et al. (5). Both methods are used to measure respiratory mechanics during flow Controlled Mechanical Ventilation (CMV) (see also ref. 61).

ponential or multi-exponential decay depending on the assumptions presented above. R_{min} can thus be determined from the transient change in pressure immediately following occlusion. Once pressure has reached a plateau, i.e., a few seconds after occlusion, R_{max} can be obtained as the ratio of the total pressure drop to the flow preceding occlusion. E_{RS} is the ratio of plateau pressure to the insufflated volume. The accuracy of the method strongly depends on the dynamic performances of the occlusion valves in the ventilator. Corrections have been proposed by the group of Bates et al. (4, 5) to take into account the finite closure time of the valves as well as valve vibrations induced by closure. The occlusion method is now integrated as an automatic process in certain ventilators to measure total flow resistance and thus includes the endotracheal tube resistance. The total tube resistance can be corrected for the endotracheal tube resistance and application of this method is now automatically performed by ventilators. The two methods have been compared by Rossi et al. (93) and they were found to be equivalent if the measured airway pressure is corrected for the internal-positive-end-expiratory pressure effects (93) and if recruitment of new lung units is not present in the early portion of inflation (92). However, the occlusion method appears preferable to the constant inspiratory flow method to quantify the difference $R_{max} - R_{min}$ because creating an occlusion is faster than suddenly generating a constant flow.

Another type of method proposed to assess respiratory mechanics in mechanically ventilated patients concerns the evaluation of the expiratory phase. The single-breath method adapted by Behrakis et al. (7) consists in performing an occlusion at the end of the inspiratory phase. Measurements of the pressure-flow relationship are performed in the relaxed patient, in conditions such that elastic recoil equilibrates the expiratory resistance through airway and circuit (see fig.7). According to Gottfried et al. (36), the respiratory system can also be evaluated by a series of brief interruptions performed by means of a pneumatic valve. This technique, initially called interrupter technique, assumes an homogeneous respiratory system and steady state behavior in spite of the rapidity of interruptions. Thus, applicability of this technique has been questioned in patients with acute respiratory failure precisely because these patient have highly inhomogeneous lung. Moreover, interrupter technique requires measurement of the flow preceding interruption in dynamic conditions which is much more difficult than it is during constant flow (46).

Complete evaluation of the respiratory system requires extension of the range of tested frequencies far above normal breathing frequencies. This may be done with the forced oscillation technique which has been studied by a wide number of groups (see the review by Peslin and Fredberg (83)). This method describes the frequency response of the respiratory system in terms of the real and imaginary parts. The frequency dependence of the real part is particularly sensitive to redistribution of flow within parallel elements when the degree of parallel inhomogeneities is high. Routine application of the forced oscillation technique to mechanically ventilated patients has not been performed yet mostly because irrelevant results were obtained when an endotracheal tube was connected. Indeed a number of fluid mechanical aspects such as kinetic energy losses at the junction between the tube and the trachea (75), and interaction with the insufflated gas flow (64) may all affect impedance measurements, rendering difficult application of the forced oscillation method to mechanically ventilated patients.

FLOW CONTROLLED MODES

Most modern intensive care ventilators are equipped with the flow controlled mode in which a constant flow can be delivered independently of whether mechanical characteristics of patients are normal or strongly altered. This depends first on the performances of flow generators which initially consisted of pistons driven by motors. To evaluate constant

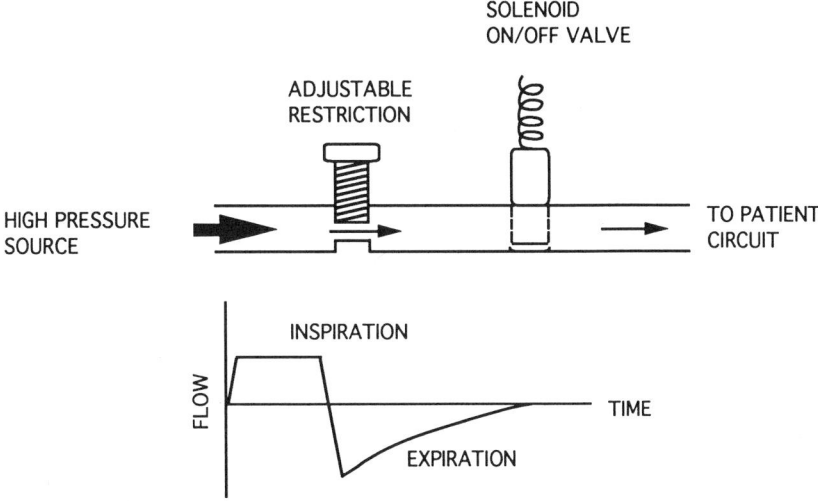

Figure 13. Orifice flow (or adjustable restriction in a tube) classically used as a constant flow generator (modified from Spearman et al. 107).

flow ventilators, Peslin (82) has proposed a method based on the simple concept of resistance internal to the ventilator which describes the negative pressure dependence of flow. When the internal resistance is high, the generated flow is less pressure dependent. In generators with moving pieces, the performance essentially depends on the power of the motor. In fluidic generators, the performance depends on the dominant conservation process.

To illustrate the nature of the interaction between ventilator and patient, we describe below the principle of a classical fluidic flow generator with high internal resistance: the orifice flow. Then, we present the mechanical characteristics of the CMV mode.

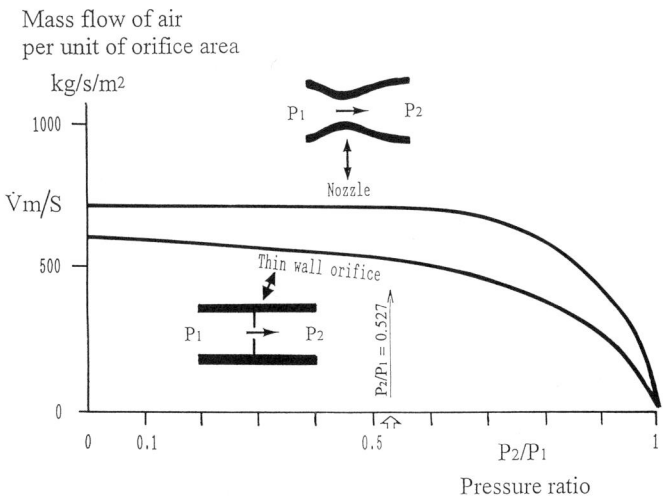

Figure 14. Typical relationship between mass flow per unit of surface area, $\dot{V}m/S$, and downstream to upstream pressure ratio, P_2/P_1, through two types of orifice: nozzle and thin wall orifice (from 46). Mass flow rate is independent on pressure ratio for the nozzle only and for pressure ratio below $P_2/P_1 = 0.527$.

Fluidic Flow Generation

Pneumatic flow generators generally comprise a source of compressed gas (3 to 5 Bars) connected to an adjustable resistance which is constituted by an orifice whose size may be adjusted as shown in figure 13.

If the size of the orifice is sufficiently small, i.e., the resistance sufficiently elevated, this system is supposed to deliver a flow independent of downstream pressure. The latter reflects the patient's mechanical conditions. However, classical results of fluid dynamics demonstrate that the flow through an orifice decreases as downstream pressure increases. This decrease requires a flow compensation usually provided by a servo-mechanism whose performances will be improved if the flow generator is optimized according to the following criterion.

Application of Bernoulli theorem to incompressible fluids at the passage through orifices, such as a thin wall hole, a nozzle or a Venturi, leads to parabolic relationships between flow ($\dot{V}$) and pressure drop ($P_1 - P_2$), or pressure ratio (P_2/P_1) through the orifice:

$$\dot{V} = \xi.S.[[2(P_1 - P_2)/\rho]^{0.5} = \xi.S.(2P_1/\rho)^{0.5}[1 - (P_2/P_1)]^{0.5}$$

where S is the minimal area of the passage, ρ the gas density at the upstream pressure P_1. P_2 is the pressure downstream the orifice. ξ is a correction coefficient currently varying from 0.5 to 1, depending on the inner shape of the passage. Values even slightly higher than 1 might be obtained if the orifice is terminated by a slow diverging portion of tube. Due to the parabolic shape of the relationship between flow and pressure ratio, the decay in flow across the orifice is slow when pressure ratio P_2/P_1 is much smaller than 1, as shown by above equation.

However, due to gas compressibility, above equation does not apply to gases as soon as pressure ratio markedly differs from 1. Note that this latter case applies during mechanical ventilation since the gas source available in Intensive Care Units is 2-3 bars (2000-3000 hPa) while airway pressure never exceeds 100 hPa.

The mass flow rate, $\dot{V}m$, defined as the product of density and volumetric flow ($\dot{V}m = \rho \dot{V}$) is the pertinent quantity to describe compressible flow through orifices. By contrast to other types of orifices, a nozzle minimizes frictional losses and will create a mass flow independent of downstream pressure below a critical value of the pressure ratio ($P_2/P_1 < 0.527$ in air). The maximum mass flow possible (in air) through a nozzle is given by the following equation (46):

$$\dot{V}m = 0.0404.S.P_1/(T_1)^{0.5}$$

in which $\dot{V}m$ (in kg/s) depends on the cross-sectional area of the throat, S (in m^2), and on the gas physical conditions upstream the orifice : absolute pressure P_1 (in Pascal) and absolute temperature T_1 (**in** °K). Most important, $\dot{V}m$ does not depend on the pressure difference across the nozzle in the range $P_2/P_1 < 0.527$ (in air) which may be understood as a flow limitation effect analogous to the well-known "waterfall" effect. This flow limitation effect results from the coupling between mass conservation - which predicts that maximum velocity necessarily occurs in the nozzle throat - and energy conservation associated to a reversible thermodynamic process - which predicts that gas velocity in the throat cannot exceed the local speed of sound.

In classical textbooks on respirators (107), the flow limitation property of nozzles has been mistakenly extended to any kind of orifice shape, while only nozzles behave in this way (see fig. 3). Moreover, a constant mass flow ($\dot{V}m$) does not necessarily mean constant volumetric flow ($\dot{V}$). For instance, an increase in maximum pressure from 0 to 100 hPa would

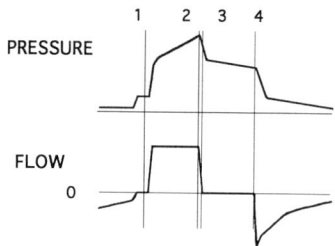

Figure 15. Typical pressure and flow signals measured at airway opening during (Flow) Controlled Mechanical Ventilation including end-inspiratory and end-expiratory pauses.

result in a 10% increase in gas density and thus a 10%-decrease in volumetric flow ($\dot{V}$) if the source pressure is maintained constant (P_1). Incidentally, the downstream-pressure-dependence effect would be reinforced in the subsonic flow regime.

Mechanical Effects of Flow (or Volume) Controlled Ventilation

Assuming that inner resistance of the constant flow ventilator is much larger than respiratory resistance of the ventilated patient and that the patient does not develop any inspiratory muscle activity, CMV signals resemble typical signals presented in fig.15.

The latter provides pressure P_1 which give an estimate of auto-PEP. Peak airway pressure (P_2) which comprises the resistive and the elastic component, is not anymore considered as a reliable index of the risks associated to a mechanical ventilation at high airway pressure: hemodynamic alteration, risk of barotrauma, risk of overinflation pulmonary edema. The pressure measured immediately after airway interruption (P_3) represents the alveolar pressure, whereas the pressure at the end of the end-inspiratory pause corresponds to the pressure of the relaxed system (P_4). An estimate of airway resistance including the endotracheal tube (R or R_{max} depending on the model used) can be obtained from $(P_2-P_4)/\dot{V}$, whereas static compliance can be obtained from $V_T/(P_4-P_1)$ (from 42).

By principle, the CMV mode requires a totally passive respiratory system. The machine controls the inspiratory cycle in terms of flow and time while only time is controlled during the expiratory cycle. The monitoring of a patient's mechanical parameters is facilitated during CMV by a perfect control of the input signal (the constant flow) and by the absence of any inspiratory activity. To obtain mechanical parameters in spite of a certain degree of patient inspiratory activity, an esophageal catheter allowing measurement of pleural pressure has to be used (see section « PHYSIOLOGICAL EFFECTS OF INSPIRATORY PRESSURE SUPPORT »). In the absence of any patient's activity, the esophageal pressure allows estimation of the mechanical properties of the chest wall.

Absence of inspiratory activity most often requires sedation and curarization of the patient, in order to inhibit any spontaneous breathing reflex. If complete inhibition of inspiratory activity has potential advantages such as reducing oxygen consumption, it renders hazardous the period of weaning from CMV because respiratory drive insufficiency caused by sedation has been identified as a cause of weaning failure Some mechanically ventilated patients remain dependent on their machine and cannot be discharged from the Intensive Care Unit (42).

During the expiratory cycle, the airway pressure represents the resistance of the expiratory circuit (tube and valve). Although the duration of expiration is ventilator-limited, expiratory duration is supposed to be sufficient to guarantee a complete exhalation, i.e., lung volume decreases down to the relaxation volume (called Functional Residual Capacity, FRC) whereas, at the same time, alveolar pressure reaches atmospheric pressure. However, a complete exhalation is not always possible in clinical conditions because, due to elevated expiratory resistance (airways + ventilator circuit), time constant ($\tau = R/E$) for expiration happens to be longer than the time fixed on the machine for expiration. In a simple R-E model as described above, it can be shown that the time necessary to expire 95% of the tidal volume roughly represent three time constants. Since airway resistance may easily be increased by 4 in altered lungs, expiratory time should also be increased by 4, which does not let enough time for oxygenation. Thus, in many clinical conditions, dynamic hyperinflation is responsible for some residual positive alveolar pressure at the end of expiration, called auto-PEP or intrinsic PEP, and the lung volume at the end of expiration is clearly above the relaxation volume (FRC). It should however be noted that expiration is not always entirely passive. For instance, patients with Chronic Obstructive Pulmonary Disease (COPD) activate abdominal muscles or intercostal muscles (76) which contributes to increase expiratory flow. The resulting effect is an increase in abdominal pressure and secondarily an increase in pleural pressure which might be mistakenly interpreted as an auto-PEP related effect.

A second mechanism contributing to auto-PEP is bronchial collapses occurring during expiration, i.e., when the outer airway wall pressure (pleural pressure) is higher than the inner airway pressure. In such a case, transmural pressure is negative. This phenomenon is known to happen in every subject during forced expiration maneuver (see the chapter by Shin et al. included in this book). In normal flow conditions, bronchial collapses can however be observed in emphysematous patients because alteration of their bronchoalveolar structure results in an increase in lung compliance and thus a reduction in lung elastic recoil. Bronchial collapse is responsible for flow limiting phenomena which are often detectable by a plateau appearing on the expiratory flow curve (42). In clinical practice, these phenomena cannot be seen by a pressure manometer placed at airway opening, unless an interruption maneuver is performed at the end of expiration (see P_1 in fig. 15).

The above characterization of the respiratory system during CMV does not really apply to highly inhomogeneous lung structures for reasons explained above. In such a case, ventilation is distributed preferentially in regions with relatively low time constants. It has sometimes been thought that the inspiratory flow waveform, e.g., decelerating versus square, was able to influence the ventilation distribution. This is why most modern ventilators include an adjustment of flow waveform (square, sinusoidal, adjustable gradient). However, in most of intensive care situations, these effects are small (13). By contrast, ventilation distribution may be altered by the reduction in diaphragmatic displacement, (by at least 50%), caused by factors associated with CMV: supine position, sedation, muscular paralysis, small amplitude of tidal volumes, variations in wall compliance. It is now considered that diaphragmatic inactivity associated to CMV mode is responsible for a dramatic alteration in ventilation distribution. This factor constitutes one of the two mechanical factors responsible for deleterious CMV complications. The other mechanical factor is pulmonary barotrauma which depends on the pressure gradient between alveolus and pulmonary interstitium. The latter is initially an alveolar disruption with penetration of gas through the small bronchovascular sheath separating alveoli from bronchiole, venule and arteriole. Then, gas migrates toward the mediastinal area, cervical, subcutaneous and peritoneal spaces (108). Lastly, among a number of non-mechanical factors responsible for CMV complications, an important one is the respiratory tract infection which depends on the duration of intubation (108).

Ventilation by High Frequency Oscillations: High Frequency Oscillation (HFO) ventilation has been intensely studied in the 1980s as a potential means to avoid barotraumatisms (18, 55). With HFO, both inspiratory and expiratory phases are active. Oscillatory volumes, ranging from 1 to 3 ml/kg, are clearly smaller than the anatomical dead space constituted by conducting airways, by contrast with CMV or spontaneous breathing for which tidal volumes are clearly higher than dead space volume (113). This is why alveolar pressure amplitude was thought to be smaller during HFO than during CMV, thus resulting in a potential reduction of barotraumatism during HFO. The low level of oscillatory volumes is compensated by the high rate at insufflation-exhalation frequency, ranging from 1 to 60 Hz, i.e., more than one order of magnitude greater than the spontaneous breathing frequency. A variety of devices have been proposed to generate the to-and-fro oscillations: piston pumps, loud speakers, and linear magnetic motors. O_2 supply and CO_2 elimination with this technique are obtained by means of a fresh gas flow, or bias flow, circulating across the circuit connecting the patient to the HFO device. Two different types of bias flow have been used (55, 56, 113): first, a low impedance bias flow, allowing the patient to breathe spontaneously throughout, but absorbing a certain part of the generated oscillatory volume; Second, a high impedance bias flow allowing a better control of the oscillatory volume entering the patient. This high impedance system has been used in most HFO studies investigating gas transport during HFO. With the high impedance bias flow, the patient is forced to follow the machine, i.e., spontaneous breathing activity is not tolerated. It is worth noting that HFO presents the same limitations as CMV considering the lack of compatibility with patient's breathing activity. To avoid intubation, High Frequency Oscillations may be delivered transthoracically (18, 49, 103) although transmission of oscillations through thoracic structure appeared problematic in adult humans and particularly in patients with Chronic Obstructive Pulmonary Disease (COPD) (85).

The difference between HFO and CMV resides in the involved gas transport mechanisms. Gas transport by direct bulk convection appears of minor importance during HFO, while it plays a major role during both CMV and spontaneous breathing. It is equivalent to

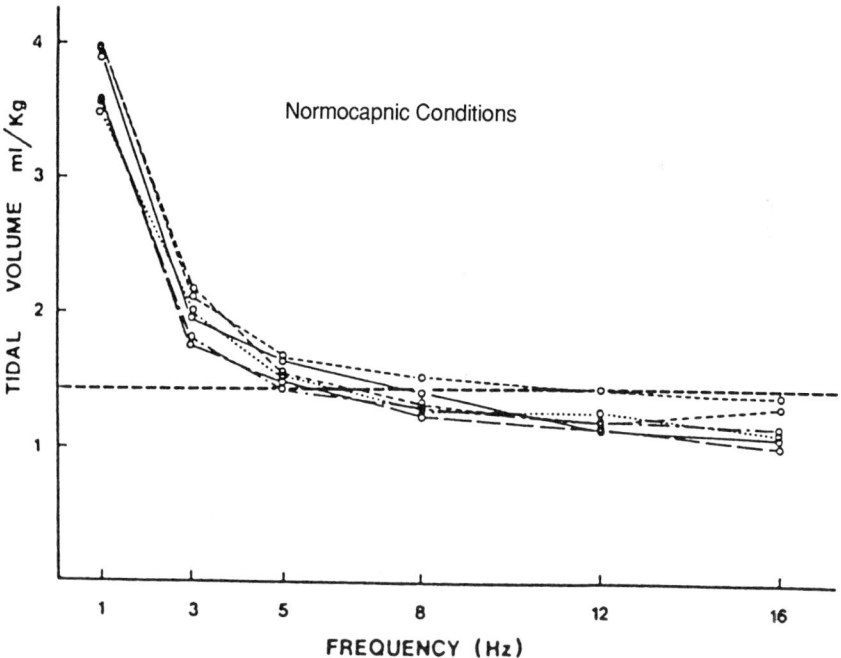

Figure 16. Tidal volume versus frequency relationship to maintain normal partial pressure of CO2 in the arterial blood of anesthetized and paralyzed normal rabbits ventilated by HFO (external oscillations) (from 87). Horizontal dotted line represents the mean dead space (in ml/kg) calculated at 1Hz by applying Bohr's equation (77). This result shows that it is actually very difficult to ventilate by HFO with oscillatory volumes smaller than 80% of the dead space volume.

say that the physiological concept of direct alveolar ventilation, classically used to describe gas exchange fails to explain gas exchange during HFO (18, 48, 55, 87, 94, 114).

$$\dot{V}_A = (V_T - V_D).f$$

$\dot{V}_A$: alveolar ventilation, V_T: tidal volume, V_D: anatomical dead space, f: frequency.

A number of theoretical studies has permitted identification of a variety of gas transport mechanisms that could play a complementary role during HFO (18, 55). These mechanisms are: (i) convective mechanisms such as convective streaming, *pendelluft*, augmented dispersion, alveolar ventilation by out of phase HFO, cardiogenic oscillations, or (ii) diffusive mechanisms such as pure molecular diffusion or augmented dispersion. These different mechanisms critically depend on the flow regime which reinforces the interest of performing fluid mechanical studies to understand oscillatory flow in complex branching geometry (21, 35, 43, 47, 51, 105, 110). Three zones of interest have been identified in the airways (113). They are essentially based on flow regime near peak flow. Upper airways, trachea and large airways constitute the first zone where turbulent flow predominates and a privileged site for convective gas transport mechanisms and augmented dispersion. Intermediate and small airways constitute the second zone characterized by laminar flow along with convective streaming, out of phase HFO, and augmented dispersion. The alveolar region where convection is negligible constitutes the third zone where gas transport occurs by molecular diffusion, *pendelluft*, cardiac oscillations. Predictive models of HFO gas transport are more or less based on the previous partition of airway flow and related gas transport mechanisms (56, 81, 100, 112). Predictive models show that the upper part of the respiratory tract appears the factor limiting the rate of gas exchange during HFO. Consistently, the endotracheal tube was also shown to alter HFO efficiency (95). Moreover, these models predict that gas transport efficiency depends on the square of the oscillatory tidal volume while it remains proportional to frequency. Experimental studies, mostly performed in small animals, have confirmed these predictions (87, 114). It means that, among a variety of dispersive mechanisms whose efficiency is undoubtedly enhanced at high frequency compared to normal breathing frequency, direct convection remains an effective mode of gas transport compared to other modes. This partly explains why normal gas exchange remains difficult to achieve by HFO at tidal volumes below 80% of dead space volume, even though oscillatory frequency is strongly increased (see fig. 16).

Some other factors have limited the generalization of HFO ventilation in adults. For different reasons, certain factors: (i) the delivered oscillatory volume (17), (ii) alveolar pressure variation, appeared difficult to accurately control due to resonant phenomena (32, 48). Moreover, increase in mean airway pressure (and lung volume) was susceptible to decrease gas transport efficiency during ventilation by HFO, secondary to the concomitant increase of the volume of conducting airways (117). In addition, control of lung volume appeared difficult during HFO because dynamic hyperinflation resulting from asymmetries of inspiratory and expiratory impedances (especially the asymmetries related to expiratory flow limitation phenomena (106)), is not detectable at airway opening. These uncontrolled factors may explain the variability of early HFO results. Later results obtained in conditions as controlled as possible, have shown that HFO was not superior to CMV except in a few cases. The results of a multicenter study (The HIFI group study (109)) in premature infants ventilated for respiratory failure with fully controlled HFO modalities did not reveal any advantage of HFO over CMV. However, an increasing use of HFO can be observed in pediatric intensive care units even though results have not been published yet. Villar et al. (113) have suggested that regular re-expansion of the lungs by positive pressure maneuvers were not sufficient in the HIFI group study compared to what was done during successful animal HFO studies. This contributes to explain the variability observed concerning the HFO results obtained in vivo. Control of ventilation parameters in neonates remains particularly difficult since, in addition to the problems mentioned above, there is a variable leak between the endotracheal tube and the trachea, whose importance depends on patient's pulmonary mechanics (31).

In adults, it is reasonable to consider that HFO ventilation and CMV failed to succeed for a common reason which is the absence of compatibility of high impedance ventilators with spontaneous muscle activity. From this point of view, High Frequency Jet Ventilation (96, 113) which is sometimes assimilated to HFO, does not actually behave as a high impedance generator but pertains to the category of low impedance generators whose principle is described below.

PRESSURE SUPPORT MODES

Muscle activity during mechanical ventilation has been initially taken into account by permitting isolated cycles of spontaneous breathing in between periods of controlled cycles. This mode called Synchronized Intermittent Mandatory Ventilation (SIMV) (26), along with the extensive use of pressure controlled modes (42, 107) reveals a progressive but fundamental evolution in the field of artificial ventilation. After SIMV, a further degree of autonomy has been permitted with partial ventilatory support modes, in which a large, but variable, part of the ventilation is controlled by the patient. Pressure Support (PS) modes pertain to this category (13, 67).

Pressure Support modes are characterized by a systematic detection of inspiratory activity which allows the synchronization of pressure assistance cycle by cycle (53, 68). Detection of the inspiratory effort is not a new problem. Let us take the example of the application of a Continuous Positive Airway Pressure (CPAP) (with continuous flow circuit) which has been early used in an attempt to improve recruitment of collapsed alveoli and better ventilation-perfusion matching. Adaptation of the CPAP mode to ventilators required to design inspiratory valves capable to detect inspiratory effort, i.e., most often a small pressure decay (from atmospheric pressure or from PEEP level), in a circuit temporary closed by valves and deliver flow on demand as long as patient activity is detected. However, in comparison to continuous flow circuits, the demand-flow mechanism of several ventilators has been shown to perform poorly (42). Indeed, certain valves offered an abnormally elevated delay to open, causing an additional inspiratory effort for the patient (22). By contrast, in PS devices as well as in most recent demand-flow systems, detection of the patient's inspiratory effort is made in an open circuit by precisely measuring the onset of inspiratory flow, or by precisely comparing flow at entrance and flow at the exit of a bias flow in which the patient breathes (97). It is more and more admitted that early detection of inspiratory activity is crucial because inspiratory muscle activity develops even before inspiratory flow becomes effective (12).

Another important feature of PS is the type of function used to control airway pressure in the system. Different modes of Pressure Support corresponding to different types of control functions have been proposed: (i) the Inspiratory Positive Airway Pressure (IPAP) in which the pressure waveform is almost constant (53, 67), (ii) the Proportional Assist Ventilation (PAV) in which the generated pressure is proportional to patient demand and basically includes a first component proportional to flow and a second component proportional to volume (118). Both systems (IPAP and PAV) have in common to offer to the patient the quantity of fresh air, or oxygen, that he desires, whereas they are supposed to maintain the desired function of pressurization throughout the whole inspiration. They sensibly differ in terms of degree of the patient's autonomy since IPAP is supposed to maintain a constant level of pressurization at airway opening whatever the flow-demand, while PAV attempts to adjust the pressure level on the flow-demand, at each instant. Compared to CMV, these different modes require a higher degree of control because internal impedance of generator devices tends to zero. Moreover, servo-controlled mechanisms may happen to be not sufficient if motor systems used to generate pressure are flow-limited. It is equivalent to say that inner impedance of PS machines is not always negligible compared to patient impedance.

To illustrate this problem, we describe below the performance of a fluidic system, the turbulent confined jet, that has been shown to give satisfactory clinical results without the adjunction of a servo-mechanism (12).

Fluidic Pressure Generation

Early pressure generators did not offer satisfactory steady and dynamic performances (73, 107). The desired level of pressurization was not maintained above a certain flow value, i.e., when inspiratory assistance was really needed by the patient to overcome flow resistance. This might explain why early pressure modes were rapidly abandoned to the benefit of the CMV mode which were more reliable.

To pressurize an unlimited quantity of gas at a pressure in the range 5-30 mbars, the turbulent jet confined in an open tube appears to satisfy the performances required to be used in the Intensive Care Unit. This system was used in a non-optimized fashion in early pressure generators (107), as well as to mix air and oxygen for oxygen therapy (20). It is also used during HFJV (96) but the actual geometrical conditions (trachea or large airways) are difficult to control. Turbines constitute another important type of pressure generator which are often used in home devices. Confined turbulent jet systems are often confused with a Venturi system (107). Their principle is however quite different (44, 74).

Due to the high shear stress between the jet and the surrounding air, the turbulent confined jet system consumes most of the kinetic energy in friction. By contrast, energy losses remain negligible in the Venturi system (99) which rather resembles the flow of an incompressible fluid through a Venturi orifice (see section Fluidic Flow Generation).

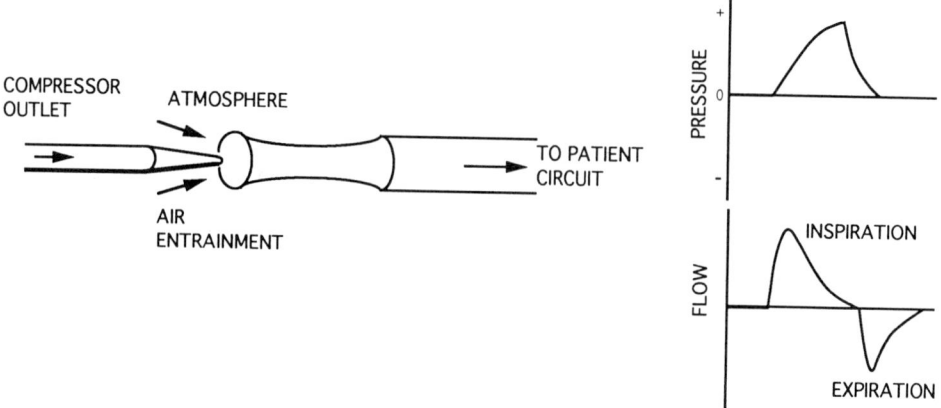

Figure 17. Turbulent jet in a tube with entry opened to atmosphere, classically used as a non-ideal pressure generator (modified from 107). Gas injection is responsible for air entrainment from atmosphere and a flow-dependent pressure generation in the patient circuit.

Jets developing in free atmosphere drag along a large amount of surrounding gas flow as a result of the preservation of the cross-sectional momentum flux in the axial jet direction (88). In contrast to free jets, turbulent jets developing in confined environments, i.e., bounded by lateral walls, generate a longitudinal pressure gradient opposed in sign to the main flow (88). As a result of the preservation of the cross-sectional averaged thrust (sum of the momentum flux and of the pressure force) between entry and exit of mixing chamber, there is an inverse relationship between the amount of entrained gas and the generated pressure (3, 23). Thus, confined turbulent jets are not pure pressure generators but behave closely like them. Their pressure performances may however be improved by an appropriate choice of the geometry of the chamber where the jet mixes with ambient (or entrained) air.

Only a few previous engineering studies on turbulent confined jets specifically focused on the estimate of the pressure generated at the exit (3, 10, 23, 39, 40). To our knowledge, none of these studies looked at the flow-dependence of the generated pressure in these systems. Elementary models presented below which describe confined jet mixing suggest that chamber geometry, injector size as well as velocity of injected flow should all play a role in the flow dependence of generated pressure.

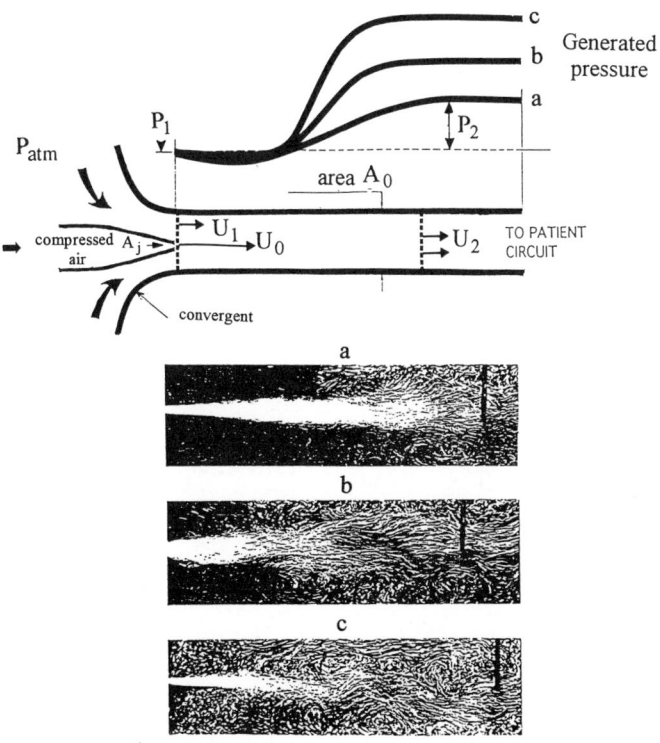

Figure 18. Mixing of a turbulent jet with ambient air in a tube (constant circular cross-section) produces a positive longitudinal pressure gradient opposed to the main flow (from (3, 23)). Generated pressure is related to the decrease in velocity of the surrounding fluid and to the eddy production resulting from high shear stress between jet and ambient air. The stronger eddy production, the higher the generated pressure. Flow patterns shown in a, b, c, correspond to different values of the Craya-Curtet number [Ct is an increasing function of the ratio between flow at the exit $(U_2.A_0)$ and jet flow $(U_1.A_j)$], Ct = 0.075 (a); 0.27 (b); 0.48 (c). The smaller Ct, the higher the pressure generated, as a result of the preservation of the pressure plus momentum integral (total thrust) at any cross section.

PRESSURE-FLOW RELATIONSHIP IN TURBULENT CONFINED JETS

We consider a turbulent jet (U_0: jet velocity; A_j: jet cross-section area) developing in a cylindrical tube, or mixing chamber, with one entry opened-to-atmosphere (see Fig. 18). U_1, is the average velocity of the secondary stream (or entrained flow). U_2 is the velocity at the exit of the chamber. P_1 and P_2 are the static pressure respectively at the entry and at the exit of the chamber. A_0 and A are respectively the entry and the exit cross-sectional areas of the tube.

Three types of mixing chamber were compared (19): (i) constant cross-section area ($A=A_0$), (ii) convergent-divergent tube with identical entry and exit area ($A=A_0$) and a minimal area located in middle of the tube, (iii) divergent tube with an exit cross section area larger than entry cross-section ($A>A_0$). Neglecting wall friction along the tube wall as well as forces acting on the lateral tube walls, the equations for mass and momentum conservation for an incompressible fluid can be written:

$$\rho.U_0.A_j + \rho.U_1.(A_0 - A_j) = \rho.U_2.A$$

$$P_1.A_0 + \rho.U_0^2.A_j + \rho.U_1^2.(A_0 - A_j) = P_2.A + \rho.U_2^2.A$$

Since inspiratory flow can become zero, U_1 may be positive or negative. Depending on the sign of U_1, the generated pressure, P_2, has different expressions.

1. case $U1 \geq 0$: Assuming smooth and rounded entrance conditions, Bernoulli equation gives:

$$P_1 = P_{atm} - \frac{1}{2}\rho.U_1^2$$

The pressure resulting from the transformation of momentum flux from the entry to the exit of the chamber may be understood as a sum of two effects: a constant area effect ($\beta=A/A_0=1$) and a complementary effect specifically resulting from the area divergence ($\beta>1$):

$$P_2 - P_{atm} = (P_2 - P_{atm})_{\beta=1} + (P_2 - P_{atm})_{\beta>1}$$

$$(P_2 - P_{atm})_{\beta=1} = \frac{\rho}{2}\left[\frac{\sigma.(2-3\alpha)}{(1-\sigma)^2}U_0^2 - \frac{2.\sigma.(1-2\sigma)}{(1-\sigma)^2}.U_0.U_2 - \frac{\sigma^2+(1-\sigma)^2}{(1-\sigma)^2}.U_2^2\right]$$

This equation shows that the generated pressure P_2 in a constant area tube is a decreasing parabolic function of velocity U_2, which also depends on the area ratio $\sigma = a/A_0$ and on jet velocity U_0; the smaller σ or U_0, the less flow-dependent is the generated pressure.

$$(P_2 - P_{atm})_{\beta>1} = \left[-\frac{\beta-1}{\beta}.P_{atm} + (\beta-1).\frac{1-2\sigma}{(1-\sigma)^2}.\frac{\rho U_2^2}{2}\right]$$

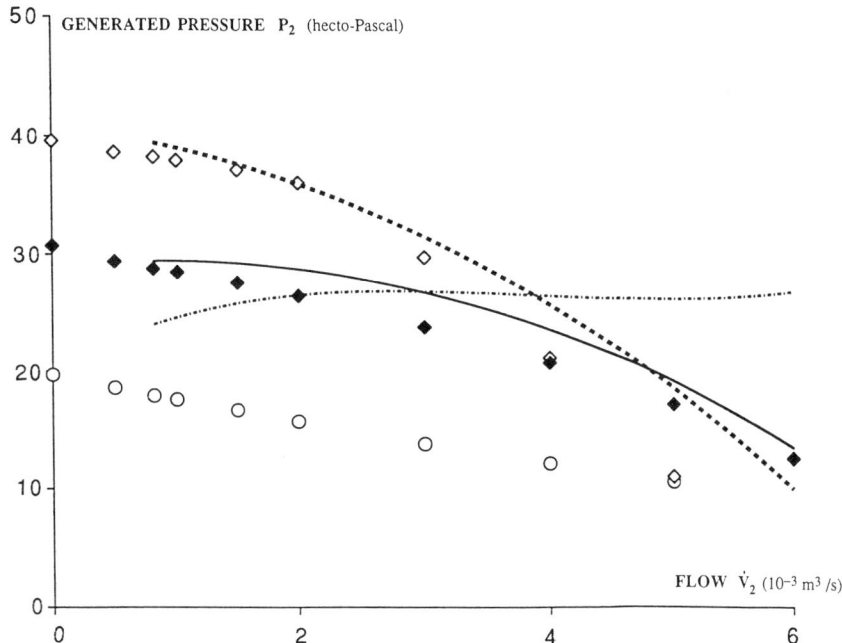

Figure 19. Relationship between generated pressure, P_2 in hPa, and inspiratory flow V_2 ($= U_2.A$) in Liter/second (10^{-3} m³/s), at the exit of chambers with axisymmetrical turbulent confined jet: comparison between experimental results and theoretical results obtained with an integral method by Hill (39) (method not presented in this chapter), and for U_1 (= velocity of entrained air) positive. Three different shapes of mixing chambers were tested (19): (i) constant and circular cross-section (continuous line: Hill's method, closed diamonds: experiments), (ii) convergent-divergent (dotted line: Hill's method, open diamonds: experiments), (iii) totally divergent with 2.5°-half angle (dashed-dotted line: Hill's method, open circles: experiments).

This equation indicates that the specific contribution of the divergence is an increasing parabolic function of the velocity U_2. As U_2 increases, the contribution of the divergence to the generated pressure, P_2, increases and may counterbalance the decay in generated pressure predicted by the straight tube contribution. These results suggest that divergent chambers theoretically provide ideal pressure generators to the extent that friction phenomena related to divergence of the chamber do not become significant (see Fig. 19).

Note on Fig.19 that the largest discrepancy between experimental and theoretical results is obtained for the totally divergent chamber although the flow-dependence of generated pressure is the smallest compared to two other chambers tested. Note also that the smallest flow-dependence of generated pressure is obtained for the divergent chamber. Further studies are necessary to understand the results obtained with the divergent tube.

2. Case $U_1 < 0$: When a part of injected flow flows back toward atmosphere, $P_1 = P_{atm}$. Thus:

$$P_2 - P_{atm} = \frac{\sigma}{1-\sigma} \cdot \rho \cdot (U_0 - U_2)^2$$

This equation indicates that P_2 is also a second degree decreasing function of U_2.

It is easy to verify the continuity between functions P_2-U_2 and their derivatives, at the point $U_2 = \sigma.U_0$, i.e., $U_1 = 0$.

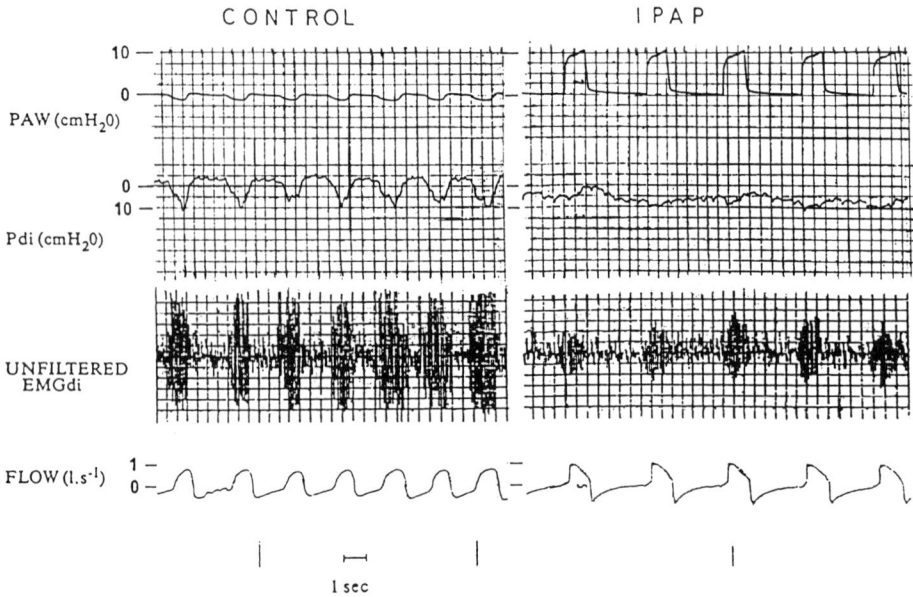

Figure 20. Airway pressure, transdiaphragmatic pressure, electromyographic activity of the diaphragm, and ventilatory pattern, during Spontaneous Breathing (left) and Inspiratory Assistance by Positive Airway Pressure at 10 hPa (right). Note the drastic reductions in both diaphragmatic activity and frequency in spite of the increase in tidal volume allowed by IPAP (from 12).

Such an axisymmetrical turbulent jet confined in a straight-divergent chamber has recently been used as pressure generator in an Inspiratory Assistance Device (12). Non-axial turbulent jets have also been used to generate a positive pressure near the tracheal extremity of endotracheal tubes (44). The principle is similar to that reported above, except that non-axial incidence of jets and/or friction between jets and with the wall, dissipate an even higher portion of kinetic energy. These jets, issued from capillaries extruded in the ETT wall, generate a positive pressure gradient capable to counterbalance the endotracheal tube resistance. For a sufficient injected flow, these jets are able to generate a positive pressure which can be used as a positive pressure ventilator usable in the frequency controlled mode or in the assist mode (9). The theoretical derivation above explains why, during HFJV, airway pressure which is created by jets developing invasively into the airways, is difficult to control (96).

Physiological Effects of Inspiratory Pressure Support

Inspiratory Positive Airway Pressure (IPAP) is now available in most modern ICU ventilators and constitutes a relatively new mode, alternative to CMV or applicable in complement to CMV. IPAP has initially been applied through an endotracheal tube, e.g., to facilitate weaning from CMV (11, 28). IPAP aims at reducing the inspiratory work of breathing cycle by cycle, and notably the additional inspiratory work caused by the endotracheal tube in patients breathing spontaneously. The diminution in work of breathing has been assessed by various studies in which esophageal pressure, and thereby work of breathing, electromyogram of the diaphragm and/or accessory inspiratory muscles were measured (11, 16, 53, 67, 69). There is an inverse linear relationship between respiratory frequency and the level of inspiratory assistance, and a direct linear relationship between tidal volume and the IPAP level (11, 59, 67). Above a certain IPAP level, e.g., 10 hPa in normal subjects, muscle activity has been shown to fall significantly (except for the initial triggering effort) both in healthy subjects (1) and in patients (11). At high levels of IPAP, IPAP level and respiratory compliance are the major determinants of tidal volume. Moreover, the sensitivity of ventilation to the concentration in inspiratory CO_2 has been found to disappear at high levels of IPAP, while ventilation is highly sensitive to CO_2 at low level of IPAP (101). This result has been used in awake normal subjects to assess static compliance of the respiratory system during cycles of pressure support for which relaxation of inspiratory muscles was ascertained (1). This method has not yet been applied to evaluate patients in the course of pressure support, although this could be done if the ventilator was able to generate mechanical cycles in which pressure level is high enough to ensure the patient's relaxation.

Classical applications of non invasive techniques have been initially limited to home ventilation. Incidentally, a pressure-limited ventilation is highly preferable during non-invasive ventilation because, above 25 hPa of pressure assistance, air might penetrate into the stomach. Application of non-invasive techniques as part of the treatment of patients in the acute stage of disease is relatively recent (Meduri et al. (71)). Non invasive application of IPAP during aggravation of chronic obstructive pulmonary disease has been very recently proposed and clinically evaluated by Brochard et al. (12). The pressure generator used in that study was a turbulent confined jet. Detection of inspiratory effort was made on the flow signal. In a first study (12), physiological effects of IPAP were evaluated in 11 patients with acute exacerbation of COPD after 45 minutes of assistance. Partial pressure of carbon dioxide was found to fell significantly down to the basal value of the chronic stage, whereas partial pressure of oxygen rose significantly. Importantly, the improvement in gas exchange, permitted by a significant increase in tidal volume, was accompanied by a marked reduction in respiratory rate. In the same study, several other signs of fatigue diminution were observed such as a tendency to pH normalization, decrease in diaphragmatic activity and reduction in transdiaphragmatic pressure swings similar to the results reported in intubated patients (11, 16). The therapeutic efficacy of the method has been evaluated against a matched historical-control group of patients treated conventionally (oxygenotherapy, antibiotherapy, bronchodilators). Results showed that, in the group treated with IPAP, only 1 over 13 patients needed tracheal intubation, while, in the control group, 11 over 13 patients needed tracheal intubation. Accordingly, the length stay in the intensive care unit was significantly reduced, i.e., 7±3 days in the group treated with IPAP, vs 19±13 days in the conventional group.

These results have been confirmed by the results of a recent multicentric study performed in 85 acute COPD patients (14). In this randomized study, the conventional treatment was either associated with inspiratory assistance (at least 6 hours a day), or without it. The use of non-invasive ventilation significantly reduced the need for endotracheal intubation from 74% to 26%. Moreover, complications (48% vs 16%), in-hospitality mortality rate (29% vs 9%), and also total length in the hospital (35±33 vs 23±17 days) were all significantly reduced in the group treated with IPAP. These two studies demonstrate that inspiratory pressure support via a face mask constitutes an alternative treatment to conventional mechanical ventilation to reduce hypercapnia in severe respiratory deficiency. But this conclusion is not fully consistent with some other clinical studies performed with different machines and in other experimental conditions. A number of methodological factors and primarily the quality of the IPAP device appear important. Indeed, depending on the gravity of the illness, the quality of the device, its rapidity to respond may become determinant for the success of the PS method. Beyond the quality of the Pressure Support delivered by the pressure generator, and/or the servo-control, another cause of the failure of the PS method reported in some other randomized studies is massive leaks occurring via the mouth when a nasal mask was used (29). Moreover, because PS mode is compatible with patient activity, the success of the treatment with non invasive IPAP strongly depends on the tolerance and acceptability of the technique by the patient. Moreover, due to the interaction between the patient and the machine, the technique requires from the clinician a substantial degree of learning and training. Although complete evaluation of PS modes is not yet finished, there is a now common agreement to consider that non invasive inspiratory pressure assistance reduces the complications associated with mechanical ventilation, mortality, and can shorten the length of the hospital stay.

CONCLUSION AND PERSPECTIVES

Most of the ventilatory modes mentioned in this chapter are available in modern mechanical ventilators which means that a wide panel of modalities is now offered to the clinician to mechanically treat respiratory deficiency. In terms of improvement of mechanical ventilation, gas transport efficiency through the airways constitutes certainly an important goal, but not the only one. This point may be illustrated by the decade of intensive studies on gas transport during HFO while clinical applications of HFO remained limited to experimental protocols (except in pediatric ICU).

In addition to gas transport improvement, minimizing dead space appears an important objective, especially in severely hypercapnic patients (13). Several attempts have been made during both flow and pressure controlled modes by insufflating, continuously or in a certain phase of ventilation, fresh gas in the lower part of the trachea. At low injected flow, the aim of tracheal gas injection is to reduce the quantity of CO_2 resident in airways at the end of expiration (13). At higher injected flow, tracheal gas insufflation avoids oxygen desaturation caused by aspiration of mucus secretions in ETT (15), counterbalances ETT airway pressure drop (44), and generates a distal Pressure Support (9). Constant Flow Insufflation (113) even allows the maintenance of life in anesthetized and paralyzed animals with however a smaller efficiency in humans.

A second important objective of mechanical ventilation as reflected by the past evolution is to maintain the pressure in the respiratory system below a level that would cause barotraumatisms or alveolar disruption secondary to excessive lung distension (volotraumatism). This objective has been in part achieved with pressure controlled modes, which in principle are pressure limited. This objective also concerns the classical CMV mode for which a recent modification, called Permissive Hypercapnia, has consisted in using only small tidal volumes, the idea being to tolerate a controlled increase in arterial CO_2 partial pressure instead of risking alveolar barotraumatisms (54); This mode can only be used in patients who have no cerebral or vascular disease.

The third important objective that can be given to mechanical ventilation is to remain compatible with a variable degree of patient's inspiratory activity. This evolution has been permitted by the Pressure Support modes recently proposed although the criteria to optimize pressure application and patient's effort detection are not totally known today. For instance, the therapeutic benefit of Proportional Assist Ventilation (118) compared to Inspiratory Positive Airway Pressure (11, 12, 16, 53, 67, 71) is not clearly established. In any case, the compatibility of mechanical ventilation with patient's activity permitted by the PS modes appears an irreversible evolution as it reduces the risks inherent to CMV. It is reasonable to consider that future improvements of assistance modes will require a precise and continuous measurement and analysis of patient's activity and state. This means a better monitoring and control of mechanical parameters whose variation could otherwise compromise the expected effect of partial pressure assistance. In this perspective, tests performed in the course of mechanical ventilation and initially developed for CMV are not directly applicable to PS. To improve monitoring during PS, the first task is to generate reliable input functions as was done for CMV several years ago. Then, maximum pressure (PS level) and form of the generated pressure signal might be adjusted to the continuous change in patient's mechanical parameters. At the same time, non-invasive methods of evaluation have to be developed and integrated to the monitoring and to the control functions of PS modes.

Such a control with complex processes requires considerable expertise (13). Preliminary study performed in ICU patients (25) suggests that intelligent monitoring may assist the clinical staff to make medical decisions, especially in critically ill patients. These systems might be able to handle in real time a huge mass of information about patient's status, to diagnose observed situations and to construct action plans with prompt reaction in alarming cases. Future systems could integrate several sorts of knowledge including fluid and tissue mechanics, in order to build a comprehensive view of patient evolution in normal and pathological conditions. This ultimate goal reinforces the importance of a pluridisciplinary approach of mechanical ventilation along with a considerable place for biomechanics.

ACKNOWLEDGMENTS

The authors gratefully thank D. Touchard for technical assistance especially in clinical studies, and Drs M. Dojat, F. Lofaso, B. Louis for their expertise and helpful discussions in the course of the present research program.

LIST OF ABBREVIATIONS

- ARDS : Acute Respiratory Distress Syndrome
- CMV : Controlled Mechanical Ventilation
- COPD : Chronic Obstructive Pulmonary Disease
- CPAP : Continuous Positive Airway Pressure
- ETT : Endotracheal tube
- FRC : Functional Residual Capacity
- HFO : High Frequency Oscillation (ventilation by HFO)
- ICU : Intensive Care Unit
- IPAP: Inspiratory Positive Assistance Pressure
- PAV : Proportional Assist Ventilation
- PEEP: Positive End-Expiratory Pressure
- PS : Pressure Support
- SB : Spontaneous Breathing
- SIMV: Synchronized Intermittent Mandatory Ventilation
- ZEEP : Zero End-Expiratory Pressure

REFERENCES

1. Azarian, R., Lofaso, F., Zerah, F., Lorino, H., Atlan, A., Isabey, D., and Harf, A., 1993, Assessment of the respiratory compliance in awake subjects using pressure support, *Eur. Respir. J.*, 6: 552-558.
2. Bachofen, H., and Hildebrandt, J., 1971, Area analysis of pressure-volume hysteresis in mammalian lungs, *J. Appl. Physiol.*, 30: 493-497.
3. Barchilon, M., and Curtet, R., 1965, Some details of the structure of an axisymmetrical confined jet with backflow, *Fluids Eng. Conf., A.S.M.E. Pap.*, 64-FE-23: 1-17.
4. Bates, J.H.T., Baconnier, P., and Milic-Emili, J., 1988, A theoretical analysis of interrupter technique for measuring respiratory mechanics, *J. Appl. Physiol.*, 64: 2204-2214.

5. Bates, J.H.T., Rossi, A., and Milic-Emili, J., 1985, Analysis of the behavior of the respiratory system with constant inspiratory flow, *J. Appl. Physiol.*, 58: 1840-1848.
6. Bayliss, L.E., and Robertson, G.W. 1939, The visco-elastic properties of the lungs, *Q. J. Exp. Physiol.*, 29: 27-47.
7. Behrakis, P.K., Higgs, B.D., Baydur, A., Zin, W.A., and Milic Emili, J., 1983, Respiratory mechanics during halothane anesthesia and anesthesia-paralysis in humans, *J. Appl. Physiol.* 55: 1085-1092.
8. Ben Fabry, Guttman J., Eberhard, L., and Wolff, G., 1994, Automatic compensation of endotracheal tube resistance in spontaneously breathing patients, *Technology and Health Care*, 1: 281-291.
9. Beydon, L., Isabey, D., Boussignac, G., Bonnet, F., Duvaldestin, P., and Harf, A., 1991, Pressure support ventilation using a new tracheal gas injection tube, *Br. J. Anaesth.*, 67: 795-800.
10. Binder, G., and Kian, K., 1983, Confined jets in a diverging duct, *Proceed. 4th Symp. on Turbulent Shear Flows*, Kalrsruhe, p. 7.18-7.23.
11. Brochard, L., Harf, A., Lorino, H., Lemaire, F., 1989, Inspiratory pressure support prevents diaphragmatic fatigue during weaning from mechanical ventilation, *Am. Rev. Respir. Dis.*, 139: 513-521.
12. Brochard, L., Isabey, D., Piquet, J., Amaro, P., Mancebo, J., Messadi, A., Brun-Buisson, C., Rauss, A., Lemaire, F., Harf, A., 1990, Reversal of acute exacerbations of chronic obstructive lung disease by inspiratory assistance with a face mask. *New Eng. J. Med.*, 323: 1523-1530.
13. Brochard, L., and Mancebo, J., 1994, *Réanimation-Ventilation artificielle-Principes et applications*, Arnette, Paris, 405 p.
14. Brochard, L., Mancebo, J., Wyzocki, M., Lofaso, F. Conti, G., Rauss, A., Simmoneau, G., Benito, S., Gasparetto, A., Lemaire, F., Isabey, D., and Harf, A., 1995, Efficacy of non-invasive ventilation for treatment of acute exacerbation of chronic obstructive lung disease. Results of a multicenter randomized trial. Submitted to *New Eng. J. Med.*
15. Brochard, L., Mion, G., Isabey, D., Bertrand, C., Messadi, A.A., Mancebo, J., Boussignac, G., Vasile, N., Lemaire, F., and Harf, A., 1991, Constant-flow insufflation prevents arterial desaturation during endotracheal suctioning, *Am. Rev. Respir. Dis.*, 144: 395-400.
16. Brochard, L., Pluskwa, F., and Lemaire, F., 1987, Improved efficacy of spontaneous breathing with inspiratory pressure support, *Am. Rev. Respir. Dis.*, 136: 411-415.
17. Brusasco, V., Beck, K.C., Crawford, M., and Rehder, K., 1986, Resonant amplification of delivered volume during high-frequency ventilation, *J. Appl. Physiol.*, 60: 885-892.
18. Chang, H.K., and Harf, A., 1984, High-frequency ventilation: A review, *Respir. Physiol.*, 57: 135-152.
19. Chaofan, S., Pigeot, J., and Isabey, D., 1995, Génération de pression par jet turbulent confiné: application en assistance respiratoire, *Arch. Int. Physiol. Biochim. Biophys.*, in press.
20. Cohen, J.L., Demers, R.R., and Saklad, M., 1977, Air-entrainment oxygen masks: a performance evaluation. *Respir. Care*, 22: 277-282.
21. Corieri, P., Benocci, C., Paiva, M., and Riethmuller, M., 1991, Numerical and experimental investigation of lung bifurcation flows, in: *NATO ASI Series*, Plenum Press, New York.
22. Cox, D., Tinloi, S.F., and Farrimond, J.G., 1988, Investigation of the spontaneous modes of breathing of different ventilators, *Intensive Care Med.*, 14: 532-537.
23. Curtet, R., and Ricou, F.P., 1964, On the tendency of self-preservation in axisymmetric ducted jets, *J. Basic Eng.*, 777-787.
24. D'Angelo, E., Prandi, E., Tavola, M., Calderini, E., and Milic Emili, J., 1994, Chest wall interrupter resistance in anesthetized paralyzed human, *J. Appl. Physiol.*, 77: 883-887.
25. Dojat, M., Brochard, L., Lemaire, F., and Harf, A., 1992, A knowledge-based for assisted ventilation of patients in intensive care, *Int. J. Clin. Monit. Comp.*, 9: 239-250.
26. Downs, J., Klein, E., Desautels, D., Modell, J., Kirby, R., 1973, Intermittent mandatory ventilation: a new approach to weaning patients from mechanical ventilators, *Chest*, 64: 331-335.
27. Drazen, J.M., Loring, S.H., and Ingram, R.H., Jr, 1976, Localisation of airway constriction using gases of varying density and viscosity, *J. Appl. Physiol.*, 41: 396-399.
28. Fiastro, J.F., Habib, M.P., and Quan, S.F., 1988, Pressure support compensation for inspiratory work due to endotracheal tubes and demand continuous positive airway pressure, *Chest*, 93: 499-505.
29. Foglio, C. Vittaca, M., Quadri, A., Scalvini, S., Marangoni, S., and Ambrosino, N., 1992, Acute exacerbations in severe COLD patients. Treatment using positive pressure ventilation by nasal mask, *Chest*, 101: 533-538.
30. Frank, N.R., Mead, J., and Whittenberger, J.L., 1971, Comparative sensitivity of four methods for measuring changes in respiratory flow resistance in man, *J. Appl. Physiol.*, 31: 934-938.
31. Fredberg, J.J., Glass, G.M., Boyton, B.R., and Frantz I.D. III, 1987, Factors influencing mechanical performance of neonatal high-frequency ventilators, *J. Appl. Physiol.*, 62: 2485-2490.

32. Fredberg, J.J., Keefe, D.H., Glass, G.M., Castile, R.G., and Frantz III, I.D.,1984, Alveolar pressure nonhomogeneity during small-amplitude high-frequency oscillation, *J. Appl. Physiol.*, 57: 788-800.
33. Fredberg, J.J., and Stamenovic, D., 1989, On the imperfect elacticity of lung tissue, *J. Appl. Physiol.*, 67: 2408-2419.
34. Froese, A.B., and Bryan, A.C., 1974, Effects of anesthesia and paralysis on diaphragmatic mechanics in man, *Anesthesiology*, 41: 242-255.
35. Gavriely, N., Solway, J., Loring, S.H., Butler, J.P., Slustky, A.S., and Drazen, J.M., 1985, Pressure-flow relationships of endotracheal tubes during high-frequency ventilation, *J. Appl. Physiol.* 59: 3-11.
36. Gottfried, S.B., Rossi, A., Higgs, B.D., Calverley, P.M.A., Zocchi, L., Bozic, C., and Milic Emili, J., 1985, Noninvasive determination of respiratory system mechanics during mechanical ventilation for acute respiratory failure, *Am. Rev. Respir. Dis.*, 131: 414-420.
37. Hildebrandt, J., 1970, Pressure-volume data of cat lung interpreted by a plastoelastic, linear viscoelastic model, *J. Appl. Physiol.*, 27: 365-372.
38. Hoppin, F.G., Stothert, J.C., Greaves, I.A., Lai, Y.L., Hildebrandt, J., 1986, Lung recoil: elastic and rheological properties, In: *Handbook of Physiology. The Respiratory System (Section 3, vol. III)*, Macklem, P.T., and Mead, J., American Physiological Society, Bethesda, p. 195-215.
39. Hill, P.G., 1967, Incompressible jet mixing in convergent-divergent axisymmetric ducts, *J. Basic Eng.*, 89: 210-220.
40. Hill, B.J., 1973, Two-dimensional analysis of flow in jet pumps, *J. Hydraul. Division*, : 1009-1026.
41. Ingram, I., Jr., and Pedley, T.J., 1986, Pressure-flow relationships in the lungs, In: *Handbook of Physiology. Respiration (section 3, vol. III)*, Macklem, P.T., and Mead, J., American Physiological Society, Bethesda, p. 277-293.
42. Iotti, G., Brochard, L., and Lemaire, F., 1992, Mechanical ventilation and weaning. In: *Care of the critically ill patient* (2nd ed.), Tinker, J. and Zapol, W.M., Springer-Verlag, London, chapt. 29, p. 457-477.
43. Isabey, D., 1982, Steady and pulsatile flow distribution in a multiple branching network with physiological applications, *J. Biomech.* 15: 395-404.
44. Isabey, D., Boussignac, G., and Harf, A., 1989, Effect of air entrainment on airway pressure during endotracheal gas injection, *J. Appl. Physiol.*, 67: 771-779.
45. Isabey, D., and Chang, H.K., 1981, Steady and unsteady pressure-flow relationships in central airways, *J. Appl. Physiol.*, 51: 1388-1348.
46. Isabey, D., and Harf, A., 1995, Basic physical principles for ventilators and ventilatory modes. In: *Acute Respiratory Failure in Chronic Obstructive Pulmonary Disease*, Derenne, J.P., Marcel Dekker Inc., New York.
47. Isabey, D., Chang, H.K., Delpuech, C., Harf, A., and Hatzfeld, C., 1993, Dependence of central airway resistance on frequency and tidal volume: a model study, *J. Appl. Physiol.*, 61: 113-126.
48. Isabey, D., Harf, A., Chang, H.K., 1985, Pressure change and gas mixing induced by oscillations in a closed system, *J. Biomech. Eng.*, 107: 68-76.
49. Isabey, D., and Piquet, J., 1989, The ventilatory effect of external oscillation, *Acta Anaesthesiol. Scand.*, 90: 87-92.
50. Jaffrin, M.Y., and Kesic, P., 1974, Airway resistance: a fluid mechanical approach, *J. Appl. Physiol.*, 36: 354-361.
51. Jan, D.L., Shapiro, A.H., and Kamm, R.D., 1989, Some feature of oscillatory flow in a model bifurcation, *J. Appl. Physiol.*, 67: 147-159.
52. Jonson, B., Beydon, L., Brauer, K., Mansson, C., Valind, S., and Grytzell, H., 1993, Mechanics of respiratory system in healthy anesthetized humans with emphasis on viscoelastic properties, *J. Appl. Physiol.*, 75: 132-140.
53. Kacmarek, R., 1988, The role of pressure support ventilation in reducing work of breathing, *Respir Care*, 33: 99-120.
54. Kacmarek, R., and Hickling, K.G., 1993, Permissive Hypercapnia, *Respir. Care*, 38: 373-387.
55. Kamm, R.D., Slutsky, A.S., Drazen, J.M., 1984, High-frequency ventilation. *C.R.C. Crit. Rev. Biomed. Eng.*, 9: 347-379.
56. Khoo, M.C.K., Slutsky, A.S., Drazen, J.M., Solway, J., Gavriely, N., and Kamm, R.D., 1984, Gas mixing during high-frequency ventilation: an improved model, *J. Appl. Physiol.*, 57: 493-506.
57. Lemaire, F., 1986, *La ventilation artificielle*, Masson, Paris.
58. Ligas, J.R., 1990, Lung tissue mechanics: historical overview, in: *Respiratory Biomechanics*, Eipstein, M.A.F., and Ligas, J.R., Springer-Verlag, New-York, p. 3-18.
59. Lofaso, F., Isabey, D., Lorino, H., Harf, A., and Scheid, P., 1992, Respiratory response to positive and negative inspiratory pressure in humans, *Respir. Physiol.*, 89: 75-88.

60. Lofaso, F., Louis, B., Brochard, L., Harf, A., and Isabey, D., 1992, Use of Blasius resistance formula to estimate the effective diameter of endotracheal tubes, *Am. Rev. Respir. Dis.* 146: 974-979.
61. Lorino, A.M., and Harf, A., 1991, Measurement of respiratory elastance and resistance in mechanically ventilated patients. In: *Adult respiratory distress syndrome*, Zapol, W.M. and Lemaire, F., Marcel Dekker Inc., New York.
62. Lorino, A.M., and Harf, A., 1993, Techniques for measuring respiratory mechanics: an analytic approach with a viscoelastic model, *J. Appl. Physiol.*, 74: 2373-2379.
63. Louis, B., Glass, G., and Fredberg, J.J., 1994, Pulmonary airway area by the two-microphone acoustic reflection method, *J. Appl. Physiol.*, 76: 2234-2240.
64. Louis, B. and Isabey, D., 1993, Interaction of oscillatory and steady turbulent flows in airway tubes during impedance measurement, *J. Appl. Physiol.*, 74: 116-125.
65. Ludwig, M.S., Dreshaj, I., Solway, A., Munoz, A., and Ingram, R.H., Jr, 1987, Partitioning of pulmonary resistance during constriction in the dog: effects of volume history, *J. Appl. Physiol.*, 62: 807-815.
66. Macklem, P., and Mead, J., 1967, Resistance of central and peripheral airways measured by a retrogade catheter, *J. Appl. Physiol.*, 22: 395-401.
67. McIntyre, N.R., 1986, Respiratory function during pressure support ventilation, *Chest*, 89: 677-683.
68. Marini, J.J., 1990, Strategies to minimize breathing effort during mechanical ventilation. In: *Mechanical ventilation - Critical care clinics*, Tobin, M.J., W.B. Saunders Company, Philadelphia, vol.6, n°3, p. 635-661.
69. Marini, J.J., Capps J.S., Culver, B.H., 1985, The inspiratory work of breathing during assisted mechanical ventilation, *Chest*, 87: 612-618.
70. Mead, J., 1961, Mechanical properties of the lungs, *Physiol. Rev.*, 41: 281-330.
71. Meduri, G.U., Conoscenti, C.C., Menashe, P., and, Nair, S., 1989, Non invasive face mask ventilation in patients with acute respiratory failure, *Chest*, 95: 865-870.
72. Mörch, E.T., 1985, History of mechanical ventilation. In: *Clinical Applications of Ventilatory Support*, Kirby, R.B., Churchill Livingstone, New York, chapt.1, p.1-61.
73. Mushin, W.W., Rendell-Baker, L., Thompson P.W., Mapleson, W.W., and Hillard, E.K., 1980, *Automatic ventilation of the lungs* (3rd ed.), Blackwell Scientific Publications, Oxford, p. 62-131.
74. Nahum, A., Sznajder, J.I., Solway, J., Wood, L.D.H., and Schumater, P.T., 1988, Pressure, flow, and density relationships in airway models during constant-flow ventilation, *J. Appl. Physiol.*, 64: 2066-2073.
75. Navajas, D., Farre, R., Rotger, M., and Canet, J., 1989, Recording pressure at the distal end of the endotracheal tube to measure respiratory impedance, *Eur. Respir. J.*, 2: 178-184.
76. Ninane, V., Rypens, F., Yernault, J.C., and De Troyer, A., 1992, Abdominal muscle use during breathing in patients with chronic airflow obstruction, *Am. Rev. Respir. Dis.*, 146: 16-21.
77. Otis, A.B., 1964, Quantitative relationships in steady-state gas exchange, In: *Handbook of Physiology. Respiration (section 3, vol. I)*, Fenn, W.O., and Rahn, H., American Physiological Society, Washington, p. 681-698.
78. Otis, A.B., McKerrow, C.B., Bartlett, R.A., Mead, J., McIlroy, M.B., Selverstone, N.J., and Radford, E.P., Jr., 1956, Mechanical factors in distribution of pulmonary ventilation, *J. Appl. Physiol.*, 8: 427-443.
79. Pedley, T.J., Drazen, J.M., 1986, Aerodynamic theory. In: *Handbook of Physiology. Respiration (section 3, vol. III)*, Macklem, P.T., and Mead, J., American Physiological Society, Bethesda, p. 41-54.
80. Pedley, T.J., Schroter, R.C., and Sudlow, M.F., 1971, Flow and pressure drop in systems of repeatedly branching tubes, *J. Fluid Mech.*, 46: 365-383.
81. Permutt, S., Mitzner, W., and Weinmann, G., 1985, Model of gas transport during high-frequency ventilation, *J. Appl. Physiol.*, 58: 1956-1970.
82. Peslin, R.L., 1969, The physical properties of ventilators in the inspiratory phase. *Anesthesiology*, 30: 315-324.
83. Peslin, R., and Fredberg, J.J., 1986, Oscillation mechanics, In: *Handbook of Physiology. The Respiratory System (section 3, vol. III)*, Macklem, P.T., and Mead, J., American Physiological Society, Bethesda, p. 277-293.
84. Petty, T.L.,1990, A historical perspective of mechanical ventilation. In: *Mechanical ventilation - Critical care clinics*, Tobin, M.J., W.B. Saunders Company, Philadelphia, vol.6, n°3, p.489-504.
85. Pimmel, R.L., Tsai, M.J., Winter, D.C., and Bromberg, P.A., 1978, Estimating central and peripheral respiratory resistance, *J. Appl. Physiol.*, 45: 375-380.
86. Piquet, J., Brochard, L., Isabey, D., De Cremoux, H., Chang, H.K., Bignon, J., and Harf, A., 1987, High frequency chest wall oscillation in patients with chronic air-flow obstruction, *Am. Rev. Respir. Dis.*, 136: 1355-1359.
87. Piquet, J., Isabey, D., Chang, H.K., and Harf, A., 1985, Stable normocapnia during High frequency body surface oscillation in rabbits, *Am. Rev. Respir. Dis.*, 132: 104-108.

88. Rajaratnam, N., 1976, *Developments in water science. Turbulent jets.* Amsterdam: elsevier scientific publishing company.
89. Rehder, K. and Marsh, M.H., 1986, Respiratory mechanics during anesthesia and mechanical ventilation. In: *Handbook of Physiology. Respiration (section 3, vol. III)*, Macklem, P.T., and Mead, J., American Physiological Society, Bethesda, p. 737-752.
90. Rodarte, J.R., Rehder, K., 1986, Dynamics of respiration. In: *Handbook of Physiology. Respiration (section 3, vol. III)*, Macklem, P.T., and Mead, J., American Physiological Society, Bethesda, p. 131-144.
91. Rohrer, F., 1915, Der strömungswiderstand in den menschlichen atemwegen und der einfluss der unregelmässigen verzweigung des bronchialsystems auf den atmungsverlauf verschiedenen lungenbezirken, *Pfluegers Arch. Gesamte Physiol. Menschen Tiere*, 162: 225-229.
92. Rossi, A., Gottfried, Higgs, B.D., B., Zocchi, L., Grassino, A., and Milic-Emili, J.,1985, Respiratory mechanics in mechanically ventilated patients with respiratory failure, *J. Appl. Physiol.*, 58:1849-1858.
93. Rossi, A., Gottfried, B., Zocchi, L., Higgs, B.D., Lennox, S., Calverley, P.M.A., Begin, P., Grassino, A., and Milic-Emili, J.,1985, Measurement of static compliance of the respiratory system in patients with acute respiratory failure during mechanical ventilation, *Am. Rev. Respir. Dis.*, 131:672-677.
94. Rossing, T., Slutsky, A.S., Lehr, J., Drinker, P.A., Kamm, R.G., and Drazen, J.M., 1981, Tidal volume and frequency dependence of carbon dioxyde elimination by high-frequency ventilation, *N. Engl. J. Med.*, 305: 1375-1379.
95. Rossing, T.H., Solway, J., Saari, A.F., Gavriely, N., Slutsky, A.S., Lehr, J.L., and Drazen, J.M., 1984, Influence of the endotracheal tube on CO2 transport during high-frequency ventilation, *Am. Rev. Respir. Dis.*, 129: 54-57.
96. Rouby, J.J., Simmoneau, G., and Benhamou, D., Sartène, R., Sardnal, F., Deriaz, H., Duroux, P., and Viars, P., 1985, Factors influencing pulmonary volumes and CO2 elimination by high-frequency jet ventilation, *Anesthesiology*, 63: 473-482.
97. Sassoon, C.S.H., Giron, A.E., Ely, E.A., and Light, R.W., 1989, Inspiratory work of breathing on flow-by and demand-flow continous positive airway pressure, *Crit. Care Med.*, 17: 1108-1114.
98. Sassoon, C.S.H., Kees Mahutte, and C., Light, R.W., 1990, Ventilator Modes: old and new. In: *Mechanical ventilation - Critical care clinics*, Tobin, M.J., W.B. Saunders Company, Philadelphia, vol.6, n°3, p. 605-634.
99. Scacci, R.P., 1979, Air entrainment masks: Jet mixing is how they work; the Bernoulli and Venturi principles are how they don't, *Respir. Care*, 24: 928-931.
100. Scherer, P.W., Haselton, F.R., Seybert, J.R., 1984, Gas transport in branched airways during high-frequency ventilation, *Ann. Biomed. Eng.* 12: 385-405.
101. Scheid, P., Lofaso, F., Isabey, D., and Harf, A., 1994, Respiratory response to inhaled CO2 during positive inspiratory pressure in humans, *J. Appl. Physiol.*, 77: 876-882.
102. Schroter, R.C., and Sudlow, M.F., 1969, Flow patterns in models of the human bronchial airways, *Respir. Physiol.*, 7: 341-355.
103. Shabtai, Y., and Gavriely, N., 1989, Frequency and amplitude effects during high-frequency vibration ventilation in dogs, *J. Appl. Physiol.*, 66: 1127-1135.
104. Slutsky, A.S., Berdine, G.B., and Drazen, J.M., 1980, Steady flow in a model of human central airways, *J. Appl. Physiol.*, 49: 417-423.
105. Snyder, B., Dantzker, D.R., and Jaeger, M., 1981, Flow partitioning in symmetric cascades of branches, *J. Appl. Physiol.*, 51: 598-606.
106. Solway, J., Rossing, T.H., Saari, A.F., and Drazen, J.M., 1986, Expiratory flow limitation and dynamic pulmonary hyperinflation during high-frequency ventilation, *J. Appl. Physiol.*, 60: 2071-2078.
107. Spearman, C.B., and Sanders, H.G., Jr., 1985, Physical principles and functional designs of ventilators. In: *Clinical Applications of Ventilatory Support*, Kirby, R.B., Churchill Livingstone, New York, chapt. 2, p.63-104.
108. Suter, P.M., 1992, Complications of mechanical ventilation. In: *Care of the critically ill patient* (2nd ed.), Tinker, J. and Zapol, W.M., Springer-Verlag, London, chapt. 30, p. 478-489.
109. The HIFI Study Group, 1989, High-Frequency oscillatory ventilation compared with conventional mechanical ventilation in the treatment of respiratory failure in preterm infants, *New Engl. J. Med.*, 320: 88-93.
110. Thiriet, M., Graham, J.M.R., and Issa, R.I., 1992, A pulsatile developing flow in a bend, *J. Phys. III*, 2: 995-1013.
111. Van Surell, C., Louis, B., Lofaso, F., Beydon, L., Brochard, L., Harf, A., Fredberg, J.J., and Isabey, D., 1994, Acoustic method to estimate the longitudinal area profile of endotracheal tubes, *Am. J. Respir. Crit. Care Med.*, 149: 28-33.

112. Venegas, J.G., Hales, C.A., and Strieder, D.J., 1986, A general dimensionless equation of gas transport by high-frequency ventilation, *J. Appl. Physiol.*, 60: 1025-1030.
113. Villar, J., Winston, B., and Slutsky, A.S., 1990, Non-conventional techniques of ventilatory support. In: *Mechanical ventilation - Critical care clinics*, Tobin, M.J., W.B. Saunders Company, Philadelphia, vol.6, n°3, p. 579-603.
114. Weinmann, G.G., Mitzner, W., and Permutt, S., 1984, Physiological dead space during high-frequency ventilation in dogs, *J. Appl. Physiol.*, 57: 881-887.
115. West, J.B., 1979, *Respiratory Physiology, The essentials (2nd ed.)*, The Williams and Wilkins company, Baltimore.
116. Wright, P.E., Marini, J.J., and Bernard, G.G., 1989, *In vitro* versus *in vivo* comparison of endotracheal tube airflow resistance, *Am. Rev. Respir. Dis.*, 140: 10-16.
117. Yamada, Y., Venegas, J.G., Strieder, D.J., and Hales, C.A., 1986, Effects of mean airway pressure on gas transport during high-frequency ventilation in dogs, *J. Appl. Physiol.*, 61: 1896-1902.
118. Younes, M., 1992, Proportional assist ventilation, a new approach to ventilatory support. *Am. Rev. Respir. Dis.*, 145:114-120.
119. Zin, W.A., Rossi, A., and Milic Emili, J., 1983, Model analysis of respiratory responses to inspiratory resistive loads, 55: 1565-1573.

17

BIOMECHANICS OF LYMPH TRANSPORT

G. W. Schmid-Schönbein and F. Ikomi

Department of Bioengineering, and Institute for Biomedical Engineering
University of California, San Diego
La Jolla, California, 92093-0412

INTRODUCTION

While lymphatics were probably observed by members of ancient schools, the first documented description of lymphatics in the dog mesentery is by the Italian Asellius in 1622. About 1652 the Swedish scientist Olaus Rudbeck at the University of Uppsala and independently the Danish anatomist Thomas Bartholin gave the first descriptions of lymphatics as a unidirectional transport system which carries fluid from the interstitial tissue compartment via a network of lymphatic conduits and nodes to the thoracic ducts and back into the venous circulation (Grotte, 1979). In addition to interstitial fluid, the lymphatics carry colloid particles of different sources as well as blood cells, such as lymphocytes. With the exception of the brain, virtually all organs have lymphatics. Man generates several liters of lymph fluid every day. While the flow rates in individual lymphatics are low compared with flow rates in the vascular system, the amount of fluid being transported by each lymphatic is still large. Even relative short interruption of the lymph flows lead to rapid swelling of lymphatics and the interstitial space. In spite of a colorful research history older than three hundred years, no consensus has been reached to explain the mechanism(s) by which interstitial fluid is carried into the terminal endings of the lymphatics and by which lymph fluid is retained and transported inside the initial lymphatics. Once lymphatic fluid reaches the contractile segments of the lymphatic system, fluid propulsion is achieved by peristaltic smooth muscle contractions and lymphatic valves to prevent reflow. While many aspects of peristaltic lymph pumping have been explored, the mechanism of lymph formation and transport in regions without lymphatic smooth muscle is, however, without even the most basic understanding. It will be the focus of the present discussion.

LYMPH VESSEL ORGANIZATION

According to the wall structure of individual lymphatic channels, the lymphatic system can be divided into two segments. The terminal segment of lymphatics, which includes the blind endings in the tissue, has only a single endothelial cell positioned on a discontinuous basement membrane which is attached via anchoring filaments to connective tissue fibers and parenchymal cells. We will refer to these as *initial lymphatics*. Initial lymphatics may or may not have intralymphatic valves, but importantly they do not have a lymphatic smooth muscle. Initial lymphatics have irregular lumen cross sections, including collapsed lumen cross sections. There is no documented evidence that initial lymphatics are contractile so to serve as a peristaltic pump. Since their lumen cross section is not circular, no significant circumferential hoop stress can be supported by the lymphatic wall, as for example in arterioles. Initial lymphatics are compressed and expanded by local tissue deformation and in many organs are positioned strategically in the adventitia of arterioles, bronchi, or mucosal smooth muscle, as well as other tissue structures with their own smooth muscle. The exact display of initial lymphatics differs among organs.

In contrast, the contractile segment of lymphatics serves to transport fluid towards lymph nodes and into the central right and left thoracic ducts. In the following they will be referred to as *collecting lymphatics*. Collecting lymphatics have a smooth muscle media and a circular cross section. They exhibit peristaltic contractions and contain lymphatic valves. Smooth muscle in collecting lymphatics has features that resemble smooth muscle in arterioles, including an innervation (Ohhashi, et al., 1982), myogenic response (Hargens and Zweifach, 1977), and endothelial - smooth muscle communication (Ohhashi and Takahashi, 1991).

PROPOSALS FOR LYMPH FORMATION

Lymphatic fluid appears to be derived largely from interstitial fluid after filtration of plasma across the microvascular endothelium. The capillary filtration process is described in first approximation by an equilibrium of hydrostatic and colloid oncotic pressures as described by the celebrated Starling relationship. The relative sparse layout of the lymphatic network in most organs requires that directed convective flow serves to transport fluid in the interstitium into the lymphatics. Convective transport requires a potential gradient such as fluid pressure or oncotic pressure gradients (Zweifach and Silberberg, 1979). This has led many investigators to assume that Starling forces may also be involved in lymph formation and transport of interstitial fluid into the initial lymphatics (Aukland and Reed, 1993).

In spite of repeated attempts, no experimental evidence has been advanced to show the presence of a *steady* pressure drop from the interstitial fluid space into the initial lymphatics, to provide a steady filling mechanism for the terminal lymphatic network (Clough and Smaje, 1978). Especially puzzling in this respect have been measurements of a negative fluid pressures in the interstitium (Guyton, et al., 1971) which in the light of close to zero intralymphatic pressures (Zweifach and Prather, 1975) would be expected to lead to actual reversal of fluid flow out of the lymphatics. Casley-Smith has proposed that osmotic pressures may serve to aspirate fluid from the interstitial space into the lymphatics (Casley-Smith, 1982).

Early investigators assumed that filling of the terminal lymphatic network is achieved by peristaltic pumping. This point of view was supported by experimental observations in the wing of the bat (Nicoll and Taylor, 1977), which has lymphatics that are indeed contractile in all parts, including initial lymphatic bulbs with a smooth muscle media (Hogan and Unthank, 1986). The bat wing, however, appears to be relativly specialized in this respect. It is the only organ where contractile initial lymphatics exist, and even other organs of the bat have non-contractile initial lymphatics. In man and several mammalian laboratory species (mouse, rat, rabbit, cat, dog) no contractile initial lymphatics have yet been encountered (Schmid-Schönbein, 1990).

In some organs the network of lymphatics has been reconstructed. For example, in the small intestine (Unthank and Bohlen, 1988) or in skeletal muscle (Skalak, et al., 1984) the entire lymphatic network inside the tissue parenchyma is made up of noncontractile initial lymphatics; collecting lymphatics are encountered only at locations where lymphatics leave the organ. Thus, major regions in these organs have no lymphatic smooth muscle, there is no evidence for peristaltic pumping, and therefore other mechanisms for lymphatic filling must be present.

Another lymph pump mechanisms was proposed by Reddy in the form of the retrograde pump (Reddy, 1986). This mechanism assumes peristaltic pumping by the collecting lymphatics. It is assumed that during expansion of contractile lymphatics a negative pressure develops in the initial lymphatics, which serves to aspirate fluid from the interstitial space into the initial lymphatics. Reversal of fluid flow in the collecting lymphatics is prevented by closure of intralymphatic valves. The elastic recoil during the expansion phase of peristalsis is provided by anchoring filaments of the endothelium to adjacent connective tissue. There is, however, no experimental confirmation of this pump mechanism (Zweifach and Prather, 1975).

EXPANSION AND COMPRESSION OF INITIAL LYMPHATICS

The lack of a contractile apparatus in the wall of initial lymphatics requires involvement of other tissue structures positioned adjacent to the lymphatics to expand and to compress the initial lymphatics. In this respect it is interesting to observe that in skeletal or heart muscle as well as in other organs the initial lymphatics are positioned in the adventitia of the arterioles/arteries (Skalak, et al., 1984). The paired arterioles and lymphatics deform in a reciprocal pattern. Relaxation of smooth muscle with expansion of the arterioles leads to compression of the adjacent initial lymphatic while constriction of the arterioles causes expansion of the lymphatic lumen (Intaglietta and Gross, 1982). Thus the arteriolar smooth muscle serves to pump lymph fluid. Similar anatomical arrangement are observed in lung bronchi or mucosal smooth muscle in the intestine.

One advantage of this arrangement is that a relatively strong arteriolar smooth muscle serves to expand and fill the initial lymphatics, so that even negative tissue fluid pressures will not significantly impede expansion and compression of initial lymphatics. During constriction of the arterioles, vascular smooth muscle contraction serves to expand the initial lymphatics, a mechanism which can readily generate 50 to 100 mmHg transmural pressures. During relaxation of the arterioles, the intra-arterial pressure serves to expand the arteriolar lumen and at the same time compress the initial lymphatics. In the absence of a lymphatic obstruction, the outflow resistance along the lymphatics is relatively low and the pressure in the initial lymphatics does not exceed a few mmHg (Wen, et al., 1994). During lymphatic obstruction, the intralymphatic pressure may reach considerably higher values and limit the expansion and compression of the initial lymphatics.

In the presence of arteriolar-lymphatic pairs, vasomotion as well as pulse pressure serves to expand and compress the initial lymphatics in a rhythmic fashion. During pressure pulsation, amplitude and frequency of the arteriolar expansion are different from the values encountered during vasomotion. The elastic expansion of the arterioles, and therefore the compression of the initial lymphatics, is determined by their elastic properties. The amplitude of the elastic diameter expansions of arterioles during pulse pressure tends to be considerably smaller than the amplitudes during vasomotion. The frequency is determined by the heart rate which is one order of magnitude faster than typical vasomotor frequencies (1 cycle about every 30 sec). An example in this regard are the experiments by Parsons and McMaster (McMaster and Parsons, 1938, Parsons and McMaster, 1938). These pioneers demonstrated that lymphatics in the dermis of the ear skin were able to carry an interstitial tracer injected at the tip of the ear towards the base only if pulsatile perfusion was present. Perfusion of the

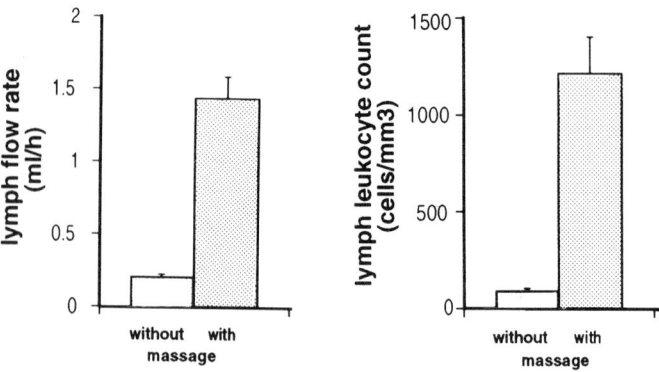

Figure 1. Mean and standard error of lymph flow rate (left) and lymph leukocyte counts (right) in afferent collecting lymph duct of rabbit hind leg prior to the popliteal node. The hind leg was rotated passively at a frequency of 0.03 Hz. Gentle superficial massage was applied over the dorsal skin in the region of the initial lymphatics that feed into the cannulated collecting lymphatic duct. Lymph flow rates without massage are significantly lower than with massage. In the absence of massage, lymph flow rates are lower without leg motion than with the leg motion (see also Figure 4). Adapted from (Ikomi, et al., 1995).

ear with steady pressure, but at the same mean pressure, did not lead to a significant removal of the interstitial tracer. Elevation of mean and pulse pressure in the intestine leads to increased lymph flow (Laine and Granger, 1983).

Furthermore, there are organ dependent mechanisms that serve to expand and compress the initial lymphatics. Studies in different tissues have shown that tissue motion enhances lymph flow, including studies in man (Olszewski and Engeset, 1980). Passive or active deformation of skeletal muscle fibers is accompanied by deformation of the adjacent lymphatics (Mazzoni, et al., 1990) and enhancement of lymph flow rates (Garlick and Renkin, 1970). In the hoof of the sheep the lymph flow in the collecting ducts is sensitive with respect to walking (McGeown, et al., 1988, McGeown, et al., 1988).

In the dog hind quarter, vibratory movement of the skin enhances lymph flow (Ohhashi, et al., 1991). In the dermis, initial lymphatics are positioned relatively close under the surface of the skin. It is apparent that external massage serves to compress and expand the soft and highly flexible initial lymphatics and increase lymph flow (Figure 1) (Ikomi, et al., 1995). Besides deliberate massage of the skin, even soft stroking of an animal fur may be sufficient to change the position of the hair follicles and may control the volume of the initial lymphatics that are positioned at the base of the follicles. Any change of the lumen volume is sufficient to promote the required cycle of expansion and compression of the initial lymphatics, irrespective of the order in which the cycle is completed. Periodic tension in the

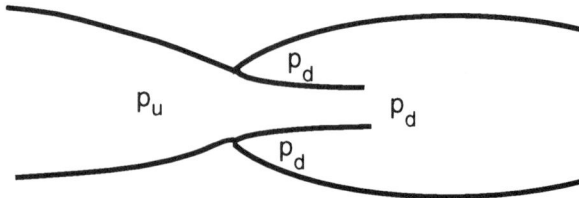

Figure 2. Schematic of a lymphatic vessel cross section containing a valve. p_u and p_d are the fluid pressures in the lymphatic lumen upstream and downstream of the valve, respectively. The downstream pressure p_d is almost uniform throughout the valve sinuses because there is no significant flow in the sinuses.

skin due to skeletal muscle motion during exercise may provide another mechanism sufficient for compression of the skin and induction of the expansion and compression cycle.

INTRALYMPHATIC VALVES

In contrast to the valves in the heart which close due to inertial fluid forces during fluid deceleration, the Reynolds number in the lymphatics is small, so that viscous forces dominate and inertial forces are negligible. Closure of lymphatic valves requires therefore another mechanism. The valves in the small lymphatics tend to be in form of bileaflets. The leaflets are highly flexible; they are attached by buttresses to the wall of the lymphatics forming relatively long funnel shapes in the lumen of the lymphatics. The valve leaflets and buttresses are made of a bilayer of lymphatic endothelial cells with a layer of collagen sandwiched in between.

The valve leaflets are closed by viscous pressure gradients along the length of the bileaflet funnel (Mazzoni, et al., 1987). When the pressure upstream of a lymphatic valve, p_u, is greater than the pressure downstream, p_d, flow is forward through the valve down a negative viscous pressure gradient (Figure 2). Since there is little flow in the valve sinuses, the downstream pressure, p_d, persists in the pockets of the leaflets so that the pressure drop, $p_u - p_d > 0$ pushes the leaflets open.

In contrast, during flow reversal the upstream pressure falls below the downstream pressure, $p_u < p_d$, a positive pressure gradient is set up in the funnel formed by the leaflets. As the pressure between the leaflets falls below p_d, a negative pressure drop, $p_u - p_d < 0$, develops across the leaflets since the downstream pressure p_d is relatively uniform in the valve sinuses. Thus the valve leaflets are pushed shut, preventing retrograde lymph flow. The bileaflets will readily close their lumen during lymph flow reversal in spite of relatively irregular lymph lumen cross sections, especially in the initial lymphatics.

ENDOTHELIAL MICROVALVES IN INITIAL LYMPHATICS

The endothelium in initial lymphatics has a structure which differs from vascular endothelium. Most striking is the lack of a continuous basement membrane and localized attachments via anchoring filaments. Instead of a continuous interendothelial tight junction in vascular endothelium, the interendothelial junctions in the wall of the initial lymphatics form an interdigitated structure which can provide a one way valve system for fluid movement. Depending on the degree of lymphatic distension, initial lymphatics exhibit macroscopic openings sufficient for unhindered passage of macromolecules (Castenholz, 1984). Interendothelial adhesion proteins, as seen in vascular endothelial junctions, are only sparsely and irregularly distributed, and the junctions can be opened during fluid inflow from the interstitium, during in-plane stretch of the lymphatic endothelium by overinflation of the lymphatics or by edema. The openings that can be formed by the interendothelial junctions are large, so that even colloids and cells enter the initial lymphatics (Figure 2). Interdigitations between neighboring endothelial cells tend to pull the junctions into a closed position (Schmid-Schönbein, et al., 1995). During compression a positive fluid pressure inside the initial lymphatic serves to compress the interendothelial junctions and prevent fluid return into the interstitium (Figure 3). During expansion a negative fluid pressure inside the initial lymphatics serves to open the interendothelial junctions in the wall and permit inflow of interstitial fluid.

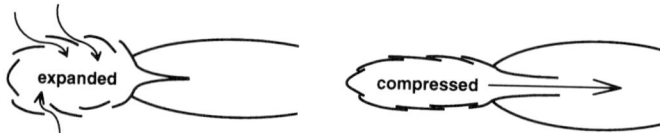

Figure 3. Schematic of initial lymphatic expansion with opening of interendothelial junctions and closure of intralymphatic valves, and during compression closure of interendothelial junctions and opening of intralymphatic valves. During expansion, fluid from the interstitial space enters the initial lymphatics, during compression it is transported away inside the lymphatic lumen.

LYMPH TRANSPORT MECHANISMS

To explore in more detail the influence of initial lymph compression and expansion, we measured quantitatively the amount of lymph flow in the hind leg of New Zealand rabbits during a controlled leg motion. The leg was rotated so that the toes followed a vertical circle with radius of 8 cm. The resultant lymph flow in an afferent lymphatic channel prior to entry into the popliteal node without rotation was small but finite (Figure 4). Without passive leg motion, lymph flow depends on pressure pulsation, since arrest of the heart beat leads to cessation of lymph flow. Lymph flow can also be maintained for several hours with active skin motion even after arrest of the heart. Increasing rotational frequencies yielded progressively elevated lymph flow rates. For almost two orders of magnitude of leg rotational frequencies, the experimental results yield a nonlinear relationship between the frequency of rotation and lymph flow rate (Ikomi and Schmid-Schönbein, 1995).

CONCLUSION

The current evidence suggests that a noncontractile initial lymphatic system drains into contractile lymphatics which carry the lymph fluid into nodes, the central lymphatic ducts, and back into the venous circulation. The initial lymphatics have irregularly shaped lumen cross sections and an attenuated endothelial layer with discontinuous basement membrane and non-contiguous cell junctions. Initial lymphatics appear to have two sets of

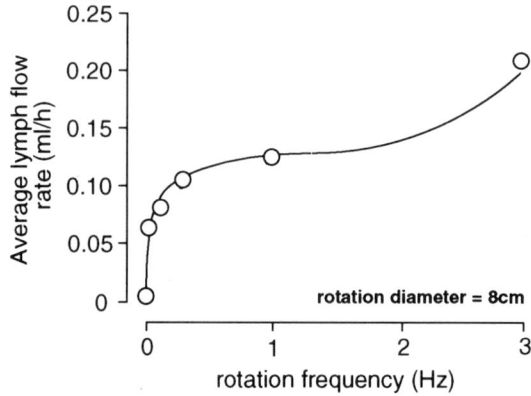

Figure 4. Average lymph flow rates in afferent collecting lymph duct of rabbit hind leg prior to the popliteal node. The hind leg was rotated in a saggital plane at different frequencies in a circle with 8 cm diameter. Adapted from (Ikomi and Schmid-Schönbein, 1995).

valves, one in their wall at the level of the endothelial junctions which prevent reflow into the interstitium during lymphatic compression, and the other one in form of the usual bileaflet intralymphatic valves which prevent reflow inside the lumen during lymphatic expansion. Lymph formation depends on periodic expansion of the initial lymphatics, while compression leads to emptying of the initial lymphatics into the contractile collecting lymphatics. Expansion and compression of initial collector lymphatics can be achieved by periodic tissue motions such as pressure pulsations, arteriolar vasomotion, skin massage, heart and skeletal muscle contraction, respiration, intestinal peristalsis, and other rhythmic tissue deformations.

ACKNOWLEDGMENTS

This work was supported by USPHS Grant HL10881 and a grant from Alliance Pharm., San Diego, CA.

REFERENCES

Aukland, K. and Reed, R. K., 1993, Interstitial-lymphatic mechanisms in the control of extracellular fluid volume. *Physiol. Rev.* 73:1-78.
Casley-Smith, J. R., 1982, Mechanisms in the formation of lymph. In: *Cardiovascular Physiology IV, International Review of Physiology*. A. C. Guyton and J. E. Hall, Ed., University Park Press, Baltimore, pp. 147-187.
Castenholz, A., 1984, Morphological characteristics of initial lymphatics in the tongue as shown by scanning electron microscopy. *Scanning Electron Microscopy* 1984:1343-1352.
Clough, G. and Smaje, L. H., 1978, Simultaneous measurement of pressure in the interstitium and the terminal lymphatics of the cat mesentery. *J. Physiol. (London)* 283:457-468.
Garlick, D. G. and Renkin, E. M., 1970, Transport of large molecules from plasma to interstitial fluid and lymph in dogs. *Am. J. Physiol.* 219:1595-1605.
Grotte, G., 1979, The discovery of the lymphatic circulation. *Acta Physiol. Scand. Suppl.* 463:9-10.
Guyton, A. C., Granger, H. J. and Taylor, A. E., 1971, Interstitial fluid pressure *Physiol. Rev.* 51:527-563.
Hargens, A. R. and Zweifach, B. W., 1977, Contractile stimuli in collecting lymph vessels. *Am. J. Physiol.* 233:H57-H65.
Hogan, R. D. and Unthank, J. L., 1986, Mechanical control of initial lymphatic contractile behavior in bat's wing. *Am. J. Physiol.* 251:H357-H363.
Ikomi, F., Hanna, G. and Schmid-Schönbein, G. W., 1995, Mechanism of colloidal particle uptake into the lymphatic system: Basic study with percutaneous lymphography. *Radiology* 196:107-113.
Ikomi, F. and Schmid-Schönbein, G. W., 1995, Lymph pump mechanics in the rabbit hind leg. *J. Appl. Physiol.* In review.
Intaglietta, M. and Gross, J. F., 1982, Vasomotion, tissue fluid flow and the formation of lymph. *Int. J. Microcirc. Clin. Exp.* 1:55-65.
Laine, G. A. and Granger, H. J., 1983, Permeability of intestinal microvessels in chronic arterial hypertension. *Hypertension* 5:722-727.
Mazzoni, M. C., Skalak, T. C. and Schmid-Schönbein, G. W., 1987, Structure of lymphatic valves in the spinotrapezius muscle of the rat. *Blood Vessels* 24:304-312.
Mazzoni, M. C., Skalak, T. C. and Schmid-Schönbein, G. W., 1990, The effect of skeletal muscle fiber deformation on lymphatic volume. *Am. J. Physiol.* 259:H1860-H1868.
McGeown, J. G., McHale, N. G. and Thornbury, K. D., 1988, Arterial pulsations and lymphformation in an isolated sheep hindlimb preparation. *J. Physiol.* 405:595-604.
McGeown, J. G., McHale, N. G. and Thornbury, K. D., 1988, Effects of varying patterns of external compression on lymph flow in the hindlimb of the anaesthetized sheep. *J. Physiol.* 397:449-457.
McMaster, P. D. and Parsons, R. J., 1938, The effect of the pulse on the spread of substances through tissues. *J. Exp. Med.* 68:377-400.
Nicoll, P. A. and Taylor, A. E., 1977, Lymph formation and flow. *Amer. Rev. Physiol.* 39:73-95.

Ohhashi, T., Kobayashi, S., Tsukahara, S. and Azuma, T., 1982, Innervation of bovine mesenteric lymphatics: From the histochemical point of view. *Microvasc. Res.* 24:377-385.

Ohhashi, T. and Takahashi, N., 1991, Acetylcholine-induced release of endothelium-derived relaxing factor from lymphatic endothelial cells. *Am. J. Physiol.* 260:H1172-H1178.

Ohhashi, T., Yokoyama, S. and Ikomi, F., 1991, Effects of vibratory stimulation and mechanical massage on micro- and lymph-circulation in the acupuncture points between the paw pads of anesthetized dogs. In: *Recent Advances in Cardiovascular Diseases.* H. Niimi and F. Y. Zhuang, Ed., National Cardiovascular Center, Osaka, pp. 125-133.

Olszewski, W. L. and Engeset, A., 1980, Intrinsic contractility of prenodal lymph vessels and lymph flow in human leg. *Am. J. Physiol.* 239:H775-H783.

Parsons, R. J. and McMaster, P. D., 1938, The effect of the pulse upon the formation and flow of lymph. *J. Exp. Med.* 68:353-376.

Reddy, N. P., 1986, Lymph circulation: physiology, pharmacology and biomechanics. *CRC Crit. Review in Biomed. Eng.* 14:45-91.

Schmid-Schönbein, G. W., 1990, Microlymphatics and Lymph Flow. *Physiol. Reviews* 70:987-1028.

Schmid-Schönbein, G. W., Kosawada, T., Skalak, R. and Chien, S., 1995, Membrane model of endothelial cells and leukocytes. A proposal for the origin of a cortical stress. *J. Biomech. Eng.* 117:171-178.

Skalak, T. C., Schmid-Schönbein, G. W. and Zweifach, B. W., 1984, New morphological evidence for a mechanism of lymph formation in skeletal muscle. *Microvasc. Res.* 28:95-112.

Unthank, J. L. and Bohlen, H. G., 1988, Lymphatic pathways and role of valves in lymph propulsion from small intestine. *Am. J. Physiol.* 254 (Gastrointest., Liver Physiol 17):G389-G398.

Wen, S., Dorffler-Melly, J., Herrig, I., Schiesser, M., Franzeck, U. K. and Bollinger, A., 1994, Fluctuation of skin lymphatic capillary pressure in controls and in patients with primary lymphedema. *Int. J. Microcirc.: Clin. Exp.* 14:139-43.

Zweifach, B. W. and Prather, J. W., 1975, Micromanipulation of pressure in terminal lymphatics in the mesentery. *Am. J. Physiol.* 228:1326-1335.

Zweifach, B. W. and Silberberg, A., 1979, The interstitial-lymphatic flow system. In: *International Review of Physiology - Cardiovascular Physiology III* A. C. Guyton and D. B. Young, Ed., University Park Press, Baltimore, pp. 215-260.

INDEX

Adventitia, 14, 177
Airways
  collapse, 37
  flow-induced oscillation, 37
  gas mixing, 35–37
    Lagrangian chaos, 37
    resonant, 37
  pressure drop, 34
    expiratory, 34
    inspiratory, 34
    oscillatory, 35
  resistance, 322, 336
  velocity profiles, 34–36
Airways, closure, 297, 302, 336
  central airways, 295, 297
  compliance, 289, 291, 295
  flutter, 306, 311
  geometry, 287, 293
  impedance, 287, 293
  network, 295
  obstruction, 296, 302
  small (peripheral) airways, 295, 299–300
  smooth muscle, 288, 297, 300, 308
  upper airway resistance, 291
  viscoelasticity, 311
  see also Collapsible tube flow; Lung ; Forced expiration, ; Obstructive disease
Algae
  bioconvection, 42, 43
  swimming, 43, 46
Anatomy, Greek, 1, 2
Aneurysm, 101, 122
  computer simulation of blood flow in, 101–103
Animation 117, 122
  frame by frame mode, 123
Aortic arch, 71, 78, 188
Aortic bifurcation, 72
Aortic stenosis, 126
Aorto-iliac arteries, 51
Apparent viscosity, blood, 165, 169
Arbitrary Lagrangian Eulerian method (ALE), 86

Arterial
  compliance, 227, 237–239, 248
  flow, 32, 33, 52
  modelling, 236
  system, 227–237, 248
  tree, 227–235
Arterial mass transfer
  in curved tubes, 110
  mathematical model, 87
Arteries, 32, 116, 177, 188
  distensibility, 141
  incremental elastic modulus, 141
  mechanical properties, 142
  muscular, 137
  radial, 143
  rat femoral, 145
  rat mesenteric 8, 146
  ulnar, 140, 143
  vertebral, 74
  wall shear stress, 32, 33
  see also vasomotion
Arteriole lymphatic pairs, 355
Arterioles, 137, 168
Asthma, 289, 296, 308
Atherogenesis, 51, 117,
Atherosclerosis, 32, 69, 79, 116, 127, 177, 190, 193
Atomic theory, Greek, 2
Atria,
  small artery, 257
  small vein, 257
Atrial contraction, 257
Auto transfusion, 219, 223

Back filtration, 217
Bacteria, 44
  bioconvection, 44, 45
  chemotaxis, 44
Basal lamina, 13, 18
Bias flow, 337
Bifurcating flows, 52
Bifurcation of arteries, 66, 127
  apex, 129

Bifurcation (cont.)
  atherosclerosis prone region, 129
  proximal regions, 129
Bifurcations, microvascular, 166
  phase separation, 166
  plasma skimming, 166
Bio-fluid dynamics, 131
Bioconvection, 42–45
Biomechanics, 5
Blasius Formula, 326
Blood cells, 8
  active motion, 9
  leukocytes, 8
  locomotion, 9
  red cells, 8
  rheology, 8
Blood circulation, 7
Blood flow, 32–33, 115, 137
Blood flow theory, 6
Blood flow velocities, 259
Blood flow, in bifurcations, 166
  in microvessels, 164
  models, 165
  regulation, 168
Blood pressure (transmural), 6, 185
Blood rheology, 165
  cross model, 89
Blood salvage, 223
Blood velocity profiles, 258
Blood viscosity, 7
Blood, mechanics of constituents, 164
Body mass, 230, 249
Body-fitted coordinate (BFC), 125
Bohr's equation, 338
Boundary conditions, 120
  inlet flow boundary, 120
  outlet boundary condition, 120
Boundary element method, 119
Bypass anastomoses
  computer simulation of blood flow in, 110

Cake formation, 201
Capillaries, 6
Capillary blood flow, 6, 8, 165
Cardiac
  contractility, 239–247
  contraction, 255
  efficiency, 245
  output, 227, 243, 249
Carotid arteries, 52, 53
Carotid artery bifurcation, 59, 73
  computer simulation of blood flow in, 103–107
Cast (vascular), 71, 117, 124
Cell
  content, 12, 22
  history, 3, 4
  membrane, 15
  morphology, 13

Cell (cont.)
  theory, 1
  ultrastructure, 13
Chaos, 154
Chemical potential gradients, 178, 180
Chemotaxis, 44,
Cholesterol, 180
Chronic Obstructive Pulmonary Disease (COPD),
     289, 296, 308, 336, 337, 345
Circulation, 32, 33
Collapsible tube flow, choke point, 288, 305, 309
  elastic jump, 304, 36
  oscillations, 306
  pressure-area relationship (law), 291, 299
  speed index, 304
  supercritical flow, 304
  wave speed, 303
  see also Airways, ; Forced Expiration ; Lung
Computational fluid dynamics (CFD), 116, 119
  commercial packages of programs, 118
  post-processor, 118–121
  pre-processor, 118–120
  solver, 118
Computational mesh, 121
Computer graphics, 121
Computer simulation
  arterial blood flow, 83, 98–111
  arterial mass transfer in curved tubes, 19,
     110
  blood flow in a carotid siphon model, 99
  blood flow in saccular aneurysms, 101
  blood flow in carotid artery bifurcation models,
     103
  blood flow in vascular graft anastomoses,
     110
  vessel mechanics and blood flow in carotid artery bifurcation, 107
Concentration polarization, 200, 206, 210
Contractile state, 239
Convective mass transport, 87, 178, 185
  in curved tubes, 110
Cork, 4
Coronary
  auto regulation, 274
  flow, 267, 276, 278
  perfusion, 249
  pressure-flow relations, 272, 274
  resistance, 273, 274
  venous system, 280
Coronary arteries, 51, 53, 255
Coronary slosh phenomenon, 259
Coronary veins, 255
Couette flow, 208
Craya-Curtet number, 341
Cytoskeleton, 8

Dean number, 36, 325
Deformation of the vessel wall, 119

Index

Diameter
  mechanical instabilities, 148
  oscillations, 137, 140
  subendocardial arterioles, 261
  subendocardial venules, 261
Diastole, 125, 232, 236, 241
Diastolic pressure, 271
Diastolic volume, 242
Diffusion, 178
Diffusivity, 201
  effective, airways, 35–36
  effective diffusivity (in blood), 207
  swimming cells, 46
Distribution volume, 179

Elastance, 257, 279, 280
  compliance, 319–320
  varying elastance, 243
Elastic vessel wall model, 90, 91
EMG, fish muscle, 42
End-systolic pressure-volume relation, 242, 247
Endothelial cell vesicles, 181
Endothelial cells, 11–14, 130
  aspects ratio, 130
  cell heights, 131
  cellular surface flow, 130
  confluent endothelial cell model, 130
  cultured endothelial cells, 130
  rigid substrate, 130
Endothelial damage, 181, 192
Endothelial microvalves, 357
Endothelial vascular regulation, 263
Endothelium mechanics, 11, 180
Endotracheal tube, 322, 346
  pressure drop, 326
  resistance, 322, 344
Epicardial coronary vessels, large and middle-sized, 267
Erythrocytes, 164
Extravascular resistance, 269, 283

Fahraeus effect, 169
Fahraeus-Lindqvist effect, 8, 169
Fibrinogen, 180, 183, 188, 190, 192
Filtration flux, 200, 203–206
Finite element mesh generation, 94
Finite element method, 91–95, 119
Finite volume method, 119
Fish swimming, 40–42
  Lighthill's theory, 40
Flight, 46
Flow
  annular, 326
  aortic, 227, 239, 245
  harmonic, 229, 234
  in models with distensible walls, 86, 107
  mass, 334
  Newtonian, 85

Flow (cont.)
  non-Newtonian inelastic, 85
  source, 245–247
  turbulent, 323, 326, 338
  undeveloped, 323
  visualization, 97
Flow limitation, see Forced Expiration ; Lung
Flow resistance, in microvessels, 164, 169
Fluid and solid interfaces, 120
Fluid flux 185, 192
Fluid-structure interaction, 86
Flux enhancement, 213, 218
Fokker-Planck Equation, 46
Forced expiration, 37, 288, 297, 302
  computational model, 287, 289, 298
  effect of gas compressibility, 304
  effect of gas physical properties, 303
  flow limiting site (FLS), 288, 303, 304, 305
  flow-volume curve, 298–302
  see also Lung ; Obstructive disease ; Collapsible tube flow
Four dimensional nature of the real flow, 117, 122
Fourier, 234
  coefficient, 229
  series, 229
Fractal dimension, 154
Frequency domain, 230, 232

Gas transport, 337
Glycocalyx, 188
Grid generation, 94
Gyrotaxis, 43

Haemodynamics
  computer simulation, 83–110
  microvascular networks, 164
Heart
  as pump, 240, 243
  rate, 227, 244, 247, 248, 249
  period
Heart contraction, 268, 276, 280
Heart failure, 126
Hematocrit, 164
  in microvessels, 170
Hemofiltration, 199
Hemolysis, 215, 219–222
Heterogeneity, in microvessel networks, 171
High density lipoprotein, 182, 189
High-frequency oscillations, lungs, 37
Hopf bifurcation, 150
Hydrodynamic instability, 43
  convective, 43, 44
  surface-tension driven, 38
Hypercapnia, 345
Hyperplasia, 70

Images of
  subendocardial arteriole, 262
  subendocardial venule, 262
Impedance
  arterial, 226–228, 230
  characteristic, 229–234, 249
  input, 229–232, 249
  modulus, 229
  phase, 229
Impulse response, 232
  in blood, 13, 17
  in cells, 12–13, 24
Inspiratory activity, 335
Instantaneous blood flow pattern, 256
Intensive care
  adult, 336, 346
  gas source, 334
  pediatric, 338
Interactive computer graphics software, 122
Intercell clefts, 181, 185, 191
Internal elastic lamella, 185, 191
Intima, 177, 181, 189, 191
Intimal thickening, 53, 66, 116
Intra myocardial pump, 276, 277, 283
Intramural blood vessels, 256
Intramural stresses
  in a carotid artery bifurcation, 105
Intramyocardial arteries, 256
Intraventricular flow field, 124
Isovolumic contractions, 227, 243

Jet
  axial, 340
  non-axial, 344

Kelvin model (cells), 14

Lagrangian chaos, 37
Laplace, 250
Laser doppler velocimetry (LDV), 54, 256
Left ventricle, 124
Lipids, 179
Low density lipoproteins, 180, 181, 189
Low wall shear stress regions, 116
Lubrication theory, 165
Lung
  Coughing, 294, 309
  elastic recoil, 297, 332, 336
  flow resistance, 295, 307
  fluid mechanics, 33–38
  forced oscillation technique, 287, 298
  forced vital capacity (FVC) maneuver, 288
  heterogeneity, 302, 308
  high frequency oscillation, 287, 325
  high frequency ventilation (HFV), 288, 298
  hysteresistivity, 328
  Inhomogeneous, 329
  normal breathing, 287

Lung (cont.)
  oscillatory flow, 295
  pseudo-elastic, plastoelastic behavior of the, 321
  respiratory muscles, 288, 293
  stress-strain relationship, 327
  supra-maximal flow, 309, 311
  surface tension effects, 38
  tissue, 326
  unsteadiness, 306
Lymphatics
  collecting lymphatics, 354
  flow rates, 353
  initial lymphatics, 354
  innervation, 354
  intestine, 354
  myogenic response, 354
  pressure, 354
  skeletal muscle, 354
  transport mechanism, 358
  valves, 357

Macro-and microscale, 133
Magnetic resonance imaging, 70–72
Mass transfer coefficient, 201
Mass transport, 119, 177, 178
Matching, 228, 248
Mathematical models, 118
  blood flow, 86–90
  vessel wall mechanics, 90, 96, 97
Maxwell element, 327
Maxwell model, 14, 138
Measurement methods
  acoustic method, 326
  during controlled ventilation, 330
  during pressure support, 345
  hydraulic method, 326
  invasive, 322,
  non invasive, 326
Mechanical properties, tissues
  fish, 41, 42
  ureter, 39,40
Media, 14, 177, 181, 189, 191
Membrane, 202, 208, 218
  fouling, 218
  resistance, 200
  separation, 199
Membrane, red blood cell, 164
Mesentery, network architecture, 161–162
Micro-organisms,
  pattern formation, 42–46
  swimming, 40–44
Microfiltration, 199, 202
Microvascular networks, architecture, 160
  hemodynamics, 164, 167
Microvessels, blood flow in, 164
Microvortices, 218
Mitosis, 190
Model building, 120

Index

Models of plasma filtration, 206
Molecular biology, 9
Molecular structure, 9
Moody diagram, 323
Moving wall problem, 120
Multi-block layered boundary CFD modeling, 133
Muscle diaphragm, 317, 344
Muscle mechanics,
  fish, 40–42
  ureter, 39
Musculo-skeletal system, 5
Myocardial infarction, 125
Myocardial mechanics, 263
Myocardial microvessels, 261
Myocardium
  blood capillaries, 283
  perfusion distribution, 276
Myogenic response, 140

Narrow tubes, blood flow in, 164
Navier Stokes equations, 86, 87, 119
  numerical solution of, 91, 92
Near wall phenomena, 133
Neddle-probe videomicroscope, 261
Network microvascular architecture, 160
  heterogeneity, 171
  methods for describing, 160
  observations, 167
  rat mesentery, 161
  simulations, 168
Newtonian viscosity, 119
Nitric oxide (NO), 145
Nitroglycerin, 260
Non-Newtonian flow, 86, 89
  in a compliant carotid artery model, 103
Non planar branching, 70–76, 79
Non-uniqueness of flow, 32
Normal heart, 125
Numerical flow visualization, 97
  of particle paths, 110
Numerical solution
  fluid-structure interaction problems, 94
  Navier-Stokes equations, 91, 92
  shell equations, 93, 94

Obstructive disease, 296, 300, 308, 336, 345
Ohm's law, 228–229
Oscillating shear, 66
Oscillatory flow, 263
Osmotic pressure, 201
Oxidised LDL, 178
Oxygen consumption, 243, 245, 248
Oxygen extraction, 268

Pacemaker cells, 140
Particle residence time, 62
Partition coefficient, 179

Pathway effect, 171
Peclet number, 37, 87, 178, 185
Peristaltic pumping, 39–40
Permeability, 179, 184
Permeate flux, see filtration flux
Physiological flow field, 131
Plasma filter, 202, 214, 219,
Plasma filtration, see plasma separation
Plasma proteins, 179, 184
Plasma separation, 200, 202–206
Plasmapheresis, see plasma separation
Platelets, 164
Platelets, 199, 205
Poiseuille's law, 7, 164
Polypropylene hollowfibers, 203, 204, 214, 223
Potential energy, 243
Pressure
  alveolar, 338
  aortic, 227, 228, 248
  arterial, 227–228
  continuous positive airway, 339
  diastolic, 243, 247
  flow relation, 227
  generator, 244
  harmonic, 229, 234
  inspiratory positive airway, 339, 344, 345
  isovolumic, 247
  positive end expiratory, 320, 336
  source, 243–247
  support physiological effect of, 344
  systolic, 238, 247
  ventricular, 241–245
  volume diagram, 243, 247
  volume relation, 240–243
Pressure pulsation, 355
Protein folding, 8
Pulmonary artery, 187, 188
Pulmonary barotrauma, 336
Pulsatile flow, 119
  computer simulation, 83–110
Pulsatile flow filtration, 213, 218
Pulsation amplitudes of arterioles, 263
Pulsed doppler velocimeter, 258
Pump function graph, 240–243

Rayleigh oscillatory boundary layer, 215
RCR model, 274–276
Real time steering, 122
Red blood cells, 4
  deformation, 164
  layer, 212
  membrane, 164, 219
  shape, 164
Reflected waves, 232
Reflection coefficient, 6, 7, 8, 232–233
Resistance, peripheral, 228, 231, 239
Respiration, 33–39

Respiratory distress
  acute respiratory distress syndrom, 321
  pneumothorax, 320
Respiratory flow, see Lung
Respiratory system
  modelized responses, 328
  visco-elastic resistance, 330
Reynolds number, 77, 87, 323
Rheology, blood, 164
Rohrer equation, 323
Rotating membrane, 208, 223

Secondary flow 32, 33, 36, 78
Septal artery, 256
Shear rate, 201, 203, 204
Shear stresses, 16, 185, 189
Shear thinning model, 89
SIMPLE, 125
Simulations, network hemodynamics, 168
Small epicardial vessels, 257
Smooth muscle cells, 179
Smooth muscle, see vascular smooth muscle
Solid mechanics, 5
Spontaneous breathing, 317, 318
Spontaneous plasmapheresis, 223
Stability analysis, 151
Stenosis, 259
Stereoscopic display, 121
Strahler order, 161
Stress
  active, 139, 149
  adaptation, 327
  concentration in cells, 12
  hoop, 141
  passive, 139, 149
  shear, 150
  total, 139, 149
Stroke volume, 247
Strouhal number, 325
Subendocardial underperfusion and ischemia, 255
Surface rendering, 128
Surfactant, lung, 38
Swimming
  fish, 40–42
  micro-organisms, 43–45
  random, 45, 46
Systole, 125, 232, 241
Systolic retrograde blood flow, 260
Systolic-diastolic interaction, 262

Taylor dispersion, 35
Taylor number, 209
Taylor vortices, 209
Tensile stress, 16
  in cell membrane, 12, 18–19, 22, 24
Tension field theory, 15–17, 23
Three dimensional configuration of the flow field, 121

Time constant, RC, 237, 248
Time scale (ventricular wall contraction), 125
Tissue compaction, 186
Tissue matrix, 178, 179, 182
Tissue porosity, 192
Tissue pressure, 274, 281
Tracer studies, 183
Transmembrane pressure, 211, 216, 221
Transmural transport, 187, 191
Transport equations, 119
Tunnel in gel, 14
Turbulence, 119
Turbulence modeling, 118

Ultrafiltration, 191, 199
Ultrasonic echotracking device, 140
Uncoupled wall motion, 124
Unsteady microfiltration, 210
Ureteral dynamics 39–40

Vaccines, 5
  anthrax, 5
  rabies, 5
Valsalva maneuver, 236
Vasa vasorum, 177, 180, 182
Vascular compliance, 185
Vascular exchange, 177
Vascular grafts, 52, 62
Vascular images, 261
Vascular smooth muscle,
  calcium channel, 151
  calcium concentration, 139, 152
  calcium regulation, 140
  contractile component, 138
  ionic transport, 151
  isometric contraction, 139
  mechanics, 138
  membrane permeability, 152
  membrane potential, 152
  parallel elastic component, 139
  potassium channel, 151
  series elastic component, 139
  sliding filament theory, 139
  tone, 137, 148
  velocity of contraction, 153
Vasomotion, 355
  amplitude, 146
  arterial, 137
  correlation, 144, 146
  flow, 145
  frequency, 140
  patterns, 138, 144
  propagation, 142
  regulating mechanisms, 145
  synchronicity, 144
  theoretical models, 148
Veins, 187, 188
Velocity profiles, 55, 59, 64, 77

Index

Velocity waveforms, 255
Ventilation
    controlled mechanical, 315, 318
    complications of, 336
    high frequency, 336
    high frequency jet, 339, 341
    mechanical effects of, 335
    non-invasive, 345
    pneumatic flow generators, 334
    proportional assist, 339
    synchronized intermittent mandatory, 339
    ventilators, 332
Ventricular, filling, 240–247
Venturi, 334, 340
Vessel wall mechanics
    carotid artery bifurcation, computer simulation of blood flow, 107
    mathematical model, 90, 96–97
    shell calculation, 93
Video signal, 122–123
Virtual reality, 121
Viscoelastic
    vessel wall model, 96, 97
Viscosity, 165
    dynamic, 324
Visualization, 117–121
Voigt element, 329

Voigt model, 14
Volume
    anatomical dead space, 338
    oscillatory, 336
    relaxation, 336
    tidal, 315
Vortex, 78, 119

Wall shear rates, in microvessels, 168
Wall shear stress, 51, 57, 60, 64, 97, 127
    in carotid bifurcation models, 105, 107
    in vascular graft anastomoses, 110
    root mean square, 129
    spatial distribution, 127, 129, 131
Waterfall model, 270
Wave
    backward, 234
    forward, 234
    propagation, 7
    reflected, 235
White blood cells, 164, 170
Windkessel, 237, 238, 239, 247
    theory, 6
Womersley number, 325

Zero stress state, 22